Jordan Crafts

Microwave and RF Wireless Systems

Microwave and RF Wireless Systems

David M. Pozar

Department of Electrical and Computer Engineering
University of Massachusetts at Amherst

JOHN WILEY & SONS, INC.

New York • Chichester • Weinheim
Brisbane • Toronto • Singapore

ACQUISITIONS EDITOR Bill Zobrist
MARKETING MANAGER Katherine Hepburn
SENIOR PRODUCTION EDITOR Patricia McFadden
SENIOR DESIGNER Karin Gerdes Kincheloe
ILLUSTRATION EDITOR Gene Aiello
PRODUCTION MANAGEMENT SERVICES Ingrao Associates

This book was set in Times Roman by TechBooks.

This book is printed on acid-free paper. ♾

Library of Congress Cataloging in Publication Data

Pozar, David M.
Microwave and RF wireless systems / David M. Pozar.
p. cm.
Includes bibliographical references.
ISBN 0-471-32282-2 (cloth : alk. paper)
1. Wireless communication systems. 2. Microwave communication systems. 3. Radio frequency. 4. Mobile communication systems. I. Title.

TK5103.2.P59 2000
621.382—dc21 00-039275

Printed in the United States of America

10 9 8 7 6

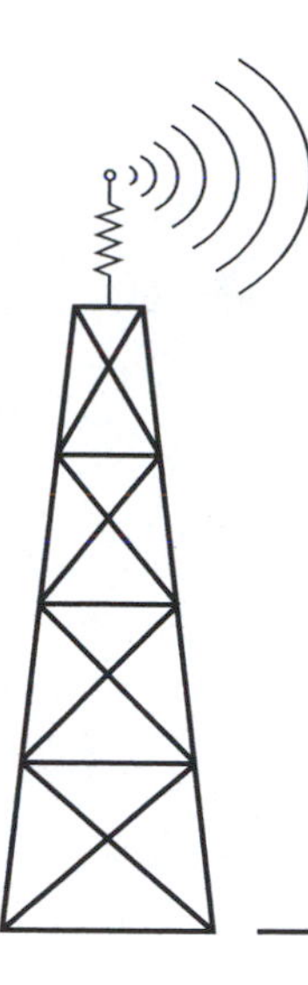

Preface

Wireless system design is one of the most exciting fields in electrical engineering today. In economic terms, wireless applications that include cellular and PCS telephony, wireless local area networks (WLANs), global positioning satellite (GPS) service, direct broadcast television service (DBS), local multipoint distribution systems (LMDS), and radio frequency identification systems (RFID) constitute a yearly market in excess of $100B, and strong growth is predicted over the long term. From a technical perspective, wireless system design involves a close integration of a variety of topics that include antennas and propagation effects, RF and microwave circuit design, noise and intermodulation effects, digital modulation methods, and digital signal processing.

The purpose of this text is to present a cohesive overview of the fundamental subjects required for the design and analysis of the RF stages of modern wireless systems, including antennas, propagation, fading, noise, receiver design, modulation methods, and bit error rates. Material is also included on the design of key components used in wireless systems, such as filters, amplifiers, mixers, oscillators, and phase-locked loops. Major wireless applications, such as cellular and PCS telephony, GPS, DBS, WLANs, and LMDS systems are described, and many design examples are given in the context of these systems. Required fundamentals on transmission lines, S parameters, impedance matching, and random processes are also included.

A key premise of this book is that a coherent understanding of wireless system performance and design can only be obtained by treating the relevant technical topics in an integrated manner. A collection of individual courses in antennas, microwave engineering, and communications engineering is unlikely to provide an understanding of the interplay between different stages and their effect on the overall performance of the system. Courses in antennas or microwave engineering, for example, generally will not discuss the effect of noise or Rayleigh fading on bit error rates in a digital radio. Similarly, a course in communications theory will probably not discuss component noise figure and intermodulation requirements for different modulation schemes and data rates. While the emphasis of this book is on the RF and microwave stages of wireless systems, we have included a chapter on modulation methods because this allows us to provide a complete characterization of a wireless system from an input data stream through the transmitter, the antennas and propagation channel, and the receiver, resulting in overall system performance measures in terms of bit error rate, data rate, or range.

There is enough material here for a full year course in RF and microwave design of wireless systems at the senior or first-year graduate student level. Prerequisites ideally would include junior-level electronics, electromagnetics, transmission lines, probability, and random variables, but Chapters 2–4 contain brief but reasonably complete reviews of these topics to the extent that they will be required later in the text. If students have a familiarity with transmission lines, S parameters, and RF circuit design, a one-semester course covering Chapters 1, 3, 4, 9, and 10 can be presented with a focus on wireless system analysis and design. Some teachers may prefer to cover the systems-oriented material in Chapters 1–4 and 9–10 first, followed by selective coverage of component design in Chapters 5–8. Other combinations are possible, depending on the background of the students and the opinions of the instructor. Much of the material on microwave circuit design presented in Chapters 2, 5, 6, and 7 was drawn from my text **Microwave Engineering,** with additional topics that include ceramic bandpass filters, stability, power amplifiers, FET mixers, and nonlinear mixer analysis.

Computer codes relevant to some of the problems and examples in the text are available on the Wiley Web site at www.wiley.com/college/pozar. These can be used for computing the complementary error function, calculating the noise figure and intermodulation point of a cascade system, determining the stability parameters of a transistor amplifier, and other applications.

ACKNOWLEDGMENTS

I would first like to thank the students who participated in our course on wireless systems at the University of Massachusetts, and who used notes and draft copies of this book for several years. Thanks also go to my colleagues in microwave engineering and the Wireless Communications Center at the University of Massachusetts. I would especially like to acknowledge my late colleague and friend, Bob McIntosh, who provided a guiding vision for many of us through the years. Several people in industry were helpful in providing photographs of wireless hardware: Lamberto Raffaelli and Earl Stewart of Arcom, Carl Marguerite and Peter Alfano of Sage Laboratories, Fred Dietrich of Globalstar, A. W. Love of Rockwell, Tuli Herscovici of Spike Technologies, and Harry Syrigos of Alpha Industries. Juraj Bartolic deserves thanks for providing a simplified derivation of the μ-parameter stability test in Chapter 6, as does Dennis Goeckel for his advice on an early version of the manuscript. Finally, I would like to thank Bill Zobrist, Jennifer Welter, and Suzanne Ingrao for their invaluable help in completing this project.

David M. Pozar
Amherst, MA

Contents

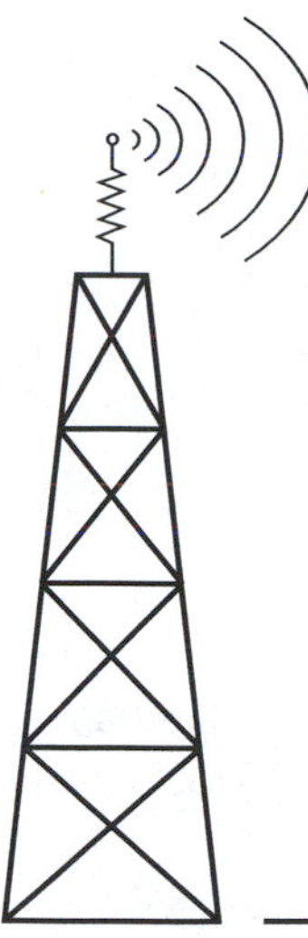

Chapter One

Introduction to Wireless Systems

In the early 1980s a marketing firm hired by AT&T to survey the potential U.S. market for its newly inaugurated cellular phone service arrived at an estimate of less than 900,000 users by the year 2000. Like many predictions of technological progress, this one turned out to be off by a wide margin—in 1998 the number of cellular subscribers in the United States was over 60 million (already an error of more than 6000 percent). It is now estimated that half of all business and personal communications will be wireless by the year 2010 [1]. Rapid growth is also occurring with other wireless systems, such as Direct Broadcast Satellite (DBS) television service, Wireless Local Area Networks (WLANs), paging systems, Global Positioning Satellite (GPS) service, and Radio Frequency Identification (RFID) systems. It is estimated that the number of consumer wireless devices will exceed 300 million by the year 2000 [2]–[3]. These systems promise to provide, for the first time in history, worldwide connectivity for voice, video, and data communications. The successes of wireless technology to date, and the technological challenges of future wireless systems, make this an exciting and rewarding field in which to work.

In this book we study the operation and design of wireless systems from the perspective of the radio frequency (RF) or microwave subsystems. These include modulators and frequency up-conversion circuits in the wireless transmitter, the transmit and receive antennas, the wireless propagation channel, and the frequency down-conversion and demodulator circuits in the wireless receiver. Generally these subsystems are analog in nature, even if the wireless system uses digital modulation techniques. We will see that noise and other characteristics of these subsystems set the ultimate limits on the performance of a wireless system, in terms of maximum data rate, operating range, power requirements, and error rates.

1.1 WIRELESS SYSTEMS AND MARKETS

In this section we give a brief introduction to some of the major wireless systems in use today. These include wireless cellular and PCS telephone systems, commercial satellite systems, wireless data networks, point-to-point radios, the global positioning system, and other wireless systems.

Classification of Wireless Systems

In the broadest sense, a wireless system allows the communication of information between two points without the use of a wired connection. This may be accomplished using sonic, infrared, optical, or radio frequency energy. While early television remote controllers used ultrasonic signals, very low data rates and poor immunity to interference make such systems a poor choice for modern applications. Infrared signals can provide moderate data rates, but the fact that infrared radiation is easily blocked by even small obstructions limits their use to short-range indoor applications such as remote controllers and local area data links. Similarly, optical signals propagating in an unobstructed environment can provide moderate to high data rates, but require a line-of-sight path, and cannot be used where foliage, fog, or dust can block the signal. For these reasons, most modern wireless systems rely on RF or microwave signals, usually in the UHF (100 MHz) to millimeter wave (30 GHz) frequency range. Because of spectrum crowding, and the need for higher data rates, the trend is to use the higher frequencies in this range, so that the majority of wireless systems today operate at frequencies ranging from about 800 MHz to a few gigahertz. RF and microwave signals offer wide bandwidths, and have the added advantage of being able to penetrate fog, dust, foliage, and even buildings and vehicles to some extent.

Historically, wireless communication using RF energy began with the theoretical work of Maxwell, followed by the experimental verification by Hertz of electromagnetic wave propagation, during the period from 1873 to 1891. Marconi built on this work to develop practical commercial radio communications systems in the early part of the 20th century. It is interesting to note that the term "wireless" dates back to this early period, and although replaced by the word "radio" for most of this century, *wireless* is again the preferred description for most of today's cellular telephone, data links, and satellite systems.

One way to categorize wireless systems is according to the nature and placement of the users. In a *point-to-point* radio system a single transmitter communicates with a single receiver. Such systems generally use high-gain antennas in fixed positions to maximize received power and minimize interference with other radios that may be operating nearby in the same frequency range. Point-to-point radios are generally used for dedicated data communications by utility companies and for connection of cellular phone sites to a central switching office. *Point-to-multipoint* systems connect a central station to a large number of possible receivers. The most common examples are commercial AM and FM broadcast radio and broadcast television, where a central transmitter uses an antenna with a broad beam to reach many listeners and viewers. Broadcast radio is similar in function to *local multipoint distribution systems* (LMDS), which are presently being deployed in urban areas to provide wireless television and Internet access to users within a small geographical area. Another example of a point-to-multipoint system is *paging*, where a central station can briefly communicate with many users over a large geographical region. *Multipoint-to-multipoint* systems allow simultaneous communication between individual users (who may not be in fixed locations). Such systems generally do not connect two users directly to each other, but instead rely on a grid of *base stations* to connect an individual user to a central switching office, which then connects to the base station of the other user. *Cellular telephone* systems and some types of *wireless local area networks* (WLANs) are examples of this type of application.

Another way to characterize wireless systems is in terms of the directionality of communication. In a *simplex* system, communication occurs only in one direction, from the transmitter to the receiver. Examples of simplex systems include broadcast radio and television. In a *half-duplex* system, communication may occur in two directions, but not simultaneously. Early mobile radios and citizens band radio are examples of duplex systems, and generally rely on a "push-to-talk" function so that a single channel can be used for both transmitting and receiving at different intervals. Some wireless data links also use half-duplex transmission. *Full-duplex* systems allow simultaneous two-way transmission and reception. Examples include cellular telephone and point-to-point radio systems. Full-duplex transmission clearly requires a *duplexing* technique to avoid interference between transmitted and received signals. This can be done by using separate frequency bands for transmit and receive (*frequency division duplexing*, FDD), or by allowing users to transmit and receive only in certain predefined time intervals (*time division duplexing*, TDD).

While most wireless systems are ground based, there is increasing interest in the development of satellite systems for voice, video, and data communications. Satellite systems offer the possibility of communication with a large number of users over wide areas, perhaps including the entire planet. Satellites in a *geosynchronous earth orbit* (GEO) are positioned approximately 36,000 km above the Earth, and remain in a fixed position relative to the surface. Such satellites are useful for point-to-point radio links between widely separated stations, and are commonly used for television and data communications throughout the world. At one time transcontinental telephone service relied heavily on such satellites, but undersea fiber optic cables have largely replaced satellites for transoceanic connections as being more economical, and avoiding the annoying delay caused by the very long round trip path between the satellite and the Earth. Another drawback of GEO satellites is that their high altitude greatly reduces the received signal strength, making it impractical for two-way communication with small transceivers. *Low earth orbit* (LEO) satellites orbit much closer to the Earth, typically in the range of 500 to 2000 km. The shorter path length allows communication between LEO satellites and handheld radios, but satellites in LEO orbits are visible from a given point on the ground for only a short time, typically from a few minutes to perhaps 20 minutes. Effective coverage therefore requires a large number of satellites in different orbital planes.

Finally, wireless systems can be grouped according to their operating frequency. The choice of operating frequency will be discussed in much more detail in a later section, but Table 1.1 lists the operating frequencies of some of the most common wireless systems.

Cellular Telephone Systems

Cellular telephone systems were proposed in the 1970s in response to the problem of providing mobile radio service to a large number of users in urban areas. Early mobile radio systems could handle only a very limited number of users due to inefficient use of the radio spectrum and interference between users. In 1976, for example, the entire mobile phone system in New York City could support only 543 users [1]. The cellular radio concept introduced by Bell Laboratories solved this problem by dividing a geographical area into non-overlapping hexagonal *cells*, where each cell has its own transmitter and receiver (*base station*) to communicate with the mobile users operating in that cell. Each cell site may allow as many as several hundred users to simultaneously communicate with other mobile users, or through the land-based telephone system.

The first cellular telephone system to offer commercial service was built by the Nippon Telephone and Telegraph company (NTT), and became operational in Japan in 1979 [4].

TABLE 1.1 Wireless System Frequencies

Wireless System	Operating Frequency
Advanced Mobile Phone Service (AMPS)	T: 824–849 MHz R: 869–894 MHz
Global System Mobile (European GSM)	T: 880–915 MHz R: 925–960 MHz
Personal Communications Services (PCS)	T: 1710–1785 MHz R: 1805–1880 MHz
US Paging	931–932 MHz
Global Positioning Satellite (GPS)	L1: 1575.42 MHz L2: 1227.60 MHz
Direct Broadcast Satellite (DBS)	11.7–12.5 GHz
Wireless Local Area Networks (WLANs)	902–928 MHz 2.400–2.484 GHz 5.725–5.850 GHz
Local Multipoint Distribution Service (LMDS)	28 GHz
US Industrial, Medical, and Scientific bands (ISM)	902–928 MHz 2.400–2.484 GHz 5.725–5.850 GHz

T/R = mobile unit transmit/receive frequency.

This was followed by the Nordic Mobile Telephone (NMT) system in Europe, which began operation in 1981. The first cellular telephone system in the United States was the Advanced Mobile Phone System (AMPS), deployed by AT&T in 1983. All of these systems use analog FM modulation and divide their allocated frequency bands into several hundred channels, each of which can support an individual telephone conversation. These early systems grew slowly at first, because of the initial costs of developing an infrastructure of base stations and the initial expense of handsets, but by the 1990s growth became phenomenal.

In 1998 there were 64 million cellular phone subscribers and over 57,000 base stations in the United States, generating annual service revenues of $30 billion with a market penetration of about 35%. Worldwide there were about 200 million cellular subscribers in 1997. While the approximately 700 million wired telephone lines far outnumber wireless telephone users, the growth rate of wireless is about 15 times that for wired lines.

In 1996 88% of all cellular telephones in the United States used the analog AMPS system, but newer digital standards have been growing in popularity and will soon replace the AMPS system. These systems are generally referred to as *Second Generation Cellular*, or *Personal Communication Systems* (PCS). Third generation PCS systems, which may include capabilities for email and Internet access, are in the planning stages.

TABLE 1.2 Major Worldwide Cellular and PCS Telephone Systems

Standard	Country	Year of Introduction	Type	Frequency Band (MHz)	Modulation	Channel Bandwidth
NTT	Japan	1979	Cellular	860–940	FM	25 kHz
NMT-450	Europe	1981	Cellular	453–468	FM	25 kHz
AMPS	United States	1983	Cellular	824–894	FM	30 kHz
E-TACS	Europe	1985	Cellular	872–950	FM	25 kHz
C-450	Germany	1985	Cellular	450–466	FM	20 kHz
NMT-900	Europe	1986	Cellular	890–960	FM	12.5 kHz
JTACS	Japan	1988	Cellular	860–925	FM	25 kHz
GSM	Europe	1990	PCS	890–960	GMSK	200 kHz
IS-54	United States	1991	PCS	824–894	DQPSK	30 kHz
NAMPS	United States	1992	Cellular	824–894	FM	10 kHz
IS-95	United States	1993	PCS	824–894	QPSK	1.25 MHz
PDC	Japan	1993	Cellular	810–1513	DQPSK	25 kHz
NTACS	Japan	1993	Cellular	843–922	FM	12.5 kHz

Personal Communications Systems

Because of the rapidly growing consumer demand for wireless telephone service, as well as advances in wireless technology, several second generation standards have been proposed for improved service in the United States, Europe, and Japan. These PCS standards all employ digital modulation methods and provide better quality service and more efficient use of the radio spectrum than analog systems. Digital systems also provide more security, preventing eavesdropping through the possible use of encryption.

PCS systems in the United States use either the IS-136 time division multiple access (TDMA) standard, the IS-95 code division multiple access (CDMA) standard, or the European Global System Mobile (GSM) system [1], [2], [4]. Many of the new PCS systems have been deployed using the same frequency bands as the AMPS system. This approach takes advantage of existing infrastructure, and facilitates the use of *dual-mode handsets* that can operate on both the older AMPS system as well as one of the newer digital PCS systems. Additional spectrum has also been allocated by the Federal Communications Commission (FCC) around 1.8 GHz, and some of the newer PCS systems use this frequency band.

Outside the United States, the Global System Mobile (GSM) TDMA system is the most widespread, being used in over 100 countries [1]. The uniformity of a single wireless telephone standard throughout Europe and much of Asia allows travelers to use a single handset throughout these regions. In contrast, the different PCS systems in the United States are incompatible. Table 1.2 lists the major cellular and PCS telephone systems that have been deployed throughout the world [1], [4].

It is interesting to compare how the development of first and second generation cellular services has differed in the United States and Europe [1]. The first U.S. cellular system, AMPS, provided a single standard allowing every cellular user in the United States and Canada to communicate within range of a base station. In the Europe of the early 1980s, however, individual countries developed their own analog cellular standards with different frequency bands and modulation methods, so that there were at least four incompatible systems in use (see Table 1.2). These situations were reversed for second generation digital systems. The organization of European countries under the European Union in the 1980s led to the establishment of GSM as a single digital PCS standard, which is now used by over 100 countries in Europe and elsewhere. In the United States, however, government policies relating to the allocation of radio spectrum, as well as the structure of the telecommunications industry and the competitive nature of R&D in the United States, has allowed the technological and economic trade-offs between CDMA, TDMA, and GSM PCS systems to be decided in the marketplace. Meanwhile, wireless telephone consumers in the United States are left to choose between an out-of-date analog system and a variety of incompatible digital systems.

Satellite Systems for Wireless Voice and Data

The key advantage of satellite systems is that a relatively small number of satellites can provide coverage to wireless users at any location, including the oceans, deserts, and mountains—areas for which it would otherwise be difficult to provide service. In principle, as few as three geosynchronous satellites can provide complete global coverage, but (as we will see in Chapter 4) the very high altitude of the geosynchronous orbit makes it difficult to communicate with handheld terminals because of very low signal strength. Satellites in lower orbits can provide usable levels of signal power, but many more satellites are then needed to provide global coverage.

There are a large number of commercial satellite systems either currently in use, or in the development stage, for wireless communications. These systems generally operate at frequencies above 1 GHz because of available spectrum, the possibility of high data rates, and the fact that such frequencies easily pass through the atmosphere and ionosphere. GEO satellite systems, such as INMARSAT and MSAT, provide voice and low-data rate communications to users with 12″ to 18″ antennas. These systems are often referred to as *very small aperture terminals* (VSATs), and in 1997 were being deployed at the rate of about 1500 per month to business users [1]. Other satellite systems operate in medium or low-earth orbits to provide mobile telephone and data service to users on a worldwide basis.

Iridium, financed by a consortium of companies headed by Motorola, was the first commercial satellite system to offer handheld wireless telephone service. It consisted of 66 LEO satellites in near-polar orbits, and connects mobile phone and paging subscribers to the public telephone system through a series of intersatellite relay links and land-based gateway terminals. The Iridium system cost was approximately $3.4 B, and it began service in 1998. Globalstar, proposed by Loral and Qualcomm, is another LEO satellite system intended for wireless telephone, fax, and paging. This system uses 48 satellites to provide global coverage, and became operational in 2000. One drawback of using satellites for telephone service is that weak signal levels require a line-of-sight path from the mobile user to the satellite. This means that satellite telephones generally cannot be used in buildings, automobiles, or even in many wooded or urban areas (the topics of *propagation, fading*, and *link loss*, which relate to this problem, will be studied in Chapter 4). This places satellite phone service at a definite performance disadvantage relative to land-based cellular and PCS wireless phone service. But an even greater problem with satellite phone service is the expense of deploying and maintaining a large fleet of LEO satellites, making it very difficult to compete economically with land-based cellular or PCS service. The typical cost of a cellular or PCS call is in the range of $0.10 to $0.20 per minute, while in 1999 the estimated cost of a call placed through the Iridium or Globalstar satellite was about $2.00 per minute. In addition, the cost of a cellular or PCS handset to new subscribers is usually minimal (or zero), while the cost of a satellite handset is several thousand dollars. For these reasons, it is hard to see how satellite telephone service can compete with land-based cellular and PCS systems in terms of either performance or cost, even though satellite systems offer (in principle) the convenience of a single phone that can be used anywhere in the world. Table 1.3 summarizes some of the current commercial voice-communication satellite systems.

TABLE 1.3 Commercial Wireless Satellite Systems

System	Organization	Number of Satellites	Orbit	Operational Date
INMARSAT-M	Inmarsat	5	GEO	1996
MSAT	AMSC, TMI	2	GEO	
Iridium	Motorola	66	LEO	1998
Globalstar	Loral, Qualcomm	48	LEO	2000
ICO Global	Hughes	10	MEO	2000
Odyssey	TRW	12	MEO	2000

In August 1999 both Iridium LLC and the ICO Global Communications companies declared bankruptcy. It remains to be seen whether Globalstar and the other large LEO systems will be financially viable, but the future of such satellite services does not look promising when land-based systems offer better performance at lower costs. A satellite from the Globalstar system is shown in Figure 1.1.

Global Positioning Satellite System

The Global Positioning Satellite system (GPS) uses 24 satellites in medium earth orbits to provide accurate position information (latitude, longitude, and elevation) to users on land, in the air, or at sea. Originally developed as the NAVSTAR system by the military, at a cost of about $12B, GPS has quickly become one of the most pervasive applications of wireless technology for consumers and businesses throughout the world. Today, GPS receivers can be

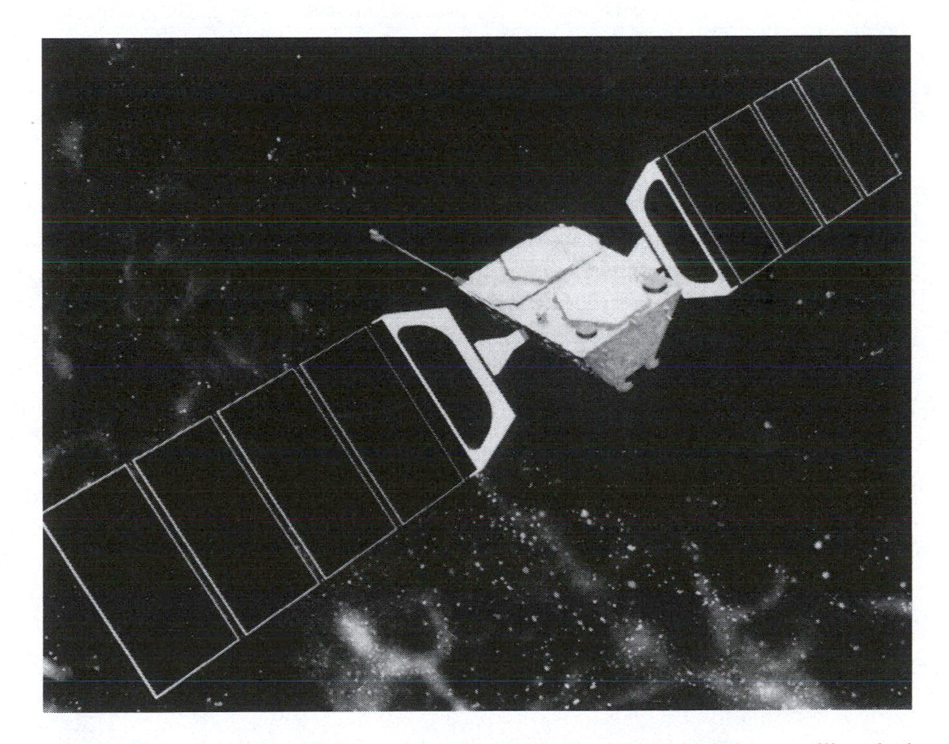

FIGURE 1.1 An artist's conception of one of the satellites used in the Globalstar satellite telephone system. (Courtesy of F. Dietrich, Globalstar, San Diego, CA.)

FIGURE 1.2 Photograph of a NAVSTAR global positioning system satellite, showing the solar panels and the L-band helix transmitting antennas. (Courtesy of Satellite and Space Division, Rockwell International, Seal Beach, CA.)

found on commercial and private airplanes, boats and ships, and ground vehicles. Advances in technology have led to substantial reductions in size and cost, so that small handheld GPS receivers can be used by hikers and sportsmen. With differential GPS, accuracies on the order of 1 cm can be achieved—a capability that has revolutionized the surveying industry [5]. A photograph of a NAVSTAR GPS satellite is shown in Figure 1.2.

The GPS positioning system operates by using triangulation with a minimum of four satellites. GPS satellites are in orbits 20,200 km above the Earth, with orbital periods of 12 hours. Distances from the user's receiver to these satellites are found by timing the propagation delay between the satellites and the receiver. The positions of the satellites (*ephemeris*) are known to very high accuracy; in addition, each satellite contains an extremely accurate clock to provide a unique set of timing pulses. A GPS receiver decodes this timing information and performs the necessary calculations in order to find the position and velocity of the receiver. The GPS receiver must have a line-of-sight view to at least four satellites in the GPS constellation, although three satellites are adequate if altitude position is known (as in the case of ships at sea). Because of the low gain antennas required for operation, the received signal level from a GPS satellite is very low—typically on the order of –130 dBm (for a receiver antenna gain of 0 dB). This signal level is usually below the noise power at the receiver, but spread spectrum techniques are used to improve the received signal-to-noise ratio.

GPS operates at two frequency bands: L1, at 1575.42 MHz; and L2, at 1227.60 MHz, transmitting spread spectrum signals with binary phase shift keying modulation. The L1 frequency is used to transmit ephemeris data for each satellite, as well as timing codes, which are available to any commercial or public user. This mode of operation is referred to as the *Course/Acquisition* (C/A) code. In contrast, the L2 frequency is reserved for military use and uses an encrypted timing code referred to as the *Protected* (P) code (there is also a P code signal transmitted at the L1 frequency). The P code offers much higher accuracy than the C/A code, and it is likely that this capability will soon be made available to all users.

The typical accuracy that can be achieved with an L1 GPS receiver is about 100 ft. Accuracy is limited by timing errors in the clocks on the satellites and the receiver, as well as some error in the assumed position of the GPS satellites. The most significant error is generally caused by atmospheric and ionospheric effects, which introduce small but variable delays in signal propagation from the satellite to the receiver. Much better accuracies can be obtained through the use of *differential* GPS, which uses a GPS receiver at a known location to provide error correction information to other nearby GPS receivers. In this way, positioning accuracies to within 1 cm can be obtained relative to the reference position. Receivers that have access to the P code can use the encrypted timing data at the L1 and L2 frequencies to correct for the atmospheric and ionospheric propagation delays, and thereby yield very accurate position information.

Wireless Local Area Networks

Wireless local area networks (WLANs) provide connections between computers over short distances. Typical indoor applications may be in hospitals, office buildings, and factories, where coverage distances are usually less than a few hundred feet. Outdoors, in the absence of obstructions and with the use of high gain antennas, ranges up to a few miles can be obtained. Wireless networks are especially useful when it is impossible or prohibitively expensive to place wiring in or between buildings, or when only temporary access is needed between computers. Mobile computers users, of course, can only be connected to a computer network by a wireless link.

In spite of their attractiveness, market penetration of WLAN products has been slow, probably due to a combination of factors that include relatively high costs, relatively slow data rates, and poor immunity to fading and interference. In 1996 the market for WLANs was about $200M, which is a negligible fraction of the several billion dollar cellular telephone industry. It is expected, however, that market growth for WLANs will soon increase substantially. A major new WLAN initiative is the *Bluetooth* standard, where very small and inexpensive RF transceivers will be used to link a wide variety of digital systems over relatively short distances. Possible Bluetooth applications include wirelessly networking printers, scanners, cell phones, notebook and desktop computers, personal digital assistants (PDAs), and even household appliances. Current Bluetooth systems operate in the ISM band at 2.4 GHz, and offer data rates up to 1 Mbps. Market projections for Bluetooth devices are in the range of several hundred million units per year.

Currently most commercial WLAN products in the United States operate in the *Industrial, Scientific, and Medical* (ISM) frequency bands, and use either frequency-hopping or direct-sequence spread spectrum techniques in accordance with IEEE Standard 802.11. Maximum bit rates range from 1 to 2 Mbps, which are much slower than the data rates that can be achieved with wired Ethernet lines. WLANs almost universally use Internet protocols (TCP/IP) for communication between computers. In Europe, the HIPERLAN standard provides for WLAN operation with data rates up to 20 Mbps.

Other Wireless Systems

Besides the wireless systems described above, there are many other applications of wireless technology. *Wireless local loop* (WLL) is similar to a cellular telephone system, but provides service over a smaller operating area. Hospitals, college campuses, factories, and office buildings can employ a WLL as a *private branch exchange* (PBX) to provide users with a single telephone number and a mobile handset with which they can communicate from any point within the operating area. The cell sizes for WLL typically range from 50 to 100 ft, and for this reason WLL is sometimes referred to as *microcellular phone service*. An interesting application of WLLs is to provide service for towns and villages in lesser developed countries that do not have wired telephone service, since installing a wireless local loop system is much more economical than installing hard-wired copper lines. For these reasons, the demand for WLL products is expected to grow rapidly in the next few years, with over 60 million WLL users predicted by the year 2000.

The *Direct Broadcast Satellite* (DBS) system provides television service from two geosynchronous satellites directly to home users with a relatively small 18″ diameter antenna. Previous to this development satellite TV service required an unsightly dish antenna as large as 6 ft in diameter. As we will see in Chapters 4 and 9, this advancement was made possible through the use of digital modulation techniques, which reduce the necessary received signal levels as compared to previous systems, which used analog modulation. The DBS system uses *quadrature phase shift keying* (QPSK) with digital multiplexing and error correction to deliver digital data at a rate of 40 Mbps. Two satellites, DBS-1 and DBS-2, located at 101.2° and 100.8° longitude, each provide 16 channels with 120 W of radiated power per channel. These satellites use opposite circular polarizations to minimize loss due to precipitation, and to avoid interference with each other (polarization duplexing), DBS-1 transmits with left-hand circular polarization (LHCP), while DBS-2 uses right-hand circular polarization (RHCP).

DBS competes directly with wired cable TV service, but within one year of its introduction in 1994, DBS sold over 1 million units to break all previous records and become the consumer electronics product with the fastest market growth in history. The initial cost of a DBS antenna and receiver was about $700, but after 2.5 million units were sold the price had dropped to about half this value. This cost reduction was the result of market competition, as well as significant economies of scale associated with large volume production rates (hundreds of thousands per month).

Local Multipoint Distribution Systems (LMDS) and *Multipoint Multichannel Distribution Systems* (MMDS) provide broadband wireless connections between a fixed base station and a cellular region of users. These systems are poised for rapid market growth because of the strong demand for the 'last mile connectivity', where wireless systems offer one of the few economical solutions to the problem of providing high data rate connections to small businesses and homes for Internet access, telephone, television, and data communications. LMDS and MMDS systems typically operate in the 2.1–2.7 GHz band, the 3.4–3.7 GHz band, or the 28 GHz millimeter wave band, and may offer two-way full-duplex data rates ranging from 50 Mbps to over 110 Mbps for each channel. These systems are sometimes referred to as *broadband fixed wireless*, because they are intended for connections between fixed, as opposed to mobile, users. Figure 1.3 shows a commercial MMDS subscriber system.

Point-to-point radios are used by businesses to provide dedicated data connections between two points. Electric utility companies use point-to-point radios for the transmission of telemetry information for the generation, transmission, and distribution of electric power between power stations and substations. Point-to-point radios are also used to connect cellular base stations to the public switched telephone network, and are generally much cheaper than running high-bandwidth coaxial or fiber-optic lines below ground. Such radios usually operate in the 18, 24, or 38 GHz bands, and use a variety of digital modulation methods to provide data rates in excess of 10 Mbps. High gain antennas are typically used to minimize power requirements and avoid interference with other users.

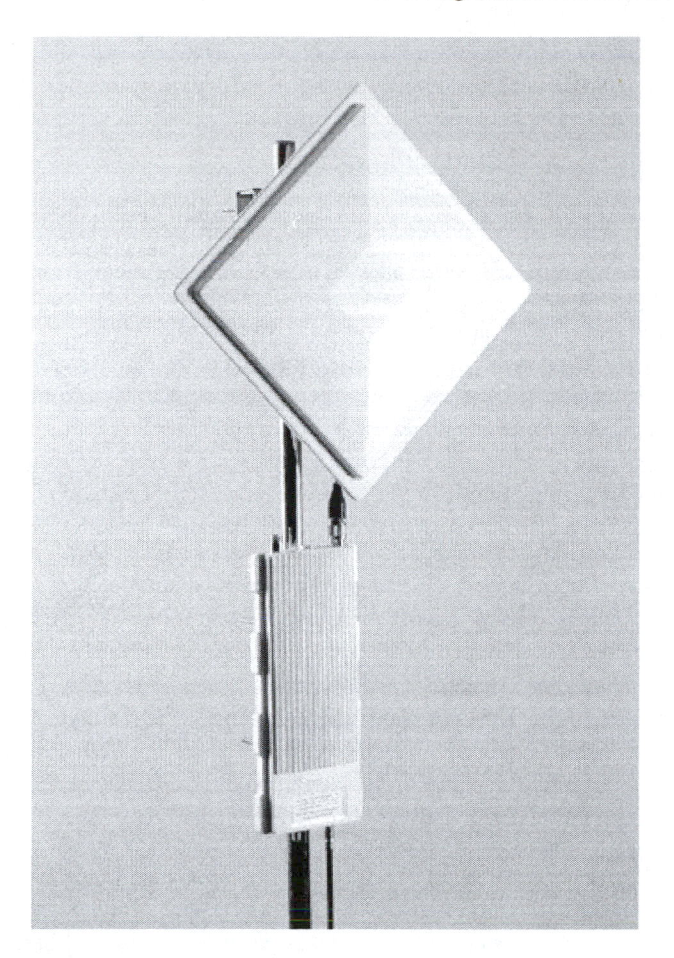

FIGURE 1.3 Photograph of the subscriber antenna and outdoor unit of an MMDS system operating at 2.4–2.6 GHz, providing a data rate of 20 Mb/sec. (Courtesy of N. Herscovici, Spike Technologies, Nashua, NH.)

Radio frequency identification (RFID) systems are used for inventory tracking, shipping, toll collection, personal security access, and other functions. Most express delivery services, for example, use handheld terminals that scan bar codes on packages and relay information to a central station. As another example, available now in several cities, automatic toll collection (ATC) uses a small transponder in an automobile that can be interrogated by an RF system mounted at the entrance to a highway or bridge. The transponder provides the vehicle's account number, which is then debited, and a monthly bill sent to the driver. RFID systems are much more specialized than cellular or WLAN systems, and use a wide range of modulation methods, operating frequencies, and duplexing schemes. It is expected that the market for RFID systems will reach \$1.5 B by the year 2000.

1.2 DESIGN AND PERFORMANCE ISSUES

In this section we discuss general considerations related to the design and performance of wireless systems. These include the choice of RF frequency, duplexing and multiple access methods, and a brief mention of some of the problems associated with propagation through the wireless channel. We will also discuss the differences between communication using a circuit-switched and a packet-switched system, and possible health hazards associated with radiated RF power.

Choice of Operating Frequency

One of the first decisions that must be made during the design of a wireless system is the operating frequency. The choice of a transmit or receive frequency is never completely free, as only small sections of the RF spectrum are available for specific applications. As listed in Appendix A, large portions of the spectrum are allocated to AM radio (550 kHz–1.6 MHz), FM radio (88–106 MHz), broadcast TV (54–88 MHz and 174–806 MHz), and a multitude of radio channels for airport, police, fire, CB, amateur, and other users. In the United States, the Federal Communications Commission (FCC) is responsible for assigning frequency spectrum to competing users. As listed previously in Table 1.1, frequency bands have been reserved for cellular and PCS telephone systems, GPS, DBS, point-to-point radios, and other major wireless applications. An important category is the *Industrial, Scientific, and Medical* (ISM) bands, which reserve three microwave frequency bands for a variety of uses not covered under other spectrum allocations. The ISM bands are used for WLANs, microwave ovens, RFID systems, and medical treatments using microwave power. For this reason, systems operating in the ISM bands are limited to a maximum of 1 W of radiated power.

Besides the availability of spectrum, other important factors influenced by the choice of operating frequency include noise, antenna gain, bandwidth, and cost. Noise power, for example, increases sharply at frequencies below 100 MHz due to a variety of sources that include lightning, ionospheric ducting, and interference from engine ignitions and other electrical equipment. At frequencies above 10 GHz, however, noise power steadily increases due to thermal noise of the atmosphere and interstellar radiation. Noise sources and noise effects are discussed in further detail in Chapters 3 and 4.

As we will see in Chapter 4, the gain of an antenna increases with frequency, for a fixed antenna size. Thus the use of higher frequencies is an advantage for point-to-point wireless systems where high antenna gain is required, as the resulting antenna will be smaller and less obtrusive. Higher gain antennas also receive less noise power from the surrounding environment.

In Chapter 9 we will see that the maximum data rate of a communications channel is determined by the available bandwidth, when noise is present. Thus a wireless system capable of high data rates will require a correspondingly high RF bandwidth, and this is easier to obtain at high frequencies than at low frequencies. For example, for a modulation method having a spectral efficiency of 1 bit per second per Hertz, a 1 Mbps data rate requires 1 MHz of bandwidth. This bandwidth could be obtained with a frequency band from 100 to 101 MHz, or from 10.000 to 10.001 GHz—the lower frequency band requires 1% fractional bandwidth, while the higher frequency band requires only 0.01% fractional bandwidth.

While most of the preceding considerations argue for the use of a high operating frequency, there are points working in favor of lower frequencies as well. One is that the efficiency of RF transistors decreases with frequency, which increases the prime power required to operate wireless transmitters and receivers. This is especially true at millimeter wave frequencies, where active device efficiencies can be as low as 30%. In addition, component cost generally increases with operating frequency, so it is much more economical to build an RF subsystem at frequencies below 1 GHz than at higher frequencies.

Finally, electromagnetic propagation characteristics vary considerably with frequency. Electromagnetic signals at frequencies above a few gigahertz propagate largely in straight line paths, thus requiring an unobstructed line-of-sight path between a wireless transmitter and receiver. At lower frequencies, however, signals can more easily pass through or around obstructions such as foliage, buildings, and vehicles. Thus lower frequencies give better propagation characteristics for wireless applications such as cellular and PCS telephone systems, while higher frequencies may be perfectly adequate for point-to-point radios and satellite systems. As a rough estimate, operating range decreases by 5% to 10% as frequency increases from 900 MHz to 2.4 GHz, and another 10% at 5 GHz.

Multiple Access and Duplexing

Because frequency spectrum is limited, and it is usually desired to accommodate as many simultaneous users as possible, several methods have been proposed for increasing the capacity of wireless channels. One such *multiple access method* is to divide the available frequency range into many narrow frequency bands. This is called *frequency division multiple access* (FDMA). The AMPS telephone system, for example, uses FDMA, dividing the 25 MHz mobile receive (869–894 MHz) and transmit (824–849 MHz) bands each into 833 channels of 30 kHz bandwidth. Another method is *time division multiple access* (TDMA), where voice or data is transmitted and received over a shared frequency band only during preassigned time intervals of very short duration, and interleaved with voice or data segments from other users. TDMA thus multiplies the number of users that can be accommodated with a single channel, but requires critical timing and range information coordinated from a central station. In practice, TDMA is often combined with frequency division duplexing to allow several users for each of several frequency bands. The third popular multiple access method is *code division multiple access* (CDMA). CDMA is a *spread spectrum* technique, whereby the relatively narrowband signal from each user is spread out in frequency using a unique *spreading code*. Several hundred signals can then occupy the same frequency band, and yet be individually recovered at the receiver with knowledge of the original spreading code.

As mentioned earlier, full-duplex wireless communication requires a duplexing method to provide transmit and receive channels that do not interfere with each other. Because of the high sensitivity of most wireless receivers, the isolation between transmitter and receiver is typically required to be on the order of 120 dB. As a practical matter, this much isolation cannot be obtained unless frequency division duplexing is used, with separate frequencies for transmit and receive. A *bandpass filter* at the input to the receiver can then be used to attenuate transmitter signals. Often it is convenient to use a single antenna for both transmit and receive, in which case a *duplexing filter* is used to pass receive frequencies from the antenna to the receiver, and transmit frequencies from the transmitter to the antenna, while providing enough attenuation between the transmit and receive bands to achieve the necessary isolation. A serious drawback of duplexing filters, however, is that they generally have several dB of insertion loss. This leads to the loss of transmit power, and increases the noise figure of the receiver. A commercial duplexing filter is shown in Figure 1.4.

In half-duplex wireless systems, as used in many TDMA telephones and wireless LANs, duplexing can be accomplished by using a *transmit/receive* (T/R) switch. This allows a single antenna to be rapidly switched between the transmitter and the receiver at the appropriate times. Electronic RF switches generally provide more than enough isolation in their off state to protect the receiver from high transmit signal levels.

Circuit Switching versus Packet Switching

Both hard-wired and wireless (cellular and PCS) telephone systems are based on centralized networks that provide a direct physical circuit between the communicating parties for the duration of the call. This is referred to as a *circuit-switching network*. The circuit-switched telephone network has proven to be extremely reliable for voice communications, with a very high *quality of service* (QoS). Circuit-switched communication systems are inefficient, however, when used for transmitting data that occurs in bursts, such as computer data, email, and telemetry data, because the physical circuit is not fully utilized. In these cases, *packet-switched* networks are preferred. In a packet-switched network,

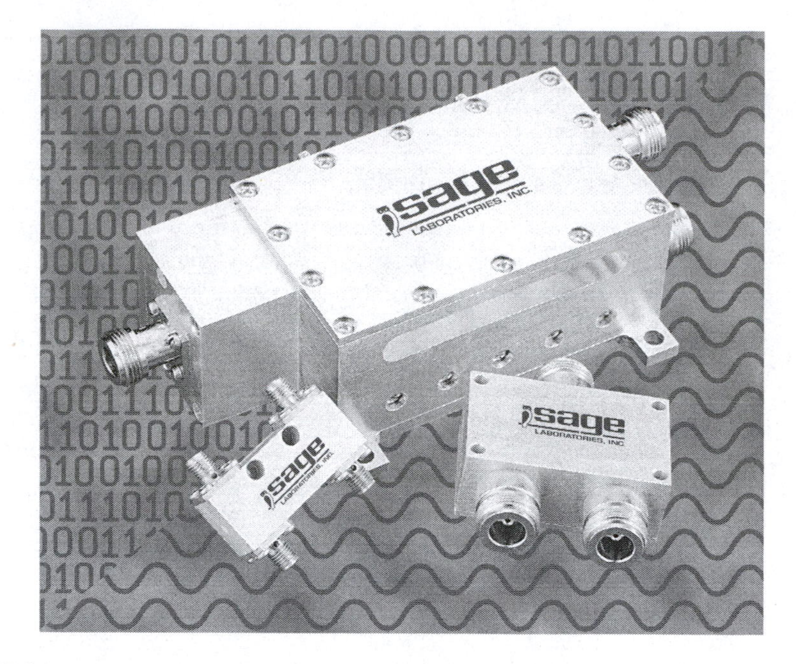

FIGURE 1.4 Photograph of a dual-band diplexer (top), a hybrid coupler (bottom left), and a two-way power divider (bottom right). These components operate over the 800–2200 MHz frequency band, providing coverage of both AMPS and PCS bands in a single component. (Courtesy of Sage Laboratories, Natick, MA.)

interconnected routers are used to provide multiple paths between any two points in the network. Messages and data are divided into packets of fixed length that are independently routed through the network from the sender to the receiver. In this way, messages and data can be multiplexed over various paths through the network, which provides efficient and robust communication links without tying up channel capacity unnecessarily.

The Internet is the most prevalent packet-switched network, and is used extensively for data, email, and multimedia communication between computers. While it is possible to use packet switching for voice communication, the fact that packet switching does not guarantee even a minimum quality of service means that Internet telephone calls often suffer from annoying delays and broken conversations. Newer protocols and standards, however, should improve this situation by implementing packet switching with priority levels that can be used for time-critical connections, such as for voice and real-time video links. In time, we can expect the majority of voice, video, and data communications to take place over packet-switched networks.

Propagation

Wireless communication is made possible by the fact that electromagnetic waves can propagate through space without the need for connecting wires or other conductors. We will see that in free space the power density of an electromagnetic wave radiated by an antenna decreases as $1/R^2$. This simple model, perhaps augmented with a factor to account for atmospheric attenuation, is usually adequate for *line-of-sight* (LOS) radio links, such as point-to-point radios and satellite communications links. In other cases, such as cellular radios in urban environments, or mobile radios in moving vehicles, the phenomena of electromagnetic energy propagation is much more complicated. Effects such as reflections from the ground, buildings, and vehicles, as well as shadowing from natural and man-made obstructions, can cause rapid variations in the amplitude of the received signal over relatively short distances or time intervals. These effects are referred to as *fading*, and are primarily due to the presence of more than one possible propagation path between the transmitter and receiver. Because different propagation paths generally have different phase (or time) delays, the superposition of signals at the receiver will involve constructive and destructive interference, leading to sharp variations in amplitude as much as 20 dB.

The large variation in received signal strength caused by fading is one of the most formidable problems facing the designer of a wireless system. Fading leads to decreased range, lower data rates, and decreased reliability and quality of service. Many of the most sophisticated techniques used in wireless communications have been developed primarily in an attempt to alleviate the degrading effects of fading. These include spread spectrum techniques, the use of antenna diversity, sophisticated modulation methods, and error-correcting codes. In all cases, such techniques increase the cost and complexity of the wireless system.

Radiated Power and Safety

Safety is a legitimate concern of users of wireless equipment, particularly in regard to possible hazards caused by radiated electromagnetic fields. The body absorbs RF and microwave energy and converts it to heat; as in the case of a microwave oven, this heating occurs within the body, and may not be felt at low power levels. Such heating is most dangerous in the brain, eyes, genitals, and stomach organs. Excessive radiation can cause cataracts, cancer, or sterility. For this reason it is important to define a safe radiation level standard, so that users of wireless equipment are not exposed to harmful power levels.

The most recent U.S. safety standard for human exposure to electromagnetic radiation is given by ANSI/IEEE Standard C95.1-1992. In the RF-microwave frequency range of 100 MHz to 300 GHz, exposure limits are set on the power density (in Watts/cm^2) as a function of frequency, as shown in Figure 1.5. The recommended safe power density limit is as low as 0.2 mW/cm^2 at the lower end of this frequency range, because fields penetrate the body more easily at low frequencies. At frequencies above 15 GHz the power density limit rises to 10 mW/cm^2, since most of the power absorption at these frequencies occurs near

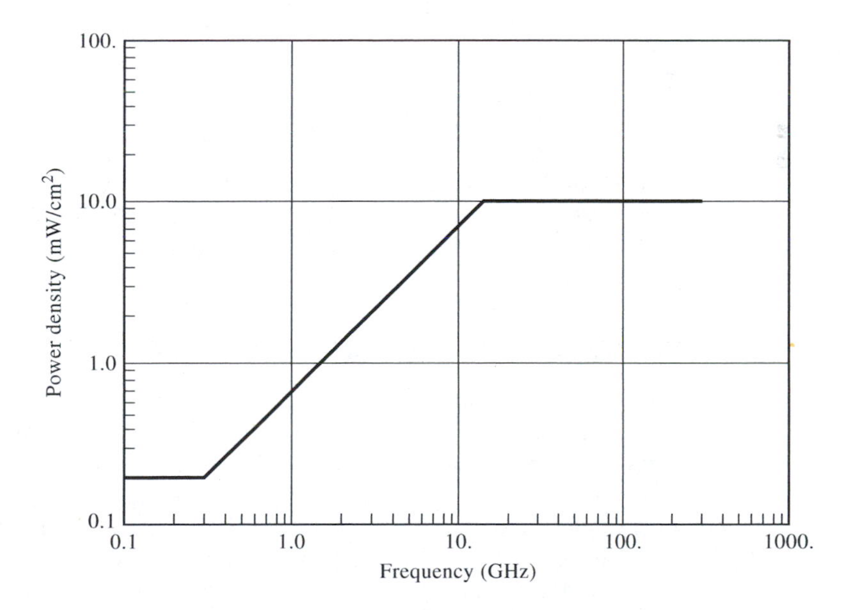

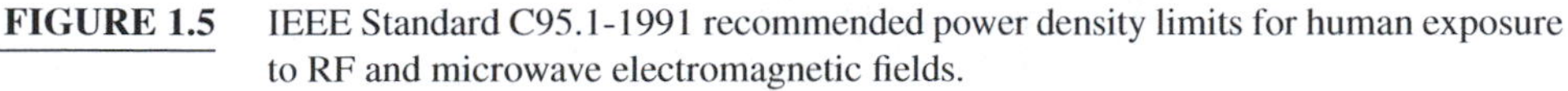

FIGURE 1.5 IEEE Standard C95.1-1991 recommended power density limits for human exposure to RF and microwave electromagnetic fields.

the skin surface. By comparison, the sun radiates a power density as high as 100 mW/cm^2 on a clear day, but the effect of this radiation is much less severe than a corresponding level of microwave frequency radiation because the sun heats the outside of the body, with much of the generated heat being reabsorbed by the air, while microwave power heats inside the body. At frequencies below 100 MHz, electric and magnetic fields interact with the body differently than at higher frequencies, and so separate limits are given for field components at these frequencies.

In addition to the above power density limits, the FCC sets limits on the total radiated power of some specific wireless equipment. Vehicle-mounted cellular phones (using an external antenna) are limited to a maximum radiated power of 3 W. For handheld cellular phones, the FCC has set an exclusionary power level of 0.76 W, below which phones are exempt from the ANSI standard on radiated power density. Most cellular phones radiate a maximum of 0.6 W, and newer PCS phones radiate even less power. Cellular and PCS base stations are limited to a total effective radiated power (see Chapter 4) of 500 W, depending on antenna height and location, but most urban base stations radiate a maximum of 10 W. Wireless products using the ISM bands are limited to a maximum radiated power of 1 W.

While other countries have different (sometimes lower) standards for RF and microwave exposure limits, most experts feel that the above limits represent safe levels with a reasonable safety margin. Some researchers, however, are concerned that health hazards may occur due to nonthermal effects of long-term exposure to even low levels of microwave radiation.

Other Issues

Although the above technical issues are critical to the performance of a wireless system, in fact the overriding consideration for the success of a given commercial wireless system is most often its cost. The cost of a system should of course include manufacturing and production costs, but also the cost of the infrastructure that is necessary to support, operate, and maintain the system. This may involve components such as base stations, antenna towers, satellite replacement costs, insurance, fees for right-of-ways for buried cables, technology licensing fees, advertising, billing, and nonrecoverable engineering costs (NRE).

Many wireless devices are portable and operate from battery power. Battery life is a critical consideration for consumers, so it is important to design for the minimization of prime power requirements through proper component selection, as well as design techniques to minimize power consumption. These may include shutting down parts of the system when their function is not required, and lowering transmit power when possible.

Finally, it should be realized that consumers will expect a wireless system to offer performance that is comparable to the wired system that it replaces. For example, consumers will not find the convenience of a cellular phone to be worthwhile if sound quality is significantly worse than with a wired phone, or if conversations are often interrupted.

1.3 INTRODUCTION TO WIRELESS SYSTEM COMPONENTS

In this section we describe the basic block diagrams for the RF stages of wireless transmitters and receivers, and provide an introductory discussion of the main RF and microwave components that are used in these systems. In later chapters we will discuss the operation and design of each of these components in much more detail, so the purpose here is simply to provide an initial broad view of the overall wireless system. In this way the reader will be able to see the larger context in which these individual components are used in practical wireless systems. Figure 1.6 shows a table of commonly used symbols that are used in block diagrams for RF and microwave components; symbols for filters are shown in Figure 1.7.

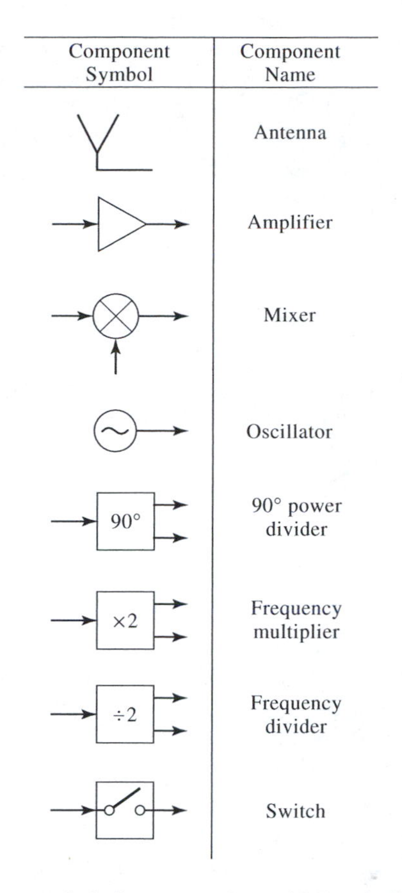

FIGURE 1.6 Block diagram symbols for commonly used RF and microwave components. (Filter symbols are shown in Figure 1.7.)

Basic Radio System

The RF stages of most wireless systems have a high degree of commonality, even though there may be many variations in practice. The block diagrams of a typical wireless transmitter and receiver are shown in Figures 1.8a,b, respectively.

The input to a wireless transmitter may be voice, video, data, or other information to be transmitted to one or more distant receivers. These data are usually referred to as

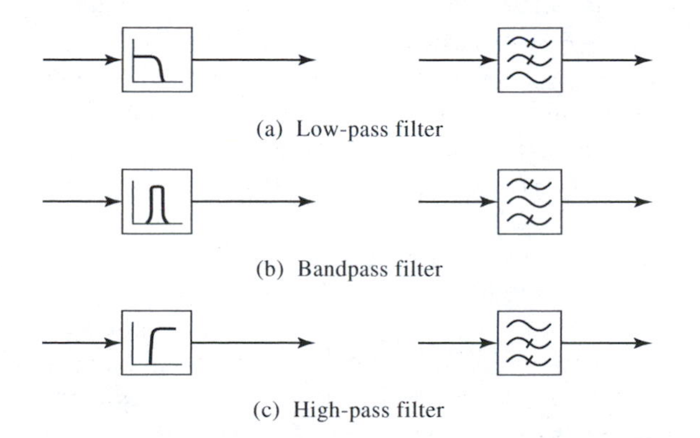

FIGURE 1.7 Symbols for filters: (a) low-pass, (b) bandpass, (c) high-pass.

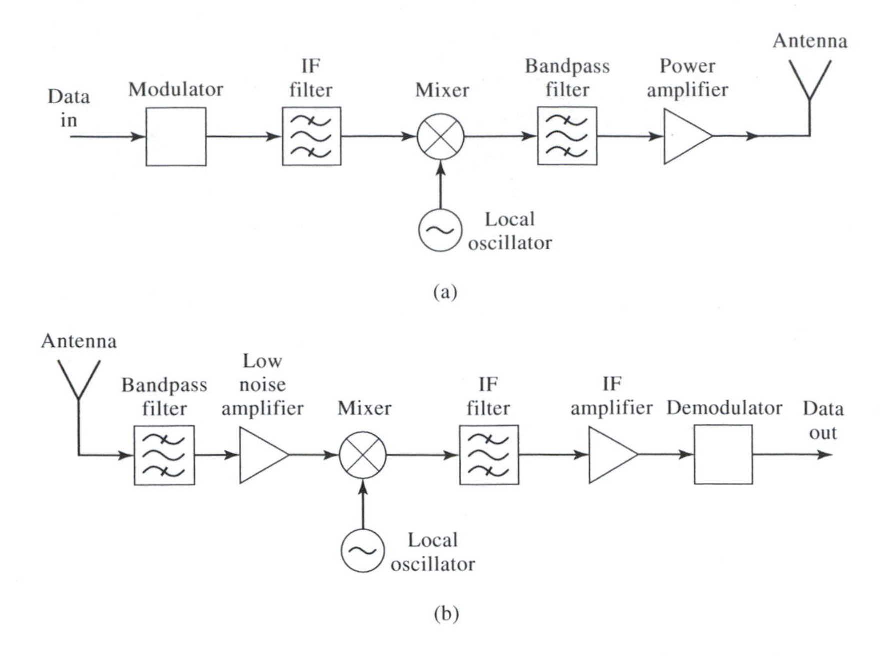

FIGURE 1.8 Block diagram of a basic radio system: (a) radio transmitter, (b) radio receiver.

the *baseband signal*. The basic function of the transmitter is to *modulate*, or encode, the baseband information onto a high frequency sine wave *carrier* signal that can be radiated by the transmit antenna. The reason for this is that signals at higher frequencies can be radiated more effectively, and use the RF spectrum more efficiently, than the direct radiation of the baseband signal. The transmitter of Figure 1.8a operates by first using the baseband data to modulate an intermediate sine wave signal. As discussed in more detail in Chapter 9, there are many possible modulation methods, both analog and digital, that function by varying either the amplitude, frequency, or phase of a sine wave. The output of the modulator is referred to as the *intermediate frequency* (IF) signal, and usually ranges between 10 and 100 MHz. The IF signal is then shifted up in frequency, or *upconverted*, to the desired RF transmit frequency using a *mixer*. The mixer operates by producing the sum and difference of the input IF signal frequency and the frequency of a separate *local oscillator* (LO). A *bandpass filter* (BPF) allows the sum frequency to pass, while rejecting the much lower difference frequency. If necessary, a *power amplifier* is used to increase the output power of the transmitter. Finally, the *antenna* converts the modulated carrier signal from the transmitter to a propagating electromagnetic plane wave.

The receiver of Figure 1.8b recovers the transmitted baseband data by essentially reversing the functions of the transmitter components. The antenna receives electromagnetic waves radiated from many sources over a relatively broad frequency range. A input bandpass filter provides some selectivity by filtering out received signals at undesired frequencies, and passing signals within the desired frequency band. The bandpass filter is followed by a *low-noise amplifier* (LNA), whose function is to amplify the possibly very weak received signal, while minimizing the noise power that is added to the received signal. Placing a bandpass filter before the LNA reduces the possibility that the sensitive amplifier will be overloaded by interfering signals of high power. Next, a mixer is used to *downconvert* the received RF signal to a lower frequency signal, again called the *intermediate frequency* (IF). When the LO is set to a frequency near that of the RF input, the output difference frequency from the mixer will be relatively low (typically less than 100 MHz), and can be easily filtered by the IF bandpass filter. A high gain IF amplifier raises the power level of the signal so that the baseband information can be recovered easily. This process is called *demodulation*, and today is usually performed with digital signal processing (DSP) circuits. As discussed in more detail in Chapter 10, this type of receiver is known as a *superheterodyne* receiver, because it uses *frequency conversion*, converting the relatively high RF carrier frequency to a lower IF frequency before final demodulation.

Antennas

As seen from the preceding discussion, the function of an antenna is to convert an RF signal from a transmitter to a propagating electromagnetic wave or, conversely, convert a propagating wave to an RF signal in a receiver. In a *transceiver*, where a transmitter and a receiver are co-located for full-duplex communications, the same antenna may be used for both transmit and receive.

The aspects of antennas that are important for wireless systems are discussed in detail in Chapter 4, along with the characteristics of the propagation channel between the transmit and receive antennas. Some of the more obvious characteristics of an antenna include operating frequency range, size, and pattern coverage. The *radiation pattern* of an antenna is a plot of the transmitted or received signal strength versus position around the antenna. It can be shown that the radiation pattern of an antenna is the same for transmitting and receiving. Wireless systems that provide broadcast-type service, such as television and AM/FM radio, require antennas with pattern coverage that is uniform in all directions. Such patterns are called *omnidirectional*, and can be obtained with wire dipole and monopole ("whip") antennas, among others. Others systems, such as point-to-point radio and DBS receivers, require antennas that radiate (or receive) power preferentially in one direction. The measure of the directionality of an antenna pattern is provided by the *directivity*, or *gain*, of the antenna; an omnidirectional antenna has low gain, while a highly directive antenna has high gain.

An important characteristic of all antennas is that there are unavoidable relationships between the operating frequency, size, and gain of an antenna [6]–[7]. Because of the nature of the electromagnetic operation of an antenna, effective radiation of a signal requires that the antenna have minimum physical dimensions on the order of the electrical wavelength ($\lambda = c/f$) at the operating frequency. This means that antenna size decreases with an increase in frequency, so that antennas at low frequencies will be very large, while antennas at microwave frequencies and higher may be very small. In addition, it can be shown that the gain of an antenna is proportional to its cross-sectional area divided by λ^2, so that high antenna gain requires an electrically large antenna. Thus a low-gain antenna used for GPS at 1.575 GHz may be as small as a few square inches, while a high gain parabolic dish antenna used in a point-to-point radio in the same frequency band may be several meters in diameter.

More sophisticated antennas are able to change the direction of their main beam electronically. Such antennas are called *phased arrays*, and in the past have generally been limited to use in military systems because of their high cost. Phased array antenna technology, however, can be very useful in commercial wireless systems because the antenna beam can be directed at a given user, while avoiding interference from other users. Such systems are called *adaptive arrays*, or sometimes *smart antennas*, and may lead to increased channel capacity for cellular and PCS telephone systems if cost reductions can be achieved.

Filters

Filters are two-port components that are used to selectively pass or reject signals on the basis of frequency. An ideal *low-pass filter* (LPF) will pass all frequency components below its *cutoff frequency*, while rejecting higher frequency components. Similarly, a *high-pass filter* (HPF) will pass frequency components above its cutoff frequency, while rejecting lower frequencies. A *bandpass filter* (BPF) passes frequency components within a narrow passband, while rejecting frequency components outside the passband. Figure 1.7 shows two sets of block diagram symbols that are commonly used to represent low-pass, bandpass, and high-pass filters.

Filters are key components in all wireless transmitters and receivers. As can be seen from the block diagrams of Figures 1.8a,b, filters are used to reject interfering signals outside the operating band of receivers and transmitters, to reject unwanted products from the outputs of mixers and amplifiers, and to set the IF bandwidth of receivers. Important filter parameters include the cutoff frequency, insertion loss, and the out-of-band attenuation rate, measured in dB per decade of frequency. Filters with sharper cutoff responses provide more rejection of out-of-band signals. Insertion loss, measured in dB, is the amount of attenuation seen by signals through the passband of the filter. Another important consideration is size and integrability with other circuit components. Today much of the front-end circuitry of receivers and transmitters in the heavily used frequency range from 800 MHz to 2 GHz can be monolithically integrated into a few integrated circuit packages. At the present time, however, it is not possible to construct high-performance bandpass filters in integrated circuit form. The inherent losses associated with RF and microwave integrated circuits leads to filters having relatively high insertion losses and low out-of-band attenuation rates. For this reason, most wireless systems today use individual "off-chip" filters that are located

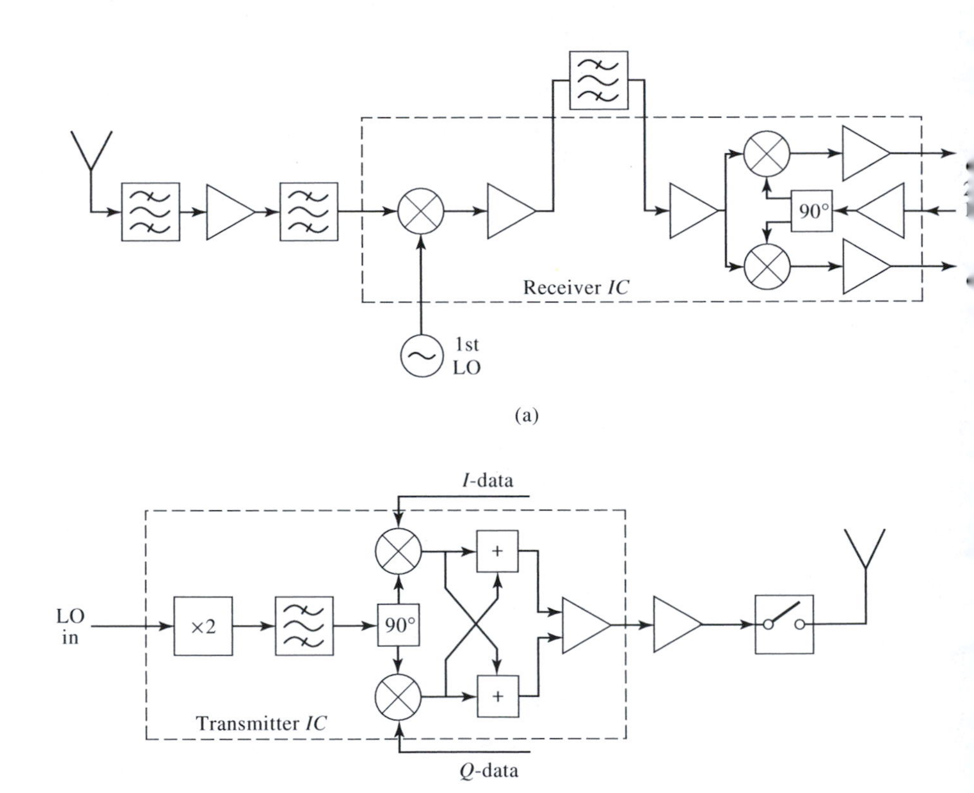

FIGURE 1.9 RF block diagrams for a 900 MHz GSM cellular telephone receiver and transmitter. Each subsystem is highly integrated with commercial RF integrated circuits, but note that the required bandpass filters are not part of the integrated circuit packages: (a) receiver block diagram, (b) transmitter block diagram.

on the circuit board, rather than fully integrated filters. This results in a larger and more costly assembly, but critical filtering performance is optimized. Figure 1.9 shows the block diagrams for a commercial GSM telephone handset with a relatively high level of integration, where the necessary bandpass filters are separate from the integrated circuit packages.

There are many technologies available for the implementation of RF and microwave filters [8], primarily dependent on frequency. In the frequency range from 800 MHz to about 4 GHz, most bandpass filters today are made with dielectric resonators, which offer small size and high Q (sharp cutoff), with reasonable insertion loss. At IF frequencies (below 100 MHz) bandpass filters using quartz crystals or surface acoustic wave (SAW) devices are very common. SAW filters have very sharp cutoffs, but suffer from the disadvantage of insertion losses that may be as high as 20 dB. At higher microwave and millimeter wave frequencies, waveguide resonators are often used for bandpass filters. Low-pass filters used in wireless systems usually have less stringent requirements than do bandpass filters, and thus are often made with simple LC networks, parallel coupled lines, or transmission line stubs [8].

Amplifiers

There are three main categories of amplifiers used in wireless systems: *low-noise amplifiers* (LNAs), used in the input stage of a receiver; *power amplifiers* (PAs), used in the output stage of a transmitter; and *IF amplifiers*, used in the IF stages of both receivers and transmitters. Important specifications for amplifiers include the *power gain* (in dB), the *noise figure*, and the *intercept points*. The noise figure of an amplifier is a measure of how much noise is added to the amplified signal by the amplifier circuitry. This is most critical in the front end of a receiver, where the input signal level is very small, and it is desired to minimize the noise added by the receiver circuitry. In addition, as we will see in Chapter 3, the noise power in a receiver is affected more by the first few components than by later components. Thus it is imperative that the first amplifier in a receiver have as low a noise figure as possible.

Because transistors are nonlinear devices, transistor amplifiers exhibit nonlinear effects. Two important phenomenon that occur in amplifiers because of these effects are *saturation* and *harmonic distortion*. At low signal levels the output power of an amplifier is linearly proportional to the input power. But because the output voltage of an amplifier cannot exceed the bias voltage level, output power gradually reaches a saturation point as input power increases. Saturation is usually only an issue with power amplifiers.

A more prevalent problem is related to the fact that harmonics of input signals are generated at the output of an amplifier, and in the case of multiple input signal frequencies some of these harmonics will lie within the passband of the amplifier. These harmonics lead to signal distortion (harmonic distortion). Generally the power level of distortion harmonics is very low but, as shown in Chapter 3, the power level of some of these distortion products increases as the cube of the input signal level. The implication of this effect is that distortion power can be significant even for input power levels well below the saturation point of an amplifier. In practice it is often desired to keep distortion levels as low as 50 to 80 dB below the output signal level.

Advances in semiconductor technology have led to the development of RF amplifiers using inexpensive silicon (Si) transistors at frequencies up to several GHz. Previously gallium arsenide (GaAs) transistors were required for frequencies at or above 1 GHz, but GaAs processing is very expensive and incompatible with silicon-based integrated circuit fabrication. This limits the level of integration that can be achieved in a wireless system, and therefore increases cost. Another semiconductor technology that is very promising is silicon germanium (SiGe), which can be used at higher frequencies than silicon, but with lower cost than gallium arsenide.

The design of transistor amplifiers is discussed in Chapter 6.

Mixers

A mixer is a three-port component that ideally forms the sum and difference frequencies from two sinusoidal inputs. This allows the important function of *frequency conversion* to be performed in superheterodyne transmitters and receivers. In the case of the transmitter shown in Figure 1.8a, the modulated baseband signal is upconverted in frequency by mixing with a high-frequency local oscillator signal. In a superheterodyne receiver the received signal is downconverted in frequency by mixing with a local oscillator to produce a difference frequency (the IF frequency). In both cases filters are required to select the desired frequency products, while rejecting undesired frequencies that are produced as a by-product of the mixing operation.

In principle, frequency conversion can be accomplished with either nonlinear devices (diodes, transistors), or time-varying elements (switches). Modern mixers generally use diodes or transistors and produce many frequencies, based on the harmonics of the input signals and their combinations, in addition to the desired sum and difference frequencies. A passive mixer (one that uses diodes) always produces an output signal (IF) of less power than the input (RF) signal, because of dissipative losses in the mixer as well as inherent losses in the frequency conversion process. This loss is characterized by the mixer *conversion loss*. Mixers that use active components (e.g., transistors) generally have lower conversion loss, and may even have conversion gain. As in the case of amplifiers, harmonic distortion and noise are also important considerations in mixer performance.

In some receivers a low-noise amplifier may not be required if the received signal level is strong enough. This cost-saving measure results in the mixer being the first component in the front end (second if a bandpass filter is used ahead of the mixer), which means that the noise and loss characteristics of the mixer will dominate the performance of the receiver. Another potential problem with this approach is that power from the local oscillator may leak backwards through the mixer and be radiated by the receiver antenna. Such radiation must be minimized in order to avoid interference with other users and other systems, so there is often a specification on the *LO-to-RF isolation* for mixers. Active mixers generally have very good isolation, because transistors are usually unilateral to a good degree. In addition, as described in more detail in Chapter 7, certain mixer circuits can yield very high isolation. Overall, however, mixer design usually involves trade-offs between noise performance, isolation, and conversion loss.

Oscillators

Oscillators are required in wireless receivers and transmitters to provide frequency conversion, and to provide sinusoidal sources for modulation. Typical transmitters and receivers may each use as many as 4–6 oscillators, at frequencies ranging from several kilohertz to many gigahertz. Often these sources must be tunable over a set frequency range, and must provide very accurate output frequencies (often to within a few parts per million).

The simplest oscillator uses a transistor with an LC network to control the frequency of oscillation. Frequency can be tuned by adjusting the values of the LC network, perhaps electronically with a varactor diode. Such oscillators are simple and inexpensive, but suffer from the fact that the output frequency is very susceptible to variations in supply voltage, changing load impedances, and temperature variations. Better frequency control can be obtained by using a quartz crystal in place of the LC resonator. A *crystal-controlled oscillator* (XCO) can provide a very accurate output frequency, especially if the crystal is in a temperature controlled environment. Crystal oscillators, however, cannot easily be tuned in frequency. A solution to this problem is provided by the *phase-locked loop* (PLL), which uses a feedback control circuit and an accurate reference source (usually a crystal controlled oscillator) to provide an output that is tunable with very high accuracy.

Phase-locked loops and other circuits that provide accurate and tunable frequency outputs are called *frequency synthesizers*. Virtually all modern wireless systems rely on frequency synthesizer circuits for the key stages of frequency conversion. Important parameters that characterize frequency synthesizers are tuning range, frequency switching time, frequency resolution, cost, and power consumption. Another very important parameter is the noise associated with the output spectrum of the synthesizer, in particular the *phase noise*. Phase noise is a measure of the sharpness of the frequency domain spectrum of an oscillator, and is critical for many modern wireless systems. Phase noise and oscillator design will be studied in more detail in Chapter 8.

Baseband Processing

Our main concern in this book is with the RF stages of transmitters and receivers, in contrast to the processing of signals between the IF stages and the baseband input or output data. Nevertheless, it is worthwhile to say a few words here about what happens to the signal in a receiver after the IF stage, in order to have a more complete view of the overall wireless system.

After down conversion to an IF signal (which may occur in two or more stages), the received signal must be demodulated. The majority of wireless systems today utilize coherent digital modulation methods (discussed in Chapter 9), for which demodulation requires a local oscillator synchronized in both frequency and phase with the down-converted carrier signal. These processes, called *carrier acquisition* and *carrier synchronization*, have traditionally been very difficult to implement, but the advent of powerful digital signal processing (DSP) chips allows these functions to be performed easily and inexpensively. Demodulated baseband data can then be obtained from the output of the DSP stage, perhaps even including error correction. If the baseband information is analog, as in the case of cellular telephones, the received digital data will be converted back to analog form with a *digital-to-analog converter* (DAC). Similarly, the transmission of baseband analog information would first involve conversion to digital form using a sampler and an *analog-to-digital converter* (ADC). Framing and multiplexing functions may also be performed on the digital baseband data.

1.4 CELLULAR TELEPHONE SYSTEMS AND STANDARDS

With the preceding introduction to wireless systems and some of the key components used in receivers and transmitters, we can now look at cellular telephone systems in more detail. Cellular telephony is the most significant, and perhaps complex, application of wireless technology, and many of the topics discussed in later chapters are of direct importance to cellular systems.

Cellular and the Public Switched Telephone Network

As discussed earlier, cellular telephony represents by far the largest commercial application of wireless technology. While there are many different standards and systems in worldwide use, all rely on the concept of cellular coverage regions for frequency reuse, and rely on circuit-switched public telephone networks to transfer calls between users.

Figure 1.10 shows how a geographical area can be covered with hexagonal-shaped cellular regions. Cellular telephone users within each cell are serviced by a base station at

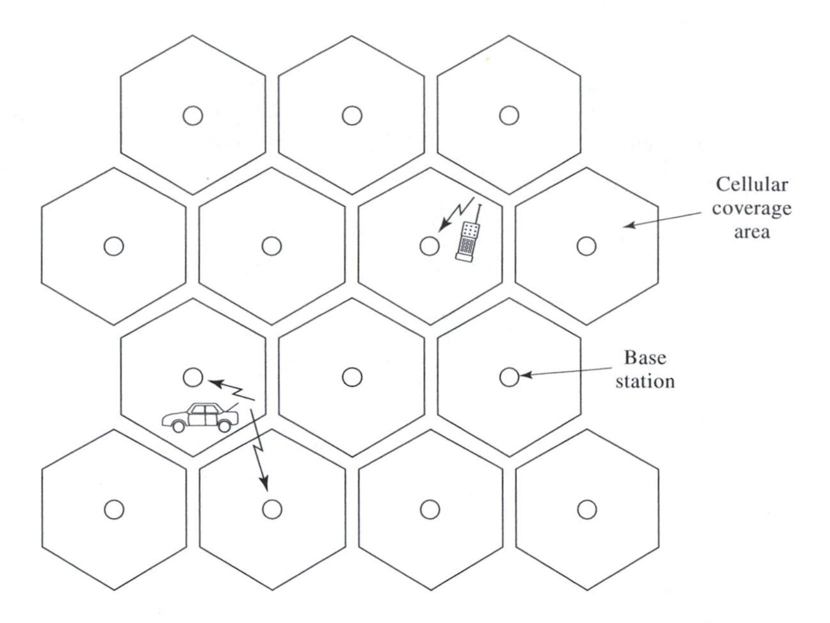

FIGURE 1.10 Layout of hexagonal cell areas and base stations for cellular radio systems.

the center of that cell. To avoid co-channel interference, adjacent cells are assigned different sets of channel frequencies. Frequencies can be reused at two different cells when there is at least one intervening cell with different frequency assignments. *Frequency reuse* is one of the key advantages of cellular radio systems because it permits more efficient utilization of valuable radio spectrum. This method is generally used for FDMA, TDMA, and CDMA multiple access systems. In the case of CDMA, further interference suppression is obtained through the use of spread spectrum techniques. (For marketing reasons, service providers often distinguish between "cellular telephone service" and PCS, but PCS still employs a cellular radio system.)

In operation, a cellular telephone user communicates with the closest base station, even though it is likely that an adjacent base station may receive a weaker signal from the same user. If the user is mobile, a *hand-off* from one base station to the next will occur when the received signal power from the closer base station becomes greater than the received signal power at the original base station. Ideally, this switchover occurs quickly and reliably, and is not noticed by the user.

All the base stations within a given geographical area are connected to a *mobile telephone switching office* (MTSO), which typically can handle several thousand simultaneous telephone calls. The MTSO provides connections to the *public switched telephone network* (PSTN), as shown in Figure 1.11. The PSTN includes high-capacity fiber-optic lines between cities, as well as transoceanic lines between countries. Local telephone exchange offices provide connections to individual private and business users, generally with twisted copper wire pairs. All cellular phone calls thus are routed through the PSTN, even when both parties are using cellular phones. In some newer PCS systems, however, callers may be connected through the base station if they are within the same cell site.

AMPS Cellular Telephone System

Like other first generation analog cellular systems, the AMPS (Advanced Mobile Phone Service) system was based on technology of the 1970s. While most other first generation

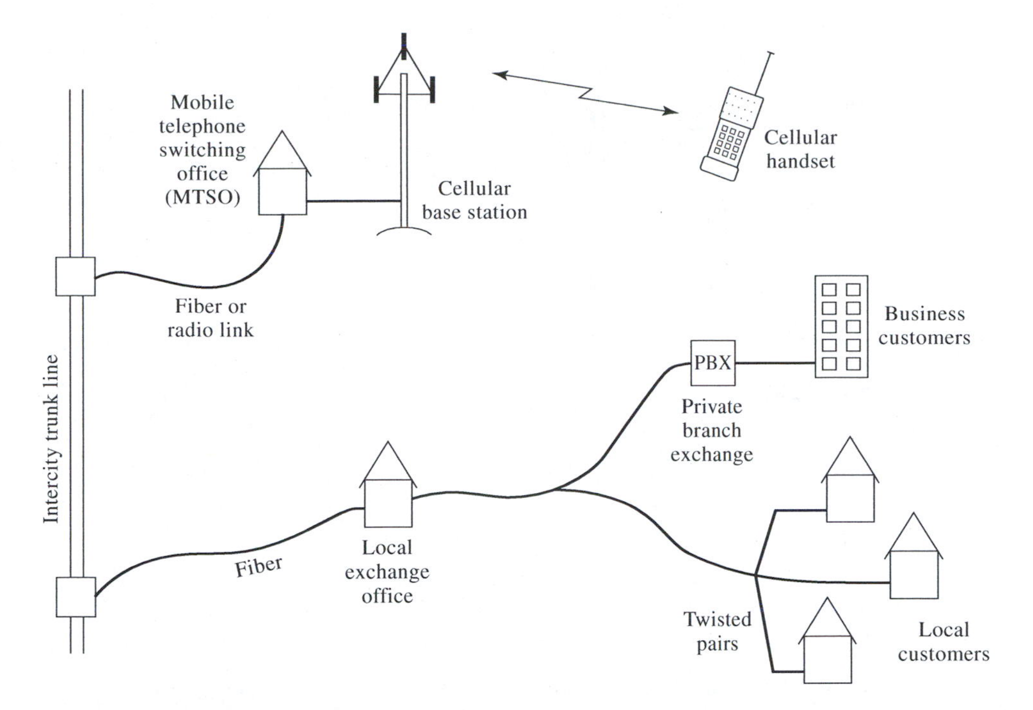

FIGURE 1.11 Pictorial diagram showing the connection of a cellular telephone and base station to the public switched telephone network.

systems have been replaced with digital cellular, for reasons discussed in Section 1.1 the U.S. AMPS system is only slowly being supplanted by newer technology.

The mobile transceiver in the AMPS system uses a transmit frequency in the range of 824–849 MHz, and a receive frequency in the range of 869–894 MHz. Both of these bands are divided into 832 channels that are 30 kHz wide. The maximum frequency deviation of the frequency modulated (FM) signal is 25 kHz, allowing a 5 kHz guard band on each channel [2], [4]. Since separate frequencies are used for transmit and receive, full duplex operation is provided (an example of frequency division duplexing). Multiple users within a cell are assigned different transmit and receive channels (frequency division multiplexing). A typical cell in the AMPS system has an area of about 10 square miles.

Figure 1.12 shows the block diagram of a typical AMPS transceiver. The receiver is dual-conversion, meaning that there are two stages of frequency down-conversion, with two mixers and two local oscillators. The desired received channel is selected at the first IF stage, which uses a phase-locked loop to provide the proper local oscillator frequency. The transmitter is single-conversion, but again a phase-locked loop is used to provide the necessary carrier frequency. The transmit-receive duplexer operates as a channel separation filter, passing the higher frequency band to the receiver circuitry while blocking the lower frequency transmitted signal; the converse operation is performed between the transmitter and antenna. The duplexer is usually a ceramic resonator filter. Early mobile transceivers used discrete components, and were large and had short battery life. Newer transceivers have most of the circuitry of Figure 1.12 integrated with one or two chips, except for the filters.

Communication between a mobile cellular phone and the base station involves four separate simplex (one-way) channels, listed as follows:

FVC—forward voice channel (base to mobile)
RVC—reverse voice channel (mobile to base)

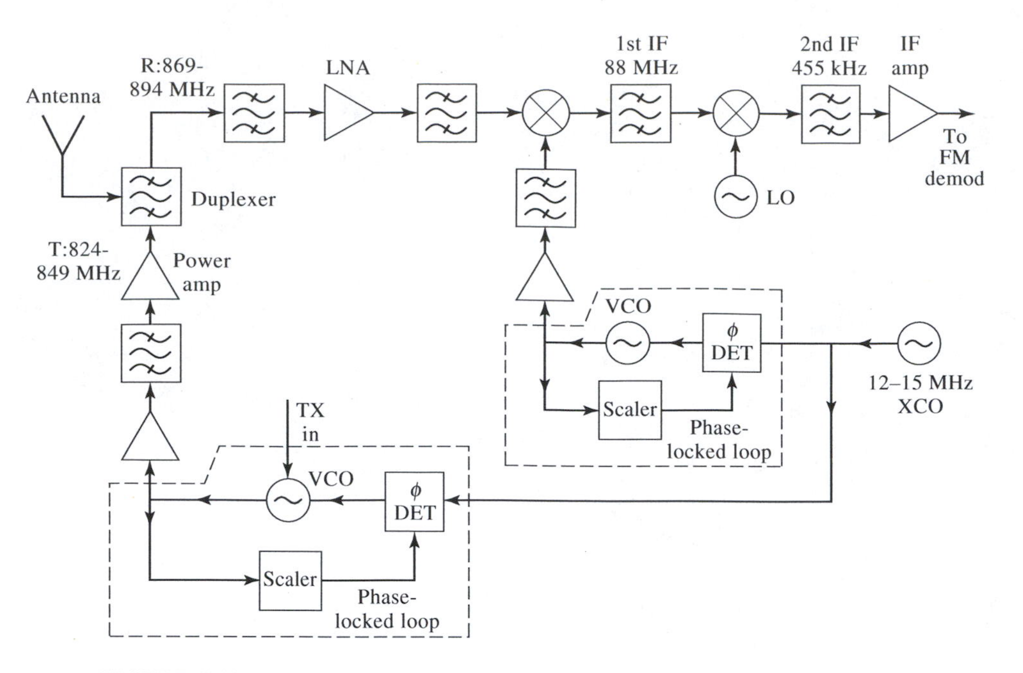

FIGURE 1.12 Block diagram of an AMPS mobile transceiver.

FCC—forward control channel (base to mobile)
RCC—reverse control channel (mobile to base)

The two voice channels are used for voice transmission and reception, while the two control channels are used for data messages that control call initiation, termination, and hand-offs. Each base station uses a dedicated FCC and RCC for all users within its cell. When an AMPS cellular telephone is first turned on, it scans the group of preassigned system FCC channels, and determines which FCC signal is strongest. This FCC is assumed to belong to the closest base station. When the signal strength of the FCC drops below a certain level, the mobile unit again scans the FCCs to pick out a new base station, and re-establishes contact over a new FCC.

When a call is placed to a mobile telephone, the MTSO sends out the request to every base station in the system, which broadcasts the called phone number over the FCCs. If the mobile phone is on, and within the coverage area of the system, it responds by sending its *mobile identification number* (MIN) over the RCC. The MTSO then sets up the call by having the base station assign an unused FVC and RVC, and sends a ring signal to the mobile phone. When a call is made from a mobile phone, a call initiation request is sent to the base station over the RCC, along with the MIN and the number being called. The MTSO makes the connection through the PSTN, and assigns voice channels with the base station. These procedures typically occur in a few seconds. Further details of cellular telephone protocols can be found in reference [4].

Digital Personal Communications System Standards

Second generation cellular telephone systems use digital modulation methods, with the primary advantage of allowing more users in a given frequency bandwidth. Additional features may include lower transmit powers, longer battery life, the use of error correcting

TABLE 1.4 International Digital PCS System Standards

PCS System	IS-54/IS-136	IS-95	GSM
Transmit Frequency (RVC)	824–849 MHz	824–849 MHz	890–915 MHz
Receive Frequency (FVC)	869–894 MHz	869–894 MHz	935–960 MHz
Duplexing Method	FDD	FDD	FDD
Multiple Access Method	TDMA	CDMA	TDMA
Channel Bandwidth	30 kHz	1.25 MHz	200 kHz
Modulation	QPSK	QPSK	GMSK
Channel Bit Rate	48.6 kbps	1,228.8 kbps	270.833 kbps
Users per Channel	3	64	8
User Bit Rate	8 kbps	1.2–9.6 kbps	13 kbps
Number of Users	2,496	15,960	992

codes for improved Quality of Service, and the possible use of encryption for privacy. As in the case of first generation analog systems, there are a multitude of competing systems proposed and in use for digital PCS. In the United States, digital PCS systems have initially been deployed using the same frequency bands as the AMPS system, but newer PCS services are using a frequency band near 1900 MHz that was recently allocated by the FCC, and auctioned to service providers in 1995. In Europe and some other countries, second generation cellular services use frequency bands at either 900 MHz or 1800 MHz. Japan currently has several PCS services operating at various bands from 800 MHz to 1500 MHz and at 1900 MHz.

Specifications for the three most commonly used digital PCS standards are listed in Table 1.4. *Interim standards* (IS) are communications standards that have been agreed upon by members of the Telecommunications Industry Association. The two competing PCS standards in the US are IS-54, a TDMA system, and IS-95, which uses CDMA. (IS-136 is an upgraded version of IS-54.) The predominant PCS system in Europe and many other countries is GSM, which uses TDMA technology. While first generation cellular systems were intended only for voice communications, most second generation systems can provide some basic data services, such as paging, fax, and low-data rate access to computer networks. Such services will be further enhanced when third generation personal communication systems are implemented.

REFERENCES

[1] National Research Council, The **Evolution of Untethered Communications**, National Academy Press, Washington, D.C., 1997.

[2] L. E. Larson, **RF and Microwave Circuit Design for Wireless Communications**, Artech House, Dedham, MA, 1996.

[3] R. Schneiderman, **Wireless Personal Communications**, IEEE Press, Piscataway, NJ, 1994.

[4] T. S. Rappaport, **Wireless Communications: Principles and Practice**, Prentice Hall, Englewood Cliff, NJ, 1996.

[5] J. Hern, **GPS–A Guide to the Next Utility**, Trimble Navigation, Sunnyvale, CA, 1989.

[6] C. A. Balanis, **Antenna Theory: Analysis and Design**, 2nd edition, Wiley, New York, 1997.

[7] W. L. Stutzman and G. A. Thiele, **Antenna Theory and Design**, 2nd edition, Wiley, New York, 1998.

[8] D. M. Pozar, **Microwave Engineering**, 2nd edition, Wiley, New York, 1998.

PROBLEMS

1.1 How many U.S. PCS/cellular telephone subscribers were there in the previous year? What was the revenue generated by this market? How many cellular users were there worldwide? Use these data in conjunction with the figures presented in the beginning of this chapter to estimate the number of U.S. subscribers for the current and the following years.

1.2 Gather data on the market size for wireless local area networks in the United States during the last five years. Plot this information, and extrapolate the curve to estimate the market for WLANs for the next three years.

1.3 Consider the frequency spectrum from 50 MHz to 2 GHz. Using data from Appendix A on frequency allocations, find the percentage of this frequency band that has been freely allocated to broadcast FM radio and television in the United States. Compare this to the percentage that is allocated to wireless systems such as cellular/PCS telephone, GPS, paging, and the ISM bands. Write a short essay discussing your opinion of this situation. Should policies regulating the frequency spectrum be changed to better serve the public?

1.4 Estimate the amount of energy required to operate a typical cellular telephone for one minute of talk time. If this phone is charged for one hour each day by using a solar panel with an area of $6'' \times 6''$, estimate the amount of talk time that can be obtained daily. (Obtain reasonable estimates for the power consumption of a typical cellular phone, the efficiency of solar cells, and the average power density of sunlight, and list your sources for this information. Show your work, and discuss your assumptions.)

1.5 Research consumer satisfaction with satellite-based telephones, and compare with the satisfaction of cellular/PCS subscribers. Consider monthly costs, availability of service, and quality of service. Can you draw any conclusions as to the long-term outcome of land-based versus satellite-based wireless telephone service?

Chapter Two

Transmission Lines and Microwave Networks

Transmission lines are essential components in modern wireless systems, being used to connect antennas to transmitters and receivers, for impedance matching in mixers and amplifiers, and as resonant elements in oscillators and filters. It is likely that the reader is already familiar with the fundamental topics of transmission line wave propagation, reflection, transmission, impedance transformation, and the Smith chart, but we will discuss this material here for completeness, and for those who may need a review of these topics. We also require some familiarity with S parameters, impedance matching techniques, and basic microwave network analysis for use in later chapters, and so these topics are also presented. Our treatment of these topics will be more than adequate for the purposes of this book, but the reader who wants to delve deeper into these topics can refer to references [1]–[2]. Finally, we note that our presentation of transmission lines and networks can be accomplished from a circuit model perspective, without recourse to electromagnetic analysis. The reader should realize, however, that transmission line theory ultimately is based on the rigorous application of electromagnetics.

2.1 TRANSMISSION LINES

The key difference between standard circuit analysis and transmission line theory is electrical size. Circuit analysis assumes that the physical dimensions of a network are much smaller than the electrical wavelength, while transmission lines may be a considerable fraction of a wavelength, or many wavelengths, in size. Thus a transmission line is a distributed-parameter network, where voltages and currents can vary in magnitude and phase over the length of the line. We begin our treatment of transmission lines with a lumped circuit model for an incremental length of line, and then study the transmission and reflection of electric waves on the line.

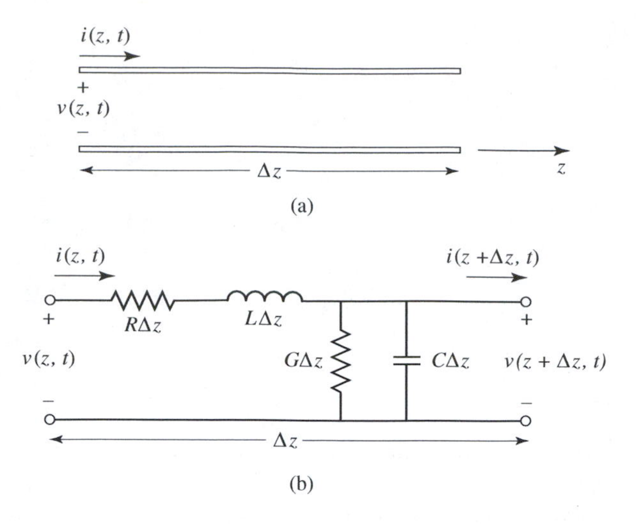

FIGURE 2.1 Voltage and current definitions and the equivalent circuit for an incremental length of transmission line. (a) Voltage and current definitions, (b) Lumped-element equivalent circuit.

Lumped Element Model for a Transmission Line

As shown in Figure 2.1a, a transmission line is often schematically represented as a two-wire line, since transmission lines usually consist of two parallel conductors. A short segment Δz of transmission line can be modeled as a lumped-element circuit, as shown in Figure 2.1b, where R, L, G, and C are per unit length quantities defined as follows:

R = series resistance per unit length, for both conductors, in Ω/m.

L = series inductance per unit length, for both conductors, in H/m.

G = shunt conductances per unit length, in S/m.

C = shunt capacitance per unit length, in F/m.

The series inductance L represents the total self-inductance of the two conductors, and the shunt capacitance C is due to the close proximity of the two conductors. The series resistance R represents the resistance due to the finite conductivity of the conductors, and the shunt conductance is due to dielectric loss in the material between the conductors. R and G, therefore, represent loss. A finite length of transmission line can be viewed as a cascade of sections of the form of Figure 2.1b.

From the circuit of Figure 2.1b, Kirchoff's voltage law can be applied to give

$$v(z,t) - R\Delta z i(z,t) - L\Delta z \frac{\partial i(z,t)}{\partial t} - v(z+\Delta z, t) = 0, \tag{2.1a}$$

and Kirchoff's current law leads to

$$i(z,t) - G\Delta z v(z+\Delta z, t) - C\Delta z \frac{\partial v(z+\Delta z, t)}{\partial t} - i(z+\Delta z, t) = 0. \tag{2.1b}$$

Dividing (2.1a) and (2.1b) by Δz and taking the limit as $\Delta z \rightarrow 0$ gives the following differential equations for the voltage and current on the line:

$$\frac{\partial v(z,t)}{\partial z} = -Ri(z,t) - L\frac{\partial i(z,t)}{\partial t}, \tag{2.2a}$$

$$\frac{\partial i(z,t)}{\partial z} = -Gv(z,t) - C\frac{\partial v(z,t)}{\partial t}. \tag{2.2b}$$

These equations are the time-domain form of the transmission line, or *telegrapher*, equations. For the sinusoidal steady-state condition, with cosine-based phasors, (2.2a) and (2.2b) simplify to

$$\frac{dV(z)}{dz} = -(R + j\omega L)I(z), \tag{2.3a}$$

$$\frac{dI(z)}{dz} = -(G + j\omega C)V(z). \tag{2.3b}$$

The solution of these differential equations leads to traveling voltage and current waves on the transmission line.

Wave Propagation on a Transmission Line

Equations (2.3a) and (2.3b) can be solved simultaneously to give a single wave equation for either $V(z)$ or $I(z)$, by eliminating either $I(z)$ or $V(z)$:

$$\frac{d^2V(z)}{dz^2} - \gamma^2 V(z) = 0, \tag{2.4a}$$

$$\frac{d^2I(z)}{dz^2} - \gamma^2 I(z) = 0, \tag{2.4b}$$

where

$$\gamma = \alpha + j\beta = \sqrt{(R + j\omega L)(G + j\omega C)}, \tag{2.5}$$

is the complex propagation constant. The imaginary part, β, of the complex propagation constant is called the *phase constant*, while the real part, α, is the *attenuation constant*. Note that the propagation constant is generally a function of frequency.

Traveling wave solutions to (2.4) can be found as

$$V(z) = V_0^+ e^{-\gamma z} + V_0^- e^{\gamma z}, \tag{2.6a}$$

$$I(z) = I_0^+ e^{-\gamma z} + I_0^- e^{\gamma z}, \tag{2.6b}$$

where the $e^{-\gamma z}$ term represents wave propagation in the $+z$ direction, and the $e^{\gamma z}$ term represents wave propagation in the $-z$ direction. Applying (2.3a) to the voltage of (2.6a) gives the current on the line:

$$I(z) = \frac{\gamma}{R + j\omega L}\left[V_0^+ e^{-\gamma z} - V_0^- e^{\gamma z}\right].$$

Comparison with (2.6b) shows that if a *characteristic impedance*, Z_0, is defined as

$$Z_0 = \frac{R + j\omega L}{\gamma} = \sqrt{\frac{R + j\omega L}{G + j\omega C}}, \tag{2.7}$$

then the voltage and current on the transmission line can be related as

$$\frac{V_0^+}{I_0^+} = Z_0 = \frac{-V_0^-}{I_0^-}.$$

Then (2.6b) can be rewritten in the following form:

$$I(z) = \frac{V_0^+}{Z_0} e^{-\gamma z} - \frac{V_0^-}{Z_0} e^{\gamma z}. \tag{2.8}$$

Converting the phasor voltage of (2.6a) to the time domain gives

$$v(z, t) = \left|V_0^+\right| \cos(\omega t - \beta z + \phi^+) e^{-\alpha z} + \left|V_0^-\right| \cos(\omega t + \beta z + \phi^-) e^{\alpha z}, \tag{2.9}$$

where $\theta^{\pm}$ is the phase angle of the complex voltage $V_0^{\pm}$. The *wavelength* of the traveling waves is defined as the distance between two successive points of equal phase on the wave at a fixed instant of time, which is seen to be

$$\lambda = \frac{2\pi}{\beta}. \tag{2.10}$$

The *phase velocity* of the wave is defined as the speed at which a constant phase point travels down the line, and is given by

$$v_p = \frac{dz}{dt} = \frac{\omega}{\beta} = \lambda f, \tag{2.11}$$

since $\omega = 2\pi f$.

Lossless Transmission Lines

The preceding results apply to general transmission lines, including loss effects, and it is seen that the propagation constant and characteristic impedance are complex. In many practical cases, however, the loss of the line is very small and can be neglected, resulting in a simplification of the above results. Thus, setting $R = G = 0$ in (2.5) gives the propagation constant as

$$\gamma = \alpha + j\beta = j\omega\sqrt{LC},$$

or,

$$\beta = \omega\sqrt{LC}, \tag{2.12a}$$

$$\alpha = 0. \tag{2.12b}$$

As expected for the lossless case, the attenuation constant α is zero. The characteristic impedance of (2.7) reduces to

$$Z_0 = \sqrt{\frac{L}{C}}, \tag{2.13}$$

which is now a real number. The general solutions for voltage and current on a lossless transmission line can then be written as

$$V(z) = V_0^+ e^{-j\beta z} + V_0^- e^{j\beta z}, \tag{2.14a}$$

$$I(z) = \frac{V_0^+}{Z_0} e^{-j\beta z} - \frac{V_0^-}{Z_0} e^{j\beta z}. \tag{2.14b}$$

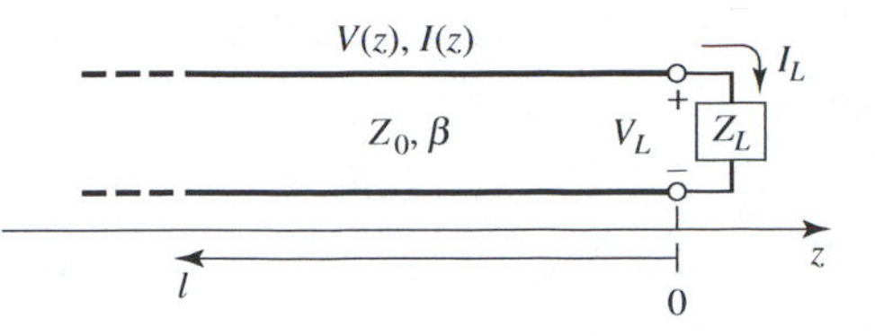

FIGURE 2.2 A transmission line terminated in a load impedance Z_L.

The wavelength on the line is

$$\lambda = \frac{2\pi}{\beta} = \frac{2\pi}{\omega\sqrt{LC}}, \tag{2.15}$$

and the phase velocity on the line is

$$v_p = \frac{\omega}{\beta} = \frac{1}{\sqrt{LC}}. \tag{2.16}$$

Terminated Transmission Lines

Figure 2.2 shows a lossless transmission line terminated in an arbitrary load impedance Z_L. This problem will illustrate wave transmission and reflection on transmission lines, which are fundamental properties of transmission line circuits.

Assume that an incident wave of the form $V_0^+ e^{-j\beta z}$ is generated from a source at $z < 0$. We have seen that the ratio of voltage to current for such a traveling wave is Z_0, the characteristic impedance. But when the line is terminated in an arbitrary load $Z_L \neq Z_0$, the ratio of voltage to current at the load must equal Z_L. Thus, a reflected wave must be generated at the load with the appropriate amplitude to satisfy this condition. The total voltage on the line can then be written as in (2.14a), as a sum of incident and reflected voltage waves. Similarly, the total current on the line consists of incident and reflected waves, as described by (2.14b).

The total voltage and current at the load are related by the load impedance, so at $z = 0$ we must have

$$Z_L = \frac{V(0)}{I(0)} = \frac{V_0^+ + V_0^-}{V_0^+ - V_0^-} Z_0.$$

Solving for V_0^- gives

$$V_0^- = \frac{Z_L - Z_0}{Z_L + Z_0} V_0^+.$$

The amplitude of the reflected voltage wave normalized to the amplitude of the incident voltage wave is defined as the *voltage reflection coefficient*, Γ:

$$\Gamma = \frac{V_0^-}{V_0^+} = \frac{Z_L - Z_0}{Z_L + Z_0}. \tag{2.17}$$

A current reflection coefficient, giving the normalized amplitude of the reflected current wave, can also be defined. But because such a current reflection coefficient is just the negative of the voltage reflection coefficient [as seen from (2.14)], we will avoid confusion by using only the voltage reflection coefficient in this book.

The total voltage and current on the line can then be written using the voltage reflection coefficient as

$$V(z) = V_0^+[e^{-j\beta z} + \Gamma e^{j\beta z}], \tag{2.18a}$$

$$I(z) = \frac{V_0^+}{Z_0}[e^{-j\beta z} - \Gamma e^{j\beta z}]. \tag{2.18b}$$

From these equations it is seen that the voltage and current on the line consist of a superposition of an incident and reflected wave; such waves are called *standing waves*. Only when $\Gamma = 0$ is there no reflected wave. To obtain $\Gamma = 0$, the load impedance Z_L must be equal to the characteristic impedance Z_0 of the line, as seen from (2.17). Such a load is then said to be *matched* to the line, since there is no reflection of the incident wave.

Now consider the time-average power flow along the line at the point z:

$$P_{\text{avg}} = \frac{1}{2}\text{Re}\{V(z)I^*(z)\} = \frac{1}{2}\frac{|V_0^+|^2}{Z_0}\text{Re}\{1 - \Gamma^* e^{-2j\beta z} + \Gamma e^{2j\beta z} - |\Gamma|^2\},$$

where (2.18) has been used. The middle two terms in the brackets are of the form $A - A^* = 2j\ \text{Im}\ \{A\}$, and so are purely imaginary. This simplifies the result to

$$P_{\text{avg}} = \frac{1}{2}\frac{|V_0^+|^2}{Z_0}(1 - |\Gamma|^2), \tag{2.19}$$

which shows that the average power flow is constant at any point on the line, and that the total power delivered to the load is equal to the incident power $(|V_0^+|^2/2Z_0)$, minus the reflected power $(|V_0^+|^2|\Gamma|^2/2Z_0)$. If $\Gamma = 0$, maximum power is delivered to the load, while no power is delivered for $|\Gamma| = 1$ (all incident power is reflected from the load). The preceding discussion assumes that the generator is matched, so that there is no re-reflection of the reflected wave from the generator at $z < 0$ (the case of a mismatched generator will be treated later).

When the load is mismatched, then, not all of the available power from the generator is delivered to the load. This "loss" is called the *return loss* (RL), and is defined (in dB) as

$$RL = -20 \log |\Gamma|\ \text{dB}, \tag{2.20}$$

so that a matched load ($\Gamma = 0$) has a return loss of ∞ dB (no reflected power), whereas a total reflection ($|\Gamma| = 1$) has a return loss of 0 dB (all incident power is reflected).

If the load is matched to the line, $\Gamma = 0$ and the magnitude of the voltage on the line is $|V(z)| = |V_0^+|$, which is a constant. For this reason such a line is sometimes said to be "flat." When the load is mismatched, however, the presence of a reflected wave leads to standing waves where the magnitude of the voltage on the line is not a constant. Thus, from (2.18a),

$$|V(z)| = |V_0^+||1 + \Gamma e^{2j\beta z}| = |V_0^+||1 + \Gamma e^{-2j\beta \ell}| = |V_0^+||1 + |\Gamma| e^{j(\theta - 2\beta\ell)}|, \tag{2.21}$$

where $\ell = -z$ is the positive distance measured from the load at $z = 0$ back toward the generator (see Figure 2.2), and θ is the phase of the reflection coefficient ($\Gamma = |\Gamma|e^{j\theta}$). This result shows that the voltage magnitude oscillates with position z along the line. The maximum value occurs when the phase term $e^{j(\theta - 2\beta\ell)} = 1$, and is given by

$$V_{\max} = |V_0^+|(1 + |\Gamma|). \tag{2.22a}$$

Similarly, the minimum value of voltage magnitude occurs when the phase term $e^{j(\theta - 2\beta\ell)} = -1$, and is given by

$$V_{\min} = |V_0^+|(1 - |\Gamma|). \tag{2.22b}$$

As $|\Gamma|$ increases, the ratio of V_{max} to V_{min} increases, so a measure of the mismatch of a line, called the *standing wave ratio* (SWR), can be defined as

$$\text{SWR} = \frac{V_{max}}{V_{min}} = \frac{1+|\Gamma|}{1-|\Gamma|}. \tag{2.23}$$

This quantity is also known as the *voltage standing wave ratio* (VSWR). From (2.23) it is seen that the SWR is a real number such that $1 \leq \text{SWR} < \infty$, where SWR $= 1$ implies a matched load.

From (2.21), it is seen that the distance between two successive voltage maxima (or minima) is $\ell = 2\pi/2\beta = \pi\lambda/2\pi = \lambda/2$, while the distance between a maximum and a minimum is $\ell = \pi/2\beta = \lambda/4$, where λ is the wavelength on the transmission line.

The reflection coefficient of (2.17) was defined as the ratio of the reflected to the incident voltage wave amplitudes at the load ($\ell = 0$), but the reflection coefficient can be generalized to any point $\ell \geq 0$ on the line as follows. From (2.14a), with $z = -\ell$, the ratio of the reflected voltage to the incident voltage is

$$\Gamma(\ell) = \frac{V_0^- e^{-j\beta\ell}}{V_0^+ e^{j\beta\ell}} = \Gamma(0)e^{-2j\beta\ell}, \tag{2.24}$$

where $\Gamma(0)$ is the reflection coefficient at $z = 0$, as given by (2.17). This form is useful when transforming the effect of a load mismatch down the line.

We have seen that the real power flow on the line is a constant, while the voltage amplitude, at least for a mismatched line, is oscillatory with position along the line. The perceptive reader may therefore conclude that the impedance seen looking into the line must vary with position, and this is indeed the case. At a distance $\ell = -z$ from the load, the input impedance seen looking toward the load is given by

$$Z_{in} = \frac{V(-\ell)}{I(-\ell)} = Z_0 \frac{V_0^+}{V_0^+}\left[\frac{e^{j\beta\ell} + \Gamma e^{-j\beta\ell}}{e^{j\beta\ell} - \Gamma e^{-j\beta\ell}}\right] = Z_0 \frac{1+\Gamma e^{-2j\beta\ell}}{1-\Gamma e^{-2j\beta\ell}}, \tag{2.25}$$

where (2.18a,b) have been used for $V(z)$ and $I(z)$. A more usable form of this result may be obtained by using (2.17) for Γ in (2.25):

$$\begin{aligned} Z_{in} &= Z_0 \frac{(Z_L + Z_0)e^{j\beta\ell} + (Z_L - Z_0)e^{-j\beta\ell}}{(Z_L + Z_0)e^{j\beta\ell} - (Z_L - Z_0)e^{-j\beta\ell}} \\ &= Z_0 \frac{Z_L \cos\beta\ell + jZ_0 \sin\beta\ell}{Z_0 \cos\beta\ell + jZ_L \sin\beta\ell} \\ &= Z_0 \frac{Z_L + jZ_0 \tan\beta\ell}{Z_0 + jZ_L \tan\beta\ell}. \end{aligned} \tag{2.26}$$

This is an important result giving the impedance at the input of a length of transmission line with an arbitrary load impedance. We will refer to this result as the *transmission line impedance equation*. Some special cases of this result will be considered next.

EXAMPLE 2.1 BASIC TRANSMISSION LINE CALCULATIONS

A load impedance of $130 + j90\ \Omega$ terminates a $50\ \Omega$ transmission line that is $0.3\ \lambda$ long. Find the reflection coefficient at the load, the reflection coefficient at the input to the line, the SWR on the line, the return loss, and the impedance seen at the input to the line.

Solution

The reflection coefficient at the load can be found from (2.17):

$$\Gamma = \frac{Z_L - Z_0}{Z_L + Z_0} = \frac{(130 + j90) - 50}{(130 + j90) + 50} = 0.598 \angle 21.8^\circ$$

Equation (2.24) can be used to transform this reflection coefficient down the line to the input, using the fact that $\beta\ell = (2\pi/\lambda)(0.3\lambda) = 108^\circ$:

$$\Gamma(\ell) = \Gamma(0)e^{-2j\beta\ell} = (0.598 \angle 21.8^\circ)e^{-2j(108^\circ)} = 0.598 \angle 165.8^\circ$$

The SWR on the line is given by (2.23):

$$\text{SWR} = \frac{1 + |\Gamma|}{1 - |\Gamma|} = \frac{1 + 0.598}{1 - 0.598} = 3.98$$

The return loss can be calculated using (2.20)

$$RL = -20 \log |\Gamma| = -20 \log(0.598) = 4.47 \text{ dB}$$

Finally, the input impedance seen at the input to the line is found from (2.26):

$$Z_{\text{in}} = Z_0 \frac{Z_L + jZ_0 \tan \beta\ell}{Z_0 + jZ_L \tan \beta\ell} = 50 \frac{(130 + j90) + j50 \tan(108^\circ)}{50 + j(130 + j90) \tan(108^\circ)}$$
$$= 12.75 + j5.8\,\Omega$$

○

Special Cases of Terminated Transmission Lines

A number of special cases of lossless terminated transmission lines are useful in practice, so it is helpful to consider the properties of such cases here.

Consider first the transmission line circuit shown in Figure 2.3, where a line is terminated in a short circuit, $Z_L = 0$. From (2.17) it is seen that the reflection coefficient for a short circuit is $\Gamma = -1$; it then follows from (2.23) that the standing wave ratio is infinite. From (2.18) the voltage and current on the short-circuited line can be written as

$$V(Z) = V_0^+[e^{-j\beta z} - e^{j\beta z}] = -2jV_0^+ \sin \beta z, \tag{2.27a}$$

$$I(Z) = \frac{V_0^+}{Z_0}[e^{-j\beta z} + e^{j\beta z}] = \frac{2V_0^+}{Z_0} \cos \beta z, \tag{2.27b}$$

which shows that $V = 0$ at the load (as expected, for a short circuit), while the current is a maximum there. From (2.26), or the ratio $V(-\ell)/I(-\ell)$, the input impedance can be found as

$$Z_{\text{in}} = jZ_0 \tan \beta\ell, \tag{2.28}$$

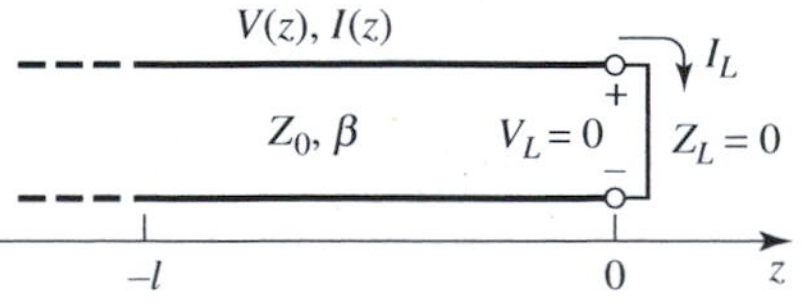

FIGURE 2.3 A transmission line terminated in a short circuit.

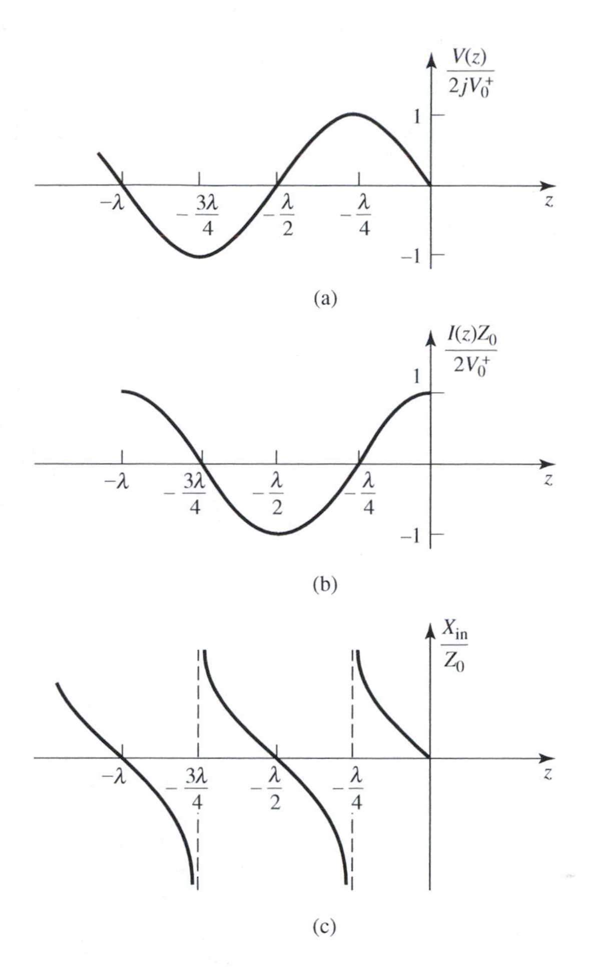

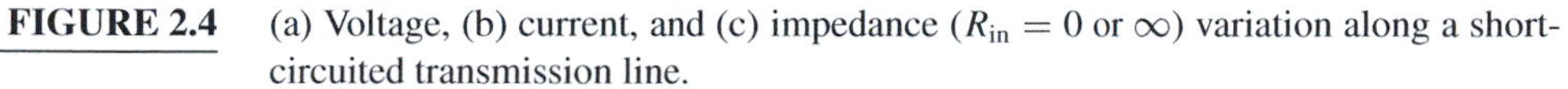

FIGURE 2.4 (a) Voltage, (b) current, and (c) impedance ($R_{in} = 0$ or ∞) variation along a short-circuited transmission line.

which is seen to be purely imaginary for any length, ℓ, and to take on all values between $+j\infty$ and $-j\infty$. For example, when $\ell = 0$ we have $Z_{in} = 0$, but for $\ell = \lambda/4$ we have $Z_{in} = +j\infty$ (open circuit). Equation (2.28) also shows that the impedance is periodic in ℓ, repeating for multiples of $\lambda/2$. The voltage, current, and input reactance for the short-circuited line are plotted in Figure 2.4.

Next consider the open-circuited line shown in Figure 2.5, where $Z_L = \infty$. Dividing the numerator and denominator of (2.17) by Z_L and allowing $Z_L \to \infty$ shows that the reflection coefficient for this case is $\Gamma = 1$, and thus the standing wave ratio is again infinite. From

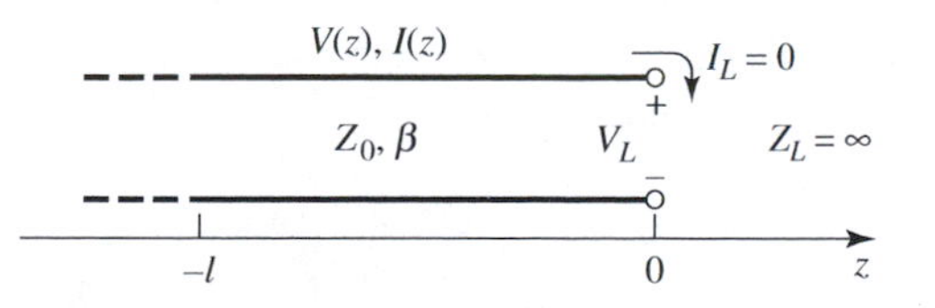

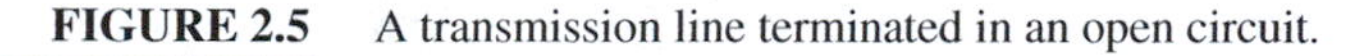

FIGURE 2.5 A transmission line terminated in an open circuit.

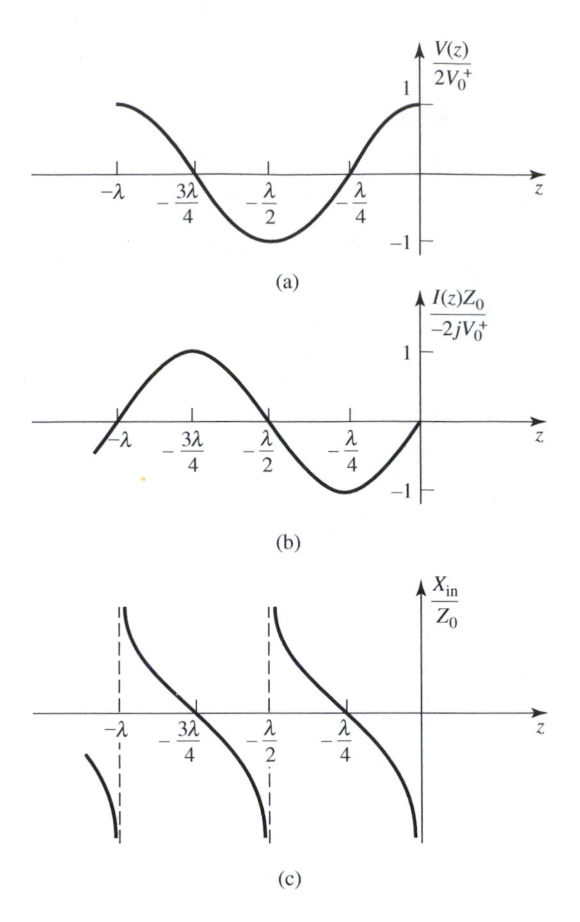

FIGURE 2.6 (a) Voltage, (b) current, and (c) impedance ($R_{in} = 0$ or ∞) variation along an open-circuited transmission line.

(2.18) the voltage and current on the open-circuited line are

$$V(Z) = V_0^+[e^{-j\beta z} + e^{j\beta z}] = 2V_0^+ \cos \beta z, \tag{2.29a}$$

$$I(Z) = \frac{V_0^+}{Z_0}[e^{-j\beta z} - e^{j\beta z}] = \frac{-2jV_0^+}{Z_0} \sin \beta z, \tag{2.29b}$$

which shows that now $I = 0$ at the load, as expected for an open circuit, while the voltage is a maximum. The input impedance can be found from (2.26), or from the ratio $V(z)/I(z)$, as

$$Z_{in} = -jZ_0 \cot \beta \ell, \tag{2.30}$$

which is also purely imaginary for any length, ℓ. The voltage, current, and input reactance of the open-circuited line are plotted in Figure 2.6.

Finally, consider terminated transmission lines with some special lengths. For example, if $\ell = \lambda/2$, (2.26) shows that

$$Z_{in} = Z_L, \tag{2.31}$$

meaning that a half-wavelength section (or any multiple of $\lambda/2$) of transmission line does not alter or transform the load impedance, regardless of the characteristic impedance of the line.

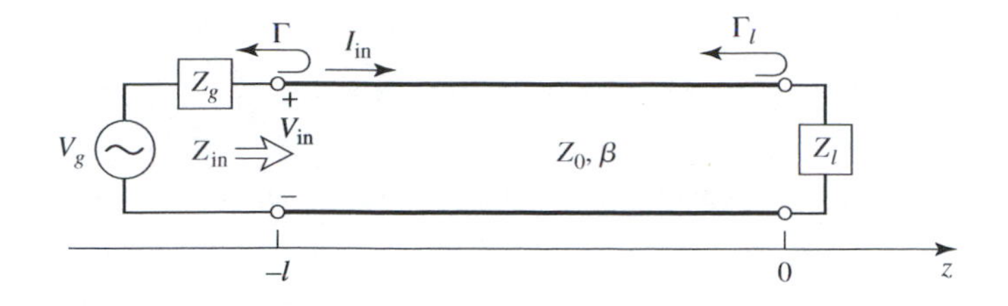

FIGURE 2.7 Transmission line circuit for mismatched load and generator.

If the line is a quarter-wavelength long or, more generally, $\ell = \lambda/4 + n\lambda/2$, for $n = 1, 2, 3, \ldots$, (2.26) shows that the input impedance is given by

$$Z_{\text{in}} = \frac{Z_0^2}{Z_L}. \tag{2.32}$$

Such a line is called a *quarter-wave transformer* because it has the effect of transforming the load impedance, in an inverse manner, depending on the characteristic impedance of the line. This provides a useful practical method of impedance matching, which we will study in more detail in a later section.

Generator and Load Mismatches

In the preceding examples of transmission line circuits it was assumed that the generator was matched to the transmission line, so that no reflections occurred at the generator. In general, however, both generator and load may present mismatched impedances to the transmission line. We will study this case here, and see that the condition for maximum power transfer from generator to load may, in some situations, require a standing wave on the line.

Figure 2.7 shows a transmission line circuit with arbitrary generator and load impedances, Z_g and Z_ℓ, which may be complex. The transmission line is assumed to be lossless, with length ℓ and characteristic impedance Z_0. This circuit is general enough to model most passive and active networks that occur in practice.

Because both the generator and load are mismatched, multiple reflections can occur on the line, since reflected waves from the load will re-reflect from the generator, and form an infinite sequence of reflections. In the steady state, the net result is a single wave traveling toward the load, and a single reflected wave traveling toward the generator. We can analyze the circuit of Figure 2.7 by first finding the input impedance looking into the terminated transmission line from the generator end. Thus, from (2.25) and (2.26),

$$Z_{\text{in}} = Z_0 \frac{1 + \Gamma_\ell e^{-2j\beta\ell}}{1 - \Gamma_\ell e^{-2j\beta\ell}} = Z_0 \frac{Z_\ell + jZ_0 \tan\beta\ell}{Z_0 + jZ_\ell \tan\beta\ell}, \tag{2.33}$$

where Γ_ℓ is the reflection coefficient of the load:

$$\Gamma_\ell = \frac{Z_\ell - Z_0}{Z_\ell + Z_0}. \tag{2.34}$$

The voltage on the line is given by (2.18a), and we can find V_0^+, the amplitude of the incident wave, from the voltage at the generator end of the line, where $z = -\ell$:

$$V(-\ell) = V_g \frac{Z_{\text{in}}}{Z_{\text{in}} + Z_g} = V_0^+ (e^{j\beta\ell} + \Gamma e^{-j\beta\ell}),$$

so that

$$V_0^+ = V_g \frac{Z_{in}}{Z_{in} + Z_g} \frac{1}{(e^{j\beta\ell} + \Gamma e^{-j\beta\ell})}. \tag{2.35}$$

This expression can be rewritten, using (2.33), as

$$V_0^+ = V_g \frac{Z_0}{Z_0 + Z_g} \frac{e^{-j\beta\ell}}{(1 - \Gamma_\ell \Gamma_g e^{-2j\beta\ell})}, \tag{2.36}$$

where Γ_g is the reflection coefficient seen looking into the generator:

$$\Gamma_g = \frac{Z_g - Z_0}{Z_g + Z_0}. \tag{2.37}$$

We can calculate the power delivered to the load as

$$P_\ell = \frac{1}{2}\mathrm{Re}\{V_{in} I_{in}^*\} = \frac{1}{2}|V_{in}|^2 \mathrm{Re}\left\{\frac{1}{Z_{in}}\right\} = \frac{1}{2}|V_g|^2 \left|\frac{Z_{in}}{Z_{in} + Z_g}\right|^2 \mathrm{Re}\left\{\frac{1}{Z_{in}}\right\}. \tag{2.38}$$

Now if $Z_{in} = R_{in} + jX_{in}$ and $Z_g = R_g + jX_g$, then (2.38) can be reduced to

$$P_\ell = \frac{1}{2}|V_g|^2 \frac{R_{in}}{(R_{in} + R_g)^2 + (X_{in} + X_g)^2}. \tag{2.39}$$

We can now use these general results to consider several special cases of load impedance, for a fixed generator impedance.

First assume the case in which the load is matched to the line, so that $Z_\ell = Z_0$. Then $\Gamma_\ell = 0$, and SWR $= 1$ on the line. The input impedance is $Z_{in} = Z_0$, and the power delivered to the load is, from (2.39),

$$P_\ell = \frac{1}{2}|V_g|^2 \frac{Z_0}{(Z_0 + R_g)^2 + X_g^2}. \tag{2.40}$$

Next, consider the case in which the generator is matched to the input impedance of a mismatched line. That is, the load impedance Z_ℓ and/or the transmission line parameters $\beta\ell$ and Z_0 are selected to make the input impedance $Z_{in} = Z_g$, so that the generator is matched to the load presented by the terminated transmission line. Then the overall reflection coefficient, Γ, seen at the input to the line is zero:

$$\Gamma = \frac{Z_{in} - Z_g}{Z_{in} + Z_g} = 0.$$

Note, however, that for this case in general $\Gamma_g \neq 0$ and $\Gamma_\ell \neq 0$, and there may be a standing wave on the line. The power delivered to the load is

$$P_\ell = \frac{1}{2}|V_g|^2 \frac{R_g}{4(R_g^2 + X_g^2)}. \tag{2.41}$$

Observe that even though the terminated line is matched to the generator, the power delivered to the load may be less than the power delivered to the load from (2.40), where the line was matched to the load, but not to the generator. This leads to the question of what is the optimum load impedance, or equivalently, what is the optimum input impedance, to achieve maximum power transfer to the load for a given generator impedance.

If we assume the generator series impedance, Z_g, is fixed, we may vary the input impedance Z_{in} until we achieve the maximum power delivered to the load. Knowing Z_{in}, it is then easy to find the corresponding load impedance Z_ℓ via an impedance transformation along the line. To maximize P_ℓ, we differentiate with respect to the real and imaginary parts of Z_{in}. Using (2.39) gives

$$\frac{\partial P_\ell}{\partial R_{in}} = 0 \rightarrow \frac{1}{(R_{in} + R_g)^2 + (X_{in} + X_g)^2} - \frac{2R_{in}(R_{in} + R_g)}{[(R_{in} + R_g)^2 + (X_{in} + X_g)^2]^2} = 0,$$

or

$$R_g^2 - R_{in}^2 + (X_{in} + X_g)^2 = 0. \tag{2.42a}$$

$$\frac{\partial P_\ell}{\partial X_{in}} = 0 \rightarrow \frac{-2X_{in}(X_{in} + X_g)}{[(R_{in} + R_g)^2 + (X_{in} + X_g)^2]^2} = 0,$$

or

$$X_{in}(X_{in} + X_g) = 0. \tag{2.42b}$$

Solving (2.42a,b) simultaneously for R_{in} and X_{in} gives $R_{in} = R_g$ and $X_{in} = -X_g$, or

$$Z_{in} = Z_g^*. \tag{2.43}$$

This condition is known as *conjugate matching*, and results in maximum power transfer to the load, for a fixed generator impedance. Under these conditions the power delivered to the load is

$$P_\ell = \frac{1}{2}|V_g|^2 \frac{1}{4R_g}, \tag{2.44}$$

which is seen to be greater than or equal to the powers of (2.40) or (2.41). Also note that the reflection coefficients Γ_ℓ, Γ_g, and Γ may be nonzero. Physically, this means that in some cases the multiple voltage reflections on a mismatched line may add in phase to deliver more power to the load than would be delivered if the line were matched (no reflections). If the generator impedance is real ($X_g = 0$), then the last two cases reduce to the same result, which is that maximum power is delivered to the load when the loaded line is matched to the generator ($R_{in} = R_g$, with $X_{in} = X_g = 0$).

Finally, note that neither matching for zero reflection ($Z_\ell = Z_0$), nor conjugate matching ($Z_{in} = Z_g^*$), necessarily yields a system with the best efficiency. For example, if $Z_g = Z_\ell = Z_0$ then both load and generator are matched (no reflections), but only half the power produced by the generator is delivered to the load (half is lost in Z_g), for a transmission efficiency of 50%. This efficiency can only be improved by making Z_g as small as possible.

2.2 THE SMITH CHART

The Smith chart, shown in Figure 2.8, is a graphical aid that is very useful for solving transmission line problems. The reader may feel that, in this day of scientific calculators and powerful computer-aided design software (CAD), graphical solutions have no place in modern engineering practice. In fact, however, the Smith chart is more than just a graphical technique. Besides being an integral part of much of the current CAD software and test equipment for microwave design, the Smith chart provides an extremely useful way of visualizing transmission line phenomenon, and so is also important for pedagogical reasons. Microwave and RF engineers can develop intuition about transmission line and impedance matching problems by learning to think in terms of the Smith chart.

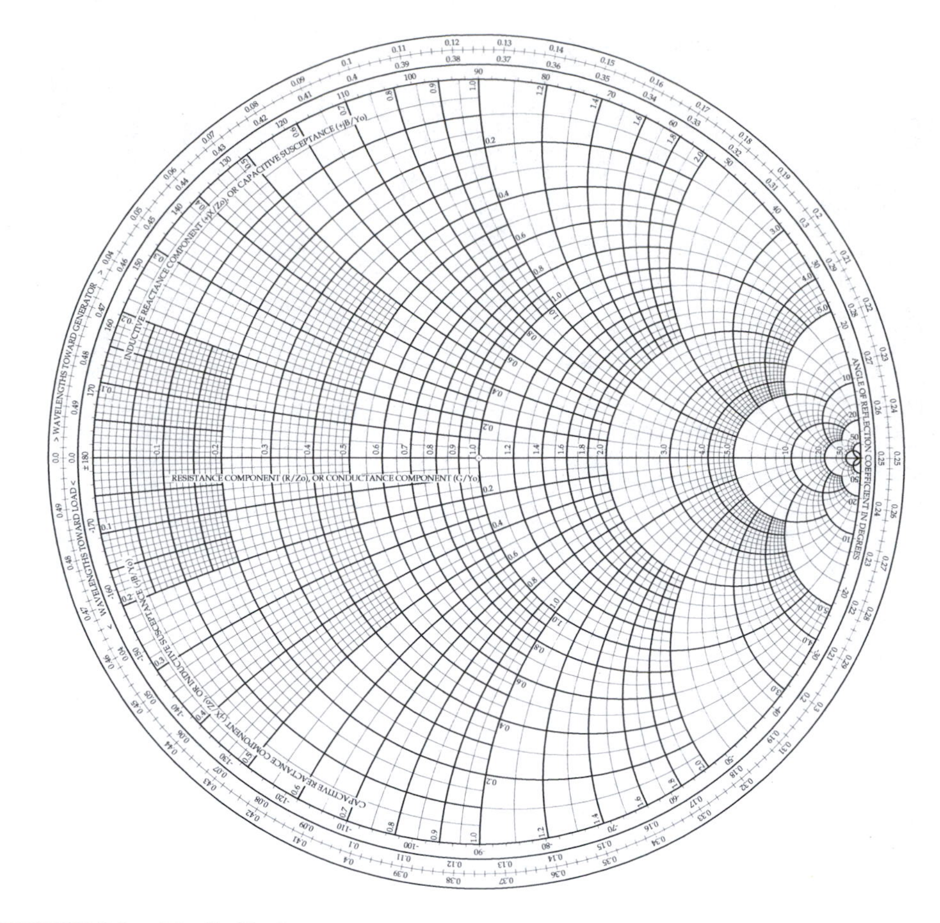

FIGURE 2.8 The Smith chart.

Derivation of the Smith Chart

At first glance the Smith chart may seem intimidating, but the key to its understanding is to realize that it is essentially a polar plot of the voltage reflection coefficient, Γ. Let the reflection coefficient be expressed in magnitude and phase (polar) form as $\Gamma = |\Gamma|e^{j\theta}$. Then the magnitude $|\Gamma|$ is plotted as a radius ($|\Gamma| \leq 1$) from the center of the chart, and the angle θ ($-180^\circ \leq \theta \leq 180^\circ$) is measured from the right-hand side of the horizontal diameter. Any passively realizable reflection coefficient can then be plotted as a unique point on the Smith chart.

The real utility of the Smith chart, however, lies in the fact that it can be used to convert from reflection coefficients to normalized impedances (or admittances), and vice versa, using the impedance (or admittance) circles printed on the chart. When dealing with impedances on a Smith chart, normalized quantities are generally used, which we will denote by lowercase letters. The normalization constant is usually the characteristic impedance of the transmission line. Thus, $z = Z/Z_0$ represents the normalized version of the impedance Z.

If a lossless line of characteristic impedance Z_0 is terminated with a load impedance Z_L, the reflection coefficient at the load can be written from (2.17) as

$$\Gamma = \frac{z_L - 1}{z_L + 1} = |\Gamma|e^{j\theta}, \tag{2.45}$$

where $z_L = Z_L/Z_0$ is the normalized load impedance. This relation can be solved for z_L in terms of Γ to give

$$z_L = \frac{1 + |\Gamma| e^{j\theta}}{1 - |\Gamma| e^{j\theta}}. \tag{2.46}$$

This complex equation (which could also be derived from (2.25) with $\ell = 0$) can be reduced to two real equations by writing Γ and z_L in terms of their real and imaginary parts. Let $\Gamma = \Gamma_r + j\Gamma_i$, and $z_L = r_L + jx_L$. Then (2.46) can be written as

$$r_L + jx_L = \frac{(1 + \Gamma_r) + j\Gamma_i}{(1 - \Gamma_r) - j\Gamma_i}.$$

The real and imaginary parts of this equation can be found by multiplying the numerator and denominator by the complex conjugate of the denominator to give

$$r_L = \frac{1 - \Gamma_r^2 - \Gamma_i^2}{(1 - \Gamma_r)^2 + \Gamma_i^2}, \tag{2.47a}$$

$$x_L = \frac{2\Gamma_i}{(1 - \Gamma_r)^2 + \Gamma_i^2}. \tag{2.47b}$$

Rearranging (2.47) gives

$$\left(\Gamma_r - \frac{r_L}{1 + r_L}\right)^2 + \Gamma_i^2 = \left(\frac{1}{1 + r_L}\right)^2, \tag{2.48a}$$

$$(\Gamma_r - 1)^2 + \left(\Gamma_i - \frac{1}{x_L}\right)^2 = \left(\frac{1}{x_L}\right)^2, \tag{2.48b}$$

which are seen to represent two families of circles in the Γ_r, Γ_i plane. Resistance circles are defined by (2.48a), and reactance circles are defined by (2.48b). For example, the $r_L = 1$ circle has its center at $\Gamma_r = 0.5$, $\Gamma_i = 0$, and has a radius of 0.5, and so passes through the center of the Smith chart. All of the resistance circles of (2.48a) have centers on the horizontal $\Gamma_i = 0$ axis, and pass through the $\Gamma = 1$ point on the right-hand side of the chart. The centers of all of the reactance circles of (2.48b) lie on the vertical $\Gamma_r = 1$ line (off the chart), and these circles also pass through the $\Gamma = 1$ point. The resistance and reactance circles are orthogonal.

Besides being useful for transforming between reflection coefficient and normalized impedance, the Smith chart can also be used to graphically solve the transmission line impedance equation of (2.25). In normalized form, this equation can be written as

$$z_{\text{in}} = \frac{1 + \Gamma e^{-2j\beta\ell}}{1 - \Gamma e^{-2j\beta\ell}}, \tag{2.49}$$

where Γ is the reflection coefficient at the load, and ℓ is the (positive) length of transmission line. We see that (2.49) is of the same form as (2.46), differing only by the phase angles of the Γ terms. Thus, if we have plotted the reflection coefficient $\Gamma = |\Gamma| e^{j\theta}$ at the load, the normalized input impedance seen looking into a length ℓ of transmission line terminates with z_L can be found by rotating the point clockwise an amount $2\beta\ell$ (subtracting $2\beta\ell$ from θ) around the center of the chart. The radius remains constant, since $|\Gamma|$ does not change with position along the line.

To facilitate such rotations, the Smith chart has scales around its periphery calibrated in electrical wavelengths, both toward and away from the "generator" (the direction away from the load). These scales are relative, so only the difference in wavelengths between two points on the Smith chart is meaningful. The scales cover a range of 0 to 0.5 wavelengths, which reflects the fact that the Smith chart automatically includes the periodicity of transmission line phenomenon. Thus, a line of length $\lambda/2$ (or any multiple) requires a rotation of $2\beta\ell = 2\pi$ around the center of the chart to transform an impedance from the load end to the input, bringing the point back to its original position.

Basic Smith Chart Operations

We can best illustrate the use of the Smith chart for basic transmission line problems through the use of an example.

EXAMPLE 2.2 USE OF THE SMITH CHART FOR BASIC TRANSMISSION LINE CALCULATIONS

A load impedance of $130 + j90\ \Omega$ terminates a $50\ \Omega$ transmission line that is $0.3\ \lambda$ long. Find the reflection coefficient at the load, the reflection coefficient at the input to the line, the SWR on the line, the return loss, and the impedance seen at the input to the line. (This is the same problem as Example 2.1.)

Solution

We begin by calculating the normalized load impedance:

$$z_L = \frac{Z_L}{Z_0} = \frac{130 + j90}{50} = 2.60 + j1.80$$

This point can be plotted on the Smith chart, as shown in Figure 2.9. Using a compass and the voltage reflection coefficient magnitude scale that is printed on most Smith charts, the reflection coefficient magnitude at the load can be read as $|\Gamma| = 0.60$. This same compass setting can then be applied to the standing wave ratio (SWR) scale to read SWR $= 3.98$, and to the return loss scale (in dB) to read RL $= 4.4$ dB. The angle of the reflection coefficient can be found by drawing a radial line from the center of the chart through the load impedance point, and reading the reflection coefficient angle on the outer scale of the chart as 21.8°. Note that these values are in close agreement with the results calculated in Example 2.1.

Now draw a circle with center at the center of the chart, and passing through the load impedance point. This circle is called a *constant SWR circle*, and it represents the locus of all possible values of reflection coefficient (and impedance) that the load can present along the line. Read the reference position of the load on the wavelengths-toward-generator (WTG) scale as 0.220λ. Moving along the line a distance of 0.3λ toward the generator brings us to the position $0.220\lambda + 0.3\lambda = 0.520\lambda$ on the WTG scale. Because the reflection coefficient repeats every 0.5λ, this is equivalent to 0.020λ. Drawing a radial line at this position gives the normalized input impedance at the intersection of this line and the SWR circle of $z_{\text{in}} = 0.255 + j0.117$. Then the input impedance at the input to the line is

$$Z_{\text{in}} = Z_0 z_{\text{in}} = 50(0.255 + j0.117) = 12.7 + j5.8\ \Omega.$$

The reflection coefficient at the input to the line still has a magnitude of $|\Gamma| = 0.60$; the phase is read from the radial line at the input position and the phase scale as 165.8°. These values are in close agreement with the results calculated in Example 2.1. ○

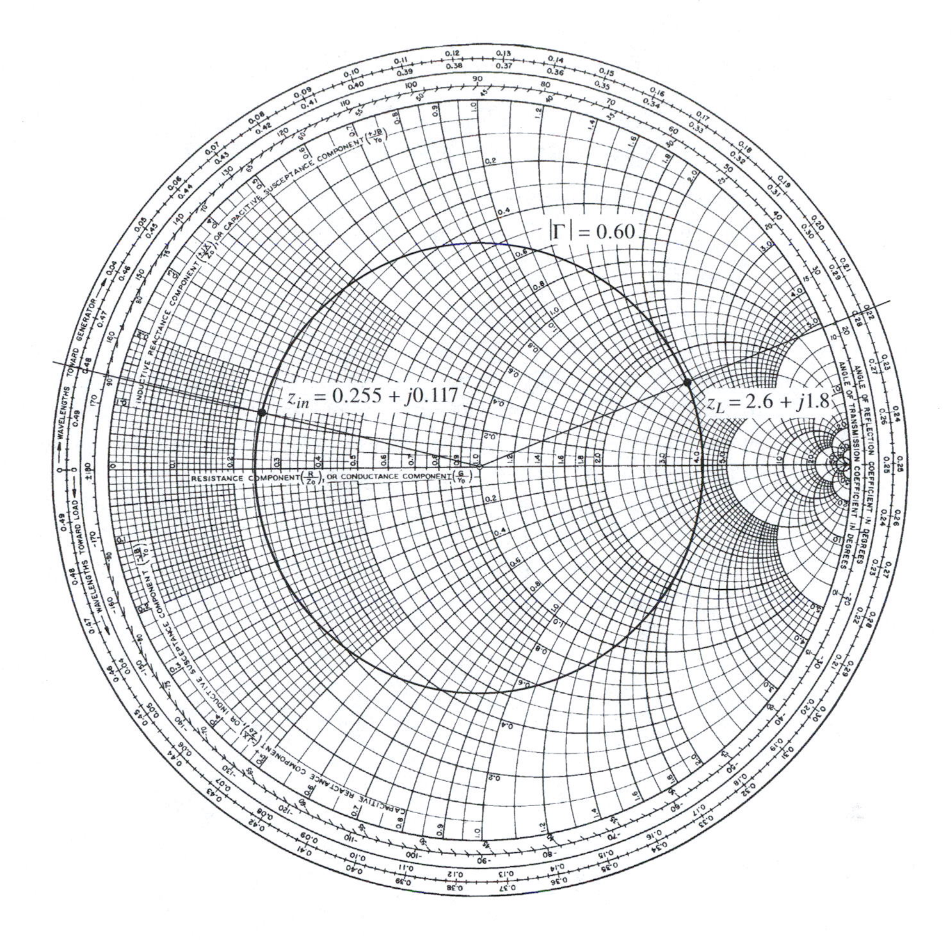

FIGURE 2.9 The Smith chart for Example 2.2.

Using The Admittance Smith Chart

Another powerful feature of the Smith chart is that it can be used with normalized admittances in the same way it is used with normalized impedances, and can be used to convert between impedance and admittance. The latter technique is based on the fact that, in normalized form, the input impedance of a load z_L connected to a $\lambda/4$ line is, from (2.32),

$$z_{\text{in}} = 1/z_L, \tag{2.50}$$

which has the effect of converting a normalized impedance to a normalized admittance.

Since a complete revolution around the Smith chart corresponds to a line length of $\lambda/2$, a $\lambda/4$ transformation is equivalent to rotating the chart by 180°; this is also equivalent to imaging a given impedance (or admittance) point across the center of the chart to obtain the corresponding admittance (or impedance) point.

The same circles can be used for either impedance or admittance, for both real and imaginary parts. As labeled on the chart, however, a positive imaginary part corresponds to an inductive reactance, or a capacitive susceptance, while a negative imaginary part corresponds to a capacitive reactance, or an inductive susceptance.

In this way, a single Smith chart can be used for both impedance and admittance calculations during the solution of a given problem. At different stages of the solution, the chart may be used as either an *impedance Smith chart* or an *admittance Smith chart.* This is often required when solving impedance matching problems with stub tuners.

EXAMPLE 2.3 SMITH CHART OPERATIONS USING ADMITTANCES

A load of $Z_L = 100 + j50\,\Omega$ terminates a 50 Ω transmission line. What are the load admittance and the input admittance if the line is 0.15λ long?

Solution

The normalized load impedance is $z_L = (100 + j50)/50 = 2 + j1$. We initially consider the Smith chart as an impedance chart, and plot the normalized load impedance point and draw the SWR circle through this point. Next, we convert to admittance by rotating $\lambda/4$ around the chart (or simply by drawing a straight line through z_L and the center of the chart to intersect the other side of the SWR circle). The chart is now considered as an admittance chart, and the normalized load admittance can be read as $y_L = 0.40 - j0.20$. (See Figure 2.10.)

To transform the load admittance to the input end of the line, first read the reference position of the load admittance on the wavelengths-toward-generator scale as 0.463λ. Adding the 0.15λ length of the line brings us to a position of

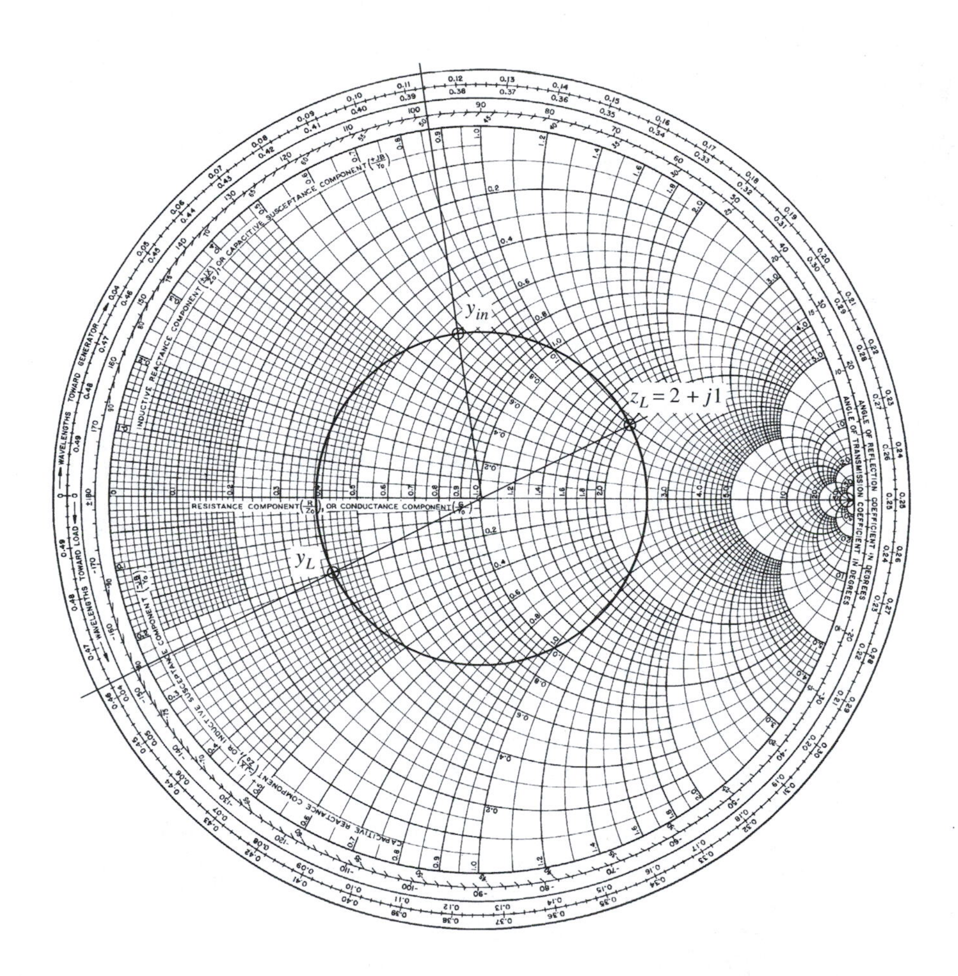

FIGURE 2.10 The Smith chart for Example 2.3.

0.613λ, or 0.113λ. The intersection of a radial line at this position with the SWR circle gives the normalized input admittance as $y_{in} = 0.60 + j0.66$. Then the actual input admittance is

$$Y_{in} = y_{in}/Z_0 = \frac{0.60 + j0.66}{50} = 0.0120 + j0.0132 \text{ S}. \qquad \bigcirc$$

2.3 MICROWAVE NETWORK ANALYSIS

In this section we show how the familiar concepts of low-frequency circuit analysis can be extended to characterize RF and microwave circuits and networks. We have seen earlier in this chapter that the distributed nature of a circuit becomes important at high frequencies, when physical dimensions become an appreciable fraction of the electrical wavelength. In addition, it is helpful to be able to view voltages and currents in terms of incident, reflected, and transmitted waves.

We begin by discussing the use of impedance and admittance matrices to describe the relationship between the total voltages and currents defined at the terminal ports of an arbitrary N-port microwave network, and show how these quantities can be decomposed into the sum of incident and reflected waves. This leads to a discussion of the scattering matrix, which gives an alternative characterization of an N-port network in terms of incident and reflected waves. The scattering matrix is central to modern RF and microwave circuit design, and will be used extensively in later chapters on amplifier and oscillator design. Finally, we will describe the transmission, or *ABCD*, matrix.

Impedance and Admittance Matrices

Consider the arbitrary N-port microwave network shown in Figure 2.11, where incident and reflected voltages, V_n^+ and V_n^-, and incident and reflected currents, I_n^+ and I_n^-, are defined at each port. Also at each port is defined a *terminal plane*, t_n, to provide a phase reference point for the voltages and currents. From (2.14), we can write the total voltage and currents at the nth port as

$$V_n = V_n^+ + V_n^-, \qquad (2.51a)$$

$$I_n = I_n^+ + I_n^-, \qquad (2.51b)$$

since the terminal plane corresponds to $z = 0$ in (2.14).

The impedance matrix $[Z]$ of the microwave network then relates these voltages and currents:

$$\begin{bmatrix} V_1 \\ V_2 \\ \cdot \\ \cdot \\ \cdot \\ V_N \end{bmatrix} = \begin{bmatrix} Z_{11} & Z_{12} & \cdot & \cdot & \cdot & Z_{1N} \\ Z_{21} & \cdot & \cdot & \cdot & \cdot & \cdot \\ \cdot & \cdot & \cdot & \cdot & \cdot & \cdot \\ \cdot & \cdot & \cdot & \cdot & \cdot & \cdot \\ \cdot & \cdot & \cdot & \cdot & \cdot & \cdot \\ Z_{N1} & \cdot & \cdot & \cdot & \cdot & Z_{NN} \end{bmatrix} \begin{bmatrix} I_1 \\ I_2 \\ \cdot \\ \cdot \\ \cdot \\ I_N \end{bmatrix},$$

or in symbolic form as

$$[V] = [Z][I]. \qquad (2.52)$$

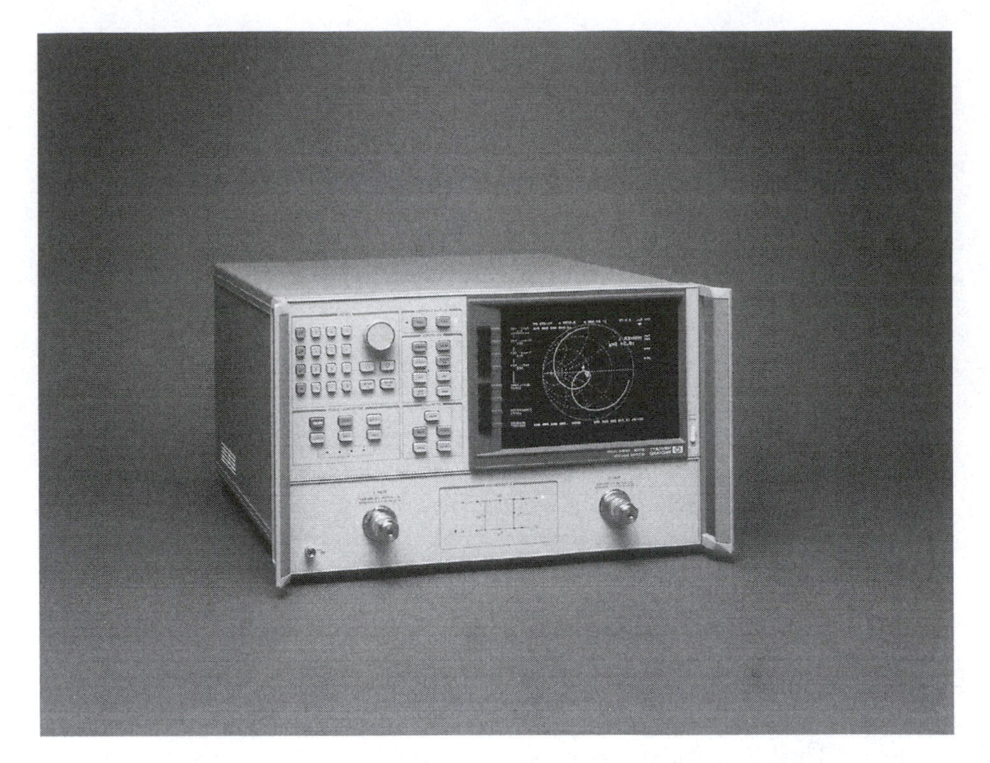

FIGURE 2.11 Photograph of the HP8720B vector network analyzer. This instrument can measure two-port scattering parameters up to 20 GHz, with built-in error correction, a synthesized source, and a color display. (Courtesy of Hewlett-Packard Company, Santa Rosa, CA.)

Similarly, we can define an admittance matrix $[Y]$ as

$$\begin{bmatrix} I_1 \\ I_2 \\ \cdot \\ \cdot \\ \cdot \\ I_N \end{bmatrix} = \begin{bmatrix} Y_{11} & Y_{12} & \cdot & \cdot & \cdot & Y_{1N} \\ Y_{21} & \cdot & \cdot & \cdot & \cdot & \cdot \\ \cdot & \cdot & \cdot & \cdot & \cdot & \cdot \\ \cdot & \cdot & \cdot & \cdot & \cdot & \cdot \\ \cdot & \cdot & \cdot & \cdot & \cdot & \cdot \\ Y_{N1} & \cdot & \cdot & \cdot & \cdot & Y_{NN} \end{bmatrix} \begin{bmatrix} V_1 \\ V_2 \\ \cdot \\ \cdot \\ \cdot \\ V_N \end{bmatrix},$$

or in symbolic form as

$$[I] = [Y][V]. \tag{2.53}$$

Of course, the $[Z]$ and $[Y]$ matrices are the inverse of each other:

$$[Y] = [Z]^{-1}. \tag{2.54}$$

Note that both the $[Z]$ and $[Y]$ matrices relate the total port voltages and currents.

From (2.52), we can see that a given matrix element Z_{ij} can be found in terms of port voltages and currents as

$$Z_{ij} = \left.\frac{V_i}{I_j}\right|_{I_k = 0 \text{ for k} \neq j}. \tag{2.55}$$

In words, (2.55) states that the ijth element of the impedance matrix can be found by driving port j with a current I_j, open-circuiting all other ports (so that $I_k = 0$ for $k \neq 0$), and measuring the open-circuit voltage at port i. Thus, Z_{ii} is the input impedance seen looking into port i when all other ports are open-circuited, and Z_{ij} is the transfer impedance between ports i and j when all other ports are open-circuited. For this reason, $[Z]$ is often called the *open-circuit impedance matrix* of the network.

Similarly, from (2.53), the ijth element of the admittance matrix can be found as

$$Y_{ij} = \left.\frac{I_i}{V_j}\right|_{V_k = 0 \text{ for k} \neq j}, \tag{2.56}$$

which states that Y_{ij} can be determined by driving port j with a voltage V_j, short-circuiting all other ports (so that $V_k = 0$ for $k \neq 0$), and measuring the short-circuit current at port i. The $[Y]$ matrix is often called the *short-circuit admittance matrix* of the network.

In general, each element of the $[Z]$ or $[Y]$ matrix may be complex. For an N-port network, the impedance and admittance matrices are $N \times N$ in size, so there are $2N^2$ independent quantities or degrees of freedom for an arbitrary N-port network. In practice, however, many networks are either reciprocal or lossless, or both. If the network is *reciprocal* (not containing any nonreciprocal media or elements such as ferrites, plasmas, or active devices), it can be shown that the impedance and admittance matrices are symmetric, so that $Z_{ij} = Z_{ji}$ and $Y_{ij} = Y_{ji}$ [1]. If the network is lossless, so that no power is dissipated in the network, then we will show that all the Z_{ij} and Y_{ij} elements are purely imaginary quantities. Either of these special cases serves to reduce the number of independent quantities or degrees of freedom that an N-port network may have.

The pure imaginary property of impedance and admittance matrix elements for lossless networks can be easily derived. If the network is lossless, the net real power delivered to the network must be zero. Thus $\text{Re}\{P_{\text{avg}}\} = 0$, where

$$\begin{aligned} P_{\text{avg}} &= \frac{1}{2}[V]^t[I]^* = \frac{1}{2}([Z][I])^t[I]^* = \frac{1}{2}[I]^t[Z][I]^* \\ &= \frac{1}{2}(I_1 Z_{11} I_1^* + I_1 Z_{12} I_2^* + I_2 Z_{21} I_1^* + \cdots) \\ &= \frac{1}{2}\sum_{n=1}^{N}\sum_{m=1}^{N} I_m Z_{mn} I_n^* \end{aligned} \tag{2.57}$$

In this expression we have used the result from matrix algebra that $([A][B])^t = [B]^t[A]^t$, where $[A]^t$ is the transpose of [A].

Since the port currents I_n are independent in (2.57), we can set all $I_n s$ equal to zero except for I_m. Then setting the real part of (2.57) to zero gives

$$\text{Re}\{I_m Z_{mm} I_m^*\} = |I_m|^2 \text{Re}\{Z_{mm}\} = 0,$$

or

$$\text{Re}\{Z_{mm}\} = 0. \tag{2.58}$$

Next, let all port currents be zero except for I_m and I_n. Then (2.57) reduces to

$$\text{Re}\{(I_n I_m^* + I_m I_n^*) Z_{mn}\} = 0,$$

since $Z_{mn} = Z_{nm}$ for a reciprocal network. But $(I_n I_m^* + I_m I_n^*)$ is a purely real quantity which is, in general, nonzero. Thus we must have that

$$\text{Re}\{Z_{mn}\} = 0. \tag{2.59}$$

Equations (2.58) and (2.59) together imply that $\text{Re}\{Z_{mn}\} = 0$ for any m, n. Thus the impedance matrix of a lossless network has purely imaginary elements. The reader can verify that the same conclusion applies to the admittance matrix as well.

The Scattering Matrix

Like the impedance or admittance matrix for an N-port network, the scattering matrix also provides a complete description of the network as seen at its N ports. While the impedance and admittance matrices relate the total voltages and currents at the ports, the scattering matrix relates the voltage waves incident on the ports to those reflected from the ports. The scattering matrix representation is especially useful at high frequencies where it is difficult to measure total voltages and currents, but easier to measure incident and reflected voltages. For some components and circuits, the scattering matrix elements can be calculated using network analysis techniques. Otherwise, the scattering parameters can be measured directly with a vector network analyzer (see photo in Figure 2.11). Once the scattering parameters of the network are known, conversion to other matrix representations can be performed, if needed.

Again consider the N-port network of Figure 2.12, where V_n^+ is the amplitude of the voltage wave incident at port n, and V_n^- is the amplitude of the voltage wave reflected from port n. The scattering matrix, or $[S]$ matrix, is defined in relation to these incident and reflected voltage waves as

$$\begin{bmatrix} V_1^- \\ V_2^- \\ \cdot \\ \cdot \\ \cdot \\ V_N^- \end{bmatrix} = \begin{bmatrix} S_{11} & S_{12} & \cdot & \cdot & \cdot & S_{1N} \\ S_{21} & \cdot & \cdot & \cdot & \cdot & \cdot \\ \cdot & \cdot & \cdot & \cdot & \cdot & \cdot \\ \cdot & \cdot & \cdot & \cdot & \cdot & \cdot \\ \cdot & \cdot & \cdot & \cdot & \cdot & \cdot \\ S_{N1} & \cdot & \cdot & \cdot & \cdot & S_{NN} \end{bmatrix} \begin{bmatrix} V_1^+ \\ V_2^+ \\ \cdot \\ \cdot \\ \cdot \\ V_N^+ \end{bmatrix},$$

or

$$[V^-] = [S][V^+]. \tag{2.60}$$

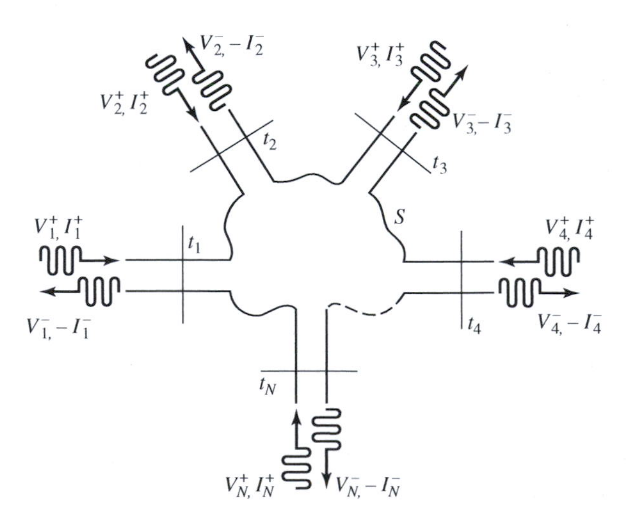

FIGURE 2.12 An arbitrary N-port microwave network.

A particular element of the $[S]$ matrix can be found as

$$S_{ij} = \left. \frac{V_i^-}{V_j^+} \right|_{V_k^+ = 0 \text{ for } k \neq j}. \tag{2.61}$$

In words, (2.61) says that S_{ij} is found by driving port j with an incident wave of voltage V_j^+, and measuring the reflected wave amplitude, V_i^-, coming out of port i. The incident waves on all ports except the jth port are set to zero, which means that all ports should be terminated with matched loads to avoid reflections from the connections (which would amount to incident waves). Thus, S_{ii} is the reflection coefficient seen looking into port i when all other ports are terminated in matched loads, and S_{ij} is the transmission coefficient from port j to port i when all other ports are terminated in matched loads.

We can now show how the $[S]$ matrix can be determined from the $[Z]$ or $[Y]$ matrix, and vice versa. First, we assume that the characteristic impedances of all ports are identical, a simplifying assumption that can be alleviated with generalized scattering parameters [1]. Then, for further convenience, we can set $Z_0 = 1$. The total voltage and current at the nth port can be written as in (2.51):

$$V_n = V_n^+ + V_n^-, \tag{2.62a}$$

$$I_n = I_n^+ + I_n^- = V_n^+ - V_n^-. \tag{2.62b}$$

Using the definition of $[Z]$ from (2.52) with (2.62) gives

$$[Z][I] = [Z][V^+] - [Z][V^-] = [V] = [V^+] + [V^-],$$

which can be rewritten as

$$([Z] + [U])[V^-] = ([Z] - [U])[V^+], \tag{2.63}$$

where $[U]$ is the *unit*, or *identity*, matrix defined as

$$[U] = \begin{bmatrix} 1 & 0 & 0 & 0 & 0 & 0 \\ 0 & 1 & & & & \cdot \\ \cdot & & \cdot & & & \cdot \\ \cdot & & & \cdot & & \cdot \\ \cdot & & & & \cdot & \cdot \\ 0 & \cdot & \cdot & \cdot & \cdot & 1 \end{bmatrix}. \tag{2.64}$$

Comparing (2.63) to (2.60) shows that

$$[S] = ([Z] + [U])^{-1}([Z] - [U]), \tag{2.65}$$

which gives the scattering matrix in terms of the impedance matrix. Note that for the special case of a one-port network, (2.65) reduces to

$$S_{11} = \frac{z_{11} - 1}{z_{11} + 1},$$

in agreement with the result for the reflection coefficient seen looking into a load with a normalized input impedance of z_{11}.

To find $[Z]$ in terms of $[S]$, rewrite (2.65) as $[Z][S] + [U][S] = [Z] - [U]$, and solve for $[Z]$ to give

$$[Z] = ([U] - [S])^{-1}([U] + [S]). \tag{2.66}$$

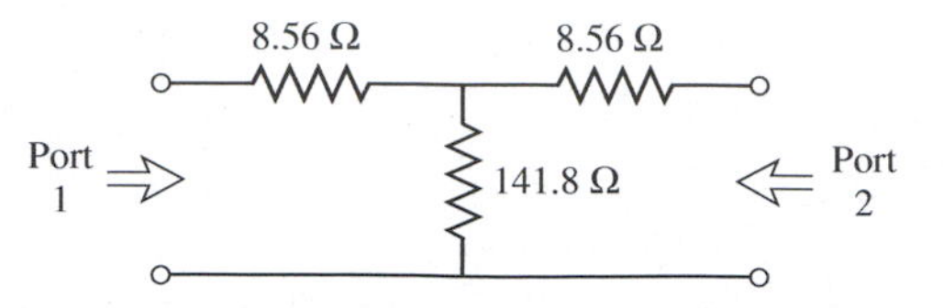

FIGURE 2.13 A matched 3 dB attenuator with a 50 Ω characteristic impedance.

Recall that we have normalized the impedance to unity, so (2.66) must be multiplied by Z_0 to recover the actual impedance. Further properties of the $[S]$ matrix, such as the symmetry of $[S]$ for reciprocal networks, and the fact that $[S]$ is unitary for lossless networks, are derived in reference [1].

EXAMPLE 2.4 EVALUATION OF SCATTERING PARAMETERS

Find the S parameters of the matched 3 dB attenuator circuit shown in Figure 2.13.

Solution

From (2.61), S_{11} can be found as the reflection coefficient seen at port 1 when port 2 is terminated in a matched load ($Z_0 = 50\ \Omega$):

$$S_{11} = \left.\frac{V_1^-}{V_1^+}\right|_{V_2^+=0} = \Gamma^{(1)}\big|_{V_2^+=0} = \left.\frac{Z_{in}^{(1)} - Z_0}{Z_{in}^{(1)} - Z_0}\right|_{Z_0 \text{ on port 2}},$$

where $Z_{in}^{(1)}$ is the input impedance seen at port 1 when port 2 is terminated with a matched load. With reference to Figure 2.13, this can be calculated as

$$Z_{in}^{(1)} = \frac{8.56 + [141.8(8.56 + 50)]}{(141.8 + 8.56 + 50)} = 50\ \Omega,$$

so $S_{11} = 0$. By symmetry of the circuit, we also have $S_{22} = 0$. The fact that $S_{11} = S_{22} = 0$ means that the input port is matched when the output port is terminated in a matched load, and vice versa. Such a network is referred to as *matched*, but it is important to realize that the ports may be mismatched if the other ports are not terminated in matched loads.

S_{21} can be found by applying an incident wave at port 1, V_1^+, and measuring the outcoming wave at port 2, V_2^-. This is equivalent to the transmission coefficient from port 1 to port 2:

$$S_{21} = \left.\frac{V_2^-}{V_1^+}\right|_{V_2^+=0}.$$

From the fact that $S_{11} = S_{22} = 0$, we know that $V_1^- = 0$ when port 2 is terminated in $Z_0 = 50\ \Omega$, and that $V_2^+ = 0$. In this case we then have that $V_1^+ = V_1$ and $V_2^- = V_2$. So by applying a voltage V_1 at port 1 and using voltage division twice we can find the voltage across the 50 Ω load resistor at port 2:

$$V_2^- = V_2 = V_1\left(\frac{41.44}{41.44 + 8.56}\right)\left(\frac{50}{50 + 8.56}\right) = 0.707V_1,$$

where 41.44 Ω is the resistance resulting from the parallel combination of the 50 Ω load and the 8.56 Ω and 141.8 Ω resistors in series. Thus, $S_{12} = S_{21} = 0.707$.

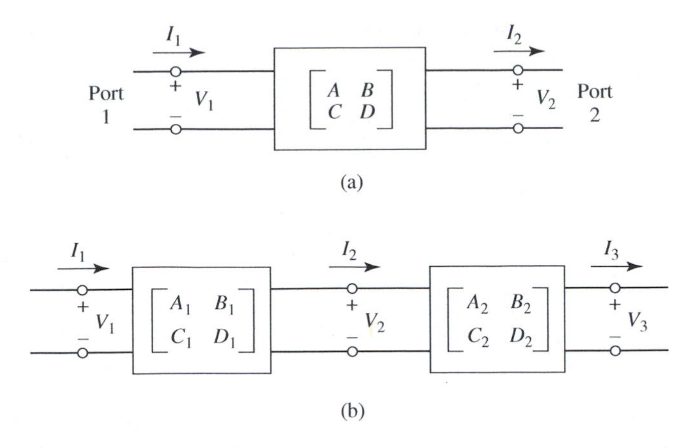

FIGURE 2.14 (a) A two-port network; (b) a cascade connection of two-port networks.

If the input power is $|V_1^+|^2/2Z_0$, then the output power is

$$\frac{|V_2^-|^2}{2Z_0} = \frac{|S_{21}V_1^+|^2}{2Z_0} = \frac{|S_{21}|^2|V_1^+|^2}{2Z_0} = \frac{|V_1^+|^2}{4Z_0},$$

which is one-half of the input power, as expected for a 3 dB attenuator. ○

The Transmission (ABCD) Matrix

The Z, Y, and S parameter representations can be used to characterize a microwave network with an arbitrary number of ports, but in practice many microwave networks consist of a cascade connection of two or more two-port networks. In this case it is convenient to define a 2×2 *transmission*, or *ABCD*, matrix, for each two-port network. We will then see that the *ABCD* matrix of the cascade connection of two or more two-port networks can be easily found by multiplying the *ABCD* matrices of the individual two-ports.

The *ABCD* matrix is defined for a two-port network in terms of the total voltages and currents as shown in Figure 2.14a and the following relations between these quantities:

$$V_1 = AV_2 + BI_2,$$
$$I_1 = CV_2 + DI_2,$$

or in matrix form as

$$\begin{bmatrix} V_1 \\ I_1 \end{bmatrix} = \begin{bmatrix} A & B \\ C & D \end{bmatrix} \begin{bmatrix} V_2 \\ I_2 \end{bmatrix}. \tag{2.67}$$

It is important to note from Figure 2.14a that a change in the sign convention of I_2 has been made from our previous definitions, which had I_2 as the current flowing *into* port 2. The convention that I_2 flows *out* of port 2 will be used when dealing with *ABCD* matrices so that in a cascade network I_2 will be the same current that flows into the adjacent network, as shown in Figure 2.14b. Then the left-hand side of (2.67) represents the voltage and current at port 1 of the network, while the right-hand side of (2.67) represents the voltage and current at port 2.

In the cascade connection of two two-port networks shown in Figure 2.14b, we have that

$$\begin{bmatrix} V_1 \\ I_1 \end{bmatrix} = \begin{bmatrix} A_1 & B_1 \\ C_1 & D_1 \end{bmatrix} \begin{bmatrix} V_2 \\ I_2 \end{bmatrix}, \tag{2.68a}$$

$$\begin{bmatrix} V_2 \\ I_2 \end{bmatrix} = \begin{bmatrix} A_2 & B_2 \\ C_2 & D_2 \end{bmatrix} \begin{bmatrix} V_3 \\ I_3 \end{bmatrix}. \tag{2.68b}$$

Substituting (2.68b) into (2.68a) gives

$$\begin{bmatrix} V_1 \\ I_1 \end{bmatrix} = \begin{bmatrix} A_1 & B_1 \\ C_1 & D_1 \end{bmatrix} \begin{bmatrix} A_2 & B_2 \\ C_2 & D_2 \end{bmatrix} \begin{bmatrix} V_3 \\ I_3 \end{bmatrix}, \tag{2.69}$$

which shows that the *ABCD* matrix of the cascade connection of the two networks is equal to the product of the *ABCD* matrices representing the individual two-ports. Note that the order of multiplication of the matrices must be the same as the order in which the networks are arranged, since matrix multiplication is not commutative.

The usefulness of the *ABCD* matrix representation is further enhanced by the fact that a library of *ABCD* matrices for elementary two-port networks can be compiled, and applied in building-block fashion to more complicated microwave networks that consist of cascades of these simpler two-ports. Table 2.1 lists a number of useful two-port networks and their ABCD matrices.

The *ABCD* parameters can be derived in terms of the Z, Y, or S parameters for a given network. To establish conversion from the impedance matrix, for example, we first change the sign convention for the current at port 2 in the impedance matrix definition of (2.52) to be consistent with that of the *ABCD* matrix:

$$V_1 = I_1 Z_{11} - I_2 Z_{12}, \tag{2.70a}$$

$$V_2 = I_1 Z_{21} - I_2 Z_{22}. \tag{2.70b}$$

Then from (2.67) we have that

$$A = \left.\frac{V_1}{V_2}\right|_{I_2=0} = \frac{I_1 Z_{11}}{I_1 Z_{21}} = \frac{Z_{11}}{Z_{21}}, \tag{2.71a}$$

$$\begin{aligned} B = \left.\frac{V_1}{I_2}\right|_{V_2=0} &= \left.\frac{I_1 Z_{11} - I_2 Z_{12}}{I_2}\right|_{V_2=0} = Z_{11} \left.\frac{I_1}{I_2}\right|_{V_2=0} - Z_{12} \\ &= Z_{11}\frac{Z_{22}}{Z_{21}} - Z_{12} = \frac{Z_{11}Z_{22} - Z_{12}Z_{21}}{Z_{21}} \end{aligned} \tag{2.71b}$$

$$C = \left.\frac{I_1}{V_2}\right|_{I_2=0} = \frac{I_1}{I_1 Z_{21}} = \frac{1}{Z_{21}}, \tag{2.71c}$$

$$D = \left.\frac{I_1}{I_2}\right|_{V_2=0} = \frac{I_2 Z_{22}/Z_{21}}{I_2} = \frac{Z_{22}}{Z_{21}}. \tag{2.71d}$$

If the network is reciprocal, then $Z_{21} = Z_{12}$, and (2.71) can be used to show that $AD - BC = 1$.

TABLE 2.1 The ABCD Parameters of Some Useful Two-Port Circuits

Circuit	*ABCD* Parameters	
Z	$A = 1$ $C = 0$	$B = Z$ $D = 1$
Y	$A = 1$ $C = Y$	$B = 0$ $D = 1$
Z_0, β l	$A = \cos \beta l$ $C = jY_0 \sin \beta l$	$B = jZ_0 \sin \beta l$ $D = \cos \beta l$
$N : 1$	$A = N$ $C = 0$	$B = 0$ $D = \frac{1}{N}$
Y_3, Y_1, Y_2	$A = 1 + \frac{Y_2}{Y_3}$ $C = Y_1 + Y_2 + \frac{Y_1 Y_2}{Y_3}$	$B = \frac{1}{Y_3}$ $D = 1 + \frac{Y_1}{Y_3}$
Z_1, Z_2, Z_3	$A = 1 + \frac{Z_1}{Z_3}$ $C = \frac{1}{Z_3}$	$B = Z_1 + Z_2 + \frac{Z_1 Z_2}{Z_3}$ $D = 1 + \frac{Z_2}{Z_3}$

2.4 IMPEDANCE MATCHING

The basic idea of *impedance matching*, or *tuning*, is illustrated in Figure 2.15, which shows an impedance matching network placed between a load impedance and a transmission line. The matching network is ideally lossless, to avoid unnecessary loss of power, and is usually designed so that the impedance seen looking into the matching network is Z_0, the characteristic impedance of the feed line. Then reflections are eliminated on the transmission line to the left of the matching network, although there will be multiple reflections between the matching network and the load.

Impedance matching is important in wireless systems for several reasons:

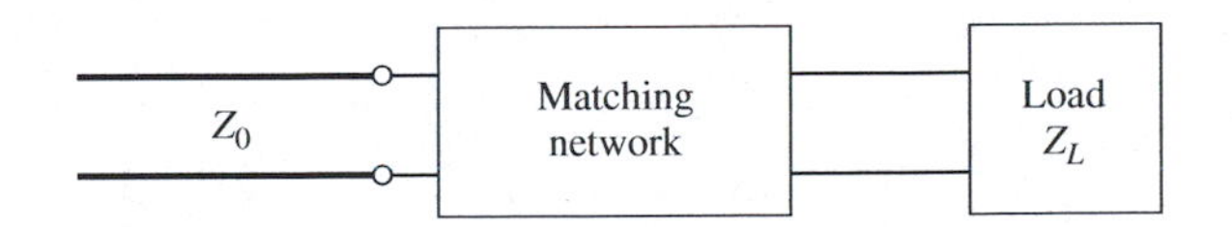

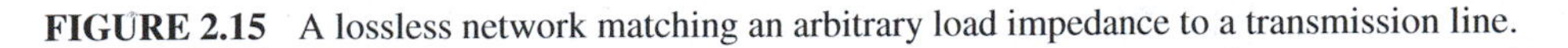

FIGURE 2.15 A lossless network matching an arbitrary load impedance to a transmission line.

- Maximum power is delivered to a load when it is matched to the feed line (assuming the generator is matched).
- Impedance matching sensitive receiver circuitry (antenna, low-noise amplifier, mixer) improves the signal-to-noise ratio of the system, and hence the maximum data rate.
- Impedance matching in transmitting system minimizes the required RF power, thus minimizing prime power (maximizing battery life, reducing risk of radiation hazard).

As long as the load impedance, Z_L, has a positive real part, a matching network can always be found. Many types of matching networks are available for practical use [1], but here we will limit our discussion to the design and performance of a few basic matching methods. These include the quarter-wave transformer, lumped element matching networks, and single-stub tuning. These techniques will be used in later chapters when we discuss the design of amplifiers and oscillators.

The Quarter-Wave Transformer

As mentioned in Section 2.1, the quarter-wave transformer is a simple and useful circuit for matching a real load impedance to a transmission line. An additional feature of the quarter- wave transformer is that it can be extended to multisection designs in a methodical manner, to provide broader bandwidth [1]. If only a narrow band impedance match is required, a single-section transformer may suffice. Although the quarter-wave transformer can only match a real load impedance, a complex load impedance can always be transformed to a real impedance by using an appropriate length of transmission line between the load and the transformer.

Here we will analyze the frequency performance of the quarter-wave transformer as a function of load mismatch. The circuit is shown in Figure 2.16, where the characteristic impedance of the matching section is given by

$$Z_1 = \sqrt{Z_0 Z_L}, \tag{2.72}$$

where Z_L is a real load impedance. At the design frequency, f_0, the electrical length of the matching section is $\lambda_0/4$, but at other frequencies the electrical length is different, and a perfect match is no longer obtained. We will now derive an approximate expression for the mismatch versus frequency.

The input impedance seen looking into the matching section is

$$Z_{\text{in}} = Z_1 \frac{Z_L + jZ_1 t}{Z_1 + jZ_L t}, \tag{2.73}$$

where $t = \tan \beta\ell = \tan\theta$, and $\beta\ell = \theta = \pi/2$ at the design center frequency, f_0. The reflection coefficient seen at the input to the transformer is then

$$\Gamma = \frac{Z_{\text{in}} - Z_0}{Z_{\text{in}} + Z_0} = \frac{Z_1(Z_L - Z_0) + jt\left(Z_1^2 - Z_0 Z_L\right)}{Z_1(Z_L + Z_0) + jt\left(Z_1^2 + Z_0 Z_L\right)}. \tag{2.74}$$

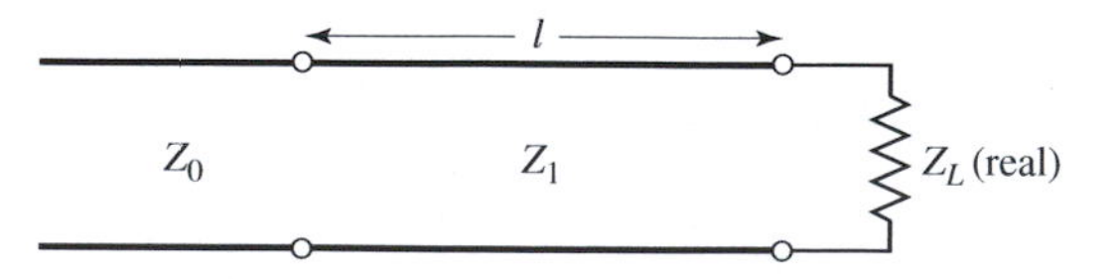

FIGURE 2.16 A quarter-wave matching transformer. $\ell = \lambda/4$ at the design frequency f_0.

Since, from (2.72), $Z_1^2 = Z_0 Z_L$, (2.74) reduces to

$$\Gamma = \frac{Z_L - Z_0}{Z_L + Z_0 + j2t\sqrt{Z_0 Z_L}}. \tag{2.75}$$

Then the reflection coefficient magnitude is

$$\begin{aligned}|\Gamma| &= \frac{|Z_L - Z_0|}{\sqrt{(Z_L + Z_0)^2 + 4t^2 Z_0 Z_L}} \\ &= \frac{1}{\sqrt{(Z_L + Z_0)^2/(Z_L - Z_0)^2 + 4t^2 Z_0 Z_L/(Z_L - Z_0)^2}} \\ &= \frac{1}{\sqrt{1 + 4Z_0 Z_L/(Z_L - Z_0)^2 + 4Z_0 Z_L t^2/(Z_L - Z_0)^2}} \\ &= \frac{1}{\sqrt{1 + \left[4Z_0 Z_L/(Z_L - Z_0)^2\right] \sec^2\theta}}\end{aligned} \tag{2.76}$$

since $1 + t^2 = 1 + \tan^2\theta = \sec^2\theta$.

Now if we assume that the frequency is near the design frequency, f_0, then $\ell \cong \lambda_0/4$ and $\theta \cong \pi/2$. Then $\sec^2\theta \gg 1$, and (2.76) simplifies to

$$|\Gamma| \cong \frac{|Z_L - Z_0|}{2\sqrt{Z_0 Z_L}} |\cos\theta|, \text{ for } \theta \text{ near } \pi/2. \tag{2.77}$$

This result gives the approximate mismatch of the quarter-wave transformer near the design frequency, as shown in Figure 2.17.

If we set a maximum value, Γ_m, of the reflection coefficient magnitude that can be tolerated, then we can define the bandwidth of the matching transformer as

$$\Delta\theta = 2\left(\frac{\pi}{2} - \theta_m\right), \tag{2.78}$$

since the response of (2.76) is symmetric about $\theta = \pi/2$, and $\Gamma = \Gamma_m$ at $\theta = \theta_m$ and at $\theta = \pi - \theta_m$. Equating Γ_m to the exact expression for reflection coefficient magnitude in (2.76) allows us to solve for θ_m:

$$\frac{1}{\Gamma_m^2} = 1 + \left(\frac{2\sqrt{Z_0 Z_L}}{Z_L - Z_0} \sec\theta\right)^2,$$

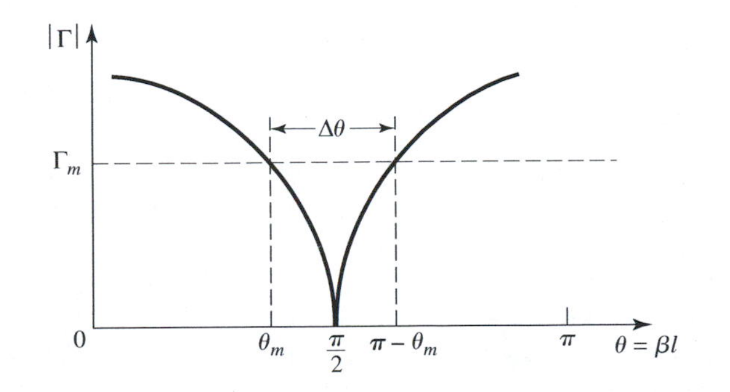

FIGURE 2.17 Approximate behavior of the reflection coefficient magnitude for a quarter-wave transformer operating near its design frequency.

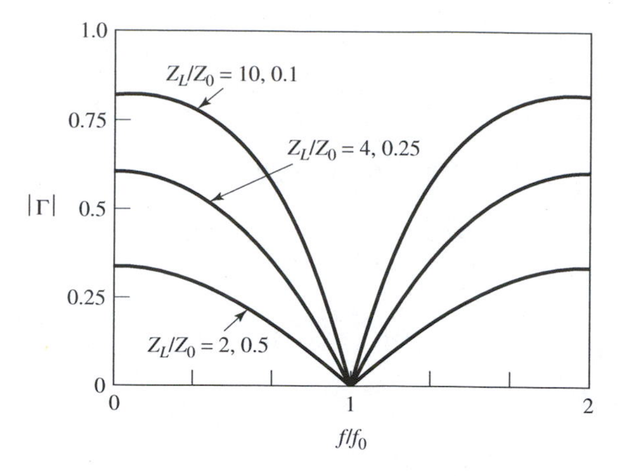

FIGURE 2.18 Reflection coefficient magnitude versus normalized frequency for a quarter-wave transformer with various load mismatches.

or

$$\cos\theta_m = \frac{\Gamma_m}{\sqrt{1-\Gamma_m^2}} \frac{2\sqrt{Z_0 Z_L}}{|Z_L - Z_0|}. \tag{2.79}$$

If we assume TEM lines, then we have that

$$\theta = \beta\ell = \frac{2\pi f}{v_p}\frac{v_p}{4f_0} = \frac{\pi f}{2f_0},$$

where v_p is the phase velocity for the transmission line. Therefore the fractional bandwidth is, from (2.79),

$$\begin{aligned} \frac{\Delta f}{f_0} &= \frac{2(f_0 - f_m)}{f_0} = 2 - \frac{2f_m}{f_0} = 2 - \frac{4\theta_m}{\pi} \\ &= 2 - \frac{4}{\pi}\cos^{-1}\left[\frac{\Gamma_m}{\sqrt{1-\Gamma_m^2}} \frac{2\sqrt{Z_0 Z_L}}{|Z_L - Z_0|}\right] \end{aligned} \tag{2.80}$$

Fractional bandwidth is usually expressed as a percentage, $100\Delta f/f_0$ %. Note that the bandwidth of the transformer increases as Z_L becomes closer to Z_0 (a less mismatched load). Figure 2.18 shows a plot of the reflection coefficient magnitude versus normalized frequency for various mismatched loads. Note the trend of increased bandwidth for smaller load mismatch.

Matching Using L-Sections

Another popular type of impedance matching network is the *L-section*, which uses two reactive elements to match an arbitrary load impedance to a transmission line. This technique is used extensively in lower frequency circuit design, and has the advantage over the quarter-wave transformer in that the load impedance need not be real.

There are two possible configurations for an L-Section matching network, as shown in Figure 2.19. If the normalized load impedance, $z_L = Z_L/Z_0$, is inside the $1 + jx$ circle on the Smith chart, then the circuit of Figure 2.19a should be used. If the normalized load impedance is outside the $1 + jx$ circle, the circuit of Figure 2.19b should be used. The $1 + jx$ circle is the resistance circle on the impedance Smith chart for which $r = 1$.

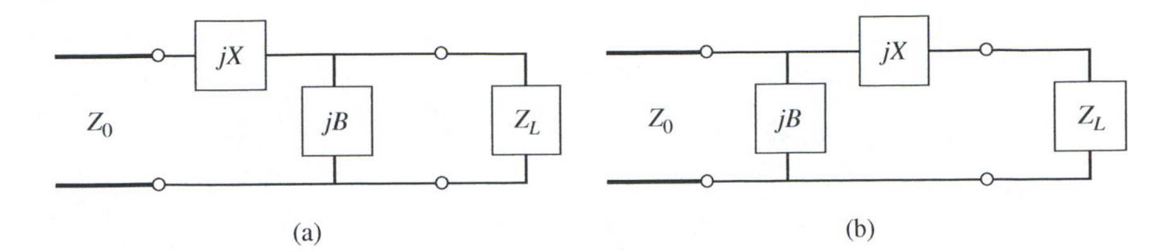

FIGURE 2.19 L-section matching networks. (a) Network for Z_L inside the $1 + jx$ circle. (b) Network for Z_L outside the $1 + jx$ circle.

In either of the configurations of Figure 2.19, the reactive elements may be either inductive or capacitive, depending on the load impedance. If the frequency is relatively low and/or the circuit size is electrically small, lumped-element inductors or capacitors can be used. At higher frequencies, however, it is difficult to implement lumped element capacitors and inductors, so in this case tuning methods using transmission line stubs may be preferred.

While analytic solutions for the required values of series reactance jX and shunt susceptance jB are available [1], it is often convenient in practice to use the Smith chart to find these values for a given load impedance. This procedure is best illustrated with an example.

EXAMPLE 2.5 L-SECTION IMPEDANCE MATCHING

Design an L-section matching network to match a series RC load having an impedance $Z_L = 200 - j100\ \Omega$, to a $100\ \Omega$ line, at a frequency of 500 MHz.

Solution

The normalized load impedance is $z_L = 2 - j1$, which is plotted on the Smith chart of Figure 2.20a. This point is inside the $1 + jx$ circle, so we will use the matching circuit of Figure 2.19a. Since the first element from the load is a shunt susceptance, it is helpful to convert to a load admittance y_L by drawing the SWR circle through the load impedance, and a straight line from the load through the center of the chart, as shown in Figure 2.20a. Now, after we add the shunt susceptance jB and convert back to impedance, we want to be on the $1 + jx$ circle, so that we can add a series reactance jX to match the load. This means that the shunt susceptance jB must move us from y_L to the $1 + jx$ circle on the admittance Smith chart. Thus, we construct the rotated $1 + jx$ circle as shown in Figure 2.20a (center at 0.333). Then we see that adding a normalized susceptance of $jb = j0.3$ will move us along a constant conductance circle to $y = 0.4 + j0.5$ (this choice is the shortest distance from y_L to the shifted $1 + jx$ circle). Converting back to impedance leaves us at $z = 1 - j1.2$, indicating that the addition of a series reactance $x = j1.2$ will bring us to the center of the chart, to complete the solution.

The matching circuit thus consists of a shunt capacitor and a series inductor, as shown in Figure 2.20b. At a frequency of $f = 500$ MHz, the shunt capacitor has a value of

$$C = \frac{b}{2\pi f Z_0} = 0.92 \text{ pF},$$

and the series inductor has a value of

$$L = \frac{x Z_0}{2\pi f} = 38.8 \text{ nH}.$$

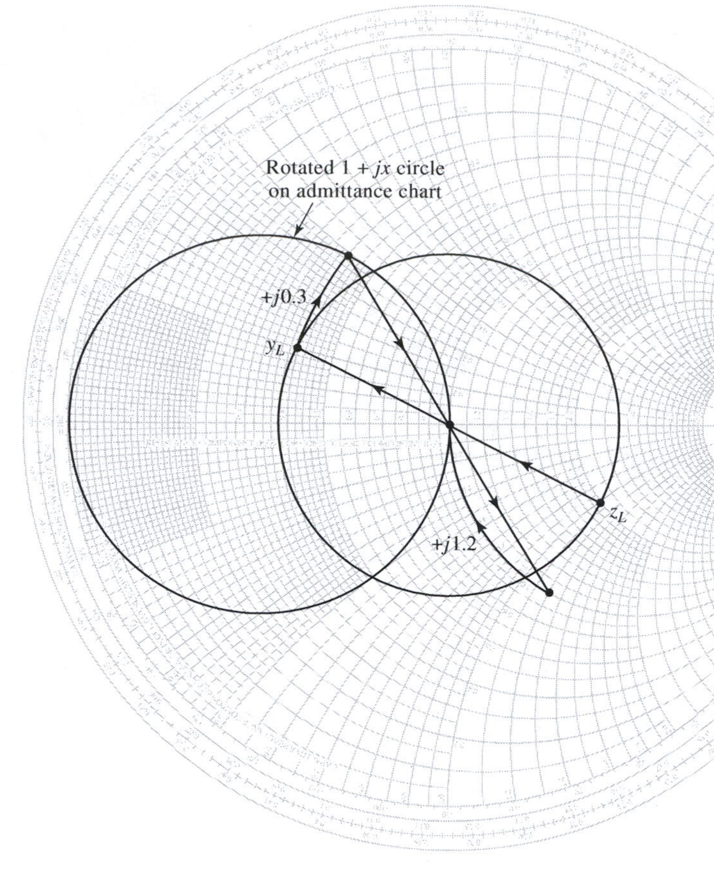

(a)

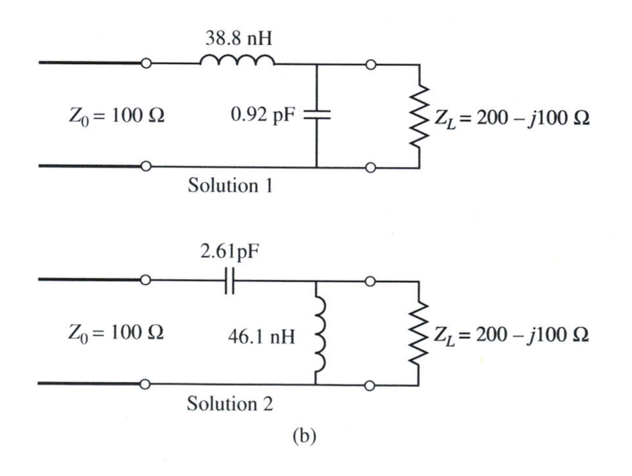

(b)

FIGURE 2.20 Solution to Example 2.5. (a) The Smith chart for the L-section matching networks. (b) The two possible L-section matching circuits. (c) Reflection coefficient magnitudes versus frequency for the matching circuits of (b).

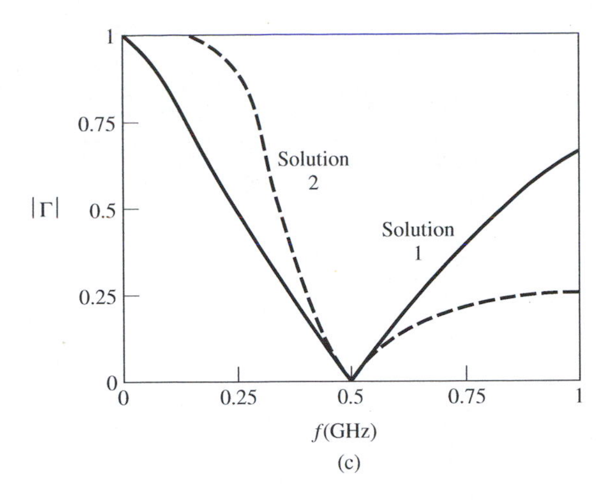

FIGURE 2.20 (*Continued*)

There is also a second solution for this tuning problem. If instead of adding a shunt susceptance of $b = 0.3$, we use a shunt susceptance of $b = -0.7$, we will move to a point on the lower half of the rotated $1 + jx$ circle, to $y = 0.4 - j0.5$. Then converting to impedance and adding a series reactance of $x = -1.2$ leads to a match as well. This matching circuit is also shown in Figure 2.20b, and is seen to have the positions of the inductor and capacitor reversed from the first matching network. At a frequency of 500 MHz, the capacitor for this solution has a value of

$$C = \frac{-1}{2\pi f x Z_0} = 2.61 \text{ pF},$$

while the inductor has a value of

$$L = \frac{-Z_0}{2\pi f b} = 46.1 \text{ nH}.$$

Figure 2.20c shows the resulting reflection coefficient magnitudes versus frequency for these two matching networks, assuming that the load impedance of $Z_L = 200 - j100\,\Omega$ at 500 MHz consists of a 200 Ω resistor and a 3.18 pF capacitor in series. There is not a substantial difference in bandwidth for these two solutions, but in other cases the difference may be more significant. ○

Single-Stub Tuning

Finally, we consider a matching technique that uses a single open-circuited or short-circuited length of transmission line (a *stub*), connected either in parallel or in series with the transmission feed line at a certain distance from the load, as shown in Figure 2.21. Such a tuning circuit is convenient from a microwave fabrication aspect, since lumped elements are not required, and the necessary transmission lines can easily be etched in planar circuit form. Single-stub tuning networks are used extensively in transistor amplifier and oscillator circuits, as discussed in Chapters 6 and 8.

In single-stub tuning the distance, d, from the load to the stub position, and the value of the shunt susceptance (or series reactance) provided by the stub, are adjustable parameters. These two degrees of freedom can be used to match an arbitrary load impedance to any feed

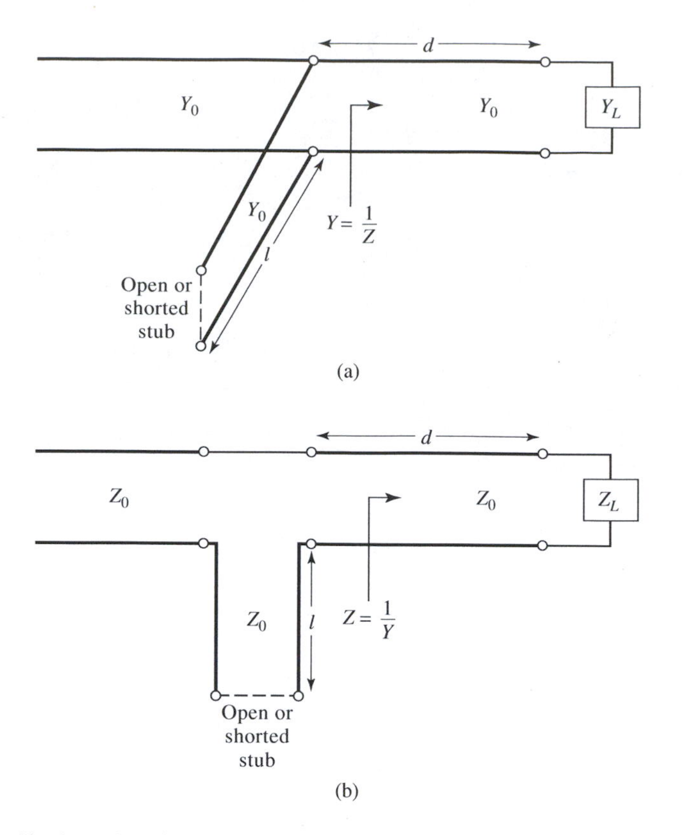

FIGURE 2.21 Single-stub tuning circuits. (a) Shunt stub. (b) Series stub.

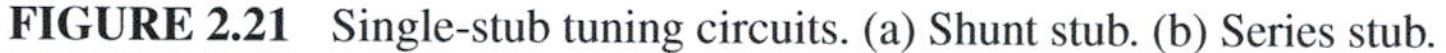

line (assuming the load impedance has a positive real part). For the shunt-stub case, the basic idea is to select d so that the admittance, Y, seen looking into the line at distance d from the load is of the form $Y_0 + jB$, where $Y_0 = 1/Z_0$. Then the stub susceptance is chosen as $-jB$, resulting in a matched condition. For the series-stub case, the distance d is selected so that the impedance, Z, seen looking into the line at a distance d from the load is of the form $Z_0 + jX$. Then the stub reactance is chosen as $-jX$, resulting in a matched condition.

As discussed in Section 2.1, the proper length of an open- or short-circuited transmission line can provide any desired value of reactance or susceptance. For a given susceptance or reactance, the difference in lengths of an open- or short-circuited stub is $\lambda/4$. For transmission line media such as microstrip or stripline, open-circuited stubs are easier to fabricate since a short-circuiting via hole is not required.

Analytic solutions for both shunt- and series-stub tuning circuits can be derived [1], but Smith chart solutions are usually accurate enough for practical work, and have the advantage of being quick and providing an intuitive view of the matching procedure. We will illustrate the method with an example for a shunt stub tuner.

EXAMPLE 2.6 SINGLE-STUB SHUNT TUNING

For a load impedance $Z_L = 20 - j15\ \Omega$, design two single-stub shunt tuning networks to match this load to a $50\ \Omega$ line.

Solution

Begin by plotting the normalized load impedance, $z_L = 0.4 - j0.3$, as shown on the Smith chart of Figure 2.22. Since we are using a shunt stub, it is convenient

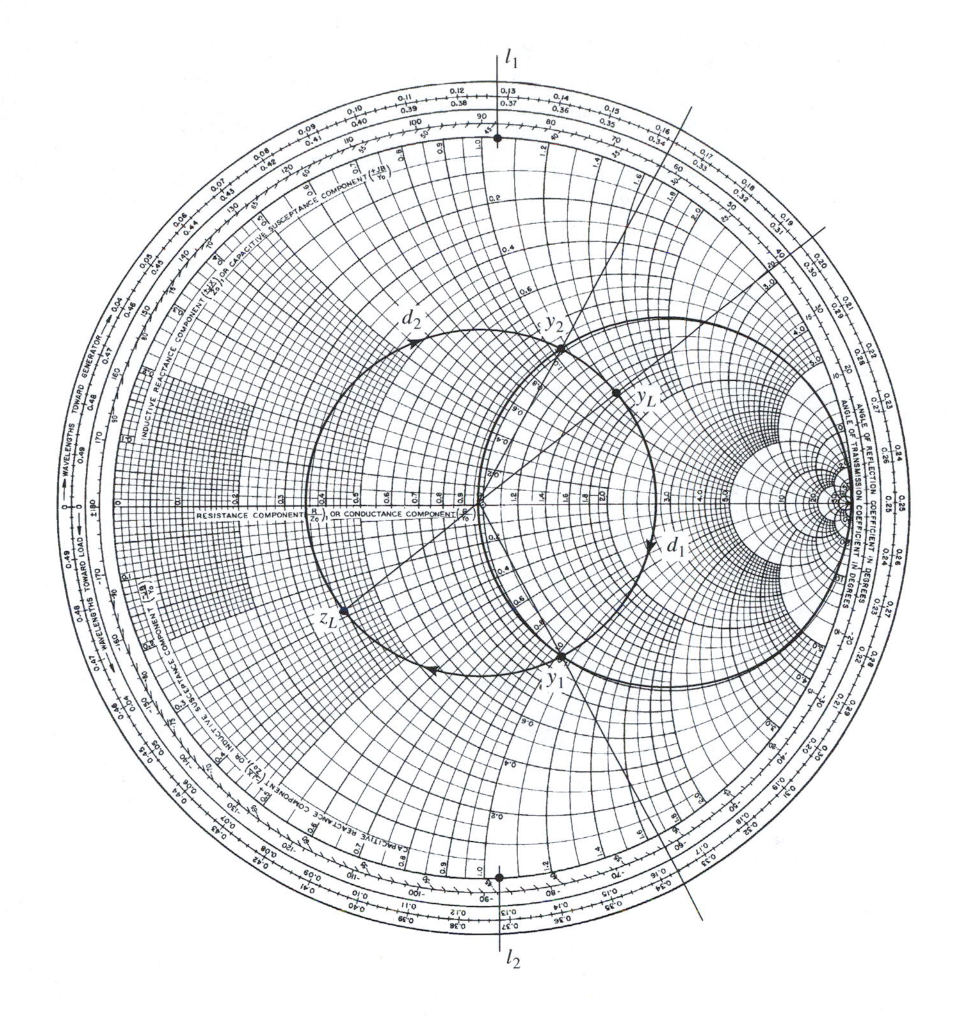

FIGURE 2.22 The Smith chart solution for Example 2.6.

to work with an admittance chart, and so we convert to a load admittance of $y_L = 1.6 + j1.2$ by plotting a SWR circle through the load impedance and drawing a diameter. Note that the SWR circle intersects the $1 + jb$ circle at two points, denoted by y_1 and y_2 in Figure 2.22. Thus the distance d, from the load to the stub, is given by either of these two intersections. Reading the WTG scale, we have

$$d_1 = 0.336 - 0.196 = 0.140\lambda,$$
$$d_2 = (0.5 - 0.196) + 0.164 = 0.468\lambda.$$

Of course, there are an infinite number of distances, d, on the SWR circle that intersect the $1 + jb$ circle. Usually, however, it is desired to keep the matching stub as close as possible to the load to improve the bandwidth of the match, and to minimize losses caused by a possibly large standing wave ratio on the line between the stub and the load.

At the two intersection points, the normalized admittances are

$$y_1 = 1 - j1.061,$$
$$y_2 = 1 + j1.061.$$

Thus, the first tuning solution requires a stub with a susceptance of $j1.061$. The length of an open-circuited stub that gives this susceptance can be found on the Smith chart by starting at $y = 0$ (the open circuit) and moving along the outer edge of the chart (since $g = 0$) toward the generator to the $j1.061$ point. The required stub length is then $\ell_1 = 0.130\lambda$. Similarly, the required open-circuit stub length for the second solution is $\ell_2 = 0.370\lambda$. ○

REFERENCES

[1] D. M. Pozar, **Microwave Engineering**, 2nd edition, Wiley, New York, 1998.
[2] D. K. Cheng, **Field and Wave Electromagnetics**, 2nd edition, Addison-Wesley, Reading, MA, 1989.

PROBLEMS

2.1 A transmission line has the following per unit length parameters: $L = 0.3\ \mu$H/m, $C = 450$ pF/m, $R = 5\ \Omega$/m, and $G = 0.01$ S/m. Calculate the complex propagation constant and characteristic impedance of this line at 880 MHz. Recalculate these quantities in the absence of loss ($R = G = 0$).

2.2 A lossless transmission line of length 0.3λ is terminated with a load impedance as shown below. Find the reflection coefficient at the load, the SWR on the line, the return loss, and the input impedance to the line.

2.3 A lossless transmission line of characteristic impedance Z_0 is terminated with a load impedance of 150 Ω. If the SWR on the line is measured to be 1.6, find the two possible values for Z_0.

2.4 A wireless transmitter is connected to an antenna having an input impedance of $80 + j40\ \Omega$ through a 50 Ω coaxial cable. If the 50 Ω transmitter can deliver 30 W when connected to a matched load, how much power is delivered to the antenna?

2.5 (a) Calculate the SWR and return loss for reflection coefficient magnitudes of 0.01, 0.1, 0.25, 0.5, and 0.75. (b) Calculate the SWR and reflection coefficient magnitudes for return losses of 1 dB, 3 dB, 10 dB, 20 dB, and 30 dB.

2.6 The transmission line circuit shown below has $V_g = 10$ v rms, $Z_g = 50\ \Omega$, $Z_0 = 50\ \Omega$, $Z_L = 60 - j40\ \Omega$, and $\ell = 0.6\lambda$. Compute the power delivered to the load using three different techniques:
(a) find Γ and compute

$$P_L = \left(\frac{V_g}{2}\right)^2 \frac{1}{Z_0}(1 - |\Gamma|^2);$$

(b) find Z_{in} and compute

$$P_L = \left|\frac{V_g}{Z_g + Z_{\text{in}}}\right|^2 \text{Re}\{Z_{\text{in}}\}; \text{ and}$$

(c) find V_L and compute

$$P_L = \left|\frac{V_L}{Z_L}\right|^2 \text{Re}\{Z_L\}.$$

Discuss the rationale for each of these methods. Which of these methods can be used if the line is not lossless?

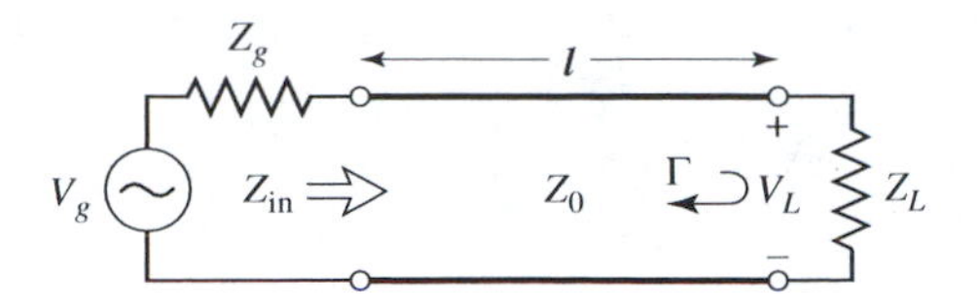

2.7 For a purely reactive load impedance of the form $Z_L = jX$, show that the reflection coefficient magnitude $|\Gamma|$ is always unity. Assume the characteristic impedance is real.

2.8 Consider the transmission line circuit shown below. Compute the incident power, the reflected power, and the power transmitted into the infinite 75 Ω line. Show that power conservation is satisfied.

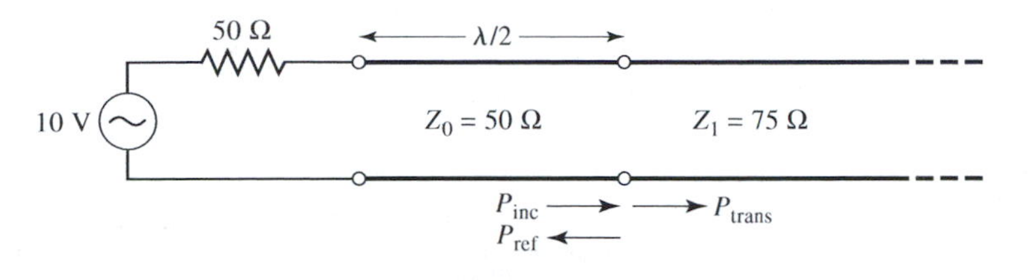

2.9 A load impedance of $Z_L = 60 + j30\ \Omega$ is to be matched to a $Z_0 = 50\ \Omega$ line using a length ℓ of lossless line of characteristic impedance Z_t. Find values for the required (real) Z_t and ℓ.

2.10 For the circuit shown below, find the power delivered to the load and the power dissipated in the generator impedance for a load impedance of $Z_l = 30 + j40\ \Omega$. What value of load impedance will result in maximum power delivered to the load? What is this power?

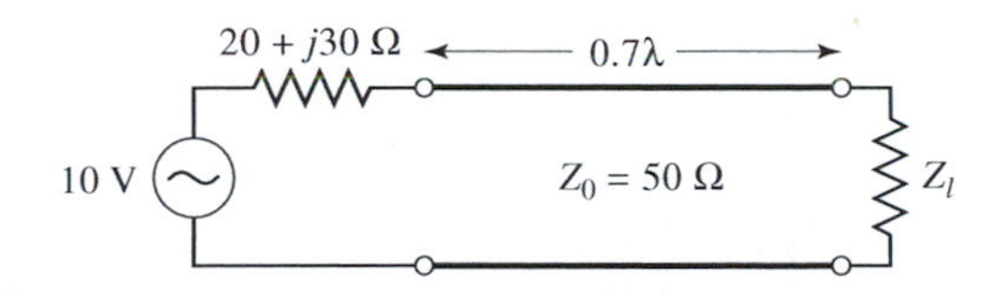

2.11 Consider the transmission line circuit below. Use the Smith chart to find the SWR on the line, the return loss, the reflection coefficient at the load, the load admittance, the input impedance to the line, the distance from the load to the first voltage minimum, and the distance from the load to the first voltage maximum.

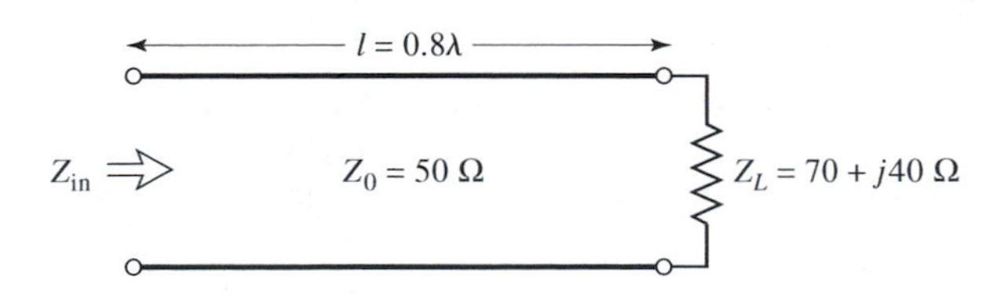

2.12 Use the Smith chart to find the shortest lengths of a short-circuited 50 Ω transmission line stub to give the following input impedance:

(a) $Z_{in} = 0$

(b) $Z_{in} = \infty$

(c) $Z_{in} = j50\ \Omega$

(d) $Z_{in} = -j50\ \Omega$

2.13 Repeat Problem 2.12 for an open-circuited length of 50 Ω line.

2.14 Derive the $[Z]$ and $[Y]$ matrices for the two-port networks shown below.

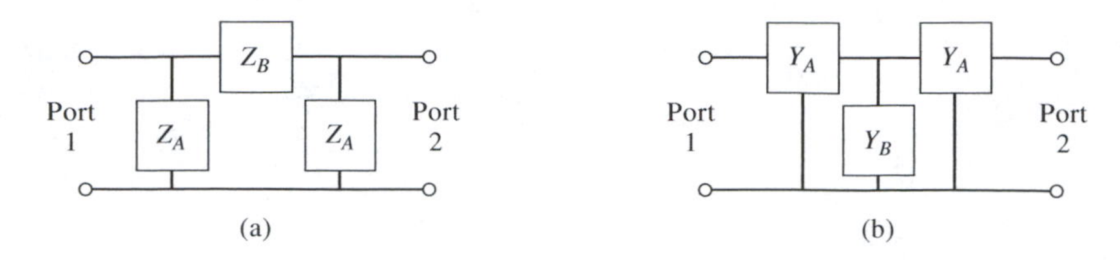

2.15 A particular two-port network is driven at both ports so that the port voltages and currents have the following values:

$$V_1 = 5.0\ \angle 45^\circ \qquad I_1 = 0.1\ \angle 45^\circ$$
$$V_2 = 3.0\ \angle -45^\circ \qquad I_2 = 0.2\ \angle 90^\circ$$

Determine the incident and reflected voltages at both ports, if the characteristic impedance is 50 Ω.

2.16 A particular two-port network is driven at port 1 with a matched generator, and terminated at port 2 with a matched load. If the total voltage and current at port 1 are measured to be $V_1 = 1.314\ \angle 12.4^\circ$ V and $I_1 = 15.4\ \angle -21.5^\circ$ mA, and the total voltage at port 2 is measured to be $V_2 = 0.8\ \angle 90^\circ$ V. If the characteristic impedance is 50 Ω, find S_{11} and S_{21}.

2.17 A three-port network has the scattering matrix given below.
(a) What is the return loss at each port, when all other ports are terminated in matched loads?
(b) What is the insertion loss and phase between ports 2 and 3, when all ports are matched?
(c) What is the return loss at port 1 when ports 2 and 3 are terminated in short circuits?

$$[S] = \begin{bmatrix} 0.1\ \angle 90^\circ & 0.4\ \angle 180^\circ & 0.4\ \angle 180^\circ \\ 0.4\ \angle 180^\circ & 0.2\ \angle 0^\circ & 0.6\ \angle 45^\circ \\ 0.4\ \angle 180^\circ & 0.6\ \angle 45^\circ & 0.2\ \angle 0^\circ \end{bmatrix}$$

2.18 Verify the *ABCD* parameters for the first three networks shown in Table 2.1.

2.19 Derive expressions giving the impedance matrix parameters in terms of the *ABCD* parameters.

2.20 Use *ABCD* matrices to find the voltage V_L across the load resistor in the circuit shown below.

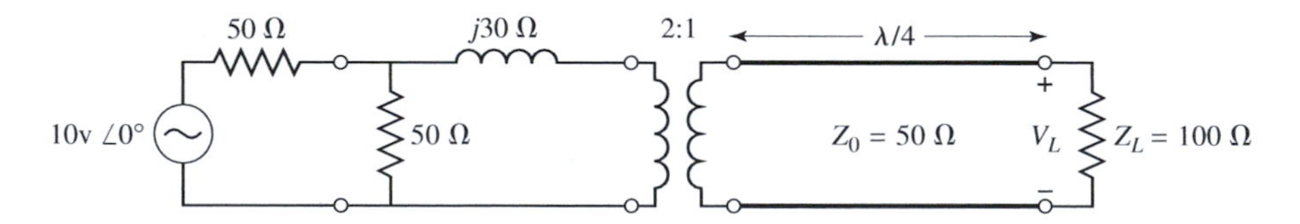

2.21 Design a quarter-wave transformer to match a 350 Ω load to a 100 Ω line. What is the percent bandwidth for this matching circuit, for an SWR ≤ 2?

2.22 In the circuit shown below, a load impedance of $Z_L = 100 + j200\ \Omega$ is to be matched to a 50 Ω feed line, using a length, ℓ, of lossless transmission line of characteristic impedance, Z_1. Find Z_1 and ℓ. Determine, in general, what type of load impedances can be matched using such a circuit.

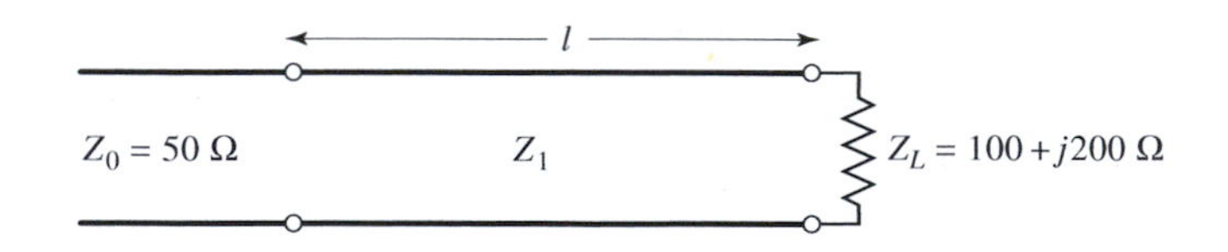

2.23 Design two lossless L-section matching networks for each of the following normalized load impedances:

(a) $z_L = 0.5 - j0.8$
(b) $z_L = 1.6 + j0.8$

2.24 A load impedance of $Z_L = 100 - j150\ \Omega$ is to be matched to a 50 Ω line using a single shunt-stub tuner. Find two solutions using open-circuited stubs.

2.25 Repeat Problem 2.24 using short-circuited stubs.

2.26 A load impedance of $Z_L = 15 + j50\ \Omega$ is to be matched to a 100 Ω line using a single series stub tuner. Find two solutions using open-circuited stubs.

2.27 Repeat Problem 2.26 using short-circuited stubs.

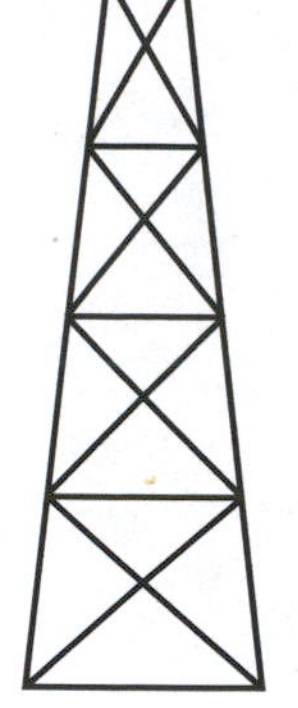

C h a p t e r T h r e e

Noise and Distortion in Microwave Systems

The effect of noise is one of the most important considerations when evaluating the performance of wireless systems because noise ultimately determines the threshold for the minimum signal level that can be reliably detected by a receiver. Noise is a random process associated with a variety of sources, including thermal noise generated by RF components and devices, noise generated by the atmosphere and interstellar radiation, and man-made interference. Noise is omnipresent in RF and microwave systems, with noise power being introduced through the receive antenna from the external environment, as well as generated internally by the receiver circuitry. In our later study of modulation methods, we will see that parameters such as signal-to-noise ratio, bit error rates, dynamic range, and the minimum detectable signal level are all directly dependent on noise effects.

Our objective in this chapter is to present a quantitative overview of noise and its characterization in RF and microwave systems. Since noise is a random process, we begin with a review of random variables and associated techniques for the mathematical treatment of noise and its effects. Next we discuss the physical basis and a model for thermal noise sources, followed by an application to basic threshold detection of binary signals in the presence of noise. The noise power generated by passive and active RF components and devices can be characterized equivalently by either noise temperature or noise figure, and these parameters are discussed in Section 3.4, followed by the propagation and accumulation of noise power through a cascade of two-port networks. A more detailed treatment of the noise figure of general passive networks is given in Section 3.5. Finally, we consider the problem of dynamic range and signal distortion in general nonlinear systems. These effects are important for large signal levels in mixers and amplifiers, and can thus be viewed as complementary to the effect of noise, which is an issue for small signal levels.

3.1 REVIEW OF RANDOM PROCESSES

In this section we review some basic principles, definitions, and techniques of random processes that we will need in our study of noise and its effects in wireless communications systems. These include basic probability, random variables, probability density functions, cumulative distribution functions, autocorrelation, power spectral density, and expected values. We assume the reader has had a beginning course in random variables, and so will not require a full exposition of the subject. References [1]–[3] should be useful for a more thorough discussion of the required concepts.

Probability and Random Variables

Probability is the likelihood of the occurrence of a particular event, and is written as $P\{\text{event}\}$. The probability of an event is a numerical value between zero and unity, where zero implies the event will never occur, and unity implies the event will always occur. Probability events may include the occurrence of an equality, such as $P\{x = 5\}$, or events related to a range of values, such as $P\{x \leq 5\}$.

In contrast to the actual terminology, a *random variable* is neither random nor a variable, but is a function that maps sample values from a random event or process into real numbers. Random variables may be used for both discrete and continuous processes. Examples of discrete processes include tossing coins and dice, counting pedestrians crossing a street, and the occurrence of errors in the transmission of data. Continuous random variables can be used for modeling smoothly varying real quantities such as temperature, noise voltage, and received signal amplitude or phase. We will be primarily concerned with continuous random variables.

Consider a continuous random variable X, representing a random process with real continuous sample values x, where $-\infty < x < \infty$. Since the random variable X may assume any one of an uncountably infinite number of values, the probability that X is exactly equal to a specific value, x_0, must be zero. Thus, $P\{X = x_0\} = 0$. On the other hand, the probability that X is less than a specific value of x may be greater than zero: $0 \leq P\{X \leq x_0\} \leq 1$. In the limit as $x_0 \to \infty$, $P\{X \leq x_0\} \to 1$, as the event becomes a certainty.

We will not adopt any particular notation for random variables in this book, as it should be clear from the context which variables are random and which are deterministic. In most cases, the only random variables we will encounter will be associated with noise voltages, and typically denoted as $v_n(t)$, or $n(t)$.

The Cumulative Distribution Function

The *cumulative distribution function* (CDF), $F_X(x)$, of the random variable X is defined as the probability that X is less than or equal to a particular value, x. Thus

$$F_X(x) = P\{X \leq x\}. \tag{3.1}$$

It can be shown that the cumulative distribution function satisfies the following properties:

$$(1)\quad F_X(x) \geq 0 \tag{3.2a}$$

$$(2)\quad F_X(\infty) = 1 \tag{3.2b}$$

$$(3)\quad F_X(-\infty) = 0 \tag{3.2c}$$

$$(4)\quad F_X(x_1) \leq F_X(x_2) \text{ if } x_1 \leq x_2 \tag{3.2d}$$

The last property is a statement that the cumulative distribution function is a monotonic (nondecreasing) function. The definition in (3.1) shows that the result of (3.2d) is equivalent to the statement that

$$P\{x_1 < x \leq x_2\} = F_X(x_2) - F_X(x_1).$$

The Probability Density Function

The *probability density function* (PDF), $f_x(x)$, of a random variable X is defined as the derivative of the CDF:

$$f_X(x) = \frac{dF_X(x)}{dx}. \tag{3.3}$$

Since the CDF is monotonically nondecreasing, $f_X(x) \geq 0$ for all x. The PDF may contain delta functions, as in the case of discrete random variables, for which the CDF is a "stair-step" type of function.

By the fundamental theorem of calculus, (3.3) can be inverted to give the following useful result:

$$P\{x_1 < X \leq x_2\} = F_X(x_2) - F_X(x_1) = \int_{x_1}^{x_2} f_X(x)\, dx. \tag{3.4}$$

In addition, since $F(-\infty) = 0$ from (3.2c), (3.4) reduces to the following result that directly relates the CDF to the PDF:

$$F_X(x) = \int_{-\infty}^{x} f_X(u)\, du. \tag{3.5}$$

Finally, because $F(\infty) = 1$ from (3.2b), (3.5) leads to the fact that the total area under a probability density function is unity:

$$\int_{-\infty}^{\infty} f_x(x)\, dx = 1. \tag{3.6}$$

These results can be extended to cases of two random variables. Thus, the joint CDF associated with random variables X and Y is defined as

$$F_{XY}(x, y) = P\{X \leq x \text{ and } Y \leq y\}. \tag{3.7}$$

Then the joint PDF is calculated as

$$f_{xy}(x, y) = \frac{\partial^2}{\partial x \partial y} F_{xy}(x, y). \tag{3.8}$$

Similar to the result of (3.4), the probability of x and y both occurring in given ranges is found from

$$P\{x_1 < X \leq x_2 \text{ and } y_1 < Y \leq y_2\} = \int_{x_1}^{x_2} \int_{y_1}^{y_2} f_{xy}(x, y)\, dx\, dy. \tag{3.9}$$

The individual probability density functions for X and Y can be recovered from the joint PDF by integration over one of the variables:

$$f_y(y) = \int_{-\infty}^{\infty} f_{xy}(x, y)\, dx, \tag{3.10a}$$

$$f_x(x) = \int_{-\infty}^{\infty} f_{xy}(x, y)\, dy. \tag{3.10b}$$

For the special case where the random variables X and Y are statistically independent, the joint PDF is the product of the PDFs of X and Y:

$$f_{xy}(x, y) = f_x(x) f_y(y). \tag{3.11}$$

Some Important Probability Density Functions

For reference, we list here some of the probability density functions that we will be using in this book. The most basic is the PDF of a uniform distribution, defined as a constant over a finite range of the independent variable:

$$f_x(x) = \frac{1}{b-a} \quad \text{for} \quad a \leq x \leq b. \tag{3.12a}$$

The constant 1/(b−a) is chosen to properly normalize the PDF according to (3.6). Many random variables have gaussian statistics, with the general gaussian PDF given by

$$f_x(x) = \frac{1}{\sqrt{2\pi\sigma^2}} e^{-(x-m)^2/2\sigma^2} \quad \text{for} \quad -\infty < x < \infty, \tag{3.12b}$$

where m is the mean of the distribution, and σ^2 is the variance.

In our study of fading and digital modulation we will encounter the Rayleigh PDF, given by

$$f_r(r) = \frac{r}{\sigma^2} e^{-r^2/2\sigma^2} \quad \text{for} \quad 0 \leq r < \infty. \tag{3.12c}$$

The reader can verify that these each satisfy the normalization condition of (3.6).

Expected Values

Since random variables are nondeterministic, we cannot predict with certainty the value of a particular sample from a random event or process, but instead must rely on statistical averages such as the mean, variance, and standard deviation. We denote the *expected value* of the random variable X as $\bar{x}$, or $E\{X\}$. The expected value is also sometimes called the mean, or average value. For discrete random variables the expected value is given as the sum of the N possible samples, x_i, weighted by the probabilities of the occurrence of that sample:

$$\bar{x} = E\{X\} = \sum_{i=1}^{N} x_i P\{X = x_i\}. \tag{3.13a}$$

This result directly generalizes to the case of continuous random variables:

$$\bar{x} = E\{X\} = \int_{-\infty}^{\infty} x f_x(x)\, dx. \tag{3.13b}$$

It is easy to show that the process of taking the expected value of a random variable is a linear operation, and that the following two properties therefore apply (assume X and Y are random variables, with c a constant):

$$(1) \quad E\{cX\} = cE\{X\} \tag{3.14a}$$

$$(2) \quad E\{X + Y\} = E\{X\} + E\{Y\}. \tag{3.14b}$$

We will also often be interested in finding the expected value of a function of a random variable. If we have a random variable x, and a function $y = g(x)$ that maps values from x

to a new random variable y, then the expected value of y can be found for the discrete case as

$$\bar{y} = E\{y\} = E\{g(x)\} = \sum_{i=1}^{N} g(x_i)P\{x = x_i\}. \tag{3.15a}$$

For the case of a continuous random variable this becomes

$$\bar{y} = E\{y\} = E\{g(x)\} = \int_{-\infty}^{\infty} g(x) f_x(x)\,dx. \tag{3.15b}$$

The result of (3.15b) can be used to find higher-order statistical averages, such as the *nth moment* of the random variable, X:

$$\overline{x^n} = E\{x^n\} = \int_{-\infty}^{\infty} x^n f_x(x)\,dx. \tag{3.16}$$

The *variance*, σ^2, of the random variable X is found by calculating the second moment of X after subtracting the mean of X:

$$\begin{aligned}\sigma^2 &= E\{(x - \bar{x})^2\} = \int_{-\infty}^{\infty} (x - \bar{x})^2 f_x(x)\,dx \\ &= E\{x^2 - 2x\bar{x} + \bar{x}^2\} = \overline{x^2} - \bar{x}^2.\end{aligned} \tag{3.17}$$

The *root-mean-square* (rms) value of the distribution is σ, the square root of the variance. If a particular zero-mean random voltage is represented by the random variable x, the power delivered to a 1 Ω load by this voltage source will be equal to the variance of x.

The expected value of a function of two random variables involves the joint PDF:

$$\overline{g(x, y)} = E\{g(x, y)\} = \int_{-\infty}^{\infty} g(x, y) f_{xy}(x, y)\,dx dy. \tag{3.18}$$

This result can be applied to the product of two random variables, x and y, by letting the function $g(x, y) = xy$. For the special case of independent random variables the joint PDF is the product of the individual PDFs by (3.11), so (3.18) reduces to

$$\overline{xy} = E\{xy\} = \int_{-\infty}^{\infty} x f_x(x)\,dx \int_{-\infty}^{\infty} y f_y(y)\,dy = E\{x\}E\{y\}. \tag{3.19}$$

Autocorrelation and Power Spectral Density

An important characteristic of both deterministic and random signals is how rapidly their sample values vary with time. This characteristic can be quantified with the *autocorrelation function*, defined for a complex deterministic signal, $x(t)$, as the time average of the product of the conjugate of $x(t)$ and a time-shifted version, $x(t + \tau)$:

$$R(\tau) = \int_{-\infty}^{\infty} x^*(t)x(t + \tau)\,dt. \tag{3.20}$$

It can be shown that $R(0) \geq R(\tau)$, and $R(\tau) = R(-\tau)$. Also, $R(0)$ is the normalized energy of the signal.

For stationary random processes, such as noise processes, the autocorrelation function is defined as

$$R(\tau) = E\{x^*(t)x(t + \tau)\}. \tag{3.21}$$

Because of the relation between the time variation of a signal and its frequency spectrum, we can also characterize the variation of random signals by examining the spectra of the autocorrelation function in the frequency domain. For stationary random processes, the *power spectral density* (PSD), $S(\omega)$, is defined as the Fourier transform of the autocorrelation function:

$$S(\omega) = \int_{-\infty}^{\infty} R(\tau)e^{-j\omega\tau}\,d\tau. \tag{3.22a}$$

The inverse transform can be used to find the autocorrelation from a known PSD:

$$R(\tau) = \frac{1}{2\pi}\int_{-\infty}^{\infty} S(\omega)e^{j\omega\tau}\,d\omega. \tag{3.22b}$$

For a noise voltage, the power spectral density represents the noise power density in the spectral (frequency) domain, assuming a 1 Ω load resistor. If $v(t)$ represents the noise voltage, the power delivered to a 1 Ω load can be found as

$$P_L = \overline{v^2(t)} = E\{v^2(t)\} = R(0) = \frac{1}{2\pi}\int_{-\infty}^{\infty} S_v(\omega)\,d\omega = \int_{-\infty}^{\infty} S_v(2\pi f)\,df\ W, \tag{3.23}$$

where $S_v(\omega)$ is the PSD of $v(t)$. The last equality follows from a change of variable with $\omega = 2\pi f$. Writing this integral in terms of f (in Hz) is convenient because $S_v(\omega)$ has dimension W/Hz, and therefore appears as a power density relative to frequency in Hertz.

EXAMPLE 3.1 OPERATIONS WITH RANDOM VARIABLES

Consider a sinusoidal voltage source, $V_0 \cos\omega_0 t$, which is randomly sampled in time to form a random process $v(t) = V_0\cos\theta$, where $\theta = \omega_0 t$ is a random variable representing the sample time. Assume θ is uniformly distributed over the interval $0 \le \theta < 2\pi$, since the cosine function is periodic with period 2π. Find the mean of the sample voltages, the average power delivered to a 1 Ω load, the autocorrelation function of the random process $v(t)$, and the power spectral density.

Solution

The PDF for the random variable θ is $f_\theta(\theta) = \frac{1}{2\pi}$, for $0 \le \theta < 2\pi$. Then we can calculate the average voltage as

$$\overline{v(t)} = E\{v(t)\} = \int_0^{2\pi} v(t)f_\theta(\theta)\,d\theta = \frac{1}{2\pi}\int_0^{2\pi}\cos\theta\,d\theta = 0.$$

The average power delivered to a 1 Ω load is given by the variance of $v(t)$:

$$P_L = \overline{v^2(t)} = E\{v^2(t)\} = \int_0^{2\pi} v^2(t)f_\theta(\theta)\,d\theta = \frac{V_0^2}{2\pi}\int_0^{2\pi}\cos^2\theta\,d\theta = \frac{V_0^2}{2}W.$$

The autocorrelation can be calculated using (3.21):

$$\begin{aligned} R_v(\tau) &= E\{v(t)v(t+\tau)\} = V_0^2 E\{\cos\omega_0 t\cos\omega_0(t+\tau)\} \\ &= \frac{V_0^2}{2\pi}\int_0^{2\pi}\cos\theta\cos(\theta+\omega_0\tau)\,d\theta = \frac{V_0^2}{4\pi}\int_0^{2\pi}[\cos\omega_0\tau + \cos(2\theta+\omega_0\tau)]\,d\theta \\ &= \frac{V_0^2}{2}\cos\omega_0\tau \end{aligned}$$

Note that $R_v(0) = V_0^2/2$, which is the variance of $v(t)$. The power spectral density is found using (3.22a):

$$\begin{aligned} S_v(\omega) &= \int_{-\infty}^{\infty} R_v(\tau)e^{-j\omega\tau}\,d\tau = \frac{V_0^2}{4}\int_{-\infty}^{\infty}\left[e^{-j(\omega-\omega_0)\tau} + e^{-j(\omega+\omega_0)\tau}\right]d\tau \\ &= \frac{\pi V_0^2}{2}[\delta(\omega-\omega_0) + \delta(\omega+\omega_0)] \end{aligned}$$

This result shows that power is concentrated at $\omega = \omega_0$ and its image at $-\omega_0$. The total power can also be calculated by integrating the PSD over frequency, as in (3.23):

$$P_L = \frac{1}{2\pi}\int_{-\infty}^{\infty} S_v(\omega)\,d\omega = \frac{V_0^2}{4}\int_{-\infty}^{\infty}[\delta(\omega-\omega_0)+\delta(\omega+\omega_0)]\,d\omega = \frac{V_0^2}{2}.$$

This result agrees with the earlier result obtained as the variance using the PDF. ○

3.2 THERMAL NOISE

Thermal noise, also known as Nyquist, or Johnson, noise, is caused by the random motion of charge carriers, and is the most prevalent type of noise encountered in RF and microwave systems. Thermal noise is generated in any passive circuit element that contains loss, such as resistors, lossy transmission lines, and other lossy components. It can also be generated by atmospheric attenuation and interstellar background radiation, which similarly involve random motion of thermally excited charges. Other sources of noise include *shot noise*, due to the random motion of charge carriers in electron tubes and solid-state devices; *flicker noise*, also occurring in solid-sate devices and vacuum tubes; *plasma noise*, caused by random motions of charged particles in an ionized gas or sparking electrical contacts; and *quantum noise*, resulting from the quantized nature of charge carriers and photons. Although these other types of noise differ from thermal noise in terms of their origin, their characteristics are similar enough that they can generally be treated in the same way as thermal noise.

Noise Voltage and Power

Figure 3.1a shows a resistor of value R at temperature T degrees Kelvin (K). The electrons in the resistor are in random motion, with a kinetic energy that is proportional to the temperature, T. These random motions produce small random voltage fluctuations across the terminals of the resistor, as illustrated in Figure 3.1b. The mean value of this voltage is zero, but its nonzero rms value in a narrow frequency bandwidth B is given by

$$V_n = \sqrt{4kTBR}, \tag{3.24}$$

where

$k = 1.380 \times 10^{-23}$ J/K is Boltzmann's constant
T is the temperature, in degrees Kelvin (K)
B is the bandwidth, in Hz
R is the resistance, in Ω

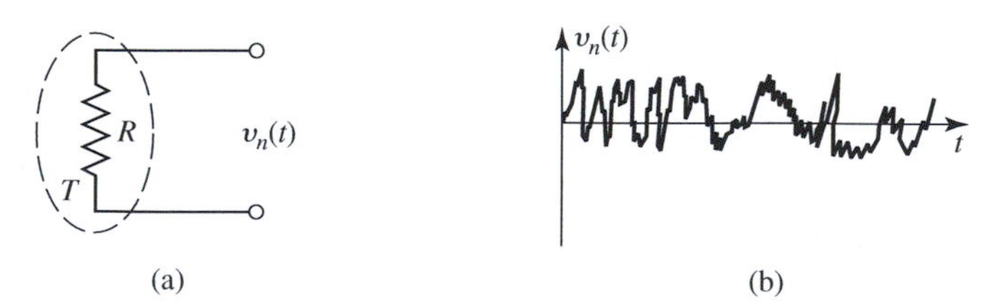

FIGURE 3.1 (a) A resistor at temperature T produces the noise voltage $v_n(t)$. (b) The random noise voltage generated by a resistor at temperature T.

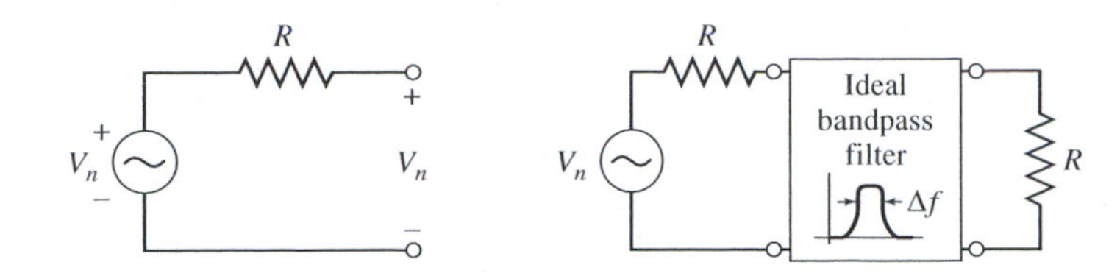

FIGURE 3.2 (a) The Thevenin equivalent circuit for a noisy resistor. (b) Maximum power transfer of noise power from a noisy resistor to a load over a bandwidth B.

The result in (3.24) is known as the Rayleigh–Jeans approximation, and is valid for frequencies up through the microwave band [4].

The noisy resistor can be modeled using a Thevenin equivalent circuit as an ideal (noiseless) resistor with a voltage generator to represent the noise voltage, as shown in Figure 3.2a. The *available noise power* is defined as the maximum power that can be delivered from the noise source to a load resistor. As shown in Figure 3.2b, maximum power transfer occurs when the load is conjugately matched to the source. Then the available noise power can be calculated as

$$P_n = \left(\frac{V_n}{2}\right)^2 \frac{1}{R} = \frac{V_n^2}{4R} = kTB, \tag{3.25}$$

where V_n is the rms noise voltage of the resistor. This is a fundamental result that is useful in a wide variety of problems involving noise. Note that the noise power decreases as the system bandwidth decrease. This implies that systems with smaller bandwidths collect less noise power. Also note that noise power decreases as temperature decreases, which implies that internally generated noise effects can be reduced by cooling a system to low temperatures. Finally, note that the noise power of (3.25) depends on absolute bandwidth, but not on the center frequency of the band. Since thermal noise power is independent of frequency, it is referred to as *white noise*, because of the analogy with white light and its makeup of all other visible light frequencies. It has been found experimentally, and verified by quantum mechanics, that thermal noise is independent of frequency for $0 < f < 1000$ GHz.

The noise power of (3.25) can also be represented in terms of the power spectral density according to (3.23). Since the power given by (3.25) is independent of frequency, the power spectral density must also be independent of frequency, and so we have that,

$$S_n(\omega) = \frac{P_n}{2B} = \frac{kT}{2} = \frac{n_0}{2}. \tag{3.26}$$

This is known as the *two-sided power spectral density* of thermal noise, meaning that the frequency range from $-B$ to B (Hz) is included in the integration of (3.23). This is the conventional definition as used in communication systems work. The notation defined in (3.26), where $n_0/2 = kT/2$ is the two-sided power spectral density for white noise, will be used throughout this book. (Note that n_0 is a constant, with the subscript 'zero'. This should not be confused with the notation $n_o(t)$, which we will often use to denote a noise output signal. The subscript for this latter notation is 'oh', and will always be written as a function of time.)

Since the power spectral density of thermal noise is constant with frequency, its autocorrelation must be a delta function according to (3.22b):

$$R(\tau) = \frac{1}{2\pi}\int_{-\infty}^{\infty} \frac{n_0}{2} e^{-j\omega\tau}\, d\omega = \frac{n_0}{2}\delta(\tau). \tag{3.27}$$

By the central limit theorem, the probability density function of white noise is gaussian

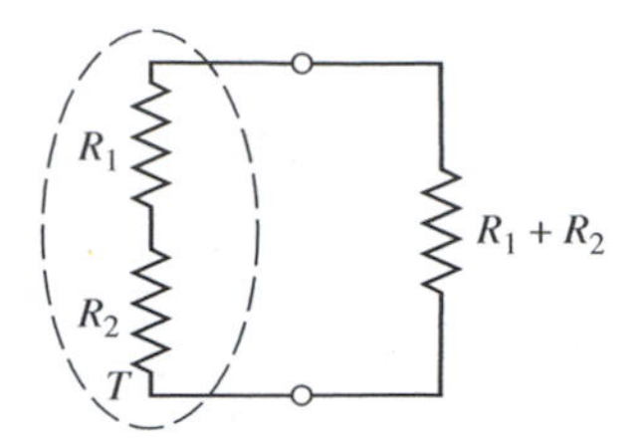

FIGURE 3.3 Circuit for Example 3.2.

with zero mean:

$$f_n(n) = \frac{1}{\sqrt{2\pi\sigma^2}} e^{-n^2/2\sigma^2}. \tag{3.28}$$

where σ^2 is the variance of the gaussian noise. Thermal noise having a zero mean gaussian PDF is known as *white gaussian noise*. Since the variance of the sum of two independent random variables is the sum of the individual variances (see Problem 3.5), and the variance is equivalent to power delivered to a 1 Ω load, the noise powers generated by two independent noise sources add in a common load. This is in contrast to the case of deterministic sources, where voltages add.

Note that (3.27) is not completely consistent with (3.28), since (3.27) indicates that $R(0)$, the variance of white noise, is infinite while (3.28) implies a finite variance. This problem arises because of the mathematical assumption that white noise has a constant power spectral density, and therefore infinite power. In fact, as we discussed earlier, thermal noise has a constant PSD only over a finite, but very wide, frequency band. We can resolve this issue if we understand our use of the concept of white noise to actually mean a bandlimited PSD having a finite frequency range, but broader than the system bandwidth with which we are working.

EXAMPLE 3.2 CALCULATION OF NOISE POWER

Two noisy resistors, R_1 and R_2, at temperature T, are shown in Figure 3.3. Calculate the available noise power from these sources by considering the individual noise power from each resistor separately. Next, consider the resistors as equivalent to a single resistor of value $R_1 + R_2$, and verify that the same available noise power is obtained. Assume a bandwidth B for the system.

Solution

The equivalent noise voltage from each resistor is found from (3.24):

$$V_{n1} = \sqrt{4kTBR_1}$$
$$V_{n2} = \sqrt{4kTBR_2}.$$

For maximum power transfer, the load resistance should be $R_1 + R_2$. Then the noise power delivered to the load from each noise source is

$$P_{n1} = \left(\frac{V_{n1}}{2}\right)^2 \frac{1}{R_1 + R_2} = \frac{kTBR_1}{R_1 + R_2}$$
$$P_{n2} = \left(\frac{V_{n2}}{2}\right)^2 \frac{1}{R_1 + R_2} = \frac{kTBR_2}{R_1 + R_2}.$$

So the total available noise power is

$$P_n = P_{n1} + P_{n2} = kTB.$$

Considering the two resistors as a single resistor of value $R_1 + R_2$, with a load resistance of $R_1 + R_2$, gives an available noise power of

$$P_n = kTB,$$

in agreement with the first result. ○

3.3 NOISE IN LINEAR SYSTEMS

In a wireless radio receiver, both desired signals and undesired noise pass through various stages, such as RF amplifiers, filters, and mixers. These functions generally alter the statistical properties of the noise, and so it is useful to study these effects by considering the general case of transmission of noise through a linear system. We then consider some important special cases, such as filters and integrators, and the nonlinear situation where noise undergoes frequency conversion by mixing.

Autocorrelation and Power Spectral Density in Linear Systems

In the case of deterministic signals, we can find the response of a linear time-invariant system to an input excitation in the time domain by using convolution with the impulse response of the system, or in the frequency domain by using the transfer function of the system. Similar results apply to wide-sense stationary random processes, in terms of either the autocorrelation function or the power spectral density.

Consider the linear time-invariant system shown in Figure 3.4, where the input random process, $x(t)$, has an autocorrelation $R_x(\tau)$ and power spectral density $S_x(\omega)$, and the output random process, $y(t)$, has an autocorrelation $R_y(\tau)$ and power spectral density $S_y(\omega)$. If the impulse response of the system is $h(t)$, we can calculate the output response as

$$y(t) = \int_{-\infty}^{\infty} h(u)x(t-u)\,du. \tag{3.29a}$$

Similarly, a time-shifted version of $y(t)$ is

$$y(t+\tau) = \int_{-\infty}^{\infty} h(v)x(t+\tau-v)\,dv. \tag{3.29b}$$

So the autocorrelation of $y(t)$ can be found as

$$\begin{aligned} R_y(\tau) = E\{y(t)y(t+\tau)\} &= \int_{-\infty}^{\infty}\int_{-\infty}^{\infty} h(u)h(v)E\{x(t-u)x(t+\tau-v)\}\,du\,dv \\ &= \int_{-\infty}^{\infty}\int_{-\infty}^{\infty} h(u)h(v)R_x(\tau+u-v)\,du\,dv. \end{aligned} \tag{3.30}$$

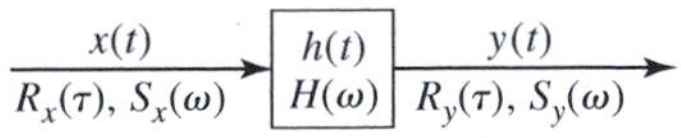

$x(t)$ → [$h(t)$ / $H(\omega)$] → $y(t)$

$R_x(\tau)$, $S_x(\omega)$ | $R_y(\tau)$, $S_y(\omega)$

FIGURE 3.4 A linear system with an impulse response $h(t)$ and transfer function $H(\omega)$. The input is a random process $x(t)$, having autocorrelation $R_x(\tau)$ and *PSD* $S_x(\omega)$. The output random process is $y(t)$, having autocorrelation $R_y(\tau)$ and *PSD* $S_y(\omega)$.

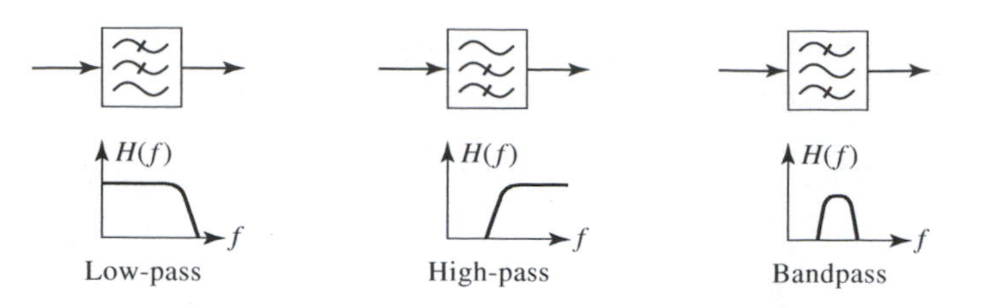

FIGURE 3.5 System symbols and frequency responses for low-pass, high-pass, and band-pass filters.

This result shows that the autocorrelation of the output is given by the double convolution of the autocorrelation of the input with the impulse response: $R_y(\tau) = h(\tau) \otimes h(-\tau) \otimes R_x(\tau)$. We can derive the equivalent result in terms of power spectral density by taking the Fourier transform of both sides of (3.30), in view of (3.22a):

$$\int_{-\infty}^{\infty} R_y(\tau)e^{-j\omega\tau}\,d\tau = \int_{-\infty}^{\infty}\int_{-\infty}^{\infty} h(u)h(v)\int_{-\infty}^{\infty} R_x(\tau + u - v)e^{-j\omega\tau}\,d\tau\,du\,dv.$$

Now perform a change of variable to $\alpha = \tau + u - v$, so that $d\alpha = d\tau$. Then we obtain

$$\int_{-\infty}^{\infty} R_y(\tau)e^{-j\omega\tau}d\tau = \int_{-\infty}^{\infty} h(u)e^{j\omega u}\int_{-\infty}^{\infty} h(v)e^{-j\omega v}\int_{-\infty}^{\infty} R_x(\alpha)e^{-j\omega\alpha}\,d\alpha\,du\,dv.$$

Since $H(\omega)$ is the Fourier transform of $h(t)$

$$H(\omega) = \int_{-\infty}^{\infty} h(t)e^{-j\omega t}dt, \tag{3.31}$$

the above simplifies to the following important result:

$$S_y(\omega) = |H(\omega)|^2 S_x(\omega). \tag{3.32}$$

We will now demonstrate the utility of these results with several applications.

Gaussian White Noise through an Ideal Low-pass Filter

As we will see in Chapters 5, 9, and 10, filters play an important role in wireless receivers and transmitters. The main function of a filter is to provide *frequency selectivity*, by allowing a certain range of frequencies to pass, while blocking other frequencies. Figure 3.5 shows the symbols and associated idealized frequency responses for low-pass, high-pass, and bandpass filters. Here we examine the effect of an ideal low-pass filter on noise.

Figure 3.6 shows white noise passing through a low-pass filter. The filter has a transfer function, $H(f)$, as shown, with a cutoff frequency of Δf. Note that the transfer function is defined for both positive and negative frequency, since we will be using the two-sided power spectral density. Our usual notation will be to use lowercase letters, such as $n_i(t)$ and

FIGURE 3.6 White noise passing through an ideal low-pass filter, and the transfer function of the filter.

$n_o(t)$, for noise and signal voltages in the time domain, and capital letters, such as N_i and N_o, for average powers of noise and signals.

Since the input noise is white, the two-sided power spectral density of the input noise is constant, as given in (3.26):

$$S_{n_i}(f) = \frac{n_0}{2} \text{ (all f)}. \tag{3.33}$$

Then from (3.32) the output power spectral density is given by

$$S_{n_o}(f) = |H(f)|^2 S_{n_i}(f) = \begin{cases} \frac{n_0}{2} & \text{for } |f| < \Delta f \\ 0 & \text{for } |f| > \Delta f \end{cases}. \tag{3.34}$$

The output noise power is then

$$N_0 = (2\Delta f) S_{n_0}(f) = \Delta f n_0. \tag{3.35}$$

We see that the output noise power is proportional to the filter bandwidth.

Gaussian White Noise through an Ideal Integrator

As we will see in Chapter 9, integrators are critical components for the detection and demodulation of digital signals. Here we derive an expression for the output noise power from an integrator with white noise input; this result will be used later for the derivation of error probabilities for digital modulation in Chapter 9.

Figure 3.7 shows a noise signal, $n_i(t)$, applied to the input of an ideal integrator. The output noise signal is $n_o(t)$. The output of the integrator is the value of the integral, at time $t = T$, of the input signal. We need to find the average power of the output noise.

The transfer function of the integration operation is

$$H(\omega) = \frac{1}{j\omega}(1 - e^{-j\omega T}). \tag{3.36}$$

where T is the integration interval time. Evaluating the magnitude squared of (3.36) gives

$$\begin{aligned} |H(\omega)|^2 &= H(\omega)H^*(\omega) = \frac{(1 - e^{-j\omega T})(1 - e^{j\omega T})}{\omega^2} = \frac{2 - 2\cos\omega T}{\omega^2} \\ &= \frac{4\sin^2 \omega T/2}{\omega^2} = \frac{\sin^2 \pi f T}{(\pi f)^2} = T^2\left(\frac{\sin \pi f T}{\pi f T}\right)^2, \end{aligned} \tag{3.37}$$

since $\omega = 2\pi$f.

If we assume white noise at the input, with $S_{n_i}(\omega) = n_0/2$, then the output noise power can be calculated using (3.23) and (3.32) to give

$$\begin{aligned} N_o &= \int_{-\infty}^{\infty} \frac{n_0}{2} |H(f)|^2 \, df = \frac{n_0 T^2}{2} \int_{-\infty}^{\infty} \left(\frac{\sin \pi f T}{\pi f T}\right)^2 df \\ &= \frac{n_0 T}{2\pi} \int_{-\infty}^{\infty} \left(\frac{\sin x}{x}\right)^2 dx = \frac{n_0 T}{2} \end{aligned} \tag{3.38}$$

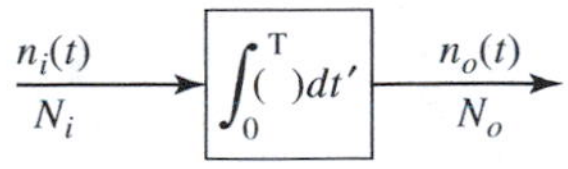

$n_i(t)$, N_i → $\int_0^T (\ )dt'$ → $n_o(t)$, N_o

FIGURE 3.7 White noise passing through an ideal integrator.

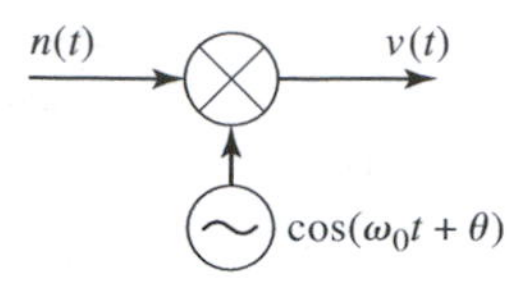

FIGURE 3.8 White noise passing through a mixer with a local oscillator signal, $\cos(\omega_0 t + \theta)$.

The integral was evaluated by using a change of variables, $x = \pi f T$, with $dx = \pi T df$, and a standard integral listed in Appendix B.

Mixing of Noise

One of the common functions of a receiver is to perform *frequency conversion*, by mixing a signal with a local oscillator to shift the original signal spectrum up or down in frequency. When noise coexists with the signal, the noise spectrum will also be shifted in frequency. While we will study mixers in detail in Chapter 7, here we idealize the function of mixing by considering it as a process of multiplication of the input signal by a local oscillator signal, as shown in Figure 3.8. We wish to find the average noise power of the output signal.

We assume that $n(t)$ is a bandlimited white gaussian noise signal with variance, or average power, $\sigma^2 = E\{n^2(t)\}$. The local oscillator signal is given by $\cos(\omega_0 t + \theta)$, where the phase, θ, is a random variable uniformly distributed on the interval $0 < \theta \leq 2\pi$, and is independent of $n(t)$. The output of the idealized mixer is

$$v(t) = n(t)\cos(\omega_0 t + \theta). \tag{3.39}$$

The average output power from the mixer can then be calculated as the variance of $v(t)$:

$$\begin{aligned} N_o &= E\{v^2(t)\} = E\{n^2(t)\cos^2(\omega_0 t + \theta)\} = E\{n^2(t)\}E\{\cos^2(\omega_0 t + \theta)\} \\ &= \frac{\sigma^2}{2\pi}\int_0^{2\pi} \cos^2(\omega_0 t + \theta)\, d\theta = \frac{\sigma^2}{2} \end{aligned} \tag{3.40}$$

This result shows that mixing reduces the average noise power by half. In this case, the factor of one-half is due to the ensemble averaging of the $\cos^2 \omega_0 t$ term over the range of random phase.

If we now consider a deterministic local oscillator signal of the form $\cos \omega_0 t$ (without a random phase), the variance of the output signal becomes

$$E\{v^2(t)\} = E\{n^2(t)\cos^2 \omega_0 t\} = \sigma^2 \cos^2 \omega_0 t. \tag{3.41}$$

The last result follows because the $\cos^2 \omega_0 t$ term is unaffected by the expected value operator, since it is no longer a random variable. (In addition, $v(t)$ is no longer stationary, and therefore does not have an autocorrelation function or power spectral density.) The variance of the output signal is now a function of time, and represents the instantaneous power of the output signal. To find the time-average output power, we must take the time-average of the variance found in (3.41):

$$N_0 = \frac{1}{T}\int_0^T E\{v^2(t)\}\, dt = \frac{\omega_0}{2\pi}\int_0^{2\pi/\omega_0} \sigma^2 \cos^2 \omega_0 t dt = \frac{\sigma^2}{2}, \tag{3.42}$$

since $T = 1/f = 2\pi/\omega_0$. We see that the same average output power is obtained whether the averaging is over the ensemble phase variation, or over time.

EXAMPLE 3.3 MIXING NOISE

Consider the complex mixing product $x(t) = n(t)e^{j\omega_0 t}$, formed by mixing noise voltage $n(t)$ with a complex exponential. If the autocorrelation and PSD of $n(t)$ are $R_n(\tau)$ and $S_n(\omega)$, find the autocorrelation and PSD of $x(t)$.

Solution

Using the definition of autocorrelation for random processes given in (3.21) we have

$$\begin{aligned} R_x(\tau) &= E\{x^*(t)x(t+\tau)\} = E\{n(t)e^{-j\omega_0 t}n(t+\tau)e^{j\omega_0(t+\tau)}\} \\ &= E\{n(t)n(t+\tau)\}e^{j\omega_0\tau} = R_n(\tau)e^{j\omega_0\tau} \end{aligned}$$

Note that $x(t)$ is still a stationary process, since it has a proper autocorrelation function. From (3.22a) the power spectral density is

$$S_x(\omega) = \int_{-\infty}^{\infty} R_x(\tau)e^{-j\omega_0\tau}d\tau = \int_{-\infty}^{\infty} R_n(\tau)e^{-j(\omega-\omega_0)\tau}d\tau = S_n(\omega-\omega_0),$$

where the last result follows by replacing ω with $\omega - \omega_0$ in $S_n(\omega) = \int_{-\infty}^{\infty} R_n(\tau)e^{-j\omega\tau}d\tau$. ○

Narrowband Representation of Noise

In many receiver circuits, signals and noise are passed through a bandpass filter. In this case, it becomes possible to represent the noise in a form that is more convenient for analysis. This is called the *narrowband representation* of noise, a result that will be very useful in Chapter 9 when analyzing the effect of noise on the demodulation of signals.

Figure 3.9 shows gaussian white noise passing through a bandpass filter with a center frequency ω_0 and bandwidth $\Delta\omega$. If the two-sided power spectral density of the input noise is $n_0/2$, then the PSD of the output noise, $n(t)$, is as shown in the figure. If $\Delta\omega \ll \omega_0$, then $n(t)$ can be represented as

$$n(t) = x(t)\cos\omega_0 t + y(t)\sin\omega_0 t, \tag{3.43}$$

where $x(t)$ and $y(t)$ are random processes, but are slowly varying due to the narrow bandwidth of the filter. To show that the above representation is valid, consider the circuit of Figure 3.10, which can be used to generate $x(t)$ and $y(t)$. Here the noise $n(t)$ is divided and mixed separately with $2\cos\omega_0 t$ and $2\sin\omega_0 t$, which produces the following results:

$$\begin{aligned} 2n(t)\cos\omega_0 t &= 2x(t)\cos^2\omega_0 t + 2y(t)\sin\omega_0 t\cos\omega_0 t \\ &= x(t) + x(t)\cos 2\omega_0 t + y(t)\sin 2\omega_0 t \\ 2n(t)\sin\omega_0 t &= 2x(t)\cos\omega_0 t\sin\omega_0 t + 2y(t)\sin^2\omega_0 t \\ &= y(t) - y(t)\cos 2\omega_0 t + x(t)\sin 2\omega_0 t. \end{aligned}$$

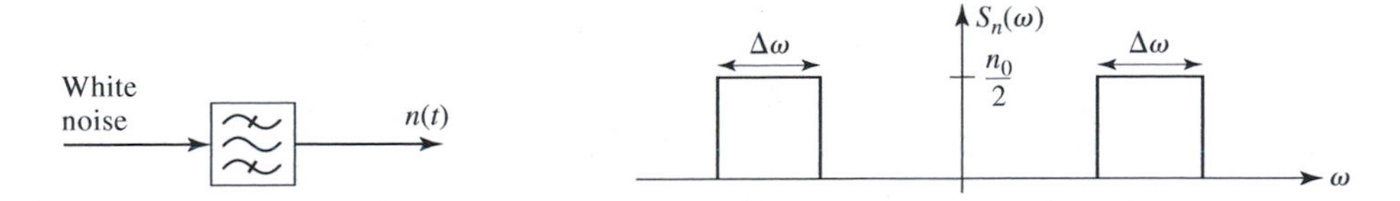

FIGURE 3.9 White noise passing through a bandpass filter, and the power spectral density of the output noise.

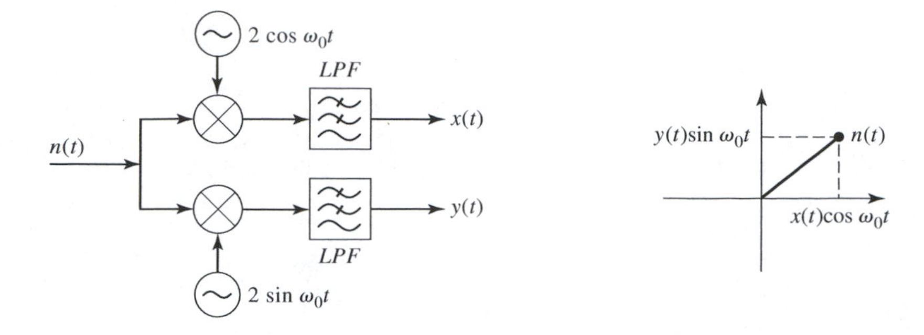

FIGURE 3.10 Circuit to generate low-pass noise $x(t)$ and $y(t)$ from a bandpass noise source $n(t)$.

After low-pass filtering with a low-pass cutoff frequency of $f_c = \Delta\omega/4\pi$, only the $x(t)$ and $y(t)$ terms will remain in the above results. Thus the output noises $x(t)$ and $y(t)$ are limited in bandwidth to $\Delta\omega/2$, and can be viewed as the bandpass noise at ω_0 shifted down to zero frequency. Since $x(t)$ and $y(t)$ represent the *in-phase* and *quadrature* components of $n(t)$, as indicated in the phasor diagram of Figure 3.10, (3.43) is also known as the *quadrature representation* of narrowband noise. We now find the statistics of $x(t)$ and $y(t)$, and show that these gaussian random processes have zero mean, the same variance as $n(t)$, and are uncorrelated. We will also find the power spectral densities of $x(t)$ and $y(t)$.

Since $E\{n(t)\}=0$, we have from (3.43) that

$$E\{n(t)\} = 0 = E\{x(t)\}\cos\omega_0 t + E\{y(t)\}\sin\omega_0 t.$$

Since $\cos\omega_0 t$ and $\sin\omega_0 t$ vary differently with time, we must have

$$E\{x(t)\} = E\{y(t)\} = 0. \tag{3.44}$$

Next, we evaluate the autocorrelation of $n(t)$ using (3.43):

$$\begin{aligned} R_n(\tau) &= E\{n(t)n(t+\tau)\} \\ &= E\{[x(t)\cos\omega_0 t + y(t)\sin\omega_0 t][x(t+\tau)\cos\omega_0(t+\tau) + y(t+\tau)\sin\omega_0(t+\tau)]\} \\ &= R_x(\tau)\cos\omega_0 t\cos\omega_0(t+\tau) + R_{xy}(\tau)\cos\omega_0 t\sin\omega_0(t+\tau) \\ &\quad + R_{yx}(\tau)\sin\omega_0 t\cos\omega_0(t+\tau) + R_y(\tau)\sin\omega_0 t\sin\omega_0(t+\tau) \end{aligned}$$

where $R_{xy}(\tau) = E\{x(t)y(t+\tau)\}$ and $R_{yx}(\tau) = E\{y(t)x(t+\tau)\}$ are the cross-correlation functions of $x(t)$ and $y(t)$. Using standard identities to expand the trigonometric products gives the following:

$$\begin{aligned} R_n(\tau) = {} & \tfrac{1}{2}R_x(\tau)[\cos\omega_0\tau + \cos\omega_0(2t+\tau)] \\ & + \tfrac{1}{2}R_{xy}(\tau)[\sin\omega_0\tau + \sin\omega_0(2t+\tau)] \\ & + \tfrac{1}{2}R_{yx}(\tau)[-\sin\omega_0\tau + \sin\omega_0(2t+\tau)] \\ & + \tfrac{1}{2}R_y(\tau)[\cos\omega_0\tau - \cos\omega_0(2t+\tau)]. \end{aligned} \tag{3.45}$$

Because $n(t)$ is a stationary process, its autocorrelation must be a function only of τ, and cannot vary with t. Thus the coefficients of $\cos\omega_0(2t+\tau)$ and $\sin\omega_0(2t+\tau)$ must vanish.

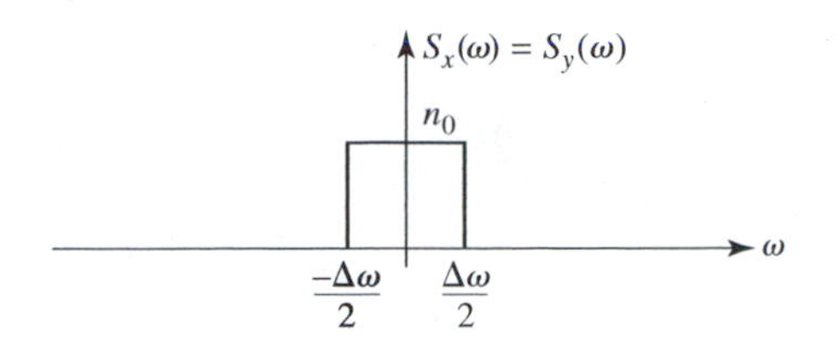

FIGURE 3.11 Power spectral density of $x(t)$ and $y(t)$.

This gives

$$R_x(\tau) = R_y(\tau) \tag{3.46a}$$

$$R_{xy}(\tau) = -R_{yx}(\tau), \tag{3.46b}$$

and then (3.45) reduces to

$$R_n(\tau) = R_x(\tau)\cos\omega_0\tau + R_{xy}(\tau)\sin\omega_0\tau = R_y(\tau)\cos\omega_0\tau - R_{yx}(\tau)\sin\omega_0\tau. \tag{3.47}$$

This result shows that $n(t)$, $x(t)$, and $y(t)$ all have the same variance, since $R_n(0) = R_x(0) = R_y(0)$.

We can also find $R_n(\tau)$ directly by evaluating the inverse Fourier transform of $S_n(\omega)$, which is shown in Figure 3.9. Since $R_n(\tau)$ is real, and symmetric about $\omega = 0$, we have

$$R_n(\tau) = \frac{1}{2\pi}\int_{-\infty}^{\infty} S_n(\omega)e^{j\omega\tau}\,d\omega = \frac{1}{2\pi}\int_{0}^{\infty} S_n(\omega)\cos\omega\tau\,d\omega = \frac{n_0}{2\pi}\int_{\omega_0-\Delta\omega/2}^{\omega_0+\Delta\omega/2}\cos\omega\tau\,d\omega$$

$$= \frac{n_0}{2\pi\tau}\left[\sin\left(\omega_0 + \frac{\Delta\omega}{2}\right)\tau - \sin\left(\omega_0 - \frac{\Delta\omega}{2}\right)\tau\right] = \frac{n_0}{\pi\tau}\sin\frac{\Delta\omega\tau}{2}\cos\omega_0\tau. \tag{3.48}$$

Comparing (3.48) to (3.47) shows that

$$R_x(\tau) = R_y(\tau) = \frac{n_0}{\pi\tau}\sin\frac{\Delta\omega\tau}{2} \tag{3.49a}$$

$$R_{xy}(\tau) = R_{yx}(\tau) = 0. \tag{3.49b}$$

The fact that $R_{xy}(\tau) = 0$ implies that $x(t)$ and $y(t)$ are statistically independent. Finally, we can find the PSD of $x(t)$ and $y(t)$ by taking the Fourier transform of $R_x(\tau)$:

$$S_x(\omega) = S_y(\omega) = \int_{-\infty}^{\infty} R_x(\tau)e^{-j\omega\tau}\,d\tau = \begin{cases} n_0 & \text{for } |\omega| < \Delta\omega/2 \\ 0 & \text{for } |\omega| > \Delta\omega/2 \end{cases} \tag{3.50}$$

where the required Fourier transform may be found in Appendix C. This power spectral density is shown in Figure 3.11, and can be viewed as the bandlimited PSD of $n(t)$ shifted up and down in frequency by the amount ω_0, and low-pass filtered. Note that the peak value of the PSD of $x(t)$ and $y(t)$ is n_0, twice that of the PSD of $n(t)$.

3.4 BASIC THRESHOLD DETECTION

We now have enough background in the topics of noise and systems to discuss an application to basic threshold detection. Threshold detection is relevant to most digital modulation schemes, and so we will see this topic again in more detail in Chapter 9. Here we evaluate the probability of error for a simple binary communications channel.

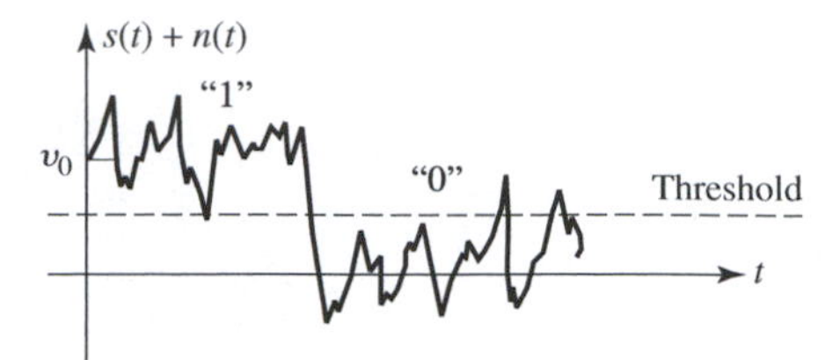

FIGURE 3.12 Input signal and noise voltage for a basic threshold detection system.

Consider a communications system where binary signals are transmitted in the presence of bandlimited white gaussian noise. Thus, the received signal, $r(t)$, can be written as the transmitted signal voltage, $s(t)$, plus a noise voltage, $n(t)$:

$$r(t) = s(t) + n(t) \tag{3.51}$$

where $n(t)$ has zero mean and variance σ^2. A sketch of a possible received voltage is shown in Figure 3.12.

When a binary "1" is transmitted the signaling voltage will be $s(t) = v_0$, and when a binary "0" is transmitted we will have $s(t) = 0$. The receiver must be designed to process the received voltage, and detect whether a "1" or a "0" has been transmitted. In the absence of noise we can simply sample the receive voltage and determine whether it is above or below a *threshold* level. In this case, the logical choice for a threshold voltage would be $v_0/2$, so that if $r(t) > v_0/2$ the receiver would detect a "1," and if $r(t) < v_0/2$ the receiver would detect a "0." This detection process can be implemented using a simple sampler and comparator circuit. In practical receivers threshold detection could incorporate matched filters or integrators to minimize the effect of noise, but here we consider only the sampling of the received signal at its maximum or minimum point.

Because the possible noise voltage amplitude ranges from $-\infty$ to ∞, the received signal may sometimes be less than the threshold when a "1" has been sent, and may be greater than the threshold when a "0" has been sent. Either of these cases will result in a detection error. In fact, there are four detection possibilities, as listed in Table 3.1.

Probability of Error

We can now find the *probability of error* for threshold detection. We define $P_e^{(1)}$ as the probability of an error in detection when a binary "1" has been transmitted, and $P_e^{(0)}$ as the probability of an error when a binary "0" has been sent. Knowing these two probabilities then defines the likelihood of all the outcomes in Table 3.1, since the probability of a correct outcome is 1 − (probability of an error).

TABLE 3.1 Possible Outcomes of Threshold Detection

Transmitted Binary Data	$s(t)$	$r(t) > v_0/2$?	Detection Outcome	Correct Detection
0	0	no	0	yes!
0	0	yes	1	error
1	v_0	yes	1	yes!
1	v_0	no	0	error

When a binary "1" is transmitted, a detection error will occur if the received signal and noise is less than the threshold level at the sampling time. For a threshold of $v_0/2$, the probability of this event is

$$P_e^{(1)} = P\{r(t) = v_0 + n(t) < v_0/2\}$$
$$= \int_{-\infty}^{v_0/2} f_r(r)\,dr = \int_{-\infty}^{v_0/2} \frac{e^{-(r-v_0)^2/2\sigma^2}}{\sqrt{2\pi\sigma^2}}\,dr \qquad (3.52)$$

where we have used (3.1), (3.5), and the gaussian probability density function given in (3.12b). Since $n(t)$ is gaussian with zero mean, the receive signal $r(t)$ is also gaussian, but with a mean value of v_0 when a binary "1" is being transmitted. The expression in (3.52) can be reduced to a standard form by using the change of variable $x = (v_0 - r)/\sqrt{2\sigma^2}$.

Then we have

$$P_e^{(1)} = \frac{1}{\sqrt{\pi}} \int_{x_0}^{\infty} e^{-x^2}\,dx, \qquad (3.53)$$

where the lower limit is

$$x_0 = \frac{v_0}{2\sqrt{2\sigma^2}}. \qquad (3.54)$$

The integral occurring in (3.53) is related to the *complementary error function*, written as

$$erfc(x) = \frac{2}{\sqrt{\pi}} \int_{x}^{\infty} e^{-u^2}\,du. \qquad (3.55)$$

Details on properties of the complementary error function, including an algorithm for calculating $erfc(x)$, can be found in Appendix D. Using the definition of (3.55) allows (3.53) to be written as

$$P_e^{(1)} = \frac{1}{2} erfc(x_0) = \frac{1}{2} erfc\left(\frac{v_0}{2\sqrt{2\sigma^2}}\right), \qquad (3.56)$$

which is our final expression for $P_e^{(1)}$. By a similar analysis we can find $P_e^{(0)}$, the probability of error when a binary "0" is sent. It is left as a problem to show that $P_e^{(0)} = P_e^{(1)}$, as might be expected from the symmetry resulting from a threshold of $v_0/2$. The result of (3.56) is dependent on the ratio v_0/σ, which can be considered a *signal-to-noise ratio* (SNR), since v_0 is the maximum signal voltage, and σ is the rms value of the noise voltage. Since *erfc* (x) decreases monotonically with x, large SNR results in lower probability of error.

A graphical interpretation of threshold detection is shown in Figure 3.13. The probability density functions are shown for the received signal and noise for the two cases of sending a binary "0" or a "1." The former has a PDF centered at $r = 0$, while the latter has

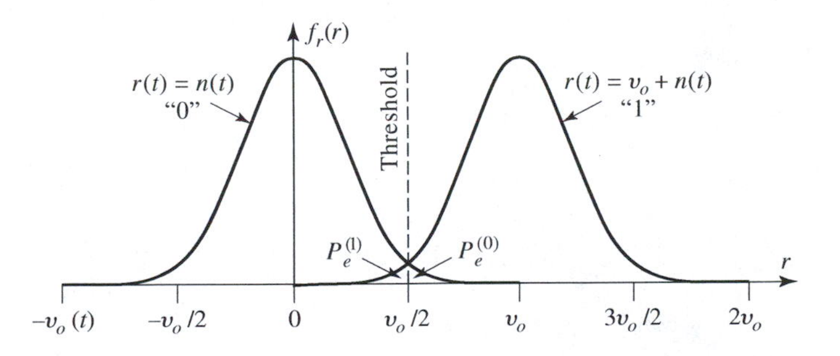

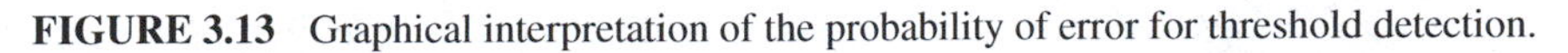
FIGURE 3.13 Graphical interpretation of the probability of error for threshold detection.

a PDF centered at $r = v_0$. The threshold of $v_0/2$ is located midway between these values. The probabilities of error are the areas of the tails of the two PDFs either above or below the threshold value.

EXAMPLE 3.4 PROBABILITY OF ERROR FOR THRESHOLD DETECTION

Calculate and plot the probability of error for threshold detection versus the signal-to-noise ratio, v_0/σ, in dB. Use a logarithmic scale for the probability of error.

Solution

Since we are dealing with voltages, the signal-to-noise ratio in dB is calculated as

$$\frac{v_0}{\sigma}(dB) = 20 \log \frac{v_0}{\sigma}.$$

Then (3.56) can be used to evaluate $P_e^{(1)}$. The algorithm of Appendix D can be used to calculate values of the complementary error function. A sample calculation follows for $v_0/\sigma = 6$ dB:

For $v_0/\sigma = 6$ dB we have a numerical value of

$$\frac{v_0}{\sigma} = 10^{6/20} = 2.0.$$

Then the argument of the complementary error function is, from (3.54)

$$x_0 = \frac{v_0}{2\sqrt{2}\sigma} = \frac{2.0}{2\sqrt{2}} = 0.707.$$

Equation (3.56) gives

$$P_e^{(1)} = \tfrac{1}{2} erfc(x_0) = \tfrac{1}{2} erfc(0.707) = \tfrac{1}{2}(0.317) = 0.159.$$

The same method can be used for other values of v_0/σ, and the result is plotted in Figure 3.14. Note that for large values of *S/N* the probability of error becomes very small. Error probabilities in the range of 10^{-5} to 10^{-8} are often desired in practice. ○

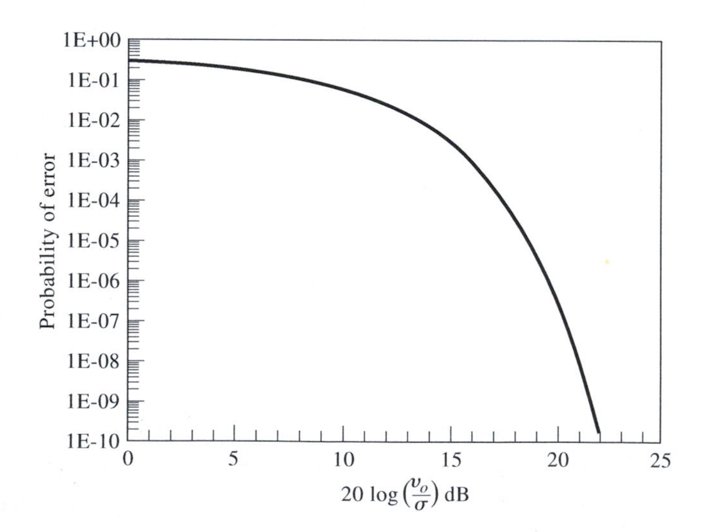

FIGURE 3.14 Probability of error versus signal-to-noise ratio for threshold detection.

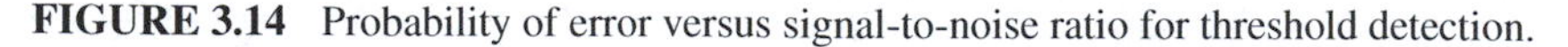

FIGURE 3.15 Equivalent noise temperature of an arbitrary white noise source.

3.5 NOISE TEMPERATURE AND NOISE FIGURE

Besides being received from the external environment by the antenna, noise is also generated by passive and active components of a wireless receiver system. In this section we will study ways of characterizing the noise properties of components such as amplifiers, mixers, couplers, and filters, and the transmission of noise through a multistage system.

Equivalent Noise Temperature

If an arbitrary noise source is white, so that its power spectral density is not a function of frequency (at least over the frequency range of interest), it can be modeled as an equivalent thermal noise source, and characterized by an *equivalent noise temperature*. This situation is illustrated in Figure 3.15, where an arbitrary white noise source of driving point impedance R delivers noise power N_o to a load resistor R. This noise source can be replaced with a noisy resistor of value R, at temperature T_e, where T_e is an equivalent temperature selected so that the same noise power is delivered to the load. Thus

$$T_e = \frac{N_o}{kB}. \tag{3.57}$$

Wireless components and receiver systems can then be characterized in terms of their equivalent noise temperature, T_e, expressed in degrees Kelvin (K). Note that $T_e \geq 0$, and may be greater or less than $T_0 = 290$ K. In addition, note that the result in (3.57) implies some fixed bandwidth, B, which is generally the bandwidth of the component or system. As an example, consider a noisy amplifier having bandwidth B and power gain G. Let the amplifier be matched to noiseless source and load resistors, as shown in Figure 3.16a. If the source resistor of Figure 3.16a is at a (hypothetical) temperature of $T_s = 0$ K, then the input noise power to the amplifier will be $N_i = 0$, and the output noise power N_o will be due only to the noise generated by the amplifier itself. We can obtain the same output noise power

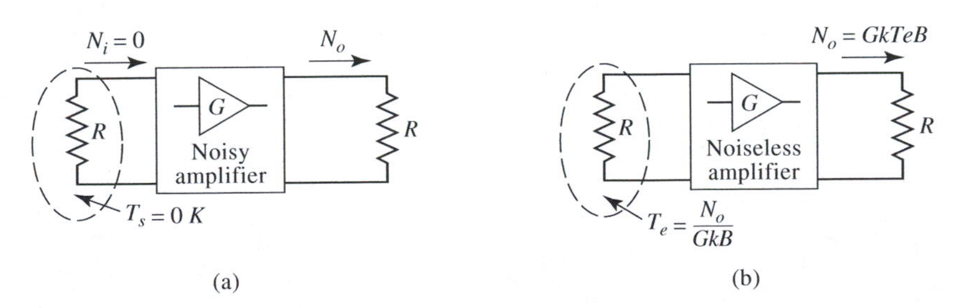

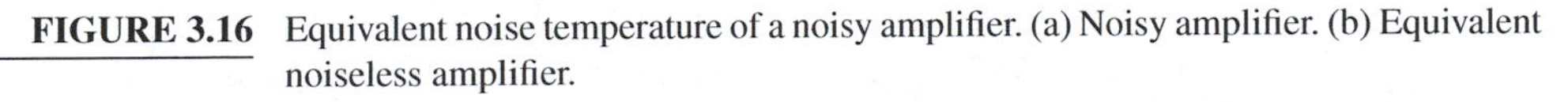

FIGURE 3.16 Equivalent noise temperature of a noisy amplifier. (a) Noisy amplifier. (b) Equivalent noiseless amplifier.

by driving an ideal noiseless amplifier with a resistor at a temperature

$$T_e = \frac{N_o}{GkB}, \tag{3.58}$$

so that the output noise power in both cases is $N_o = GkT_eB$, as illustrated in Figure 3.16b. Then T_e is the equivalent noise temperature of the amplifier. Note the important point that the equivalent noise source is applied at the *input* to the device, and that the input noise power, $N_i = kT_eB$, must be multiplied by the gain of the amplifier to obtain the output noise power. This is the convention that we will use throughout most of this book, but be aware that it is also possible to reference the equivalent noise source at the output of the component.

Measurement of Noise Temperature

A direct way to measure equivalent noise temperature is to drive the component with a known noise source, and measure the increase in output noise power. Because the thermal noise power generated by a resistor is so small, active noise sources are available for this purpose. Active noise sources use a diode or an electron tube to provide a calibrated noise power output, and are useful for laboratory tests and measurements. Active noise generators can be characterized by their equivalent noise temperature, but a more common measure of noise power for such components is the *excess noise ratio* (ENR), defined as

$$ENR(dB) = 10 \log \frac{N_g - N_0}{N_0} = 10 \log \frac{T_g - T_0}{T_0}, \tag{3.59}$$

where N_g and T_g are the noise power and equivalent temperature of the noise generator, and $N_0 = kT_0B$ and $T_0 = 290$ K are the noise power and temperature associated with a passive source (a matched load) at room temperature. Solid-state noise generators typically have ENRs ranging from 20 to 40 dB, meaning that their noise power output is 100 to 10,000 times greater than the thermal noise power from a matched load at room temperature.

The problem with direct measurement of noise temperature is that most components generate only small levels of noise, which are difficult to measure reliably. Instead, a ratio technique, called the *Y-factor method*, is often used in practice. This method is illustrated in Figure 3.17, where the device under test is connected to one of two different matched loads at different temperatures. Let T_1 be the temperature of the hotter load, and T_2 the temperature of the cooler load, and let the respective output noise powers be N_1 and N_2. Since the source noise is uncorrelated with the noise of the device under test, the total output noise powers for the two cases can be written as

$$N_1 = GkT_1B + GkT_eB \tag{3.60a}$$

$$N_2 = GkT_2B + GkT_eB, \tag{3.60b}$$

where the first term is due to the input noise power, and the second term is due to the noise generated by the device under test. This set of equations has two unknowns: T_e (the desired

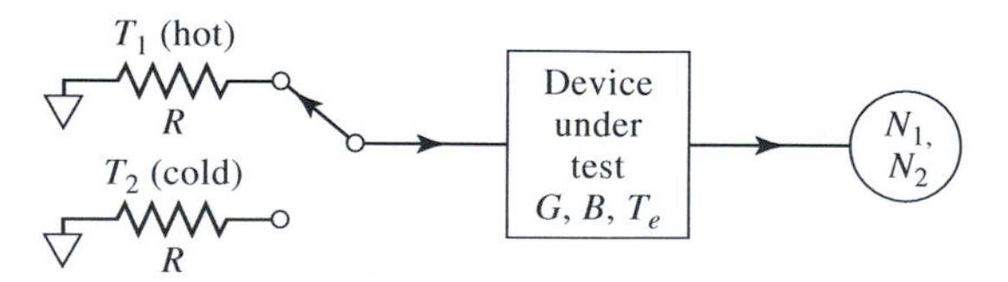

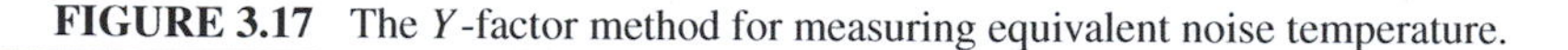

FIGURE 3.17 The Y-factor method for measuring equivalent noise temperature.

noise temperature of the device under test), and GB (the gain-bandwidth product of the amplifier). Define the Y factor as

$$Y = \frac{N_1}{N_2} = \frac{T_1 + T_e}{T_2 + T_e} \geq 1, \tag{3.61}$$

which is a ratio determined via power measurements, and does not depend on GB. Then (3.60) can be solved for the equivalent noise temperature of the device under test:

$$T_e = \frac{T_1 - YT_2}{Y - 1}. \tag{3.62}$$

To obtain accurate results with this method, the two source temperatures should not be too close together, so that Y is not close to unity. In practice, one noise source may be a resistor at room temperature, while the other is either hotter or colder, depending on whether T_e is greater or lesser than T_0. An active noise source can be used as a "hot" source, while a "cold" source can be obtained by immersing a load resistor in liquid nitrogen ($T = 77$ K), or liquid helium ($T = 4$ K).

Noise Figure

We have seen that noisy RF and microwave components can be characterized by an equivalent noise temperature. An alternative characterization is the *noise figure*, which can be viewed as a measure of the degradation in the signal-to-noise ratio between the input and output of the component. When noise and a desired signal are applied to the input of a noiseless network, both noise and signal will be attenuated or amplified by the same factor, so that the SNR will be unchanged. But if the network is noisy, the output noise power will be increased to a greater degree than the output signal power, so that the output SNR will be reduced. The noise figure, F, is a measure of this reduction in SNR, and is defined as

$$F = \frac{S_i/N_i}{S_o/N_o} \geq 1, \tag{3.63}$$

where S_i and N_i are the input signal and noise powers, and S_o and N_o are the output signal and noise powers. By definition, the input noise power must be the noise power from a matched load at $T_0 = 290$ K; that is, $N_i = kT_0B$. Noise figure is usually expressed in dB, obtained as $F(\text{dB}) = 10 \log F$. (Note: some authors define the numerical value of F as the *noise factor*, and the corresponding value in dB as the noise figure, but we will not make this distinction.)

We can establish the relation between noise figure and equivalent noise temperature by referring to Figure 3.18, which shows noise power N_i and signal power S_i being fed into a noisy two-port network. The network is characterized by a power gain G, a bandwidth B, and an equivalent noise temperature T_e. Note that the input noise power is $N_i = kT_0B$, as required by the definition of noise figure. The output signal power is $S_o = GS_i$, while

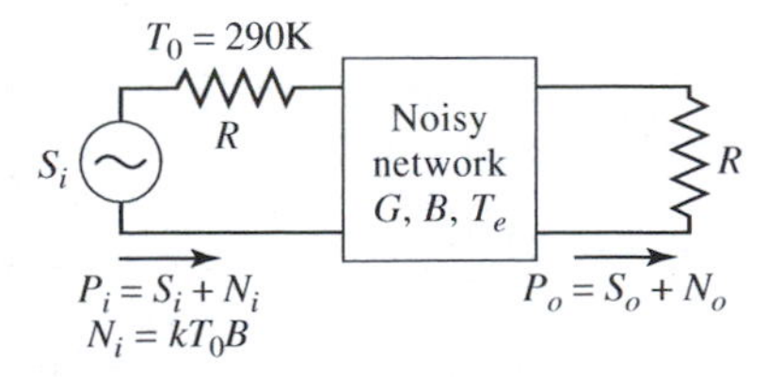

FIGURE 3.18 Relating the noise figure of a noisy network to its equivalent noise temperature.

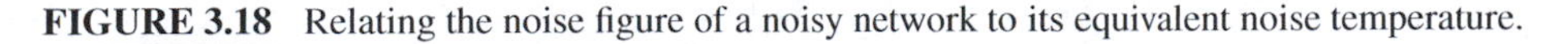

the output noise power is the sum of the amplified input noise power and the internally generated noise:

$$N_o = kGB(T_0 + T_e).$$

Using these results in (3.63) gives the noise figure as

$$F = \frac{S_i}{kT_0B}\frac{kGB(T_0 + T_e)}{GS_i} = 1 + \frac{T_e}{T_0} \geq 1. \tag{3.64}$$

This result can be solved for T_e in terms of F to give

$$T_e = (F - 1)T_0. \tag{3.65}$$

If the network were perfectly noiseless, its equivalent noise temperature would be zero, and its noise figure would be unity, or 0 dB. These results show that equivalent noise temperature and noise figure are interchangeable ways of characterizing the noise properties of RF and microwave components. In practice, mixers and amplifiers are usually specified in terms of noise figure, while antennas and receivers are often specified in terms of noise temperature.

Again referring to the two-port network of Figure 3.18, if we define N_{added} as the noise power added by the network, then the output noise power can be expressed as

$$N_o = G(N_i + N_{\text{added}}),$$

assuming that N_{added} is applied to the input of the network. Then, using (3.63) and the fact that $S_o = GS_i$, allows the noise figure to be written as

$$F = \frac{S_i/N_i}{GS_i/G(N_i + N_{\text{added}})} = 1 + \frac{N_{\text{added}}}{N_i}. \tag{3.66}$$

Since N_{added} is independent of N_i it cannot be proportional to N_i, and so the noise figure depends on the particular value chosen for the input noise power. For this reason N_i must be defined according to a fixed standard if the noise figure is to be meaningful in a general sense. The standard chosen is $N_i = kT_0B$.

Noise Figure of a Lossy Line

We can now determine the noise figure of an important practical component—the lossy transmission line (or attenuator). Figure 3.19 shows a lossy transmission line of characteristic impedance $Z_0 = R$, and held at a physical temperature T. The power gain, G, of a lossy network is less than unity; the *power loss factor*, L, can be defined as $L = 1/G > 1$.

If the input of the line is terminated with a matched load at temperature T, then the entire system is in thermal equilibrium, and the output of the line will appear as a resistor of value R, at temperature T. Thus the available noise power at the output must be $N_o = kTB$. But we can also view the output noise power as a sum of the input noise power attenuated through the lossy line, and the noise power added by the lossy line itself. In this case we

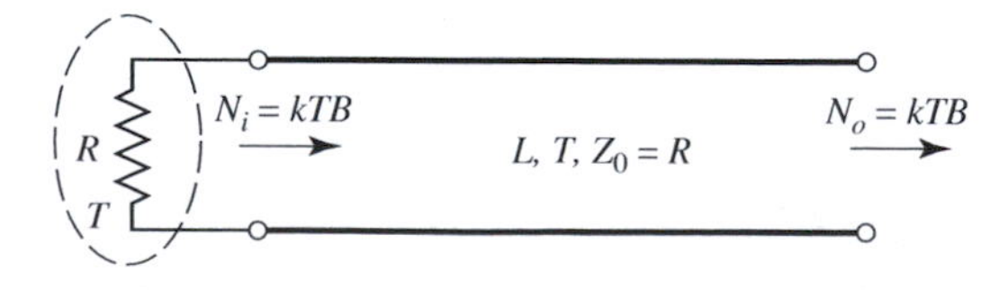

FIGURE 3.19 Determining the noise figure of a lossy line or attenuator with loss L and temperature T.

would write

$$N_o = kTB = G(kTB + N_{\text{added}}), \tag{3.67}$$

where N_{added} is the noise generated by the line, as if it appeared at the input terminals of the line. Solving (3.67) for this power gives

$$N_{\text{added}} = \frac{1-G}{G}kTB = (L-1)kTB. \tag{3.68}$$

Then (3.57) shows that the equivalent noise temperature of the lossy line (as referred to the input) is

$$T_e = \frac{N_{\text{added}}}{kB} = (L-1)T. \tag{3.69}$$

The noise figure of the lossy line can then be found using (3.64) to be

$$F = 1 + \frac{T_e}{T_0} = 1 + (L-1)\frac{T}{T_0}. \tag{3.70}$$

Note that in the limiting case of a lossless line, with $L = 1$, the above results reduce to $T_e = 0$ and $F = 1$ (0 dB) as expected, since a lossless component does not generate thermal noise. Another special case occurs when the line or attenuator is at room temperature. Then $T = T_0$, and the noise figure reduces to $F = L$. For instance, a 6 dB attenuator at room temperature has a noise figure of 6 dB. At higher temperatures, however, the noise figure will be higher.

Noise Figure of Cascaded Components

In a typical wireless receiver the input signal travels through a cascade of several different components such as filters, amplifiers, mixers, and transmission lines. Each of these stages will progressively degrade the signal-to-noise ratio, so it is important to quantify this effect to evaluate the overall performance of the receiver. If we know the noise figure (or noise temperature) of the individual stages, we can determine the noise figure of the cascade connection of stages. We will see that the most critical stage is usually the first, and that later stages generally have a progressively reduced effect on the overall noise figure. This is an important consideration for the design and layout of receiver circuitry.

Consider a cascade of two components having power gains G_1 and G_2, noise figures F_1 and F_2, and noise temperatures T_{e1} and T_{e2}, as shown in Figure 3.20a. We wish to find the overall noise figure, F, and noise temperature, T_e, of the cascade as if it were the single component of Figure 3.20b. Note that we set the input noise power to be $N_i = kT_0B$.

Using noise temperatures, the noise power at the output of the first stage is

$$N_1 = G_1kT_0B + G_1kT_{e1}B. \tag{3.71}$$

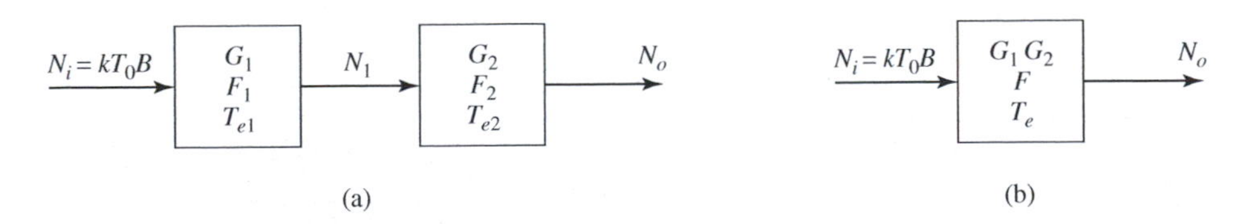

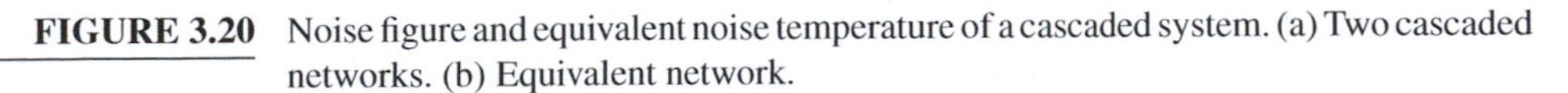

FIGURE 3.20 Noise figure and equivalent noise temperature of a cascaded system. (a) Two cascaded networks. (b) Equivalent network.

Then the noise power at the output of the second stage is

$$N_o = G_2 N_1 + G_2 k T_{e2} B$$
$$= G_1 G_2 k B \left(T_0 + T_{e1} + \frac{T_{e2}}{G_1} \right). \tag{3.72}$$

For the equivalent system of Figure 3.20b the output noise power can be written as

$$N_o = G_1 G_2 k B (T_e + T_0), \tag{3.73}$$

so comparison with (3.72) gives the noise temperature of the cascade system:

$$T_e = T_{e1} + \frac{T_{e2}}{G_1}. \tag{3.74}$$

Using (3.64) to convert noise temperature to noise figure gives the noise figure of the cascade system:

$$F = F_1 + \frac{F_2 - 1}{G_1}. \tag{3.75}$$

The above results are for two cascaded networks, but can be generalized to an arbitrary number of stages as follows:

$$T_e = T_{e1} + \frac{T_{e2}}{G_1} + \frac{T_{e3}}{G_1 G_2} + \cdots. \tag{3.76}$$

$$F = F_1 + \frac{F_2 - 1}{G_1} + \frac{F_3 - 1}{G_1 G_2} + \cdots. \tag{3.77}$$

These results show that the noise characteristics of a cascaded system are dominated by the first few stages, since the effect of later stages is reduced by the product of the gains of the preceding stages. Thus, for best overall system noise performance, the first stage of a receiver should have a low noise figure and at least moderate gain. Expense and effort are most rewarded when applied to improving the noise characteristics of the first or second stage, as opposed to later stages, since later stages have a diminished impact on overall noise performance.

EXAMPLE 3.5 ANALYSIS OF A WIRELESS RECEIVER

The block diagram of a wireless receiver front end is shown in Figure 3.21. Compute the overall noise figure of this subsystem. If the input noise power from a feeding antenna is $N_i = kT_a B$, where $T_a = 15$ K, find the output noise power in dBm. What is the two-sided power spectral density of the output noise? If we require a minimum SNR of 20 dB at the output of the receiver, what is the minimum signal voltage that can be applied at the receiver input? Assume the system is at temperature T_0, with a characteristic impedance of 50 Ω, and an IF bandwidth of 10 MHz.

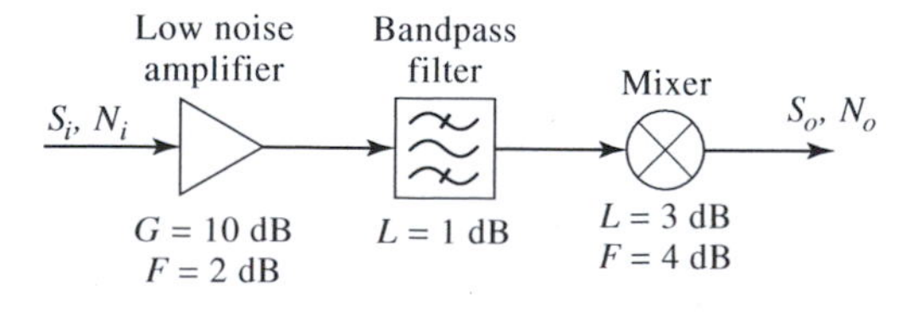

FIGURE 3.21 Block diagram of a wireless receiver front-end circuit.

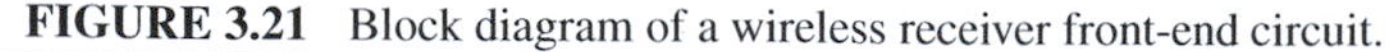

Solution

We first carry out the required conversions from dB to numerical values:

$$G = 10 \text{ dB} = 10 \qquad G = -1.0 \text{ dB} = 0.79 \qquad G = -3.0 \text{ dB} = 0.5$$
$$F = 2 \text{ dB} = 1.58 \qquad F = 1 \text{ dB} = 1.26 \qquad F = 4 \text{ dB} = 2.51$$

Then we can use (3.77) to find the overall noise figure of the system:

$$F = F_1 + \frac{F_2 - 1}{G_1} + \frac{F_3 - 1}{G_1 G_2} = 1.58 + \frac{(1.26 - 1)}{10} + \frac{(2.51 - 1)}{(10)(0.79)}$$
$$= 1.80 = 2.55 \text{ dB}$$

The best way to compute the output noise power is to use noise temperatures. From (3.65), the equivalent noise temperature of the overall system is

$$T_e = (F - 1)T_0 = (1.80 - 1)(290) = 232 \text{ K}.$$

The overall gain of the system is $G = (10)(0.79)(0.5) = 3.95$. Then we can find the output noise power as

$$N_o = k(T_a + T_e)BG = (1.38 \times 10^{-23})(15 + 232)(10 \times 10^6)(3.95)$$
$$= 1.35 \times 10^{-13} \text{ W} = -98.7 \text{ dBm}$$

From (3.26), the power spectral density of the output noise over the IF bandwidth is

$$S_n(\omega) = \frac{N_o}{2B} = \frac{1.35 \times 10^{-13} \text{ W}}{2(10 \times 10^6)} = 6.8 \times 10^{-21} \text{ W/Hz}.$$

Finally, for an output *SNR* of 20 dB = 100, the input signal power must be

$$S_i = \frac{S_o}{G} = \frac{S_o}{N_o}\frac{N_o}{G} = 100\,\frac{1.35 \times 10^{-13}}{3.95} = 3.42 \times 10^{-12} \text{ W} = -84.7 \text{ dBm}.$$

For a 50 Ω system impedance, this corresponds to an input signal voltage of

$$V_i = \sqrt{Z_0 S_i} = \sqrt{(50)(3.42 \times 10^{-12})} = 1.31 \times 10^{-5} \text{ v} = 13.1\ \mu\text{V (rms)}.$$

Note: It may be tempting to compute the output noise power from the definition of the noise figure as

$$N_o = N_i F\left(\frac{S_o}{S_i}\right) = N_i F G = kT_a BFG$$
$$= (1.38 \times 10^{-23})(15)(10 \times 10^6)(1.8)(3.95) = 1.47 \times 10^{-14} \text{ W}.$$

This is an *incorrect* result! The reason for the disparity with the earlier result is because the definition of noise figure assumes an input noise level of kT_0B, while this problem involves an input noise of kT_aB, with $T_a = 15$ K. This is a common error and suggests that when computing absolute noise powers it is often safer to use noise temperatures to avoid this confusion. ○

3.6 NOISE FIGURE OF PASSIVE NETWORKS

As we have seen in the previous section, the noise figure of an RF or microwave system can be evaluated if we know the noise figures of the individual components. In the previous section we derived the noise figure for a matched lossy line or attenuator by

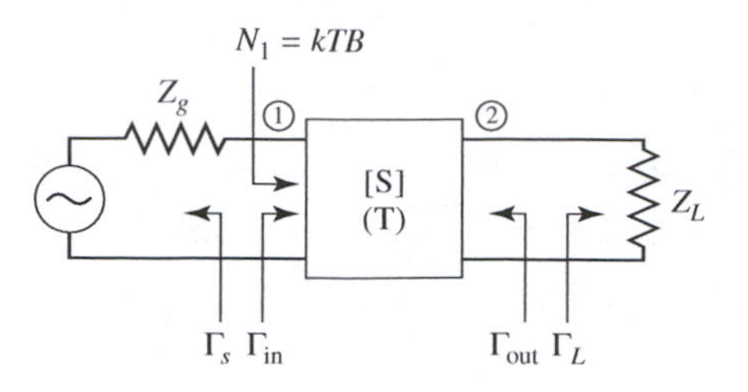

FIGURE 3.22 A passive two-port network with impedance mismatches. The network is at physical temperature T.

using a thermodynamic argument, but that method is not useful for very many circuits. Here we extend the thermodynamic method to evaluate the noise figure of general passive RF and microwave networks (networks that do not contain active devices such as diodes or transistors, which generate nonthermal noise). In addition, this method will account for the change in noise figure that occurs when a component is impedance mismatched at either its input or output port.

Generally it is easier and more accurate to find the noise characteristics of an active device, such as a diode or transistor, by direct measurement than by calculation from first principles. Once the noise parameters of a device are known, the overall noise figure of a circuit containing that device can be evaluated. This is demonstrated in Chapter 6 for the design of low-noise amplifiers.

This section requires knowledge of S-parameters and available gain, topics which are briefly treated in Chapter 2, and in more detail in reference [4]. If the reader does not have this background, he or she may skip this section without any loss of continuity for the rest of the book.

Noise Figure of a Passive Two-port Network

Figure 3.22 show an arbitrary passive two-port network, with a generator at port 1 and a load at port 2. The network is characterized by its S-parameter matrix, $[S]$. In the general case, impedance mismatches may exist at each port, and we define these mismatches in terms of the following reflection coefficients:

$$\begin{aligned}
\Gamma_s &= \text{reflection coefficient looking toward generator}\\
\Gamma_{\text{in}} &= \text{reflection coefficient looking toward port 1 of network}\\
\Gamma_{\text{out}} &= \text{reflection coefficient looking toward port 2 of network}\\
\Gamma_L &= \text{reflection coefficient looking toward load}
\end{aligned}$$

If we assume the network is at temperature T, and an input noise power of $N_1 = kTB$ is applied to the input of the network, the available output noise power at port 2 can be written as

$$N_2 = G_{21}kTB + G_{21}N_{\text{added}} \tag{3.78}$$

where N_{added} is the noise power generated internally by the network (referenced to port 1), and G_{21} is the *available gain* of the network from port 1 to port 2. As derived in reference [4], and later in Chapter 6, the available gain can be expressed in terms of the S-parameters of the network and the port mismatches as

$$G_{21} = \frac{\text{power available from network}}{\text{power available from source}} = \frac{|S_{21}|^2(1 - |\Gamma_S|^2)}{|1 - S_{11}\Gamma_S|^2(1 - |\Gamma_{\text{out}}|^2)}. \tag{3.79}$$

Also, the output mismatch can be expressed as

$$\Gamma_{\text{out}} = S_{22} + \frac{S_{12}S_{21}\Gamma_S}{1 - S_{11}\Gamma_S}. \tag{3.80}$$

Observe that when the network is matched to its external circuitry, so that $\Gamma_s = 0$ and $S_{22} = 0$, we have $\Gamma_{\text{out}} = 0$ and $G_{21} = |S_{21}|^2$, which is the gain of the network when it is matched. Also observe that the available gain of the network does not depend on the load mismatch, Γ_L. This is because available gain is defined in terms of the maximum power that is available from the network, which occurs when the load impedance is conjugately matched to the output impedance of the network.

Since the input noise is kTB, and the network is at temperature T, the network is in thermodynamic equilibrium, and so the available output noise power must be $N_2 = kTB$. Then we can solve for N_{added} from (3.78) to give

$$N_{\text{added}} = \frac{1 - G_{21}}{G_{21}} kTB. \tag{3.81}$$

Then the equivalent noise temperature of the network is

$$T_e = \frac{N_{\text{added}}}{kB} = \frac{1 - G_{21}}{G_{21}} T, \tag{3.82}$$

and the noise figure of the network is

$$F = 1 + \frac{T_e}{T_0} = 1 + \frac{1 - G_{21}}{G_{21}} \frac{T}{T_0}. \tag{3.83}$$

Note the similarity of (3.81)–(3.83) to the results in (3.68)–(3.70) for the lossy line—the essential difference is that here we are using the available gain of the network, which accounts for impedance mismatches between the network and the external circuit. We will now illustrate the use of this result with some applications to problems of practical interest.

Application to a Mismatched Lossy Line

In Section 3.5 we found the noise figure of a lossy transmission line under the assumption that it was matched to its input and output circuits. Now we consider the case where the line is mismatched to its input circuit.

Figure 3.23 shows a transmission line of length ℓ at temperature T, with a power loss factor $L = 1/G$, and an impedance mismatch between the line and the generator. Thus,

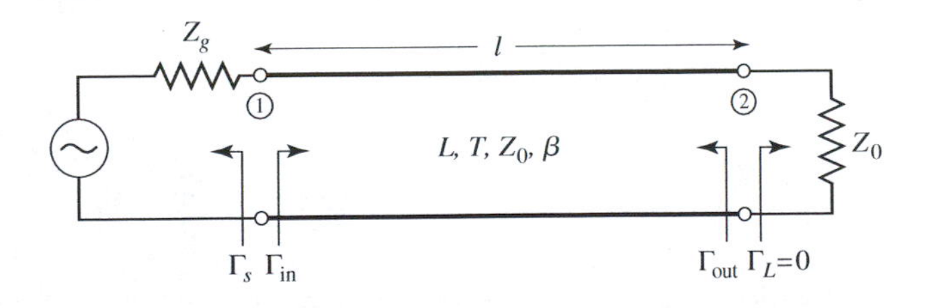

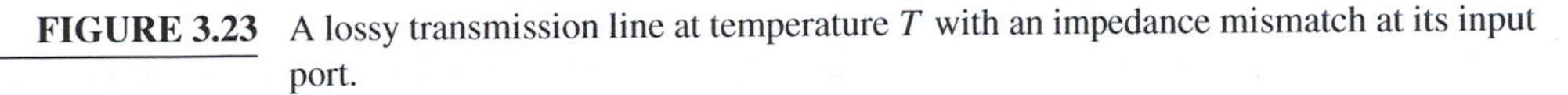

FIGURE 3.23 A lossy transmission line at temperature T with an impedance mismatch at its input port.

$Z_g \neq Z_0$, and the reflection coefficient looking toward the generator can be written as

$$\Gamma_s = \frac{Z_g - Z_0}{Z_g + Z_0} \neq 0.$$

The scattering matrix of the lossy line of characteristic impedance Z_0 can be written as

$$[S] = \begin{bmatrix} 0 & 1 \\ 1 & 0 \end{bmatrix} \frac{e^{-j\beta\ell}}{\sqrt{L}}, \tag{3.84}$$

where β is the propagation constant of the line. Using (3.80) gives the reflection coefficient looking into port 2 of the line as

$$\Gamma_{\text{out}} = S_{22} + \frac{S_{12}S_{21}\Gamma_s}{1 - S_{11}\Gamma_s} = \frac{\Gamma_s}{L} e^{-2j\beta\ell}. \tag{3.85}$$

Then the available gain, from (3.79), is

$$G_{21} = \frac{\frac{1}{L}(1 - |\Gamma_s|^2)}{1 - |\Gamma_{\text{out}}|^2} = \frac{L(1 - |\Gamma_s|^2)}{L^2 - |\Gamma_s|^2}. \tag{3.86}$$

We can verify two limiting cases of (3.86): when $L = 1$ we have $G_{21} = 1$, and when $\Gamma_s = 0$ we have $G_{21} = 1/L$.

Using (3.82) gives the equivalent noise temperature of the mismatched lossy line as

$$T_e = \frac{1 - G_{21}}{G_{21}} T = \frac{(L - 1)(L + |\Gamma_s|^2)}{L(1 - |\Gamma_s|^2)} T. \tag{3.87}$$

The corresponding noise figure can then be evaluated using (3.64). Observe that when the line is matched, $\Gamma_s = 0$ and (3.87) reduces to $T_e = (L - 1)T$, in agreement with the result for the matched lossy line given by (3.69). If the line is lossless, then $L = 1$ and (3.87) reduces to $T_e = 0$ regardless of mismatch, as expected. But when the line is lossy and mismatched, so that $L > 1$ and $|\Gamma_s| > 0$, then the noise temperature given by (3.87) is greater than $T_e = (L - 1)T$, the noise temperature of the matched lossy line. The reason for this increase is that the lossy line actually delivers noise power out of both its ports, but when the input port is mismatched some of the available noise power at port 1 is reflected from the source back into port 1, and appears at port 2. When the generator is matched to port 1, none of the available power from port 1 is reflected back into the line, so the noise power available at port 2 is a minimum.

Application to a Wilkinson Power Divider

Here we evaluate the noise figure of a power divider, which is another common component found in wireless systems. Figure 3.24 shows a Wilkinson power divider. A detailed analysis and description of this circuit are given in reference [4], but for our purposes knowledge of the scattering matrix is sufficient:

$$[S] = \frac{-j}{\sqrt{2L}} \begin{bmatrix} 0 & 1 & 1 \\ 1 & 0 & 0 \\ 1 & 0 & 0 \end{bmatrix}, \tag{3.88}$$

where L is the dissipative insertion power loss from port 1 to port 2 or 3, due to the loss of the quarter-wave transmission lines connecting those ports. The scattering matrix shows that the divider is matched at all ports, and divides input power at port 1 evenly to ports 2 and 3, when those ports are matched. The shunt resistor across ports 2 and 3 provides

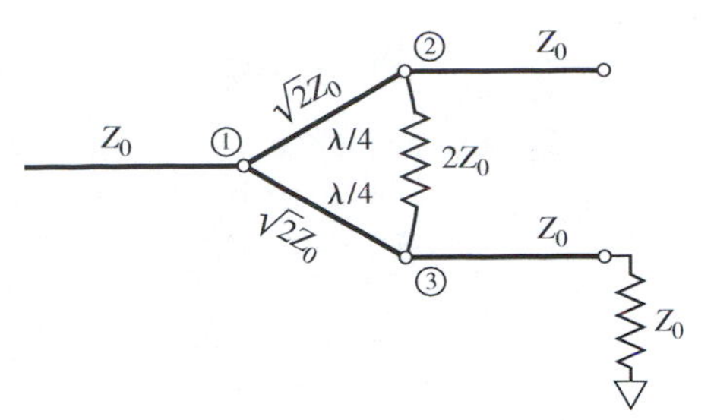

FIGURE 3.24 A Wilkinson power divider with port 3 terminated in a matched load.

isolation between those ports ($S_{23} = S_{32} = 0$), but does not dissipate input power when ports 2 and 3 are terminated with matched loads.

To evaluate the noise figure of the Wilkinson divider, we first terminate port 3 with a matched load; this converts the 3-port device to a 2-port device. If we assume a matched source at port 1, we have $\Gamma_s = 0$. Equation (3.80) then gives $\Gamma_{out} = S_{22} = 0$, and so the available gain can be calculated from (3.79) as

$$G_{21} = |S_{21}|^2 = \frac{1}{2L}. \tag{3.89}$$

Then the equivalent noise temperature of the Wilkinson divider is, from (3.82),

$$T_e = \frac{1 - G_{21}}{G_{21}} T = (2L - 1)T, \tag{3.90}$$

where T is the physical temperature of the divider. Using (3.64) gives the noise figure as

$$F = 1 + \frac{T_e}{T_0} = 1 + (2L - 1)\frac{T}{T_0}. \tag{3.91}$$

Observe that if the divider is at room temperature, then $T = T_0$ and (3.91) reduces to $F = 2L$. If the divider is at room temperature and lossless, (3.91) reduces to $F = 2 = 3$ dB. In this case the source of the noise power is the isolation resistor.

Because the divider circuit of Figure 3.24 is matched at its input and output, it is easy to obtain these same results using thermodynamic arguments. Thus, if we apply an input noise power of kTB to port 1 of the matched Wilkinson divider at temperature T, the system will be in thermal equilibrium and the output noise power must therefore be kTB. We can also express the output noise power as the sum of the input power times the gain of the divider, and N_{added}, the noise power added by the divider itself (referenced at the input to the divider):

$$kTB = \frac{kTB}{2L} + \frac{N_{added}}{2L}. \tag{3.92}$$

Solving for N_{added} gives

$$N_{added} = kTB(2L - 1), \tag{3.93}$$

so the equivalent noise temperature is

$$T_e = \frac{N_{added}}{kB} = (2L - 1)T,$$

in agreement with (3.90).

3.7 DYNAMIC RANGE AND INTERMODULATION DISTORTION

Since thermal noise is generated by any lossy component, and all realistic components have at least a small loss, the ideal linear component or network does not exist in the sense that its output response is always exactly proportional to its input excitation. Thus, all realistic devices are nonlinear at very low power levels due to noise effects. In addition, all practical components also become nonlinear at high power levels. This may ultimately be the result of catastrophic destruction of the device at very high powers or, in the case of active devices such as diodes and transistors, due to effects such as gain compression or the generation of spurious frequency components due to device nonlinearities. In either case these effects set a minimum and maximum realistic power range, or *dynamic range*, over which a given component or network will operate as desired. In this section we will study dynamic range, and the response of nonlinear devices in general. These results will be useful for our later discussions of amplifiers (Chapter 6), mixers (Chapter 7), and wireless receiver design (Chapter 10).

Devices such as diodes and transistors are nonlinear components, and it is this nonlinearity that is of great utility for functions such as amplification, detection, and frequency conversion [5]. Nonlinear device characteristics, however, can also lead to undesired responses such as gain compression and the generation of spurious frequency components. These effects may produce increased losses, signal distortion, and possible interference with other radio channels or services.

Figure 3.25 shows a general nonlinear network, having an input voltage v_i and an output voltage v_o. In the most general sense, the output response of a nonlinear circuit can be modeled as a Taylor series in terms of the input signal voltage:

$$v_o = a_0 + a_1 v_i + a_2 v_i^2 + a_3 v_i^3 + \cdots. \tag{3.94}$$

where the Taylor coefficients are defined as

$$a_0 = v_o(0) \qquad \text{(DC output)} \tag{3.95a}$$

$$a_1 = \left.\frac{dv_o}{dv_i}\right|_{v_i=0} \qquad \text{(linear output)} \tag{3.95b}$$

$$a_2 = \left.\frac{d^2v_o}{dv_i^2}\right|_{v_i=0} \qquad \text{(squared output)} \tag{3.95c}$$

and higher order terms. Thus, different functions can be obtained from the nonlinear network depending on the dominance of particular terms in the expansion. If a_0 is the only nonzero coefficient in (3.94), the network functions as a rectifier, converting an AC signal to DC. If a_1 is the only nonzero coefficient, we have a linear attenuator ($a_1 < 1$) or amplifier ($a_1 > 1$). If a_2 is the only nonzero coefficient, we can achieve mixing and other frequency conversion functions. Usually, however, practical devices have a series expansion containing many nonzero terms, and a combination of several of these effects will occur. We consider some important special cases below.

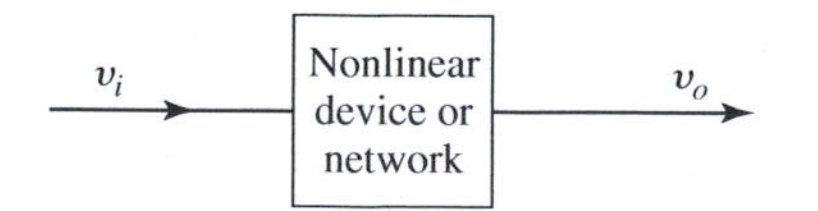

FIGURE 3.25 A general nonlinear device or network.

Gain Compression

First consider the case where a single frequency sinusoid is applied to the input of a general nonlinear network, such as an amplifier:

$$v_i = V_0 \cos \omega_0 t. \tag{3.96}$$

Then (3.94) gives the output voltage as

$$\begin{aligned} v_o &= a_0 + a_1 V_0 \cos \omega_0 t + a_2 V_0^2 \cos^2 \omega_0 t + a_3 V_0^3 \cos^3 \omega_0 t + \cdots \\ &= \left(a_0 + \tfrac{1}{2} a_2 V_0^2\right) + \left(a_1 V_0 + \tfrac{3}{4} a_3 V_0^3\right) \cos \omega_0 t \\ &\quad + \tfrac{1}{2} a_2 V_0^2 \cos 2\omega_0 t + \tfrac{1}{4} a_3 V_0^3 \cos 3\omega_0 t + \cdots. \end{aligned} \tag{3.97}$$

This result leads to the voltage gain of the signal component at frequency ω_0:

$$G_v = \frac{v_0^{(\omega_0)}}{v_i^{(\omega_0)}} = \frac{a_1 V_0 + \frac{3}{4} a_3 V_0^3}{V_0} = a_1 + \tfrac{3}{4} a_3 V_0^2, \tag{3.98}$$

where we have retained only terms through the third order.

The result of (3.98) shows that the voltage gain is equal to the a_1 coefficient, as expected, but with an additional term proportional to the square of the input voltage amplitude. In most practical amplifiers a_3 is typically negative, so that the gain of the amplifier tends to decrease for large values of V_0. This effect is called *gain compression*, or *saturation*. Physically, this is usually due to the fact that the instantaneous output voltage of an amplifier is limited by the power supply voltage used to bias the active device. Smaller values of a_3 will lead to higher output voltages.

A typical amplifier response is shown in Figure 3.26. For an ideal linear amplifier a plot of the output power versus input power is a straight line with a slope of unity, and the gain of the amplifier is given by the ratio of the output power to the input power. The amplifier response of Figure 3.26 tracks the ideal response over a limited range, then begins to saturate, resulting in reduced gain. To quantify the linear operating range of the amplifier, we define the *1 dB compression point* as the power level for which the output power has decreased by 1 dB from the ideal characteristic. This power level is usually denoted by P_1, and can be stated in terms of either input power or output power. For amplifiers P_1 is

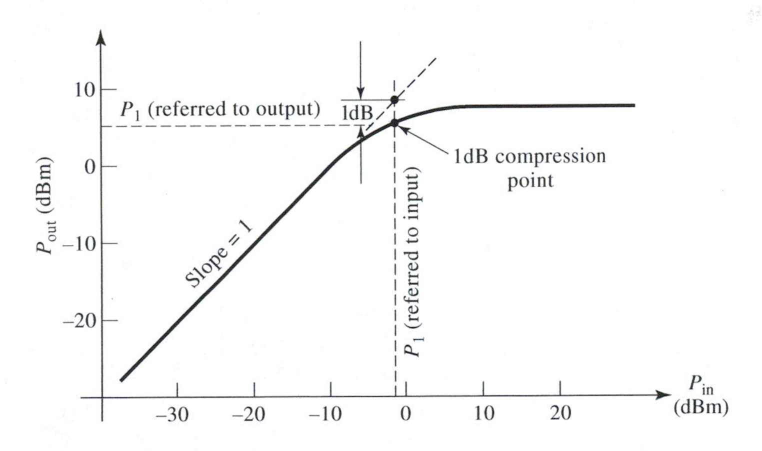

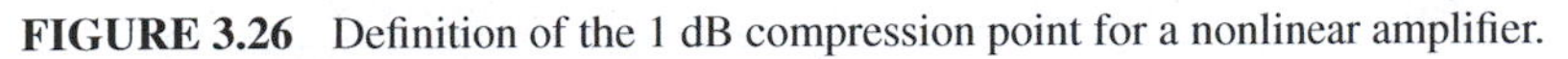
FIGURE 3.26 Definition of the 1 dB compression point for a nonlinear amplifier.

usually specified as an output power, while for mixers P_1 is usually specified in terms of input power.

Intermodulation Distortion

Observe from the expansion of (3.97) that a portion of the input signal at frequency ω_0 is converted to other frequency components. For example, the first term of (3.97) represents a DC voltage, which would be a useful response in a rectifier application. The voltage components at frequencies $2\omega_0$ or $3\omega_0$ might be useful for frequency multiplier circuits. In amplifiers, however, the presence of other frequency components will lead to signal distortion if those components are in the passband of the amplifier.

For a single input frequency, or *tone*, ω_0, the output will in general consist of harmonics of the input frequency of the form $n\omega_0$, for $n=0, 1, 2, \ldots$. Usually these harmonics lie outside the passband of the amplifier, and so do not interfere with the desired signal at frequency ω_0. The situation is different, however, when the input signal consists of two closely spaced frequencies.

Consider a *two-tone* input voltage, consisting of two closely spaced frequencies, ω_1 and ω_2:

$$v_i = V_0(\cos\omega_1 t + \cos\omega_2 t) \tag{3.99}$$

From (3.94) the output is

$$\begin{aligned} v_o &= a_0 + a_1 V_0(\cos\omega_1 t + \cos\omega_2 t) + a_2 V_0^2(\cos\omega_1 t + \cos\omega_2 t)^2 \\ &\quad + a_3 V_0^3(\cos\omega_1 t + \cos\omega_2 t)^3 + \cdots \\ &= a_0 + a_1 V_0 \cos\omega_1 t + a_1 V_0 \cos\omega_2 t + \tfrac{1}{2} a_2 V_0^2(1 + \cos 2\omega_1 t) \\ &\quad + \tfrac{1}{2} a_2 V_0^2(1 + \cos 2\omega_2 t) + a_2 V_0^2 \cos(\omega_1 - \omega_2)t + a_2 V_0^2 \cos(\omega_1 + \omega_2)t \\ &\quad + a_3 V_0^3\left(\tfrac{3}{4}\cos\omega_1 t + \tfrac{1}{4}\cos 3\omega_1 t\right) + a_3 V_0^3\left(\tfrac{3}{4}\cos\omega_2 t + \tfrac{1}{4}\cos 3\omega_2 t\right) \\ &\quad + a_3 V_0^3\left[\tfrac{3}{2}\cos\omega_2 t + \tfrac{3}{4}\cos(2\omega_1 - \omega_2)t + \tfrac{3}{4}\cos(2\omega_1 + \omega_2)t\right] \\ &\quad + a_3 V_0^3\left[\tfrac{3}{2}\cos\omega_1 t + \tfrac{3}{4}\cos(2\omega_2 - \omega_1)t + \tfrac{3}{4}\cos(2\omega_2 + \omega_1)t\right] + \cdots, \end{aligned} \tag{3.100}$$

where standard trigonometric identities have been used to expand the initial expression. We see that the output spectrum consists of harmonics of the form

$$m\omega_1 + n\omega_2, \tag{3.101}$$

with $m, n = 0, \pm 1, \pm 2, \pm 3, \ldots$. These combinations of the two input frequencies are called *intermodulation products*, and the *order* of a given product is defined as $|m| + |n|$. For example, the squared term of (3.100) gives rise to the following four intermodulation products of second order:

$2\omega_1$	(second harmonic of ω_1)	$m=2$ $n=0$	order $=2$
$2\omega_2$	(second harmonic of ω_2)	$m=0$ $n=2$	order $=2$
$\omega_1 - \omega_2$	(difference frequency)	$m=1$ $n=-1$	order $=2$
$\omega_1 + \omega_2$	(sum frequency)	$m=1$ $n=1$	order $=2$

All of these second-order products are undesired in an amplifier, but in a mixer the sum or

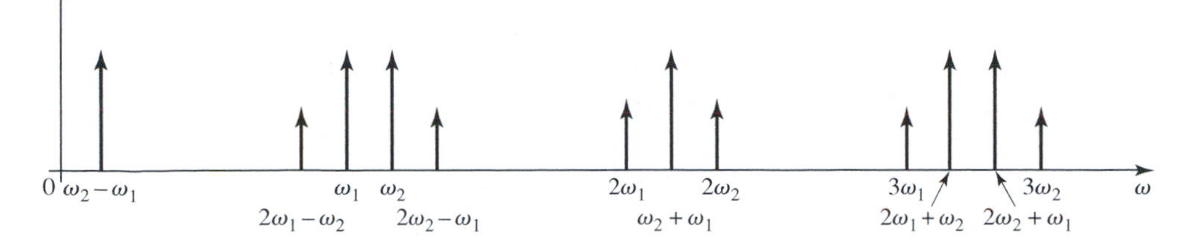

FIGURE 3.27 Output spectrum of second and third-order two-tone intermodulation products, assuming $\omega_1 < \omega_2$.

difference frequencies form the desired outputs. In either case, if ω_1 and ω_2 are close, all the second-order products will be far from ω_1 or ω_2, and can easily be filtered (either passed or rejected) from the output of the component.

The cubed term of (3.100) leads to six third-order intermodulation products: $3\omega_1, 3\omega_2, 2\omega_1 + \omega_2, 2\omega_2 + \omega_1, 2\omega_1 - \omega_2$, and $2\omega_2 - \omega_1$. The first four of these will again be located far from ω_1 and ω_2, and will typically be outside the passband of the component. But the two difference terms produce products located near the original input signals at ω_1 and ω_2, and so cannot be easily filtered from the passband of an amplifier. Figure 3.27 shows a typical spectrum of the second- and third-order two-tone intermodulation products. For an arbitrary input signal consisting of many frequencies of varying amplitude and phase, the resulting in-band intermodulation products will cause distortion of the output signal. This effect is called *third-order intermodulation distortion*.

Third-Order Intercept Point

Equation (3.100) shows that as the input voltage V_0 increases, the voltage associated with the third-order products increases as V_0^3. Since power is proportional to the square of voltage, we can also say that the output power of third-order products must increase as the cube of the input power. So for small input powers the third-order intermodulation products must be very small, but will increase quickly as input power increases. We can view this effect graphically by plotting the output power for the first- and third-order products versus input power on log-log scales (or in dB), as shown in Figure 3.28.

The output power of the first order, or linear, product is proportional to the input power, and so the line describing this response has a slope of unity (before the onset of compression). The line describing the response of the third-order products has a slope of 3. (The second-order products would have a slope of 2, but since these products are generally not in the passband of the component, we have not plotted their response in Figure 3.28.) Both the linear- and third-order responses will exhibit compression at high input powers, so we show the extension of their idealized responses with dotted lines. Since these two lines have different slopes, they will intersect, typically at a point above the onset of compression, as shown in the figure. This hypothetical intersection point, where the first-order and third-order powers are equal, is called the *third-order intercept point*, denoted P_3, and specified as either an input or an output power. Usually P_3 is referenced at the output for amplifiers, and at the input for mixers.

As depicted in Figure 3.28, P_3 generally occurs at a higher power level than P_1, the 1 dB compression point. Many practical components follow the approximate rule that P_3 is 12 to 15 dB greater than P_1, assuming these powers are referenced at the same point.

We can express P_3 in terms of the Taylor coefficients of the expansion of (3.100) as follows. Define P_{ω_1} as the output power of the desired signal at frequency ω_1. Then from

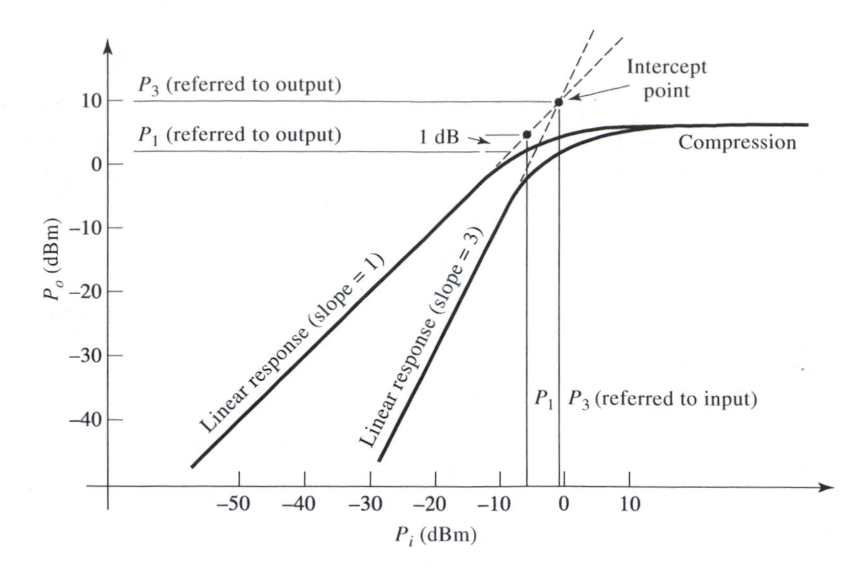

FIGURE 3.28 Third-order intercept diagram for a nonlinear component.

(3.100) we have

$$P_{\omega_1} = \tfrac{1}{2} a_1^2 V_0^2. \tag{3.102}$$

Similarly, define $P_{2\omega_1-\omega_2}$ as the output power of the intermodulation product of frequency $2\omega_1 - \omega_2$. Then from (3.100) we have

$$P_{2\omega_1-\omega_2} = \tfrac{1}{2}\left(\tfrac{3}{4} a_3 V_0^3\right)^2 = \tfrac{9}{32} a_3^2 V_0^6. \tag{3.103}$$

By definition, these two powers are equal at the third-order intercept point. If we define the input signal voltage at the intercept point as V_{IP}, then equating (3.102) and (3.103) gives

$$\tfrac{1}{2} a_1^2 V_{IP}^2 = \tfrac{9}{32} a_3^2 V_{IP}^6.$$

Solving for V_{IP} yields

$$V_{IP} = \sqrt{\frac{4a_1}{3a_3}}. \tag{3.104}$$

Since P_3 is equal to the linear response of P_{ω_1} at the intercept point, we have from (3.102) and (3.104) that

$$P_3 = P_{\omega_1}\Big|_{V_0=V_{IP}} = \frac{1}{2} a_1^2 V_{IP}^2 = \frac{2a_1^3}{3a_3}, \tag{3.105}$$

where P_3 in this case is referred to the output port. This expression will be useful in the following section.

Dynamic Range

We can define *dynamic range* in a general sense as the operating range for which a component or system has desirable characteristics. For a power amplifier this may be the

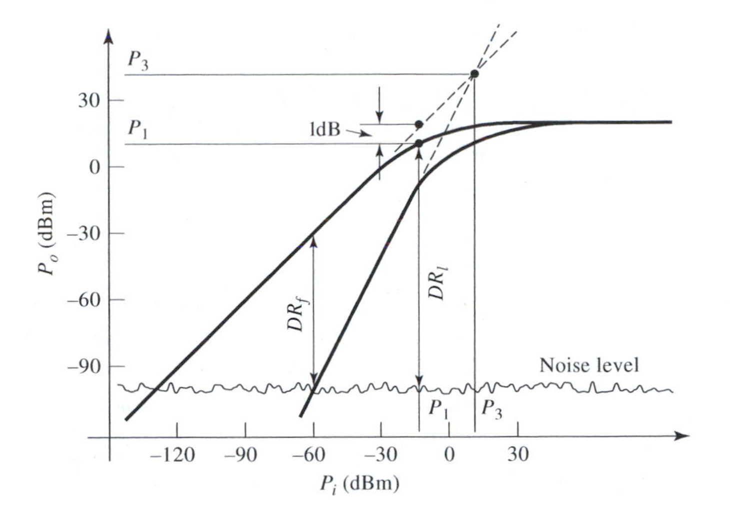

FIGURE 3.29 Illustrating linear dynamic range and spurious free dynamic range.

power range that is limited at the low end by noise and at the high end by the compression point. This is essentially the linear operating range for the amplifier, and is called the *linear dynamic range* (DR_ℓ). For low-noise amplifiers or mixers, operation may be limited by noise at the low end and the maximum power level for which intermodulation distortion becomes unacceptable. This is effectively the operating range for which spurious responses are minimal, and is called the *spurious-free dynamic range* (DR_f).

We thus compute the linear dynamic range DR_ℓ as the ratio of P_1, the 1 dB compression point, to the noise level of the component, as shown in Figure 3.29. These powers can be referenced at either the input or the output of the device. Note that some authors [6] prefer to define the linear dynamic range in terms of a minimum detectable power level. This definition is more appropriate for a receiver system, rather than an individual component, as it depends on factors external to the component itself, such as the type of modulation used, the recommended system SNR, effects of coding, and related factors.

The spurious free dynamic range is defined as the maximum output signal power for which the power of the third-order intermodulation product is equal to the noise level of the component. This situation is shown in Figure 3.29. If P_{ω_1} is the output power of the desired signal at frequency ω_1, and $P_{2\omega_1-\omega_2}$ is the output power of the third-order intermodulation product, then the spurious free dynamic range can be expressed as

$$DR_f = \frac{P_{\omega_1}}{P_{2\omega_1-\omega_2}}, \tag{3.106}$$

with $P_{2\omega_1-\omega_2}$ taken equal to the noise level of the component. $P_{2\omega_1-\omega_2}$ can be written in terms of P_3 and P_{ω_1} as follows:

$$P_{2\omega_1-\omega_2} = \frac{9a_3^2V_0^6}{32} = \frac{\frac{1}{8}a_1^6V_0^6}{\dfrac{4a_1^6}{9a_3^2}} = \frac{(P_{\omega_1})^3}{(P_3)^2}, \tag{3.107}$$

where (3.102) and (3.105) have been used. Observe that this result clearly shows that the third-order intermodulation power increases as the cube of the input signal power. Solving (3.107) for P_{ω_1}, and applying the result to (3.106) gives the spurious free dynamic range in terms of P_3 and N_o, the output noise power of the component:

$$DR_f = \left.\frac{P_{\omega_1}}{P_{2\omega_1-\omega_2}}\right|_{P_{2\omega_1-\omega_2}=N_o} = \left(\frac{P_3}{N_o}\right)^{2/3}. \tag{3.108}$$

This result can be written in terms of dB as

$$DR_f(dB) = \tfrac{2}{3}(P_3 - N_o), \tag{3.109}$$

for P_3 and N_o expressed in dB or dBm. If the output SNR is specified, this can be added to N_o to give the spurious free dynamic range in terms of the minimum detectable signal level. Finally, although we derived this result for the $2\omega_1 - \omega_2$ product, the same result applies for the $2\omega_2 - \omega_1$ product.

EXAMPLE 3.6 DYNAMIC RANGES

A receiver has a noise figure of 7 dB, a 1 dB compression point of 25 dBm (referenced to output), a gain of 40 dB, and a third-order intercept point of 35 dBm (referenced to output). If the receiver is fed with an antenna having a noise temperature of $T_A = 150$ K, and the desired output SNR is 10 dB, find the linear and spurious free dynamic ranges. Assume a receiver bandwidth of 100 MHz.

Solution

The noise power at the receiver output can be calculated as

$$\begin{aligned} N_o &= GkB[T_A + (F-1)T_o] = 10^4(1.38\times10^{-23})(10^8)[150 + (4.01)(290)] \\ &= 1.8\times10^{-8}\ \text{W} = -47.4\ \text{dBm}. \end{aligned}$$

Then the linear dynamic range is, in dB

$$DR_\ell = P_1 - N_o = 25\ \text{dBm} + 47.4\ \text{dBm} = 72\ \text{dB}.$$

Equation (3.109) gives the spurious free dynamic range as

$$DR_f = \tfrac{2}{3}(P_3 - N_o - SNR) = \tfrac{2}{3}(35 + 47.4 - 10) = 48.3\ \text{dB}.$$

Observe that $DR_f \ll DR_\ell$. ○

Intercept Point of Cascaded Components

As in the case of noise figure, the cascade connection of components has the effect of degrading (lowering) the third-order intercept point. Unlike the case of a cascade of noisy components, however, the intermodulation products in a cascaded system are deterministic (coherent), so we cannot simply add powers, but must deal with voltages.

With reference to Figure 3.30, let G_1 and P_3' be the power gain and third-order intercept point for the first stage, and G_2 and P_3'' be the corresponding values for the second stage. From (3.107) the third-order distortion power at the output of the first stage is

$$P'_{2\omega_1-\omega_2} = \frac{(P'_{\omega_1})^3}{(P_3')^2}. \tag{3.110}$$

where P'_{ω_1} is the desired signal power at frequency ω_1 at the output of the first stage. The

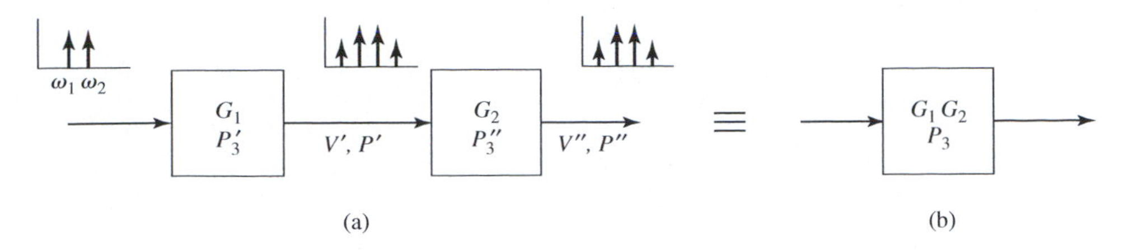

FIGURE 3.30 Third-order intercept point for a cascaded system. (a) Two cascaded networks. (b) Equivalent network.

voltage associated with this power is

$$V'_{2\omega_1-\omega_2} = \sqrt{P'_{2\omega_1-\omega_2} Z_0} = \frac{\sqrt{\left(P'_{\omega_1}\right)^3 Z_0}}{P'_3}, \tag{3.111}$$

where Z_0 is the system impedance.

The total third-order distortion voltage at the output of the second stage is the sum of this voltage times the voltage gain of the second stage, and the distortion voltage generated by the second stage. This is because these voltages are deterministic and phase-related, unlike the uncorrelated noise powers that occur in cascaded components. Adding these voltages gives the worst-case result for the distortion level, because there may be phase delays within the stages that could cause partial cancellation. Thus we can write the worst-case total distortion voltage at the output of the second stage as

$$V''_{2\omega_1-\omega_2} = \frac{\sqrt{G_2\left(P'_{\omega_1}\right)^3 Z_0}}{P'_3} + \frac{\sqrt{\left(P''_{\omega_1}\right)^3 Z_0}}{P''_3}.$$

Since $P''_{\omega_1} = G_2 P'_{\omega_1}$, we have

$$V''_{2\omega_1-\omega_2} = \left(\frac{1}{G_2 P'_3} + \frac{1}{P''_3}\right)\sqrt{\left(P''_{\omega_1}\right)^3 Z_0}. \tag{3.112}$$

Then the output distortion power is

$$P''_{2\omega_1-\omega_2} = \frac{\left(V''_{2\omega_1-\omega_2}\right)^2}{Z_0} = \left(\frac{1}{G_2 P'_3} + \frac{1}{P''_3}\right)^2 \left(P''_{\omega_1}\right)^3 = \frac{\left(P''_{\omega_1}\right)^3}{P_3^2}. \tag{3.113}$$

Thus the third-order intercept point of the cascaded system is

$$P_3 = \left(\frac{1}{G_2 P'_3} + \frac{1}{P''_3}\right)^{-1}. \tag{3.114}$$

Note that $P_3 = G_2 P'_3$ for $P''_3 \to \infty$, which is the limiting case when the second stage has no third-order distortion. This result is also useful for transferring P_3 between input and output reference points.

EXAMPLE 3.7 CALCULATION OF CASCADE INTERCEPT POINT

A low-noise amplifier and mixer are shown in Figure 3.31. The amplifier has a gain of 20 dB and a third-order intercept point of 22 dBm (referenced at output), and the mixer has a conversion loss of 6 dB and a third-order intercept point of 13 dBm (referenced at input). Find the intercept point of the cascade network.

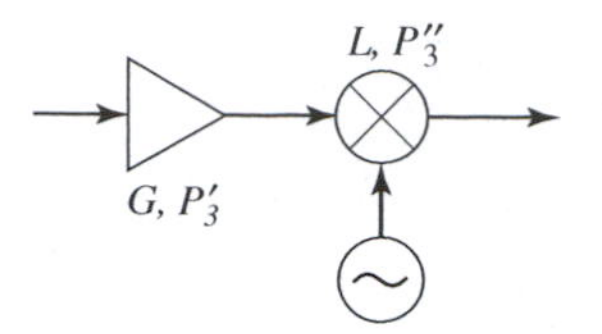

FIGURE 3.31 System for Example 3.7.

Solution

First we transfer the reference of P_3 for the mixer from its input to its output:

$$P_3'' = 13 \text{ dBm} - 6 \text{ dB} = 7 \text{ dBm (referenced at output)}$$

Converting the necessary dB values to numerical values:

$$\begin{aligned} P_3' &= 22 \text{ dBm} = 158 \text{ mW} && \text{(for amplifier)} \\ P_3'' &= 7 \text{ dBm} = 5 \text{ mW} && \text{(for mixer)} \\ G_2 &= -6 \text{ dB} = 0.25 && \text{(for mixer)} \end{aligned}$$

Then using (3.114) gives the intercept point of the cascade as

$$P_3 = \left(\frac{1}{G_2 P_3'} + \frac{1}{P_3''}\right)^{-1} = \left(\frac{1}{(0.25)(158)} + \frac{1}{5}\right)^{-1} = 4.4 \text{ mW} = 6.4 \text{ dBm},$$

which is seen to be much lower than the P_3 of the individual components. ○

Passive Intermodulation

The above discussion of intermodulation distortion was in the context of active circuits involving diodes and transistors, but it is also possible for intermodulation products to be generated by passive nonlinear effects in connectors, cables, antennas, or almost any component where there is a metal-to-metal contact. This effect is called *passive intermodulation* (PIM) and, as in the case of intermodulation in amplifiers and mixers, occurs when signals at two or more closely spaced frequencies mix to produce spurious products.

Passive intermodulation can be caused by a number of factors, such as poor mechanical contact, oxidation of junctions between ferrous-based metals, contamination of conducting surfaces at RF junctions, or the use of nonlinear materials such as carbon fiber composites or ferromagnetic materials. In addition, when high powers are involved, thermal effects may contribute to the overall nonlinearity of a junction. It is very difficult to predict PIM levels from first principles, so measurement techniques must usually be used.

Because of the third-power dependence of the third-order intermodulation products with input power, passive intermodulation is usually only significant when input signal powers are relatively large. This is frequently the case in cellular telephone base station transmitters, which may operate with powers of 30–40 dBm, with many closely spaced RF channels. It is often desired to maintain the PIM level below −125 dBm, with two 40 dBm transmit signals. This is a very wide dynamic range, and requires careful selection of components used in the high-power portions of the transmitter, including cables, connectors, and antenna components. Because these components are often exposed to the weather, deterioration due to oxidation, vibration, and sunlight must be offset by a careful maintenance program. Passive intermodulation is generally not a problem in receiver systems, due to the much lower power levels.

REFERENCES

[1] A. Leon-Garcia, **Probability and Random Processes for Electrical Engineering**, 2nd edition. Addison-Wesley, Reading, MA. 1994.

[2] B. P. Lathi, **Modern Digital and Analog Communications Systems**, 3rd edition. Oxford University Press, New York, 1998.

[3] A. Papoulis, **Probability, Random Variables, and Stochastic Processes**, 2nd edition. McGraw-Hill, New York, 1984.

[4] D. M. Pozar, **Microwave Engineering**. 2nd edition, Wiley, New York, 1998.

[5] M. E. Hines, "The Virtues of Nonlinearity—Detection, Frequency Conversion, Parametric Amplification, and Harmonic Generation," *IEEE Trans. Microwave Theory and Techniques*, vol. MTT-32, pp. 1097–1104, September 1984.

[6] G. Gonzalez, **Microwave Transistor Amplifiers: Analysis and Design**, 2nd edition, Prentice Hall, New Jersey, 1997.

PROBLEMS

3.1 Prove the four properties of cumulative distribution functions as given in (3.2a)–(3.2d).

3.2 Use the definition of expected value to prove the linearity properties of (3.14).

3.3 Evaluate the mean and variance for the uniform, gaussian, and Rayleigh probability density functions given in (3.12).

3.4 Evaluate the nth moment of a random process having a gaussian PDF with zero mean ($m = 0$). Show that $E\{x^n\} = 0$ for odd n.

3.5 Consider the random variable $z = x + y$, where x, y, and z are zero-mean gaussian random variables having variances σ_x^2, σ_y^2, and σ_z^2. If x and y are independent, show that $\sigma_z^2 = \sigma_x^2 + \sigma_y^2$.

3.6 A gaussian white noise process, X, has zero mean and variance σ^2. Evaluate the probability $P\{-\sigma < x < \sigma\}$, that is, the probability that a particular sample, x, of the process satisfies the inequality $-\sigma < x < \sigma$.

3.7 The autocorrelation function for a white noise source is $R(\tau) = \frac{1}{2}n_0\delta(\tau)$. If this source is applied to an ideal bandpass filter with the frequency response shown below, find the total noise power output.

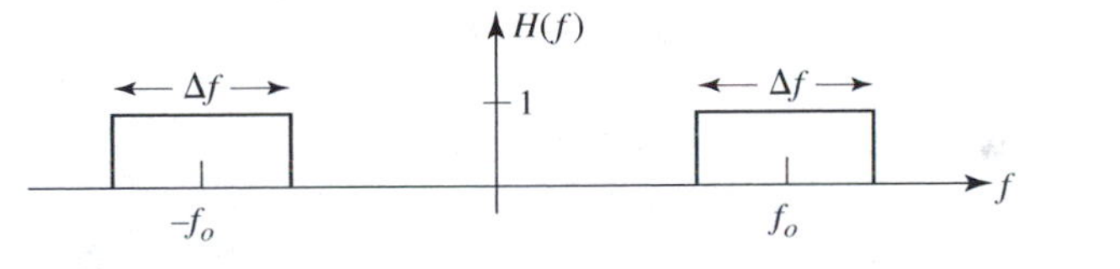

3.8 A noisy resistor of value R, at temperature T, is connected to a load resistor of value R_L. Calculate and plot the average power dissipated in the load, normalized to kTB, for $0 \leq R_L < 10R$. Prove that maximum power transfer occurs for $R_L = R$.

3.9 A gaussian white noise source with a two-sided power spectral density $S_i(f) = n_0/2$ is applied to the RC low-pass filter circuit shown below. Find the output noise power, N_o, in terms of n_0 and $f_c = 1/2\pi RC$ (the cutoff frequency of the filter).

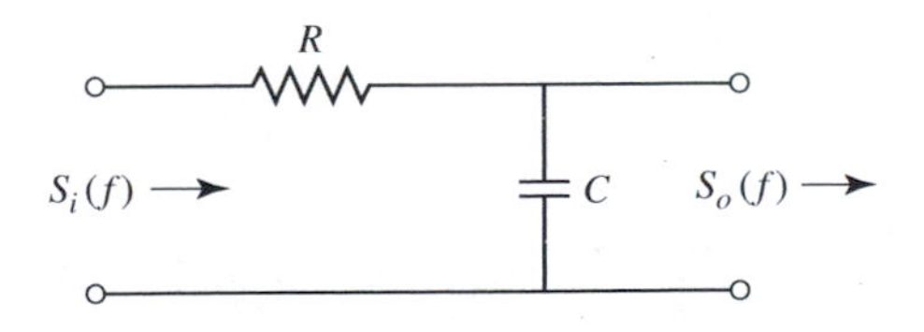

3.10 Derive the result of (3.38) for white noise passing through an ideal integrator by using the autocorrelation function, instead of the power spectral density. Find the output noise power from $N_o = E\{n_o^2(t)\}$, and evaluate $n_o(t)$ by integrating $n_i(t)$ according to the integrator operation.

3.11 Consider the random process $y(t) = x(t)\cos(\omega_0 t + \theta)$, where $x(t)$ is stationary with autocorrelation $R_x(\tau)$ and power spectral density $S_x(\omega)$. If θ is a uniformly distributed random variable over the range $0 \leq \theta < 2\pi$, find the autocorrelation and power spectral density of $y(t)$. Assume $x(t)$ and θ are independent.

3.12 Evaluate $P_e^{(0)}$, the probability of error when a binary "0" is transmitted, for the threshold detection system of Section 3.4.

3.13 For the threshold detection system of Section 3.4, determine the SNR (in dB) required for error probabilities of 10^{-2}, 10^{-5}, and 10^{-8}.

3.14 Consider a threshold detection receiver where a binary "1" is transmitted as v_0, and a binary "0" is transmitted as $-v_0$. If additive gaussian white noise is present with the received signal, choose a reasonable threshold value, and evaluate the probability of error for the two cases. Compare the SNR required for a given P_e of this system with the corresponding result derived in Section 3.4.

3.15 The Y-factor method is used to measure the equivalent noise temperature of a component, with a hot load of $T_1 = 320$ K and a cold load of $T_2 = 77$ K. If the Y-factor ratio is measured to be 0.608 dB, what is the noise figure of the component under test?

3.16 A certain transmission line has a noise figure $F = 2$ dB at a temperature of $T_0 = 290$ K. Calculate and plot the noise figure of the line (in dB) as its physical temperature ranges from 0 K to 1000 K.

3.17 An amplifier with a gain of 15 dB, a bandwidth of 200 MHz, and a noise figure of 3 dB feeds a detector/demodulator with a noise temperature of 800 K. Find the noise figure and equivalent noise temperature of the overall system.

3.18 Consider the wireless local area network receiver front end shown below, where the bandwidth of the bandpass filter is 150 MHz centered at 2.4 GHz. If the system is at room temperature, find the noise figure of the overall system. What is the output SNR if the input signal level is -85 dBm? Can the components be rearranged to give a better noise figure?

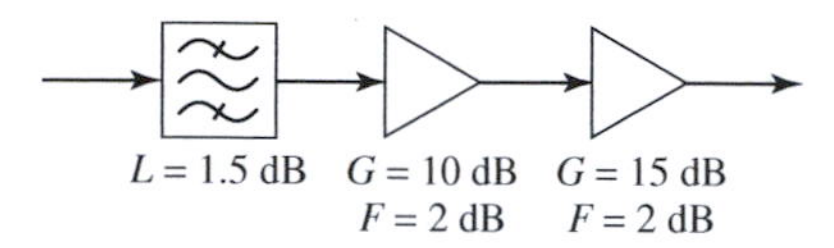

3.19 A digital PCS receiver front-end circuit is shown below. The operating frequency is 1805–1880 MHz, and the physical temperature of the system is 300 K. A noise source with $N_i = -95$ dBm is applied to the receiver input. What is the equivalent noise temperature of the source over the operating bandwidth? What is the noise figure of the amplifier? What is the noise figure of the cascaded transmission line and amplifier? What is the total noise power output of the receiver over the operating bandwidth?

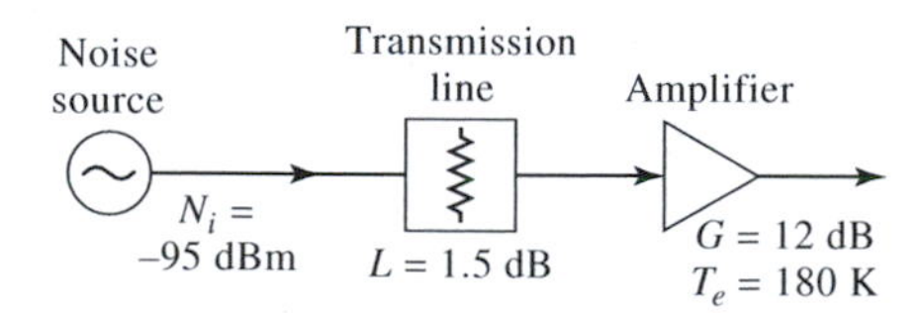

3.20 Prove that, for fixed loss $L > 1$, the equivalent noise temperature of a mismatched lossy line given in (3.87) is minimized when $|\Gamma_s| = 0$.

3.21 A matched resistive power divider is shown below, with its scattering matrix. If port 3 is terminated in a matched load Z_0 derive an expression for the noise figure from port 1 to port 2. Assume the divider is at physical te0perature T.

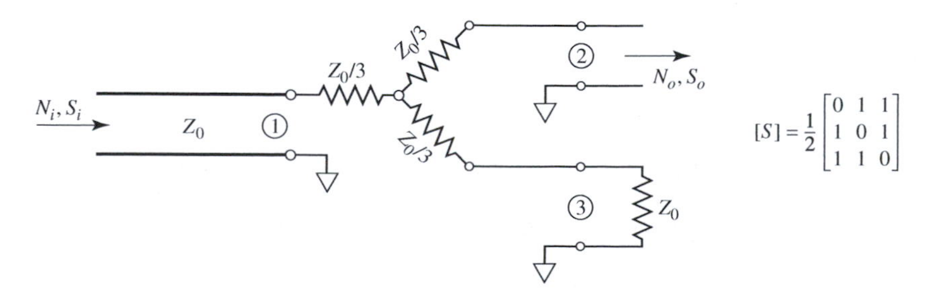

3.22 A lossless matched reactive power divider is shown below, with its scattering matrix. If port 3 is terminated in a matched load, find the noise figure of the divider and discuss the source of the resulting thermal noise. Assume the system is at physical temperature T_0.

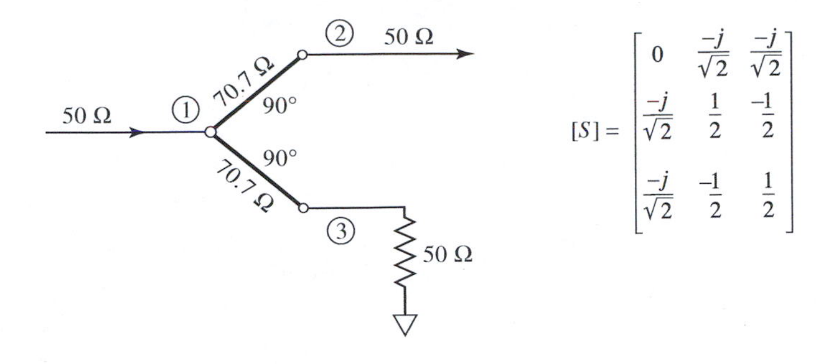

$$[S] = \begin{bmatrix} 0 & \frac{-j}{\sqrt{2}} & \frac{-j}{\sqrt{2}} \\ \frac{-j}{\sqrt{2}} & \frac{1}{2} & \frac{-1}{2} \\ \frac{-j}{\sqrt{2}} & \frac{-1}{2} & \frac{1}{2} \end{bmatrix}$$

3.23 Consider the lossless matched quadrature hybrid junction shown below, with the scattering matrix as given. Find the noise figure between ports 1 and 2, when ports 3 and 4 are terminated in matched loads. How does this result change when a dissipative power loss of L is included in each of the branch lines? Assume the system is at physical temperature T.

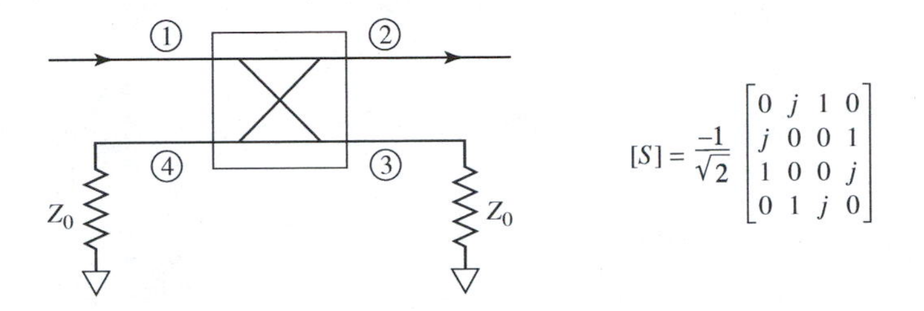

$$[S] = \frac{-1}{\sqrt{2}} \begin{bmatrix} 0 & j & 1 & 0 \\ j & 0 & 0 & 1 \\ 1 & 0 & 0 & j \\ 0 & 1 & j & 0 \end{bmatrix}$$

3.24 A lossy line at temperature T feeds an amplifier with noise figure F, as shown below. If an impedance mismatch Γ is present at the input of the amplifier, find the overall noise figure of the system.

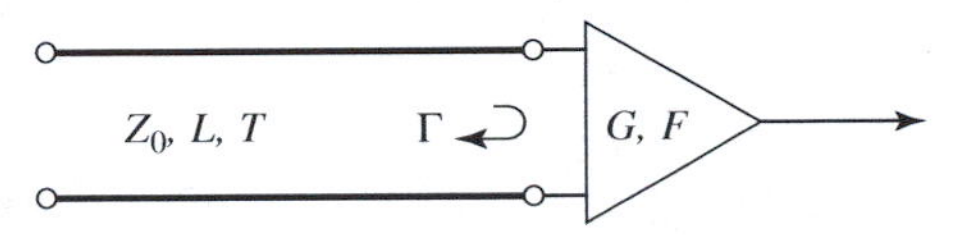

3.25 A balanced amplifier circuit is shown below. The two amplifiers are identical, each with power gain G and noise figure F. The two quadrature hybrids are also identical, with an insertion loss from the input to either output of $L > 1$ (the scattering matrix for a lossless hybrid is given in Problem 3.23). Derive an expression for the overall noise figure of the balanced amplifier. What does this result reduce to when the hybrids are lossless?

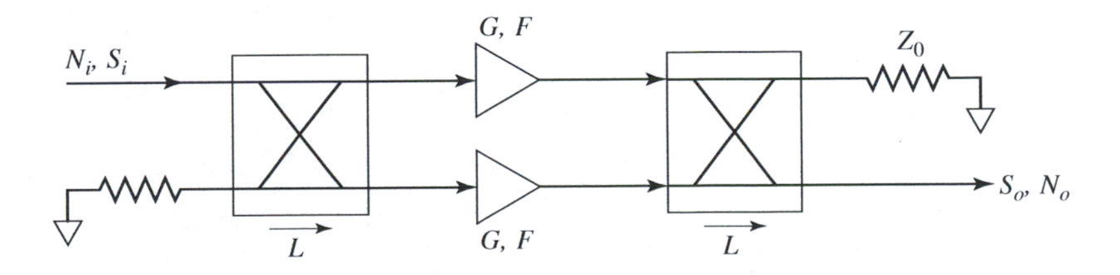

3.26 A receiver subsystem has a noise figure of 6 dB, a 1 dB compression point of 21 dBm (referenced to output), a gain of 30 dB, and a third-order intercept point of 33 dBm (referenced to output). If the subsystem is fed with a noise source with $N_i = -105$ dBm, and the desired output SNR is 8 dB, find the linear and spurious free dynamic ranges of the subsystem. Assume a system bandwidth of 20 MHz.

3.27 Find the third-order intercept point for the problem of Example 3.7 when the positions of the amplifier and mixer are reversed.

3.28 In practice, the third-order intercept point is extrapolated from measured data taken at input power levels well below P_3. For the spectrum analyzer display shown below, where ΔP is the difference in power between P_{ω_1} and $P_{2\omega_1-\omega_2}$, show that the third-order intercept point is given by $P_3 = P_{\omega_1} + \frac{1}{2}\Delta P$. Is this referenced at the input or output?

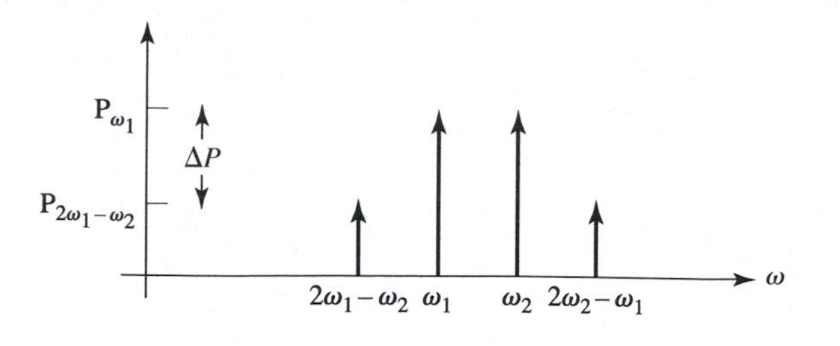

Chapter Four

Antennas and Propagation for Wireless Systems

The propagation of electromagnetic energy is the key physical phenomenon that makes wireless communications possible. In this chapter we study antennas and propagation effects, which can be viewed as forming the channel that links the transmitter to the receiver in a wireless communications system. Our goal is to characterize antennas from a systems point of view to enable the calculation of the received signal and noise powers, in terms of transmit power, range, antenna gain, efficiency, and background noise. We will also want to consider propagation effects through the environment, including reflection, diffraction, scattering, and attenuation. Such effects are often critical in mobile communications systems, especially when multipath scattering leads to signal fading.

4.1 ANTENNA SYSTEM PARAMETERS

In this section we describe the basic characteristics and parameters of antennas that are needed for our study of the wireless communications link. We are not interested here in the detailed electromagnetic theory of operation of antennas, but rather the systems aspect of the operation of an antenna in terms of its radiation patterns, directivity, gain, efficiency, and polarization. References [1]–[3] can be reviewed for a more in-depth treatment of the fascinating subject of antenna theory and design. Figure 4.1 shows some of the different types of antennas that have been developed for commercial wireless systems.

An antenna can be viewed as a device that converts a guided electromagnetic wave on a transmission line to a plane wave propagating in free space. Thus, one side of an antenna appears as an electrical circuit element, while the other side provides an interface with a propagating plane wave. Antennas are inherently bidirectional, in that they can be used for both transmit and receive functions.

Figure 4.2 illustrates the basic operation of transmit and receive antennas. The transmitter can be modeled as a Thevenin source consisting of a voltage generator and series impedance, delivering a transmit power P_t to the transmit antenna. The transmit antenna

FIGURE 4.1 Photograph of various antennas used for commercial millimeter wave wireless systems. Clockwise from top: a high-gain 38 GHz reflector antenna with radome, a 24 GHz microstrip array, a GPS antenna, a 38 GHz microstrip array, a 28 GHz microstrip array, a 24 GHz dual-beam microstrip array, and a planar LMDS array. (Courtesy of H. Syrigos, Alpha Industries, Inc., Woburn, Mass.)

radiates a spherical wave which, at large distances, approximates a plane wave, at least over a localized area. The receive antenna intercepts a portion of the propagating wave, and delivers a receive power P_r to the receiver load impedance.

Fields and Power Radiated by an Antenna

While we do not require detailed solutions to Maxwell's equations for our purposes, we do need to be familiar with the far-zone electromagnetic fields radiated by an antenna. We consider an antenna located at the origin of a spherical coordinate system, as shown in Figure 4.3. At large distances, where the localized near-zone fields are negligible, the radiated electric field of an arbitrary antenna can be expressed as

$$\bar{E}(r, \theta, \phi) = [\hat{\theta} F_\theta(\theta, \phi) + \hat{\phi} F_\phi(\theta, \phi)] \frac{e^{-jk_0 r}}{r} \text{ V/m}, \tag{4.1}$$

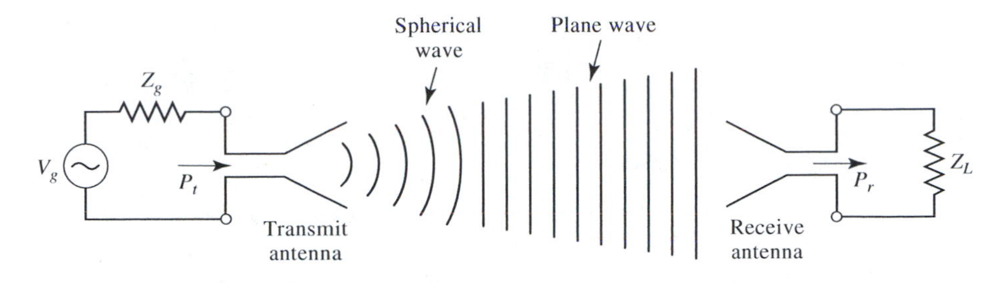

FIGURE 4.2 Basic operation of transmit and receive antennas.

where $\bar{E}$ is the electric field vector, $\hat{\theta}$ and $\hat{\phi}$ are unit vectors in the spherical coordinate system, r is the radial distance from the origin, and $k_0 = 2\pi/\lambda$ is the free-space propagation constant, with wavelength $\lambda = c/f = 3 \times 10^8/f$ (for frequency in Hz). Also defined in (4.1) are the *pattern functions*, $F_\theta(\theta, \phi)$ and $F_\phi(\theta, \phi)$. The interpretation of (4.1) is that this electric field propagates in the radial direction, with a phase variation of $e^{-jk_0 r}$ and an amplitude variation of $1/r$. The electric field may be polarized in either the θ or ϕ directions, but not in the radial direction. These are some of the well-known characteristics of *transverse electromagnetic* (TEM) waves.

Whenever we have a propagating electric field, there must be an associated magnetic field. For the TEM wave of (4.1) the magnetic fields can be found as

$$H_\phi = \frac{E_\theta}{\eta_0} \tag{4.2a}$$

$$H_\theta = \frac{-E_\phi}{\eta_0}, \tag{4.2b}$$

where $\eta_0 = 377\ \Omega$, the wave impedance of free space. Note that the magnetic field vector is also polarized only in the transverse direction. The Poynting vector for electromagnetic fields is given by the cross product of the electric and magnetic field vectors:

$$\bar{S} = \bar{E} \times \bar{H}^* \text{ W/m}^2, \tag{4.3}$$

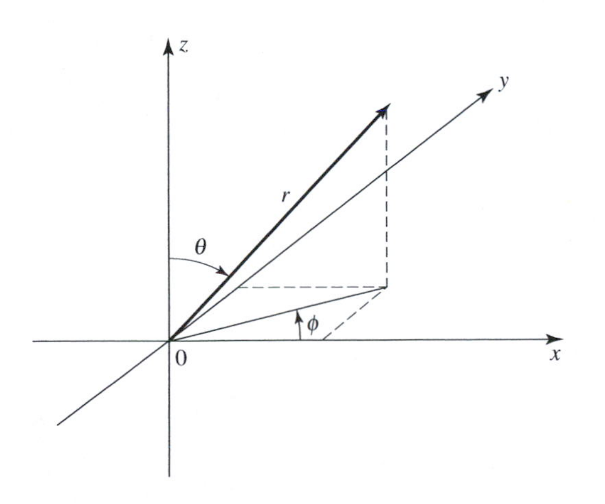

FIGURE 4.3 The spherical coordinate system.

and the time-average Poynting vector is

$$\bar{S}_{\text{avg}} = \tfrac{1}{2}\,\text{Re}\{\bar{S}\} = \tfrac{1}{2}\,\text{Re}\{\bar{E} \times \bar{H}^*\}\ \text{W/m}^2. \tag{4.4}$$

Far-Field Distance

We mentioned above that at large distances the near fields of an antenna are negligible, and the radiated electric field can be written as in (4.1). We can give a more precise meaning to this concept by defining the *far-field distance* as the distance where the spherical wave front radiated by an antenna becomes a close approximation to the ideal planar phase front of a plane wave. This approximation applies over the aperture area of the antenna, and so depends on the maximum dimension of the antenna. If we call this dimension D, then the far-field distance is defined as

$$R_{ff} = \frac{2D^2}{\lambda}\ \text{m}. \tag{4.5}$$

This result is derived from the condition that the actual spherical wave front radiated by the antenna departs less than $\pi/8 = 22.5^\circ$ from a true plane wave front over the maximum extent of the antenna. For electrically small antennas, such as short dipoles and small loops, this result may give a far-field distance that is too small; in this case, a minimum value of $R_{ff} = 2\lambda$ should be used.

EXAMPLE 4.1 FAR-FIELD DISTANCE OF AN ANTENNA

A parabolic reflector antenna used for reception with the Direct Broadcast System (DBS) is 18″ in diameter and operates at 12.4 GHz. Find the operating wavelength and the far-field distance for this antenna.

Solution

The operating wavelength at 12.4 GHz is

$$\lambda = \frac{c}{f} = \frac{3 \times 10^8}{12.4 \times 10^9} = 2.42\ \text{cm}.$$

The far-field distance is found from (4.5), after converting 18″ to 0.457 m:

$$R_{ff} = \frac{2D^2}{\lambda} = \frac{2(0.457)^2}{0.0242} = 17.3\ \text{m}.$$

The actual distance from a DBS satellite to earth is about 36,000 km, so it is safe to say that the receive antenna is in the far field of the transmitter. ○

Radiation Intensity

Next we define the *radiation intensity* of the radiated electromagnetic field as

$$\begin{aligned} U(\theta, \phi) = r^2|\bar{S}_{\text{avg}}| &= \frac{r^2}{2}\,\text{Re}\{E_\theta\hat{\theta} \times H_\phi^*\hat{\phi} + E_\phi\hat{\phi} \times H_\theta^*\hat{\theta}\} \\ &= \frac{r^2}{2\eta_0}\left[|E_\theta|^2 + |E_\phi|^2\right] = \frac{1}{2\eta_0}\left[|F_\theta|^2 + |F_\phi|^2\right]\ \text{W}, \end{aligned} \tag{4.6}$$

where (4.1), (4.2), and (4.4) were used. The units of the radiation intensity are Watts (W), or Watts per unit solid angle, since the radial dependence has been removed. The radiation intensity gives the variation in radiated power versus position around the antenna. We can find the total power radiated by the antenna by integrating the Poynting vector over the surface of a sphere of radius r that encloses the antenna. This is equivalent to integrating the radiation intensity over a unit sphere:

$$P_{\text{rad}} = \int_{\phi=0}^{2\pi} \int_{\theta=0}^{\pi} \bar{S}_{\text{avg}} \cdot \hat{r} r^2 \sin\theta \, d\theta \, d\phi = \int_{\phi=0}^{2\pi} \int_{\theta=0}^{\pi} U(\theta, \phi) \sin\theta \, d\theta \, d\phi. \tag{4.7}$$

Radiation Patterns

The *radiation pattern* of an antenna is a plot of the magnitude of the far-zone field strength versus position around the antenna, at a fixed distance from the antenna. Thus the pattern can be plotted from the pattern functions $F_\theta(\theta, \phi)$ or $F_\phi(\theta, \phi)$, versus either the angle θ (for an elevation plane pattern) or the angle ϕ (for an azimuthal plane pattern). The choice of plotting either F_θ or F_ϕ is dependent on the polarization of the antenna.

A typical antenna pattern is shown in Figure 4.4, for the particular case of a small array of dipoles often used in cellular telephone base stations. This pattern is plotted in polar form, versus the azimuth angle ϕ. The plot shows the relative variation of the radiated power of the antenna in dB, normalized to the maximum value. Since the pattern functions are proportional to voltage, the radial scale of the plot is computed as $20 \log |F(\theta, \phi)|$; alternatively, the plot could be computed in terms of the radiation intensity as $10 \log |U(\theta, \phi)|$. The plot shows that this antenna has maximum radiation in the $\phi = 0$ direction, and theoretically zero radiation at the angles $\phi = \pm 42^\circ$ and $\phi = \pm 90^\circ$.

There are several features of radiation pattern plots with which we should be familiar. First, as in the example of Figure 4.4, an antenna pattern may exhibit several distinct radiation

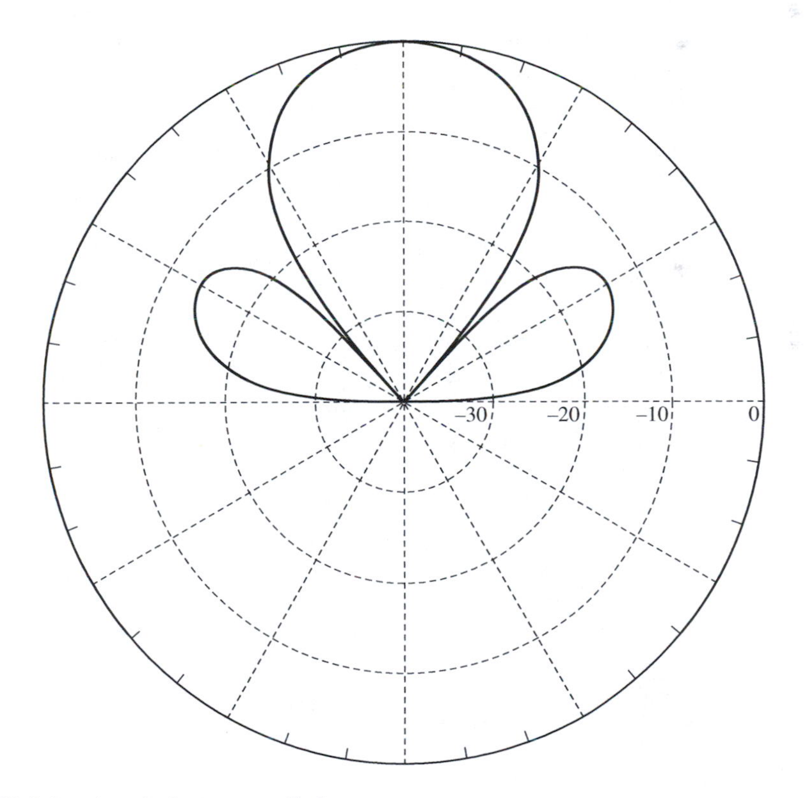

FIGURE 4.4 A typical antenna radiation pattern.

lobes, with different maxima in different directions. The lobe having the maximum value is called the *main beam*, while those lobes at lower levels are called *sidelobes*. Thus the pattern of Figure 4.4 has one main beam at $\phi = 0$, and two sidelobes located at $\phi = \pm 60°$. The level of these sidelobes is 14 dB below the level of the main beam.

A fundamental property of an antenna is its ability to focus power in a given direction, to the exclusion of other directions. Thus, an antenna with a broad main beam can transmit (or receive) power over a wide angular region, while an antenna having a narrow main beam will transmit (or receive) power over a small angular region. A measure of this focusing effect is the *3 dB beamwidth* of the antenna, defined as the angular width of the main beam at which the power level has dropped 3 dB from its maximum value (its half-power points). The 3 dB beamwidth of the pattern of Figure 4.4 is 36°.

Antennas having a constant pattern in the azimuthal direction are called *omnidirectional*, and are useful for applications such as broadcasting or for handheld cellular phones, where it is desired to transmit or receive equally in all directions. Patterns that have relatively narrow main beams in both planes are known as *pencil beam* antennas, and are useful in applications requiring point-to-point transmission, such as satellite links.

Directivity

Another measure of the focusing ability of an antenna is the *directivity*, defined as the ratio of the maximum radiation intensity in the main beam to the average radiation intensity over all space:

$$D = \frac{U_{\max}}{U_{\text{avg}}} = \frac{4\pi U_{\max}}{P_{\text{rad}}} = \frac{4\pi U_{\max}}{\int_{\theta=0}^{\pi}\int_{\phi=0}^{2\pi} U(\theta,\phi)\sin\theta\, d\theta\, d\phi}, \tag{4.8}$$

where (4.7) has been used for the radiated power. Directivity is a dimensionless ratio of power, and is usually expressed in dB as $D(\text{dB}) = 10\log(D)$.

An antenna that radiates equally in all directions is called an *isotropic* antenna. Applying the integral identity that

$$\int_{\theta=0}^{\pi}\int_{\phi=0}^{2\pi} \sin\theta\, d\theta\, d\phi = 4\pi,$$

to the denominator of (4.8) for $U(\theta,\phi) = 1$ shows that the directivity of an isotropic element is $D = 1$, or 0 dB. Since the minimum directivity of any antenna is unity, directivity is often stated as relative to the directivity of an isotropic radiator, and written as *dBi*.

Beamwidth and directivity are both measures of the focusing ability of an antenna: an antenna pattern with a narrow main beam will have a high directivity, while a pattern with a wide beam will have a lower directivity. We might therefore expect a direct relation between beamwidth and directivity, but in fact there is not an exact relationship between these two quantities. This is because beamwidth is only dependent on the size and shape of the main beam, whereas directivity involves integration of the entire radiation pattern. Thus it is possible for many different antenna patterns to have the same beamwidth, but quite different directivities due to differences in sidelobes or the presence of more than one main beam. With this qualification in mind, however, it is possible to develop approximate relations between beamwidth and directivity that apply with reasonable accuracy to a large number of practical antennas. One such approximation that works well for antennas with pencil beam patterns is the following:

$$D \cong \frac{32{,}400}{\theta_1\theta_2}, \tag{4.9}$$

where θ_1 and θ_2 are the beamwidths in two orthogonal planes of the main beam, in degrees. This approximation does not work well for omnidirectional patterns because there is a well-defined main beam in only one plane for such patterns.

EXAMPLE 4.2 PATTERN CHARACTERISTICS OF A DIPOLE ANTENNA

The far-zone electric field radiated by an electrically small wire dipole antenna on the z axis is given by,

$$E_\theta(r, \theta, \phi) = V_0 \sin\theta \frac{e^{-jk_0 r}}{r} \text{ V/m}$$
$$E_\phi(r, \theta, \phi) = 0.$$

Find the main beam position of the dipole, its beamwidth, and its directivity.

Solution

The radiation intensity for the above far field is

$$U(\theta, \phi) = C \sin^2\theta,$$

where the constant $C = V_0^2/2\eta_0$. The radiation pattern is seen to be independent of the azimuth angle ϕ, and so is omnidirectional in the azimuth plane. The pattern has a "donut" shape, with nulls at $\theta = 0°$ and $\theta = 180°$ (on the z axis), and a beam maximum at $\theta = 90°$ (the horizontal plane). The angles where the radiation intensity has dropped by 3 dB is given by the solutions to

$$\sin^2\theta = 0.5,$$

thus the 3 dB, or half-power, beamwidth is $135° - 45° = 90°$.

The directivity is calculated using (4.8). The denominator of this expression is

$$\int_{\theta=0}^{\pi}\int_{\phi=0}^{2\pi} U(\theta, \phi) \sin\theta \, d\theta \, d\phi = 2\pi C \int_{\theta=0}^{\pi} \sin^3\theta \, d\theta = 2\pi C\left(\frac{4}{3}\right) = \frac{8\pi C}{3},$$

where the required integral identity is listed in Appendix B. Since $U_{max} = C$, the directivity reduces to

$$D = \frac{3}{2} = 1.76 \text{ dB.}$$

○

Radiation Efficiency

Resistive losses, due to nonperfect metals and dielectric materials, exist in all antennas. Such losses result in a difference between the power delivered to the input of an antenna and the power radiated by that antenna. As with other electrical components, we can define the *radiation efficiency* of an antenna as the ratio of the desired output power to the supplied input power

$$e_{rad} = \frac{P_{rad}}{P_{in}} = \frac{P_{in} - P_{loss}}{P_{in}} = 1 - \frac{P_{loss}}{P_{in}}, \tag{4.10}$$

where P_{rad} is the power radiated by the antenna, P_{in} is the power supplied to the input of the antenna, and P_{loss} is the power lost in the antenna. Efficiency is always less than or equal to unity, and is commonly expressed as a percent. Radiation efficiency has been defined in (4.10) for transmit antennas, but applies as well to receiving antennas.

Note that there are other factors that can contribute to the effective loss of transmit power, such as impedance mismatch at the input to the antenna, or polarization mismatch with the receive antenna. But these losses are external to the antenna, and could be eliminated by the proper use of matching networks, or the proper choice and positioning of the receive antenna. Therefore losses of this type should not be attributed to the antenna itself, as are dissipative losses due to metal conductivity or dielectric loss within the antenna.

Gain

Recall that antenna directivity is a function only of the shape of the radiation pattern (the radiated fields) of an antenna, and is not affected by losses in the antenna. In order to account for the fact that an antenna with radiation efficiency less than unity will not radiate all of its input power, we define *antenna gain* as the product of directivity and efficiency:

$$G = e_{\text{rad}} D. \tag{4.11}$$

Thus, gain is always less than or equal to directivity. Gain can also be computed directly, by replacing P_{rad} in the denominator of (4.8) with P_{in}, since by the definition of radiation efficiency in (4.10) we have $P_{\text{rad}} = e_{\text{rad}} P_{\text{in}}$. Gain is usually expressed in dB, as $G(\text{dB}) = 10 \log (G)$.

Aperture Efficiency

Many types of antennas can be classified as *aperture antennas*, meaning that the antenna has a well-defined aperture area from which radiation occurs. Examples include reflector antennas, horn antennas, lens antennas, and array antennas. For such antennas, it can be shown that the maximum directivity that can be obtained from an aperture of area A is

$$D_{\max} = \frac{4\pi A}{\lambda^2}. \tag{4.12}$$

For example, a rectangular horn antenna having an aperture $2\lambda \times 3\lambda$ could have a maximum directivity of 24π. In practice, there are several factors that can serve to reduce the maximum directivity, such as nonideal amplitude or phase characteristics of the aperture field, aperture blockages or, in the case of reflector antennas, spillover of the feed pattern. For this reason we can define an *aperture efficiency* as the ratio of the actual directivity of an aperture antenna to the maximum directivity given by (4.12). Then we can write the directivity of an aperture antenna as

$$D = e_{ap} \frac{4\pi A}{\lambda^2}. \tag{4.13}$$

Aperture efficiency is always less than or equal to unity.

Effective Area

The above definitions of antenna directivity, efficiency, and gain were stated in terms of a transmitting antenna, but apply to receiving antennas as well. For a receiving antenna it is also of interest to determine the received power for a given incident plane wave field. This is the converse problem to finding the power density radiated by a transmitting antenna, as given in (4.4); both of these cases are illustrated in Figure 4.2. Finding received power is important for the derivation of the radio system link equation, to be discussed in the following section.

We expect that received power will be proportional to the power density, or Poynting vector, of the incident wave. Since the Poynting vector has dimensions of W/m^2, and the received power P_r has dimensions of W, the proportionality constant must have units of area. Thus we write

$$P_r = A_e S_{avg}, \tag{4.14}$$

where A_e is defined as the *effective aperture area* of the receive antenna. The effective aperture area has dimensions of m^2, and can be interpreted as the "capture area" of a receive antenna, intercepting part of the incident power density radiated toward the receive antenna. P_r in (4.14) is the power available at the terminals of the receive antenna, as delivered to a matched load.

The maximum effective aperture area of an antenna can be shown to be related to the directivity of the antenna as [1], [2]

$$A_e = \frac{D\lambda^2}{4\pi}, \tag{4.15}$$

where λ is the operating wavelength of the antenna. For electrically large aperture antennas the effective aperture area is often close to the actual physical aperture area. But for many other types of antennas, such as dipoles and loops, there is no simple relation between the physical cross-sectional area of the antenna and its effective aperture area. The maximum effective aperture area as defined above does not include the effect of losses in the antenna, which can be accounted for by replacing D in (4.15) with G, the gain, of the antenna.

Antenna Polarization

Polarization of an electromagnetic wave is defined as the orientation of the radiated electric field vector. For the case of a plane wave propagating along the z axis, the electric field may only have components in the x- or y directions, so the general expression for the electric field is of the form

$$\bar{E} = (E_{0x}\hat{x} + E_{0y}\hat{y})e^{-jk_0 z}, \tag{4.16}$$

where E_{0x} and E_{0y} are the independent amplitudes of the x and y components. If $E_{0x} = 1$ and $E_{0y} = 0$, the field is *linearly polarized* in the x direction. If $E_{0x} = E_{0y} = 1$, the field is linearly polarized in the direction 45° between the x- and y axes. In general, as long as E_{0x} and E_{0y} have the same phase, the wave will be linearly polarized.

Now consider the case in which the x and y components have equal magnitudes and a 90° phase shift, such as $E_{0x} = 1$ and $E_{0y} = j$. Then the electric field still lies in the x-y plane, but now rotates counterclockwise when viewed toward the z-axis. To see this, convert the phasor expression of (4.16) to the time domain by multiplying by $e^{j\omega t}$ and taking the real part:

$$\begin{aligned}\bar{e}(x, y, z, t) &= \text{Re}\{(\hat{x} + j\hat{y})e^{-jk_0 z}e^{j\omega t}\} = \text{Re}\left\{\hat{x}e^{j(\omega t - k_0 z)} + \hat{y}e^{j(\omega t - k_0 z + \pi/2)}\right\} \\ &= \hat{x}\cos(\omega t + k_0 z) - \hat{y}\sin(\omega t - k_0 z).\end{aligned} \tag{4.17}$$

For a given point on the z-axis, this result shows that the electric field vector rotates from the x-axis, to the $-y$-axis, ..., as time increases. Since the direction of rotation is counterclockwise when viewed toward the direction of propagation, this is referred to as *left-hand circular polarization*, or LHCP. *Right-hand circular polarization* (RHCP) can be obtained by changing the sign of the E_{0y} term.

If the magnitudes of E_{0x} and E_{0y} are not equal, or if the phase difference is not exactly 90°, then the electric field amplitude will vary as the field vector rotates around the z axis. This is called *elliptical polarization*, and is the most general case.

According to IEEE standards, the polarization of an antenna is defined as the polarization of the radiated field when the antenna is operating as a transmitter. For a dipole on the z axis, as in Example 4.2, the radiated field has only an E_θ component in the far zone. In the main beam of the antenna, which occurs at $\theta = 90°$, the electric field is vertically oriented at any position around the axis of the dipole, and so the dipole is said to be linearly polarized in the vertical direction. We will see in the next section that maximum power transfer between two antennas requires that they have the same polarization.

4.2 THE FRIIS EQUATION

In this section we derive the *Friis equation*, which is the fundamental result for radio system links. It expresses the received power in terms of transmitted power, antenna gains, range, and frequency, and thus forms the basis for all wireless system design. We will also discuss equivalent circuits for transmit and receive antennas, and a modification to the Friis equation to account for impedance or polarization mismatches.

The Friis Equation

A general radio system link is shown in Figure 4.5, where the transmit power is P_t, the transmit antenna gain is G_t, the receive antenna gain is G_r, and the received power (delivered to a matched load) is P_r. The transmit and receive antennas are separated by the distance R.

By conservation of energy, the power density radiated by an isotropic antenna ($D = 1 = 0$ dB) at a distance R is given by,

$$S_{\text{avg}} = \frac{P_t}{4\pi R^2} \text{ W/m}^2, \tag{4.18}$$

where P_t is the power radiated by the antenna. This result is deduced from the fact that we must be able to recover all of the radiated power by integrating over a sphere of radius R surrounding the antenna; since the power is distributed isotropically, and the area of a sphere is $4\pi R^2$, (4.18) follows. If the transmit antenna has a directivity greater than 0 dB, we can find the radiated power density by multiplying by the directivity, since directivity is defined as the ratio of the actual radiation intensity to the equivalent isotropic radiation intensity. Also, if the transmit antenna has losses, we can include the radiation efficiency factor, which has the effect of converting directivity to gain. Thus, the general expression for the power density radiated by an arbitrary transmit antenna is

$$S_{\text{avg}} = \frac{G_t P_t}{4\pi R^2} \text{ W/m}^2. \tag{4.19}$$

If this power density is incident on the receive antenna, we can use the concept of effective aperture area, as defined in (4.14), to find the received power:

$$P_r = A_e S_{\text{avg}} = \frac{G_t P_t A_e}{4\pi R^2} \text{ W}.$$

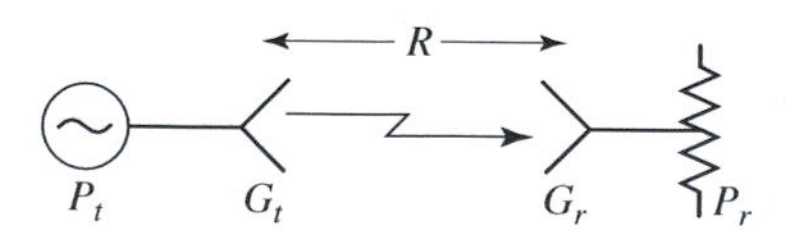

FIGURE 4.5 Basic radio system.

Next, (4.15) can be used to relate the effective area to the directivity of the receive antenna. Again, the possibility of losses in the receive antenna can be accounted for by using the gain (rather than the directivity) of the receive antenna. So the final result for the received power is

$$P_r = \frac{G_t G_r \lambda^2}{(4\pi R)^2} P_t \text{ W}. \tag{4.20}$$

This is the Friis equation, which addresses the fundamental question of how much power is received by a radio antenna. In practice, the value given by (4.20) should be interpreted as the maximum possible received power, as there are a number of factors that can serve to reduce the received power in an actual radio system. These include impedance mismatch at either antenna, polarization mismatch between the antennas, propagation effects leading to attenuation or depolarization, and multipath effects that may cause partial cancellation of the received field.

Observe in (4.20) that the received power decreases as $1/R^2$ as the separation between transmitter and receiver increases. This dependence is a result of conservation of energy. While it may seem to be prohibitively large for large distances, in fact the space decay of $1/R^2$ is much better than the exponential decrease in power due to losses in a wired communications link. This is because the attenuation of power on a transmission line varies as $e^{-2\alpha z}$(where α is the voltage attenuation constant of the line), and at large distances the exponential function always decreases faster than an algebraic dependence like $1/R^2$. Thus for long distance communications, radio links will perform better than wired links. This conclusion applies to any type of transmission line, including coaxial lines, waveguides, and even fiber optic lines. (It may not apply, however, if the communications link is land or sea-based, so that repeaters can be inserted along the link to recover lost signal power.)

EXAMPLE 4.3 COMMUNICATIONS SATELLITE LINK LOSS

A geosynchronous satellite orbiting 36,900 km above the earth has a transmit power of 2 W, a transmit antenna gain of 37 dB, and operates at 20 GHz. If the receiving station on earth has an antenna gain of 45.8 dB, find the received power. Ignore possible mismatch and propagation loss effects.

Solution

We first compute the quantity $4\pi R/\lambda = 3.09 \times 10^{10}$. Then we can evaluate the Friis equation of (4.20) using dB:

$$\begin{aligned} P_r(dB) &= G_r(dB) + G_t(dB) + P_t(dB) - 10\log\left(\frac{4\pi R}{\lambda}\right)^2 \\ &= 45.8 + 37.0 + 33.0 - 209.8 = -94.0 \text{ dBm} = 3.98 \times 10^{-10} \text{ mW}. \end{aligned}$$

○

Effective Isotropic Radiated Power

As can be seen from the Friis equation of (4.20), received power is proportional to the product $P_t G_t$. These two terms, the transmit power and transmit antenna gain, characterize the transmitter, and in the main beam of the antenna the product $P_t G_t$ can be interpreted equivalently as the power radiated by an isotropic antenna with input power $P_t G_t$. Thus, this product is defined as the *effective isotropic radiated power* (EIRP):

$$\text{EIRP} = P_t G_t \text{ W}. \tag{4.21}$$

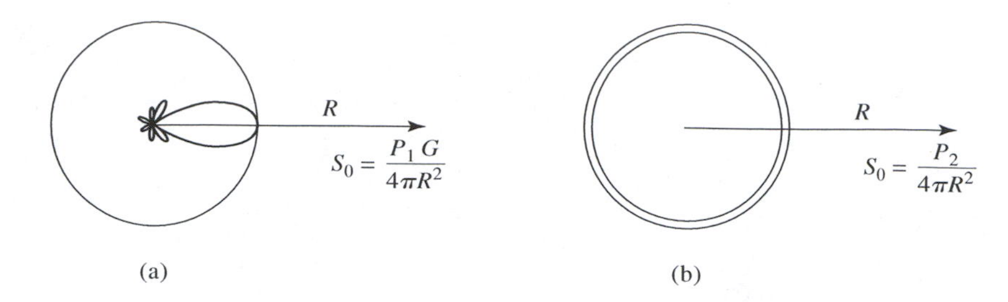

FIGURE 4.6 Effective isotropic radiated power. (a) An antenna with $G > 1$ with input power P_1 radiates a power density S_0 at distance R. (b) An isotropic antenna with input power P_2 radiates the same power density at distance R.

For a given frequency, range, and receiver antenna gain, the received power is proportional to the EIRP of the transmitter, and can only be increased by increasing the EIRP. This can be done by increasing the transmit power, or the transmit antenna gain, or both.

Figure 4.6 illustrates the meaning of EIRP. The antenna of Figure 4.6a has a gain G greater than unity, and an input power P_1. Thus, at a distance R in the main beam of the antenna, the radiated power density is given by (4.19) as

$$S_0 = \frac{P_1 G}{4\pi R^2} \text{ W/m}^2. \tag{4.22a}$$

The antenna of Figure 4.6b is isotropic, with gain $G = 1$ (0 dB), and input power P_2. We choose the input power so that the radiated power density is equal for both cases. Thus for the antenna of Figure 4.6b we have

$$S_0 = \frac{P_2}{4\pi R^2} \text{ W/m}^2. \tag{4.22b}$$

Comparison with (4.22a) shows that $P_2 = GP_1$. Both cases will then appear equivalent as seen by a receiver, but the transmitter of Figure 4.6a requires less input power by the factor $1/G$. Of course, the coverage of the antenna of Figure 4.6a is much less than that of the omnidirectional antenna of Figure 4.6b, and may not be usable in all applications.

A related quantity is the *effective radiated power* (ERP), which is defined relative to a half-wave dipole, rather than an isotropic element. ERP is sometimes used for mobile communications systems because of the prevalence of dipole antennas for base station antennas. Since the gain of a half-wave dipole is 2.2 dB, we have the following relation with EIRP:

$$\text{ERP(dB)} = \text{EIRP(dB)} + 2.2 \text{ dB}. \tag{4.23}$$

Impedance Mismatch

The derivation of the Friis equation as given in (4.20) assumed that the transmit and receive antennas were impedance matched to the transmitter and receiver, respectively. As with any RF or microwave system, an impedance mismatch will reduce the power delivered from a source to a load by the factor $(1 - |\Gamma|^2)$, where Γ is the reflection coefficient between the source and the load [4]. In a radio link there is the possibility of an impedance mismatch between the transmitter and the transmit antenna, as well as between the receive antenna and the receiver. Thus the Friis formula of (4.20) can be multiplied by the *impedance mismatch factor*, e_{imp}, defined as

$$e_{\text{imp}} = (1 - |\Gamma_t|^2)(1 - |\Gamma_r|^2), \tag{4.24}$$

to account for the reduction in received power due to impedance mismatch effects at the transmitter and receiver. In (4.24) Γ_t is the reflection coefficient at the transmitter, and Γ_r is the reflection coefficient at the receiver. Note that impedance mismatch is not included in the definition of antenna gain. This is because mismatch is dependent on the external source or load impedances to which the antenna is connected, and thus is not a property of the antenna itself. It is always possible to match an antenna to a given source or load by using an appropriate external tuning network.

Polarization Mismatch

Maximum transmission between two antennas requires that both antenna be polarized in the same direction. If a transmit antenna is vertically polarized, for example, maximum power will be delivered to a vertically polarized receiving antenna, while zero power would be delivered to a horizontally polarized receive antenna. Polarization matching of antennas is therefore critical for optimum communications system performance.

Since receive voltage is proportional to the dot product of the electric field and the polarization vector of the receive antenna, we can account for polarization mismatch effects by multiplying the Friis equation by the *polarization loss factor*, defined as

$$e_{\text{pol}} = |\hat{e}_i \cdot \hat{e}_r|^2. \tag{4.25}$$

In (4.25) $\hat{e}_i$ is a unit vector representing the polarization of the electric field of the incident plane wave. That is,

$$\bar{E}_i = \hat{e}_i E_i e^{-jk_0 r}.$$

Similarly, $\hat{e}_r$ is a unit vector representing the polarization of the electric field of the receive antenna *when it is operating as a transmit antenna.* That is,

$$\bar{E}_r = \hat{e}_r E_r e^{-jk_0 r}.$$

EXAMPLE 4.4 POLARIZATION LOSS FACTOR

A transmitting antenna is circularly polarized (LHCP), and transmits in the direction of the horizontal z axis. Find the polarization loss factor for the following receive antennas:

(a) a vertically polarized dipole
(b) a LHCP antenna
(c) a RHCP antenna

Solution
The unit vector for the polarization of the incident field is, from (4.17),

$$\hat{e}_i = (\hat{x} + j\hat{y})/\sqrt{2}.$$

For case (a) the polarization vector for a vertically polarized antenna is

$$\hat{e}_r = \hat{y},$$

so the polarization loss factor is

$$e_{\text{pol}} = |\hat{e}_i \cdot \hat{e}_r|^2 = \frac{|(\hat{x} + j\hat{y}) \cdot \hat{y}|^2}{2} = 0.5 = -3 \text{ dB}.$$

We assumed the y axis was in the vertical direction, but the same result would be obtained if the x axis were used. This result shows that the linearly polarized receive antenna captures only half the incident power from the circularly polarized transmitted wave.

For case (b) the polarization vector for an LHCP antenna transmitting along the $-z$ axis (toward the incident wave) is

$$\hat{e}_i = (\hat{x} - j\hat{y})/\sqrt{2},$$

so the polarization loss factor is

$$e_{\rm pol} = |\hat{e}_i \cdot \hat{e}_r|^2 = \frac{|(\hat{x} + j\hat{y}) \cdot (\hat{x} - j\hat{y})|^2}{4} = 1 = 0 \text{ dB}.$$

In this case the receive antenna is polarization matched to the transmit antenna. For case (c) the polarization vector for a RHCP antenna transmitting along the $-z$ axis (toward the incident wave) is

$$\hat{e}_i = (\hat{x} + j\hat{y})/\sqrt{2},$$

so the polarization loss factor is

$$e_{\rm pol} = |\hat{e}_i \cdot \hat{e}_r|^2 = \frac{|(\hat{x} + j\hat{y}) \cdot (\hat{x} + j\hat{y})|^2}{4} = 0.$$

This receive antenna is completely mismatched to the polarization of the transmitted wave. ○

Equivalent Circuits for Transmit and Receive Antennas

As discussed in the beginning of this section, an antenna appears as a circuit element at its terminals. For a transmitting antenna, power is delivered to the antenna from a generator and converted to a propagating electromagnetic wave. For a receiving antenna, input power is delivered to the antenna in the form of a plane wave, and converted to received power at the terminals of the antenna. In both cases the terminals of the antenna appear as an equivalent circuit port. Figure 4.7a shows the case of a transmitting antenna, where the transmitter is represented as a Thevenin source with voltage generator V_g and generator impedance Z_g, and the antenna impedance Z_A appears as a load. The power dissipated in this load represents power radiated by the antenna, as well as power dissipated due to loss in the antenna. The case of a receiving antenna is shown in Figure 4.7b, where the voltage generator V_A and generator impedance Z_A represents power received from the incident

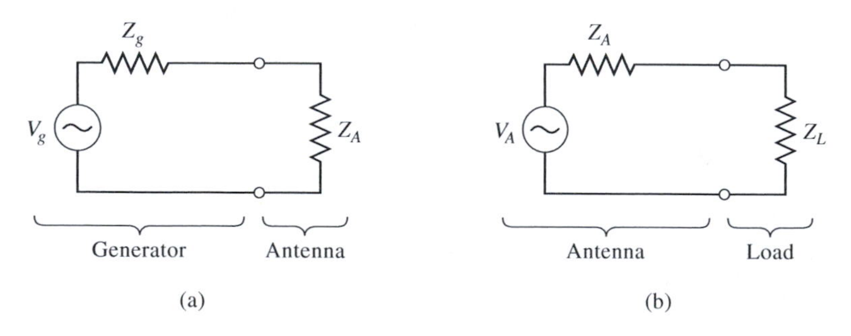

FIGURE 4.7 Equivalent circuits for (a) transmit, and (b) receive antennas.

plane wave, and Z_L represents the receiver load impedance. It is important to note that Z_A is a property of the antenna itself, and is the same for both transmit and receive operation.

4.3 ANTENNA NOISE TEMPERATURE

We have seen how noise is generated in a receiver due to lossy components and active devices, but noise can also be delivered to the input of a wireless receiver by the receive antenna. Antenna noise may be received from the external environment, or generated internally as thermal noise due to losses in the antenna itself. While noise produced within a receiver is controllable to some extent (by judicious design and component selection), the noise received from the environment by a receiving antenna is generally not controllable, and may exceed the noise level of the receiver itself. Thus it is important that we are able to characterize the noise power delivered to a radio receiver by its antenna. In this section we will study the noise characteristics of receiving antennas, and define the equivalent noise temperature of an antenna.

Background and Brightness Temperature

Consider the three situations shown in Figure 4.8. In Figure 4.8a we have the simple case of a resistor at temperature T, producing an available output noise power given by (3.25):

$$N_o = kTB, \tag{4.26}$$

where B is the system bandwidth, and k is Boltzmann's constant. In Figure 4.8b we have an antenna enclosed by an anechoic chamber at temperature T. The anechoic chamber appears as a perfectly absorbing enclosure, and is in thermal equilibrium with the antenna. Thus the terminals of the antenna are indistinguishable from the resistor terminals of Figure 4.8a (assuming an impedance-matched antenna), and therefore the antenna produces the same output noise power as the resistor of Figure 4.8a. Lastly, Figure 4.8c shows the same antenna directed at the sky. If the main beam of the antenna is narrow enough so that it sees a uniform region at physical temperature T, then the antenna again appears as a resistor at temperature T, and produces the output noise power given in (4.26). This is true regardless of the radiation efficiency of the antenna, as long as the physical temperature of the antenna is T.

In actuality an antenna typically sees a much more complex environment than the cases depicted in Figure 4.8. A general scenario of both naturally occurring and man-made

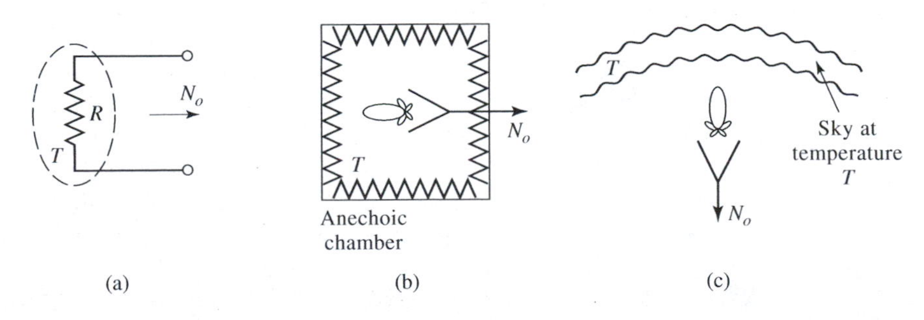

FIGURE 4.8 Illustrating the concept of background temperature. (a) A resistor at temperature T. (b) An antenna in an anechoic chamber at temperature T. (c) An antenna viewing a uniform sky background at temperature T.

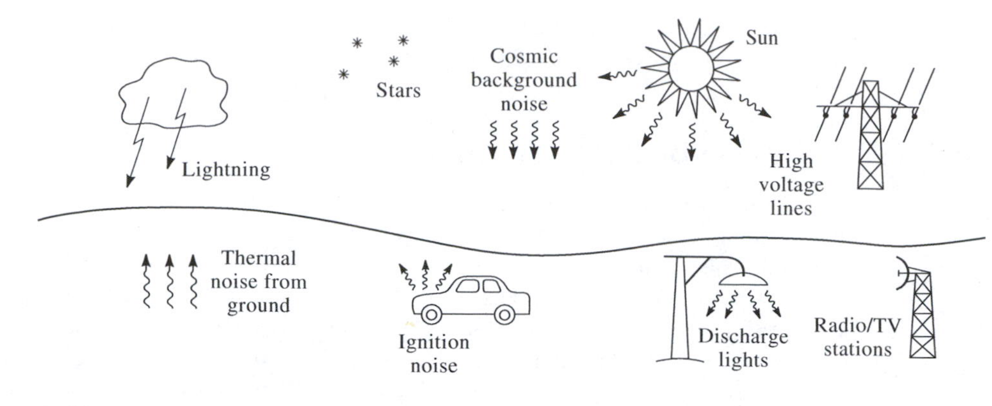

FIGURE 4.9 Natural and man-made sources of background noise.

noise sources is shown in Figure 4.9, where we see that an antenna with a relatively broad main beam may pick up noise power from a variety of origins. In addition, noise may be received through the sidelobes of the antenna pattern, or via reflections from the ground or other large objects. As in Chapter 3, where the noise power from an arbitrary white noise source was represented as an equivalent noise temperature, we define the *background noise temperature*, T_B, as the equivalent temperature of a resistor required to produce the same noise power as the actual environment seen by the antenna. Some typical background noise temperatures that are relevant at low microwave frequencies are:

- sky (toward zenith) 3–5 K
- sky (toward horizon) 50–100 K
- ground 290–300 K

The overhead sky background temperature of 3–5 K is the cosmic background radiation believed to be a remnant of the big bang at the creation of the universe. This would be the noise temperature seen by an antenna with a narrow beam and high radiation efficiency pointed overhead, away from "hot" sources such as the sun or stellar radio objects. The background noise temperature increases as the antenna is pointed toward the horizon because of the greater thickness of the atmosphere, so that the antenna sees an effective background closer to that of the anechoic chamber of Figure 4.8b. Pointing the antenna toward the ground further increases the effective loss, and hence the noise temperature.

Figure 4.10 gives a more complete picture of the background noise temperature, showing the variation of T_B versus frequency, and for several elevation angles [5]. Note that the noise temperature shown in the graph follows the trends listed above, in that it is lowest for the overhead sky ($\theta = 90°$), and greatest for angles near the horizon ($\theta = 0°$). Also note the sharp peaks in noise temperature that occur at 22 GHz and 60 GHz. The first is due to the resonance of molecular water, while the second is caused by resonance of molecular oxygen. Both of these resonances lead to increased atmospheric loss, and hence increased noise temperature. The loss is great enough at 60 GHz that the atmosphere effectively appears as a matched load at 290 K. While loss in general is undesirable, these particular resonances can be useful for remote sensing applications [4], or for using the inherent attenuation of the atmosphere to limit propagation distances for cellular communications over small regions.

When the antenna beamwidth is broad enough that different parts of the antenna pattern see different background temperatures, the effective *brightness temperature* seen by the antenna can be found by weighting the spatial distribution of background temperature by the pattern function of the antenna. Mathematically we can write the *brightness temperature*

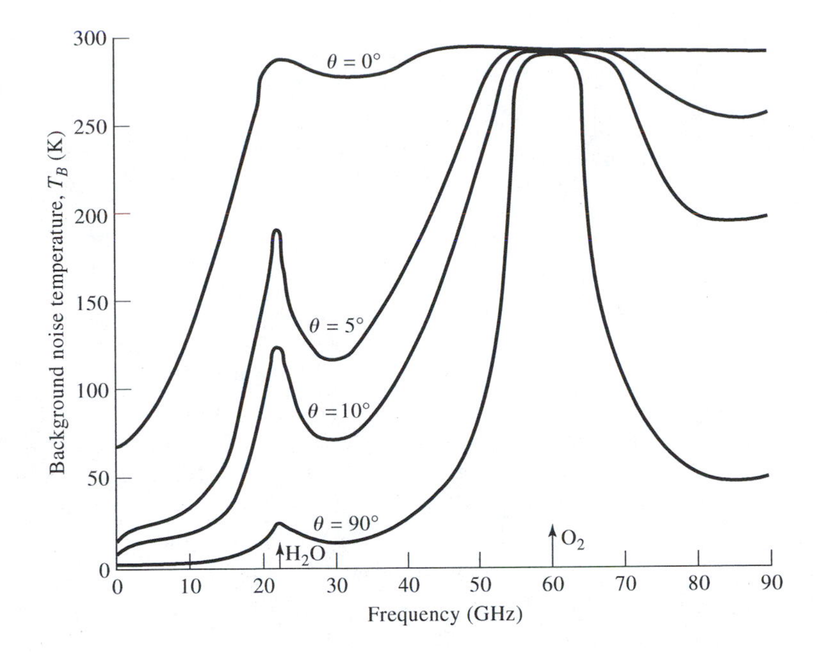

FIGURE 4.10 Background noise temperature of sky versus frequency. θ is the elevation angle measured from the horizon. Data are for sea level, with surface temperature of 15°C, and surface water vapor density of 7.5 gm/m^3.

T_b seen by the antenna as

$$T_b = \frac{\int_{\phi=0}^{2\pi} \int_{\theta=0}^{\pi} T_B(\theta, \phi) D(\theta, \phi) \sin\theta \, d\theta \, d\phi}{\int_{\phi=0}^{2\pi} \int_{\theta=0}^{\pi} D(\theta, \phi) \sin\theta \, d\theta \, d\phi}, \tag{4.27}$$

where $T_B(\theta, \phi)$ is the distribution of the background temperature, and $D(\theta, \phi)$ is the directivity (or the power pattern function) of the antenna. Antenna brightness temperature is referenced at the terminals of the antenna. Observe that when T_B is a constant, (4.27) reduces to $T_b = T_B$, which is essentially the case of a uniform background temperature shown in Figure 4.8b or 4.8c. Also note that this definition of antenna brightness temperature does not involve the gain or efficiency of the antenna, and so does not include thermal noise due to losses in the antenna.

Antenna Noise Temperature

If the antenna has dissipative loss, so that the radiation efficiency e_{rad} is less than unity, then the power available at the terminals of a receive antenna is reduced by the factor e_{rad} from that intercepted by the antenna. This applies to received noise power, as well as received signal power, so the noise temperature of the antenna will be reduced from the brightness temperature given in (4.27) by the factor e_{rad}. In addition, thermal noise will be generated by resistive losses in the antenna, and this will increase the noise temperature of the antenna.

The overall problem of a lossy antenna at physical temperature T_p viewing a background noise temperature distribution T_B can be represented by the system shown in Figure 4.11. The lossy antenna is modeled as an ideal antenna with $e_{\text{rad}} = 1$, followed by an attenuator having a power loss factor of $L \geq 1$, at physical temperature T_p. Since radiation efficiency is

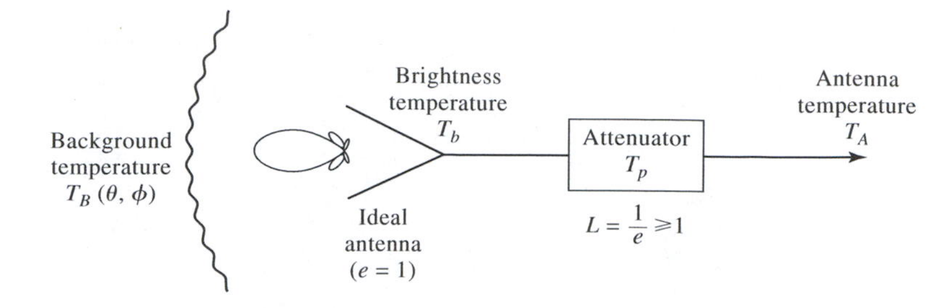

FIGURE 4.11 Illustrating the relation of background noise temperature, antenna brightness temperature, and antenna noise temperature. An antenna with dissipative losses is modeled as an ideal antenna followed by an attenuator.

the ratio of output to input power, it is clear that the relation between the radiation efficiency of the antenna and the attenuator loss factor is $L = 1/e_{\text{rad}}$. The brightness temperature seen by the ideal antenna is given by (4.27). The overall noise temperature appearing at the output terminals can be found by adding the brightness temperature seen by the antenna and the equivalent noise temperature of the attenuator as given by (3.69), with both reduced by the loss of the attenuator:

$$T_A = \frac{T_b}{L} + \frac{(L-1)}{L}T_p = e_{\text{rad}}T_b + (1 - e_{\text{rad}})T_p. \tag{4.28}$$

The resulting equivalent temperature T_A is called the *antenna noise temperature*, and is a combination of the external brightness temperature seen by the antenna and the thermal noise generated by the antenna. As with other equivalent noise temperatures, the proper interpretation of T_A is that a matched load at this temperature will produce the same available noise power as does the antenna. Note that this temperature is referenced at the output terminals of the antenna; since an antenna is not a two-port circuit element, it does not make sense to refer its equivalent noise temperature to its "input."

Observe that (4.28) reduces to $T_A = T_b$ for a lossless antenna with $e_{\text{rad}} = 1$. If the efficiency is zero, meaning that the antenna appears as a matched load and does not see the external background noise, then (4.28) reduces to $T_A = T_p$, due to the thermal noise generated by the losses. If an antenna is pointed toward a known background temperature different than T_0, (4.28) can be used to measure radiation efficiency.

Finally, it is important to realize the difference between radiation efficiency and aperture efficiency, and their effects on antenna noise temperature. While radiation efficiency accounts for resistive losses, and thus involves the generation of thermal noise, aperture efficiency does not. Aperture efficiency applies to the loss of directivity in aperture antennas, such as reflectors, lenses, or horns, due to feed spillover or suboptimum aperture excitation, and by itself does not lead to any effect on noise temperature that would not be included through the pattern of the antenna.

EXAMPLE 4.5 ANTENNA NOISE TEMPERATURE

A high-gain antenna has the idealized hemispherical elevation plane pattern shown in Figure 4.12, and is rotationally symmetric in the azimuth plane. If the antenna is facing a region having a background temperature T_B, as given in the figure below, find the antenna noise temperature. Assume the radiation efficiency is 100%.

Solution

Since $e_{\text{rad}} = 1$, (4.28) reduces to $T_A = T_b$. The brightness temperature can be computed from (4.27), after normalizing the directivity to a maximum value

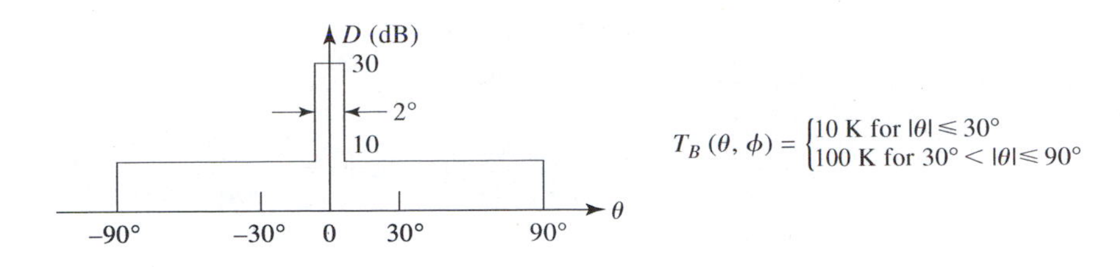

FIGURE 4.12 Idealized antenna pattern and background noise temperature for Example 4.5.

of unity:

$$T_b = \frac{\int_{\phi=0}^{2\pi} \int_{\theta=0}^{\pi} T_B(\theta, \phi) D(\theta, \phi) \sin\theta \, d\theta \, d\phi}{\int_{\phi=0}^{2\pi} \int_{\theta=0}^{\pi} D(\theta, \phi) \sin\theta \, d\theta \, d\phi}$$

$$= \frac{\int_{\theta=0}^{1°} 10 \sin\theta \, d\theta + \int_{\theta=1°}^{30°} 0.1 \sin\theta \, d\theta + \int_{\theta=30°}^{90°} \sin\theta \, d\theta}{\int_{\theta=0}^{1°} \sin\theta \, d\theta + \int_{\theta=1°}^{90°} 0.01 \sin\theta \, d\theta}$$

$$= \frac{-10 \cos\theta \Big|_0^{1°} - 0.1 \cos\theta \Big|_{1°}^{30°} - \cos\theta \Big|_{30°}^{90°}}{-\cos\theta \Big|_0^{1°} - 0.01 \cos\theta \Big|_{1°}^{10°}}$$

$$= \frac{0.00152 + 0.0134 + 0.866}{0.0102} = 86.4 \text{ K}.$$

This example shows that most of the noise power is collected through the sidelobe region of the antenna. ○

G/T

The antenna noise temperature defined in the previous section is a useful figure of merit for a receive antenna because it characterizes the total noise power delivered by the antenna to the input of a receiver. Another useful figure of merit for receive antennas is the G/T *ratio*, defined as

$$G/T(\text{dB}) = 10 \log \frac{G}{T_A} \text{ dB/K}, \tag{4.29}$$

where G is the gain of the antenna, and T_A is the antenna noise temperature. This quantity is important because the signal-to-noise ratio (*SNR*) at the input to a receiver is proportional to G/T_A. To see this, consider the *SNR* at the terminals of a receive antenna calculated using the Friis formula of (4.20) and the antenna temperature of (4.28). The signal power delivered by the receive antenna to a matched receiver input is

$$S_i = \frac{G_r G_t P_t \lambda^2}{(4\pi R)^2},$$

and the noise input to the receiver is $N_i = kT_A B$, where G_r is the receiver antenna gain, G_t is the transmitter antenna gain, P_t is the transmitter power, R is the distance between transmitter and receiver, and B is the radio system bandwidth. Then the *SNR* at the input to the receiver is

$$\frac{S_i}{N_i} = \frac{G_r G_t P_t \lambda^2}{kT_A B(4\pi R)^2} = \left(\frac{G_r}{T_A}\right) \frac{G_t P_t \, G\lambda^2}{kB(4\pi R)^2}, \tag{4.30}$$

showing that the received *SNR* is proportional to G/T of the receive antenna. Observe that only the factor G_r/T_A in (4.30) is controllable at the receiver, as the remaining factors are fixed by the transmitter design and location. Thus, for a fixed transmitter, receiver performance is optimized by maximizing G/T for the receive antenna.

G/T can often be maximized by increasing the gain of the antenna, since this increases the numerator and usually minimizes reception of noise from hot sources at low elevation angles. Of course, higher gain requires a larger and more expensive antenna, and high gain may not be desirable for applications requiring omnidirectional coverage (e.g., cellular telephones or mobile data networks), so often a compromise must be made.

Sensitive receivers used in satellite or point-to-point radio links often have the first amplifier and/or mixer mounted at the antenna in an *outdoor unit* (ODU), with a connecting transmission line at the IF or baseband frequency to an *indoor unit* (IDU). This avoids a long lossy RF transmission line before the first stage of the receiver which, by the cascade noise figure formula of (3.77), provides a significant improvement in the overall noise figure of the system. The amplifier/mixer components used in the outdoor unit are selected to have good noise figure, and are often referred to as the *low-noise block* (LNB) of the receiver. When an LNB is combined with an antenna in this manner, G/T is usually modified to include the combined noise temperature of the LNB and the antenna.

Finally, note that the dimensions given in (4.29) for $10\log(G/T)$ are not actually decibels per degree Kelvin, but this is the nomenclature that is commonly used for this quantity.

EXAMPLE 4.6 ANALYSIS OF DBS SYSTEM

The Direct Broadcast System (DBS) operates at 12.2–12.7 GHz, with a transmit carrier power of 120 W, a transmit antenna gain of 34 dB, an IF bandwidth of 20 MHz, and a worst-case slant angle (30°) distance from the geosynchronous satellite to earth of 39,000 km. The 18″ receiving dish antenna has a gain of 33.5 dB and sees an average background brightness temperature of $T_b = 50$ K, with a receiver LNB having a noise figure of 1.1 dB. The DBS system is shown in Figure 4.13. Find (a) the EIRP of the transmitter, (b) G/T for the receive antenna and LNB system, (c) the received carrier power at the receive antenna terminals, and (d) the *carrier-to-noise ratio* (*CNR*) at the output of the LNB.

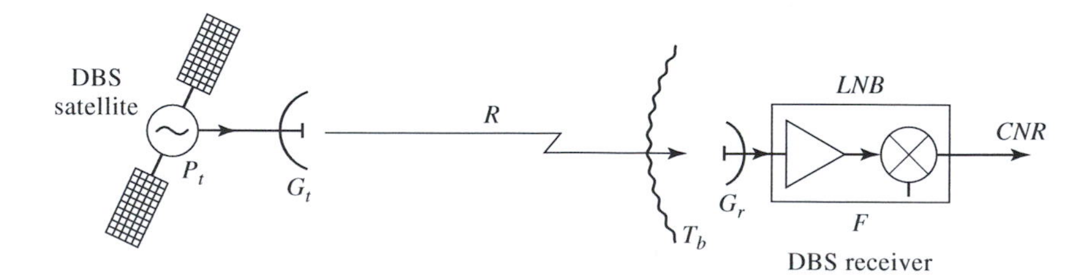

FIGURE 4.13 System diagram of the DBS system for Example 4.6.

Solution
First we convert quantities in dB to numerical values:

$$34 \text{ dB} = 2512$$
$$1.1 \text{ dB} = 1.29$$
$$33.5 \text{ dB} = 2239$$

We will take the operating frequency to be 12.45 GHz, so the wavelength is 0.0241 m.

(a) The EIRP of the transmitter is found using the definition of (4.21):

$$EIRP = P_t G_t = (120)(2512) = 3.01 \times 10^5 \text{ W} = 54.8 \text{ dBm}.$$

(b) To find G/T we first find the noise temperature of the antenna and LNB cascade, referenced at the input of the LNB:

$$T_e = T_A + T_{\text{LNB}} = T_b + (F - 1)T_0 = 50 + (1.29 - 1)(290) = 134 \text{ K}.$$

Then G/T for the antenna and LNB is

$$G/T(\text{dB}) = 10 \log \frac{2239}{134} = 12.2 \text{ dB/K}.$$

(c) The received carrier power is found from the Friis equation:

$$P_r = \frac{P_t G_t G_r \lambda^2}{(4\pi R)^2} = \frac{(3.01 \times 10^5)(2239)(0.0241)^2}{(4\pi)^2(3.9 \times 10^7)^2}$$
$$= 1.63 \times 10^{-12} \text{ W} = -117.9 \text{ dBW}.$$

(d) Then the CNR at the output of the LNB is

$$CNR = \frac{P_r G_{\text{LNB}}}{kT_e B G_{\text{LNB}}} = \frac{1.63 \times 10^{-12}}{(1.38 \times 10^{-23})(134)(20 \times 10^6)} = 44.1 = 16.4 \text{ dB},$$

where G_{LNB}, the gain of the LNB module, cancels in the ratio for the output CNR. A CNR of 16 dB is adequate for good video quality with the error-corrected digital modulation used in the DBS system. ○

4.4 BASIC PRACTICAL ANTENNAS

Although a thorough coverage of antenna design for wireless systems is far beyond the scope of this book, it will be useful for the reader to gain familiarity with some of the practical aspects of the more commonly used antennas. Thus in this section we discuss the operation and basic characteristics of dipole and loop antennas, which are used in a wide variety of wireless communications equipment. The interested reader can consult the references for more detail on antenna design. Reference [6] provides an in-depth treatment of antennas for mobile radio systems. Other types of antennas that are important in the field of wireless communications, such as microstrip antennas, reflector antennas, and arrays, are discussed in references [1]–[3] and [7].

Here we will focus on wire antennas, which include dipoles, loops, and their many variations. Dipole and loop antennas date back to the early work of Hertz, were used extensively in the pioneering wireless work of Marconi, and continue to be used in a large number of wireless systems today. Wire dipoles and loops are relatively small at the

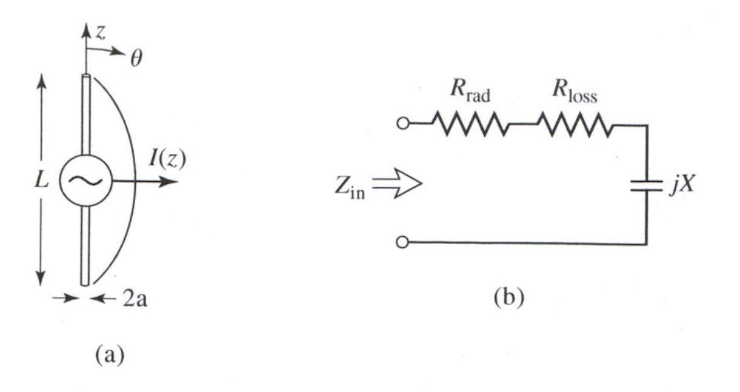

FIGURE 4.14 An electrically small dipole antenna. (a) Geometry and current distribution. (b) Equivalent circuit.

usual wireless frequencies of 900 MHz and above, and are low in cost. Dipoles and loops generally provide omnidirectional coverage, which is desired for mobile systems such as cellular telephones, pagers, and mobile data terminals.

Electrically Small Dipole Antenna

An electrically small dipole is probably the simplest type of radiating element that finds practical use. As shown in Figure 4.14a, the dipole consists of a thin wire of radius a and length L, with a feeding voltage generator located at its midpoint; an equivalent circuit for the dipole is shown in Figure 4.14b. The term "small" means that the dipole is below resonance in length, or $L < \lambda/2$. The distribution of current along the dipole goes to zero at the ends of the dipole, and is maximum at the center. With $L < \lambda/2$, there are no other maxima or minima in the current.

As shown in references [1]–[3], the far field of the small dipole can be derived as

$$E_\theta = \frac{jk_0\eta_0 I_0 L}{4\pi r}\sin\theta e^{-jk_0 r},\ E_\phi = 0, \tag{4.31}$$

where I_0 is the dipole current at the feed point. This expression applies for $0 \le \theta \le \pi$. Thus, a vertical dipole has a main beam in the horizontal plane at $\theta = 90^\circ$ and is vertically polarized. The elevation plane pattern of the dipole is shown in Figure 4.15. The $\sin\theta$ far-field pattern was already studied in Example 4.2, where we found the half-power beamwidth to be 90°, and the directivity to be 1.76 dB.

We can find the total power radiated by the dipole by integrating the average Poynting vector over all space. Thus, using (4.6)–(4.7), we have

$$\begin{aligned} P_{\text{rad}} &= \frac{r^2}{2\eta_0}\int_{\phi=0}^{2\pi}\int_{\theta=0}^{\pi}|E_\theta|^2 \sin\theta\, d\theta\, d\phi \\ &= \frac{I_0^2 L^2 k_0^2 \eta_0}{32\pi^2}\int_{\phi=0}^{2\pi}\int_{\theta=0}^{\pi}\sin^3\theta\, d\theta\, d\phi = \frac{I_0^2 L^2 k_0^2\eta_0}{12\pi} \cong \frac{40 I_0^2 L^2 \pi^2}{\lambda^2}, \end{aligned} \tag{4.32}$$

where the last step follows from using $k_0 = 2\pi/\lambda$ and $\eta_0 = 377 \cong 120\pi\ \Omega$. If we consider the equivalent circuit of the dipole as shown in Figure 4.14b, with the radiation resistance R_{rad} accounting for the total radiated power, then

$$P_{\text{rad}} = \tfrac{1}{2} I_0^2 R_{\text{rad}},$$

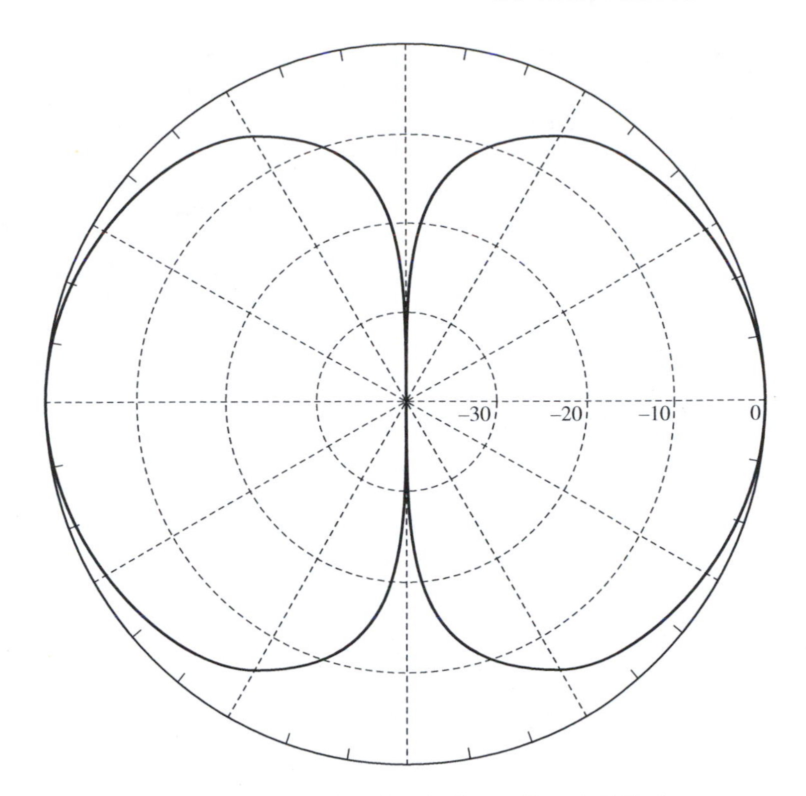

FIGURE 4.15 Elevation plane pattern of an electrically small vertical dipole antenna.

and so comparison with (4.32) gives the radiation resistance of the small dipole as

$$R_{\text{rad}} = 20\pi^2 \left(\frac{L}{\lambda}\right)^2 \ \Omega. \tag{4.33}$$

More advanced techniques can be used to find the reactive part of the dipole input impedance, which is

$$X = \frac{-60\lambda}{\pi L}\left[\ln\left(\frac{L}{a}\right) - 1\right] \Omega, \tag{4.34}$$

where a is the dipole radius. Then the input impedance of the dipole is $Z_{\text{in}} = R_{\text{rad}} + jX$. The results of (4.33) and (4.34) show that, for electrically small dipoles, the real part of the input impedance is only a few ohms, or less, while the imaginary part is often several hundred ohms, or more. This makes matching a small dipole to normal system impedances of 50 Ω difficult, and generally requires high-Q matching circuits.

So far we have assumed the dipole to be lossless. The finite conductivity of a realistic dipole, however, introduces a small amount of loss which reduces the radiation efficiency of the antenna. The effective loss resistance of the small dipole can be derived to be

$$R_{\text{loss}} = \frac{R_s L}{6\pi a} \ \Omega, \tag{4.35}$$

where R_s is the surface resistance of the dipole conductor, given by

$$R_s = \sqrt{\frac{\omega\mu_0}{2\sigma}} \ \Omega. \tag{4.36}$$

In (4.36), $\mu_0 = 4\pi \times 10^{-7}$ H/m, and σ is the conductivity of the metal, in S/m. Once we have the radiation and loss resistances, the radiation efficiency of the small dipole can be found from (4.10) as

$$e_{\text{rad}} = \frac{P_{\text{rad}}}{P_{\text{rad}} + P_{\text{loss}}} = \frac{R_{\text{rad}}}{R_{\text{rad}} + R_{\text{loss}}}. \tag{4.37}$$

EXAMPLE 4.7 IMPEDANCE AND EFFICIENCY OF A DIPOLE ANTENNA

A dipole antenna for a paging system operates at 930 MHz. If the dipole is 3.0 cm long with a radius of 0.01 cm, find the input impedance of the dipole, and its radiation efficiency. Assume a copper conductor.

Solution

The conductivity of copper is 5.8×10^7 S/m [4]. So from (4.36) the surface resistance is

$$R_s = \sqrt{\frac{\omega\mu_0}{2\sigma}} = \sqrt{\frac{(2\pi)(930 \times 10^6)(4\pi \times 10^{-7})}{2(5.8 \times 10^7)}} = 7.96 \times 10^{-3}\ \Omega$$

Then from (4.35) the loss resistance is

$$R_{\text{loss}} = \frac{R_s L}{6\pi a} = \frac{(7.96 \times 10^{-3})(3.0)}{(6\pi)(0.01)} = 0.13\ \Omega.$$

The wavelength at 930 MHz is $\lambda = 32.3$ cm, so the radiation resistance found from (4.33) is

$$R_{\text{rad}} = 20\pi^2\left(\frac{L}{\lambda}\right)^2 = 20\pi^2\left(\frac{3.0}{32.3}\right)^2 = 1.70\ \Omega.$$

and the reactance, from (4.34), is

$$X = \frac{-60\lambda}{\pi L}\left[\ln\left(\frac{L}{a}\right) - 1\right] = \frac{-60(32.3)}{\pi(3.0)}\left[\ln\left(\frac{3.0}{0.01}\right) - 1\right] = -967\ \Omega.$$

Then the radiation efficiency is

$$e_{\text{rad}} = \frac{R_{\text{rad}}}{R_{\text{rad}} + R_{\text{loss}}} = \frac{1.70}{1.70 + 0.13} = 93\%.$$

○

Half-Wave Dipole Antenna

As the length of the dipole increases, its resistance increases and its reactance becomes less capacitive. When the length is approximately $\lambda/2$, the reactance becomes zero, and the dipole is said to be *resonant*. The input impedance of a resonant dipole is about $Z_{\text{in}} = 72 + j0\ \Omega$, which is much easier to match to a transmitter or receiver.

The current distribution on a half-wave dipole has approximately a half-cosine shape, with the current zero at the ends of the dipole and a maximum at the center. As derived in references [1]–[3], the far-field pattern of the half-wave dipole is

$$E_\theta = V_0 \frac{\cos\left(\frac{\pi}{2}\cos\theta\right)}{\sin\theta}\frac{e^{-jk_0 r}}{r}, \quad E_\phi = 0. \tag{4.38}$$

This pattern is similar to the $\sin\theta$ pattern of the small dipole, but is slightly narrower. Its maximum occurs at $\theta = 90^\circ$, and the half-power beamwidth is 78°. The directivity is 2.2 dB.

The loss resistance of a half-wave dipole is comparable to that of the small dipole given in (4.35), but because the radiation resistance is much greater, the efficiency of a half-wave dipole is usually close to 100%.

Monopole Antenna

By image theory, a horizontal ground plane can be placed at the midpoint of a vertical dipole antenna without altering the fields of the dipole. This forms a *monopole* antenna, the evolution of which is shown in Figure 4.16.

Figure 4.16a shows a dipole antenna, with a feed voltage V_0 and feed current I_0. Also shown is the elevation plane pattern of the dipole. If the feed generator is bisected into two series-connected generators each having voltage $V_0/2$, as shown in Figure 4.16b, then a symmetry plane exists through the middle of the antenna, as indicated by the dotted line. In fact, since the tangential electric field is zero on this midplane, an infinitely large perfectly conducting horizontal ground plane can be inserted at this position, without changing the fields anywhere in the problem. Once the ground plane is in place, the bottom generator and dipole arm can be removed to leave a *monopole antenna*, as shown in Figure 4.16c. In this case, the fields are zero below the ground plane. The monopole antenna can be conveniently fed from a coaxial cable from below the ground plane.

The radiated fields of the dipole and the equivalent monopole are identical in the hemisphere above the ground plane, but the fields are zero below the ground plane of the monopole. Thus, for an electrically small monopole (height $< \lambda/4$), the far field can be expressed as

$$E_\theta = \frac{jk_0\eta_0 I_0 L}{4\pi r}\sin\theta e^{-jk_0 r}, \quad E_\phi = 0, \tag{4.39}$$

for $0 \leq \theta \leq \pi/2$. This field has the same maximum value as the field of (4.31) for the small dipole, but the total radiated power will be half of that for the small dipole. Therefore the directivity of the electrically small monopole will be twice the directivity of the dipole, or

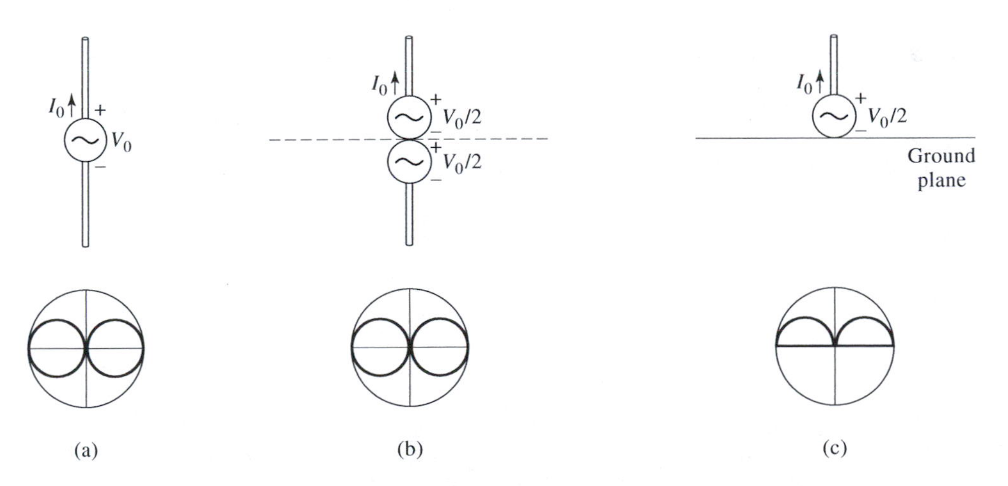

FIGURE 4.16 Evolution of a monopole antenna from a dipole. (a) Dipole antenna and radiation pattern. (b) Dipole antenna with series generators and symmetry plane. (c) Monopole antenna mounted on a ground plane.

$D = 1.76 + 3$ dB $= 4.76$ dB. Similarly, the directivity of a $\lambda/4$ monopole will be twice the directivity of a half-wave dipole, or 5.2 dB.

The impedance of the monopole is also different from the impedance of the corresponding dipole. The relation can be determined from Figure 4.16, which indicates that the terminal current is the same for each case, but the terminal voltage of the monopole is half the terminal voltage of the dipole. Therefore the input impedance of the monopole is half the impedance of the dipole:

$$Z_{\text{monopole}} = \tfrac{1}{2} Z_{\text{dipole}}. \tag{4.40}$$

Thus the input impedance of a resonant $\lambda/4$ monopole is about 36 Ω.

Sleeve Monopole Antenna

There are dozens of variations of dipole and monopole antennas that have been developed over the years. Some of these offer improvements in performance, while others are of interest for specialized applications. One of the more important variations for wireless systems is the *sleeve monopole*.

A problem with the basic dipole antenna is that it requires a balanced feed at the middle of the element. This can be done with two-wire "twin-lead" line, as in the case of broadcast television or FM radio antennas, but is not very convenient for most mobile wireless applications. A monopole antenna can be used with an unbalanced coax feed line, but requires a relatively large ground plane. This may be reasonable for vehicle-mounted antennas, but is clearly impractical for handheld wireless equipment.

The sleeve monopole, shown in Figure 4.17, solves both of these problems. The sleeve monopole consists of a coax feed line that passes through a metallic sleeve without direct contact. The center conductor of the coax is extended past the top of the sleeve to form a monopole element. The overall length of the antenna is $L + \ell \cong \lambda/4$. The effective feed point is located at the top end of the outer coax conductor and generates currents on the sleeve as well as the monopole element. The current distribution is zero at the top of the monopole and at the bottom of the sleeve, and is similar in form to the current on a monopole antenna. Thus the pattern, beamwidth, and directivity of the sleeve antenna are comparable to those of a monopole element. Because of these features, and the fact that they are inexpensive

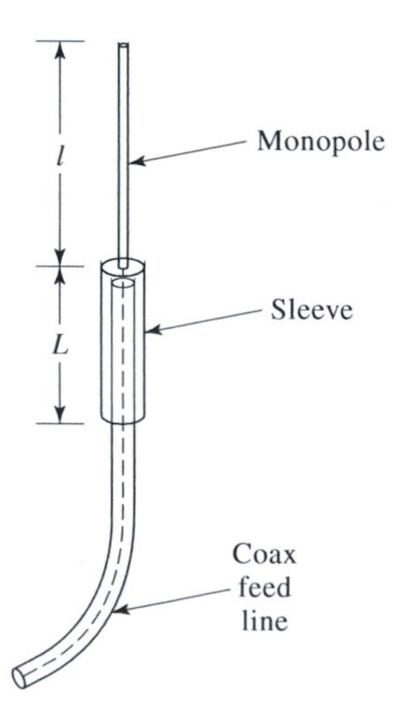

FIGURE 4.17 Geometry of a sleeve monopole antenna.

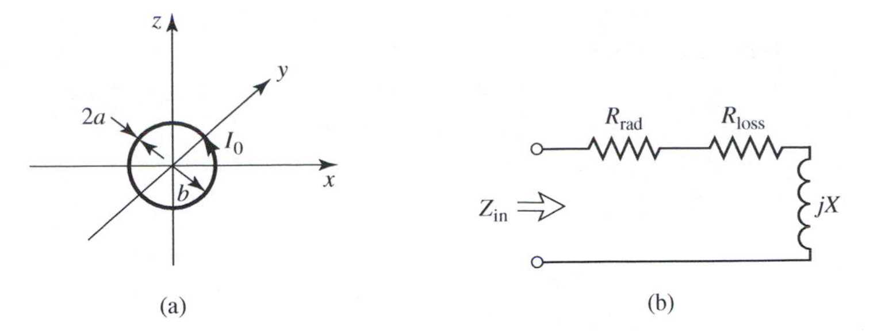

FIGURE 4.18 An electrically small loop antenna. (a) Geometry. (b) Equivalent circuit.

and compact, commercially available sleeve antennas are used in a wide variety of portable wireless systems.

Electrically Small Loop Antenna

Wire loop antennas also offer the advantages of low-cost and low-gain, and are therefore useful in many portable wireless devices. Loop antennas have an advantage over dipole-type radiators for applications where the receiver is held close to the body, in that the performance of a loop element is not degraded as much due to the high conductivity of the body [6].

The dual to the small dipole studied in the previous section is the small loop, shown in Figure 4.18a. It consists of a wire loop of radius b, carrying a uniform current I_0. The radius of the wire is a. The far field of the loop can be shown to be

$$E_\theta = 0, \qquad E_\phi = V_0 \sin\theta \frac{e^{-jk_0 r}}{r}. \tag{4.41}$$

Thus the far-field pattern of the loop antenna is the same as the pattern of the small dipole antenna (plotted in Figure 4.15), implying that the main beam occurs at $\theta = 90^\circ$, with a half-power beamwidth of 90°, and a directivity of 1.76 dB. The polarization, however, is in the ϕ direction, meaning that the loop antenna is horizontally polarized, in contrast to the vertically polarized dipole.

Using the same method as used to derive (4.33), the radiation resistance of the loop is

$$R_{\text{rad}} = 31,200\left(\frac{\pi b^2}{\lambda^2}\right)^2 \Omega. \tag{4.42}$$

This result generally gives good results for loops with $b < \lambda/10$. The reactance of the loop is

$$X = k_0\eta_0\left[\ln\left(\frac{8b}{a}\right) - 1.75\right] \Omega, \tag{4.43}$$

which is seen to be inductive, in contrast to the capacitive reactance of the small dipole. As with the small dipole, the radiation resistance of the small loop is typically only a few ohms, or less, while the reactance is several thousand ohms, or more. The equivalent circuit of the small loop antenna is shown in Figure 4.18b.

The loss resistance of the small loop antenna can be derived as

$$R_{\text{loss}} = \frac{b}{a} R_s \,\Omega, \tag{4.44}$$

where R_s is the surface resistance given by (4.36). The radiation efficiency can then be found from (4.37).

The radiation resistance of the loop antenna can be increased by using more turns; if N turns are used, the radiation resistance of (4.42) will increase by the factor N^2, although loss resistance will also increase. Loop antennas are sometimes made square or rectangular in shape. In general, an arbitrarily shaped small loop antenna of perimeter length L with an enclosed area S has radiation and loss resistances given by,

$$R_{\text{rad}} = 31{,}200\left(\frac{S}{\lambda^2}\right)^2 \ \Omega, \tag{4.45}$$

$$R_{\text{loss}} = \frac{L}{2\pi a} R_s \ \Omega. \tag{4.46}$$

EXAMPLE 4.8 IMPEDANCE AND EFFICIENCY OF A LOOP ANTENNA

A loop antenna is used for the paging system of Example 4.7, operating at 930 MHz. The loop has a diameter of 3.0 cm and is made from copper wire with a diameter of 0.02 cm. Find the input impedance of the loop, including loss effects, and the radiation efficiency.

Solution

The wavelength and surface resistance are the same as in Example 4.7. The radiation resistance is computed using (4.42) as

$$R_{\text{rad}} = 31{,}200\left(\frac{\pi b^2}{\lambda^2}\right)^2 = 31{,}200\left(\frac{\pi(0.015)^2}{(0.323)^2}\right)^2 = 1.43\ \Omega.$$

The reactance is computed from (4.43):

$$X = k_0\eta_0\left[\ln\left(\frac{8b}{a}\right) - 1.75\right] = \frac{2\pi}{0.323}(377)\left[\ln\left(\frac{8(3)}{0.02}\right) - 1.75\right] = 39{,}212\ \Omega.$$

The loss resistance is computed from (4.44) as

$$R_{\text{loss}} = \frac{b}{a}R_s = \frac{3}{0.02}(7.96 \times 10^{-3}) = 1.19\ \Omega.$$

So the input impedance of the loop is $Z_{\text{in}} = R_{\text{rad}} + R_{\text{loss}} + jX = 2.62 + j39{,}212\ \Omega$. The efficiency is

$$e_{\text{rad}} = \frac{R_{\text{rad}}}{R_{\text{rad}} + R_{\text{loss}}} = \frac{1.43}{1.43 + 1.19} = 55\%,$$

which is considerably lower than the efficiency of the dipole of Example 4.7. ○

4.5 PROPAGATION

Between the transmit and receive antennas of a radio communication channel the propagating wave is subject to a variety of effects that can alter its amplitude, phase, or frequency. Such *propagation effects* include:

- reflection (from the ground or large objects)
- diffraction (from edges and corners of terrain or buildings)

- scattering (from foliage or other small objects)
- attenuation (from rain or the atmosphere)
- Doppler (from moving users)

This list covers the most important effects for frequencies above 500 MHz, and so includes most wireless systems in which we are interested. At frequencies below about 100 MHz, however, a number of other propagation effects can be important, such as ground surface waves, atmospheric ducting, and ionospheric reflection.

Propagation effects generally have the effect of reducing the received signal power, and thus limit either the usable range or maximum data rate of a wireless system. We will discuss in some detail ground reflections, and make some qualitative comments on the subjects of path loss and attenuation. This material will be followed by the important related topic of multipath fading in the following section. We refer the reader to the literature for a more detailed discussion of radio wave propagation.

Free-space Propagation

In Section 4.2 we derived the Friis equation in (4.20), which shows that received power decreases as $1/R^2$ with distance from the transmitter. This *path loss* strictly applies only to propagation in free space, where there is no reflection, scattering, or diffraction along the path between transmitter and receiver. In practice, the Friis formula can be used when there is essentially a single *line-of-sight* (LOS) path between the transmitter and receiver. This usually implies that at least one of the link antennas has a narrow beamwidth (high gain), as in the case of point-to-point radio links, satellite-to-satellite links, and earth-to-satellite links. Figure 4.19a illustrates a typical situation where the Friis formula can be used with good accuracy.

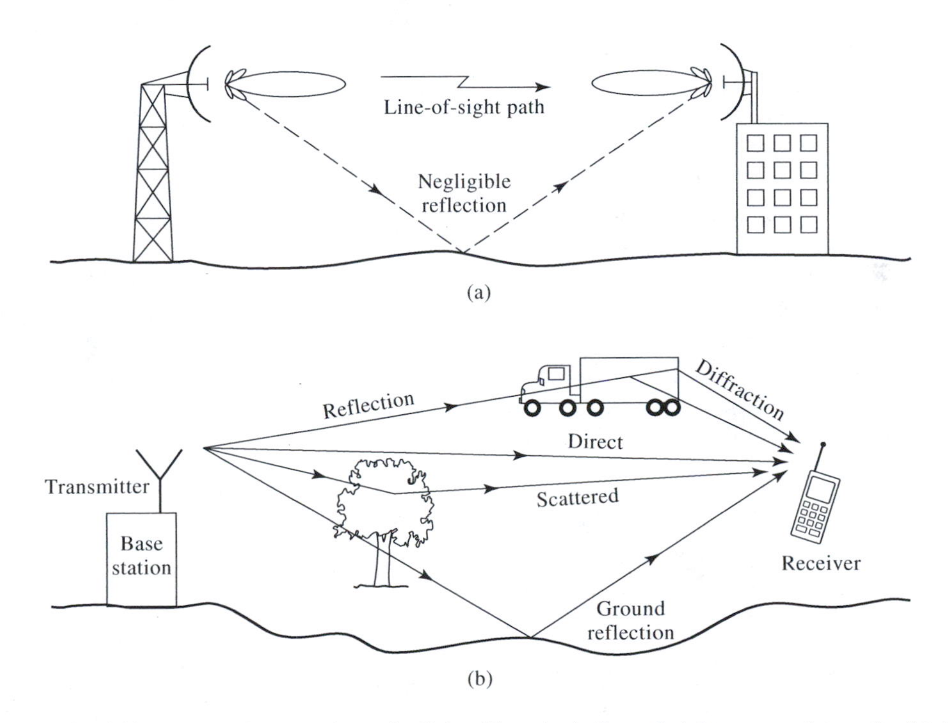

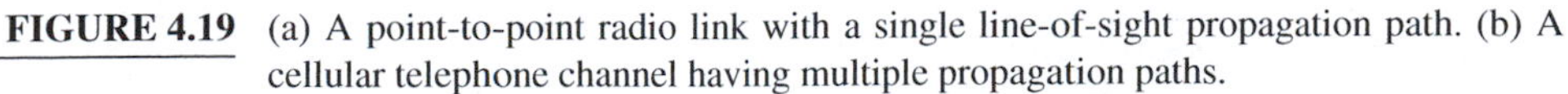

FIGURE 4.19 (a) A point-to-point radio link with a single line-of-sight propagation path. (b) A cellular telephone channel having multiple propagation paths.

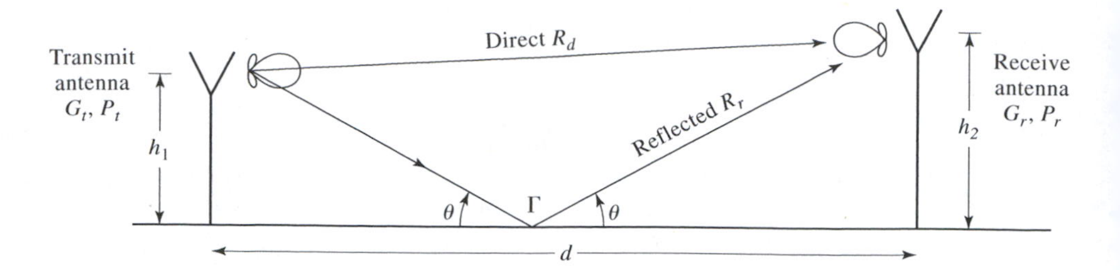

FIGURE 4.20 Geometry of a radio link with a direct propagation path and a ground reflection path.

In many practical cases reflections, scattering, and diffractions create more than a single path between transmitter and receiver, as illustrated in Figure 4.19b. *Multipath* propagation is particularly likely when the antennas have broad beams (low gain) and are in close proximity to the ground or other large reflecting structures such as buildings, vehicles, or heavy foliage. In the worst case, there may be no direct line-of-sight path between the transmitter and receiver; this is a common situation for cellular and PCS telephone users located in a building or vehicle. Communication is still possible in the presence of multiple propagation paths, even in the absence of a LOS path, but the total signal voltage at the receiver will experience varying degrees of destructive or constructive interference due to the variable phase delays that occur along different paths. The Friis formula cannot be used in these cases.

Ground Reflections

Determining the received signal power in a multipath environment is usually a very difficult problem, but we can study the essential effects of reflections by considering an LOS path with a single reflected signal. This model is useful for ground reflections, which frequently occur in practice, as well as reflections from buildings, vehicles, and other large structures.

The geometry of a propagation path with a single ground reflection is shown in Figure 4.20. The transmit antenna is located at height h_1 above a flat ground, and the receive antenna is located at height h_2. The distance between the transmitter and receiver is d, which, in practice, is much greater than h_1 or h_2. Besides the direct LOS path from the transmit antenna to the receive antenna, a reflected path also exists. By Snell's law, the incident wave is *specularly reflected* from the ground, meaning that the angle of incidence, θ, is equal to the angle of reflection.

Let the transmit antenna gain and power be G_t and P_t, and the receive antenna gain and power be G_r and P_r, respectively. Then from (4.1) and the Friis formula of (4.20), we can write the received voltage due to the direct wave as

$$V_d = \sqrt{P_r Z_0} e^{-jk_0 R_d} = \frac{\sqrt{G_t P_t G_r Z_0 \lambda^2}}{4\pi R_d} e^{-jk_0 R_d} = C \frac{e^{-jk_0 R_d}}{R_d}, \tag{4.47}$$

where R_d is the path length of the direct ray, and Z_0 is the receiver load impedance. Note that we have retained the phase of the received voltage, as given by (4.1), because the total voltage at the receiver will be the phasor sum of the voltages due to the direct and reflected signals. The constant C is used to simplify our notation and is defined as

$$C = \frac{\lambda}{4\pi} \sqrt{G_t G_r P_t Z_0}. \tag{4.48}$$

Assuming that $d \gg h_1$ and $d \gg h_2$, the direct path distance can be approximated using the Taylor expansion that $\sqrt{1+x} \cong 1 + x/2$:

$$R_d = \sqrt{d^2 + (h_2 - h_1)^2} \cong d + \frac{(h_2 - h_1)^2}{2d}. \tag{4.49}$$

Similarly, the receive voltage due to the reflected wave can be written as

$$V_r = C\Gamma \frac{e^{-jk_0 R_r}}{R_r}. \tag{4.50}$$

R_r is the path length of the reflected wave, which can be approximated as

$$R_r = \sqrt{d^2 + (h_2 + h_1)^2} \cong d + \frac{(h_2 + h_1)^2}{2d}. \tag{4.51}$$

In (4.50), Γ is the plane wave voltage reflection coefficient of the ground, which is a function of the incident angle, the frequency, the polarization of the incident wave, and the dielectric constant and conductivity of the ground [4]. Typical values for the material properties of the ground are $\varepsilon_r = 15$ and $\sigma = 0.005$ S/m at 100 MHz [8]. For vertical polarization, at angles of incidence close to grazing (small θ), the reflection coefficient will be close to -1. Thus we will make the approximation that $\Gamma = -1$.

We also assume that the gain patterns of the transmit and receive antennas are equal for the direct and reflected rays. This is a good assumption, unless the antenna gains are very high, because in practice the angle θ is very small.

Combining the direct and reflected voltages of (4.47) and (4.50) gives the total received voltage as

$$V = V_d + V_r = C\left[\frac{e^{-jk_0 R_d}}{R_d} - \frac{e^{-jk_0 R_r}}{R_r}\right]. \tag{4.52}$$

For the amplitude variation of (4.52) we can assume that $R_r \cong R_d \cong d$ with negligible error because $d \gg h_1$ and $d \gg h_2$. The phase terms, however, must be treated with greater accuracy because of the modulo 2π periodicity of the complex exponential. Thus (4.52) reduces to

$$V = C\frac{e^{-jk_0 R_d}}{R_d}\left[1 - e^{-jk_0(R_d - R_r)}\right] = C\frac{e^{-jk_0 R_d}}{R_d}[1 - e^{-2jk_0 h_1 h_2/d}]. \tag{4.53}$$

The magnitude of the last factor in (4.53) is called the *path gain factor*, F:

$$F = |1 - e^{-2jk_0 h_1 h_2/d}| = 2\left|\sin\left(\frac{k_0 h_1 h_2}{d}\right)\right|. \tag{4.54}$$

Observe that $0 \leq F \leq 2$, so that the received voltage may be doubled (power quadrupled) when the two signals are in phase, or reduced to zero in the case of complete destructive interference.

For a fixed transmitter height h_1 and distance d, the path gain factor can be plotted as a function of the receive antenna height h_1. This is usually done by defining the angle ψ as the elevation angle of the receive antenna as seen at the transmitter:

$$\tan\psi = \frac{h_2}{d}. \tag{4.55}$$

Then (4.54) can be written as

$$F = 2|\sin(k_0 h_1 \tan\psi)|. \tag{4.56}$$

In a practical situation the frequency (k_0) and transmit antenna height h_1 are usually fixed, and the receive antenna distance d and height h_2 are variable. The path gain factor F can then be plotted versus ψ as a *coverage diagram*, to give the relative field strength versus position of the receive antenna.

EXAMPLE 4.9 EFFECT OF GROUND REFLECTION

The height of a cellular telephone transmit antenna operating at 1800 MHz is 8.33 m. If the distance to the receiver is 1 km, find the smallest receiver antenna height that will maximize the receive signal voltage.

Solution

At 1800 MHz the wavelength is 0.1667 m, so $h_1 = 8.33\text{ m} = 50\lambda$. The path gain factor has a maximum when the argument of the sin function is $\pi/2$, $3\pi/2$, etc:

$$\frac{k_0 h_1 h_2}{d} = \frac{2\pi(50\lambda)h_2}{1000\lambda} = \frac{\pi h_2}{10} = \frac{\pi}{2} + n\pi, \quad \text{for } n = 0, 1, 2, \ldots.$$

So the minimum height for a maximum path gain is

$$h_2 = \frac{\pi}{2}\frac{10}{\pi} = 5\text{ m}.$$

○

Path Loss for Ground Reflections

In contrast to the $1/R^2$ path loss of the Friis formula, the received signal power in the presence of a ground reflection varies according to the path gain factor of (4.56), and is not simply a function of the separation distance. In the limiting case of very large distances, however, the path gain factor can be simplified by using the Taylor series approximation that $\sin x \cong x$:

$$F = 2\left|\sin\frac{k_0 h_1 h_2}{d}\right| \cong \frac{2k_0 h_1 h_2}{d}.$$

Applying this result to (4.53) gives the received voltage as

$$V = \frac{2Ck_0 h_1 h_2}{d}\frac{e^{-jk_0 R_d}}{R_d} \cong 2Ck_0 h_1 h_2 \frac{e^{-jk_0 R_d}}{R_d^2}. \tag{4.57}$$

Since the signal voltage decreases as $1/R^2$, the received signal power will decrease as $1/R^4$:

$$P_r = \frac{|V|^2}{Z_0} = \frac{4C^2 k_0^2 h_1^2 h_2^2}{Z_0 R_d^4} = \frac{P_t G_t G_r h_1^2 h_2^2}{R_d^4}, \tag{4.58}$$

where (4.48) was used for the constant C. This result applies when $k_0 h_1 h_2/d < 0.3$ rad, or for

$$d > \frac{k_0 h_1 h_2}{0.3} \cong \frac{20 h_1 h_2}{\lambda}. \tag{4.59}$$

Realistic Path Loss

We have seen that the Friis formula leads to a path loss factor for free-space propagation of $1/R^2$, for the decrease in received power with distance. When a ground reflection is

TABLE 4.1 Typical Path Loss Exponents for Realistic Cellular Environments

Environment	Path Loss Exponent
Free space	2
Urban	2.7–3.5
Shadowed urban	3–5
In-building LOS	1.6–1.8
In-building shadowed	4–6
Factory shadowed	2–3
Retail store	2.2
Office—soft partitions	2.4

present, the path loss can be as much as $1/R^4$ in the worst case. In practice, actual propagation paths will be more complicated than either of these idealized cases due to the possibility of multiple reflections, diffractions, and scatterings. The resulting path loss can be expressed as $1/R^n$, where the exponent may vary from $n = 2$ to 5 or 6, in the case of many lossy obstructions. The complexity of a realistic propagation environment usually requires that path loss be measured, rather than calculated from first principles. Table 4.1 gives some typical path loss exponents that have been measured; this and further data can be found in reference [8].

Attenuation

Attenuation is the decrease in signal power due to losses in the propagation path. These may be due to the atmosphere, precipitation, walls, or ceilings.

Figure 4.21 shows the attenuation of the atmosphere versus frequency, for two different altitudes. Note that atmospheric attenuation is generally negligible at frequencies below about 10 GHz, and exhibits peaks at frequencies near 22 GHz, 60 GHz, and higher

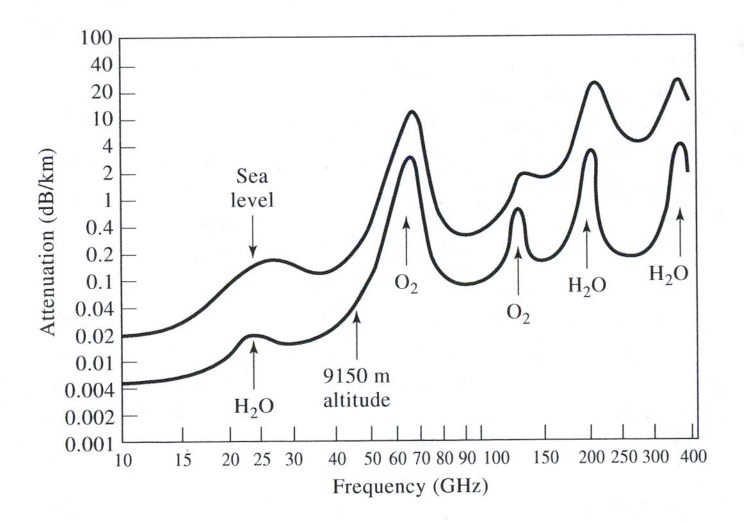

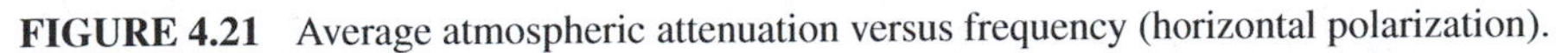

FIGURE 4.21 Average atmospheric attenuation versus frequency (horizontal polarization).

TABLE 4.2 Attenuation of Some Common Building Materials

Material	Frequency	Loss (dB)
Concrete block wall	1300 MHz	13
Sheetrock (2 × 3/8″)	9.6 GHz	2
Plywood (2 × 3/4″)	9.6 GHz	4
Concrete wall	1300 MHz	8–15
Chain link fence	1300 MHz	5–12
Loss between floors	1300 MHz	20–30
Corner in corridor	1300 MHz	10–15

frequencies. As in the case of atmospheric noise temperature shown in Figure 4.10, these peaks are due to resonances of molecular water and oxygen. Also observe that attenuation decreases with altitude, due to a decreased air and water vapor density. The effective atmospheric path length for earth-to-satellite links is about 4–5 km.

Precipitation, in the form of rain or snow, can greatly increase atmospheric attenuation at microwave frequencies. For example, at 10 GHz the attenuation for a rain rate of 5 mm/h is about 0.1 dB/km, but for a rate of 100 mm/h this increases to 3 dB/km.

Many wireless applications involve propagation into or within buildings and may depend on propagation through walls or ceilings. Different building materials lead to different attenuation rates which can be easily measured, but practical situations are often complicated by multiple reflections and multipath signals. Table 4.2 lists attenuation for some common materials; this and related data can be found in reference [8].

4.6 FADING

In many practical wireless systems there may be no direct line-of-sight path between the transmitter and receiver. A typical example is a cellular telephone system, where the signal between the base station and a mobile user may be blocked by buildings, vehicles, or similar obstructions. In such cases multiple paths between transmitter and receiver may exist due to scattering, reflection, or diffraction, and thus communication may still be possible. Because such multipath signals arrive at the receiver with different amplitudes and phases, the net received signal voltage can vary due to destructive interference. Because of long path lengths, rapid variations in received amplitude can occur over distances as short as $\lambda/2$. If a mobile user is moving, or if some of the scatterers/reflectors are moving, these variations may occur over relatively short time intervals. This effect is known as *fading*, and is one of the most significant factors affecting the performance of wireless systems [9]–[10]. Besides large amplitude variations, fading can also involve frequency modulation due to Doppler effects, and variable signal delays caused by time-varying propagation paths.

Fading is referred to as a *small-scale* effect, since it involves large variations in amplitude over small distances or time intervals. This is in contrast to *large-scale* propagation effects caused by $1/R^n$ path losses, or blockage or other effects that produce slower variations over relatively large distances. These effects are illustrated in Figure 4.22, which shows received signal power versus distance from the transmitter. Fading effects are observed as the rapid variations in amplitude versus distance, which are seen to be as severe as 20–30 dB in some cases. The large-scale variation of the received signal power is found by averaging the small-scale variations to give the dashed curve. In an ideal free-space environment, or one with a single ground reflection, the large-scale variation would monotonically decrease

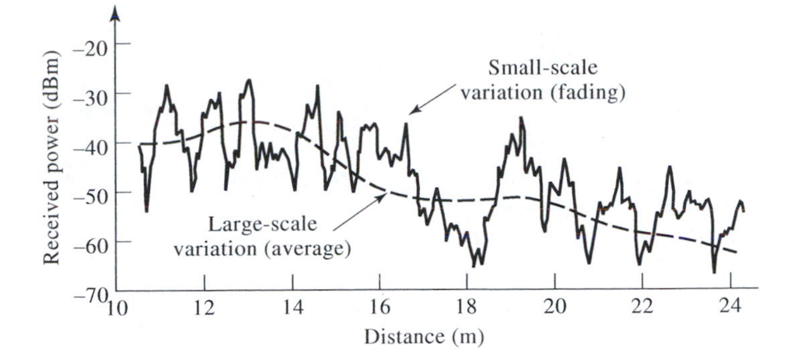

FIGURE 4.22 Typical received signal power versus distance from transmitter, showing small-scale fading effects and large-scale variations.

with distance according to $1/R^2$ or $1/R^4$. But in more realistic situations the large-scale variation may be nonmonotonic due to varying combinations of constructive and destructive interference.

There are several types of fading that occur in practice, depending on whether the signal bandwidth is greater or lesser than the channel bandwidth, and whether Doppler shift is appreciable or not. Here we consider only the case of *flat fading*, defined as the case for which Doppler effects are negligible and the signal bandwidth is less than the channel bandwidth. This means that the transfer function of the propagation channel has a constant amplitude ("flat") and a linear phase variation versus frequency over the bandwidth of the signal. We will show next that the statistics of the received signal amplitude in this case satisfy a Rayleigh probability distribution function. This is probably the most common type of fading, but nevertheless still represents an idealized view of multipath fading.

Rayleigh Fading

To study the effect of multipath fading we assume a situation with a large number of signal paths between the transmitter and receiver, but no direct line-of-sight path, and find the statistics of the received voltage envelope. Each received voltage component may have an independent amplitude and phase, so the total received voltage can be expressed as

$$V(t) = \sum_{n=1}^{N} V_n \cos(\omega t + \phi_n), \tag{4.60}$$

where V_n and ϕ_n are independent random variables describing the amplitude and phase of the nth received signal component, and ω is the RF carrier frequency that is common to each component. If we have a large number of multipath components, the amplitude can be considered to be gaussian distributed. This follows from the central limit theorem, which states that the sum of a large number of independent random variables tends to a gaussian distribution. The phases will be uniformly distributed between 0 and 2π.

Since $\cos(\omega t + \phi_n) = \cos\omega t \cos\phi_n - \sin\omega t \sin\phi_n$, we can rewrite (4.60) in terms of the in-phase and quadrature components of the carrier waveform:

$$V(t) = x(t)\cos\omega t - y(t)\sin\omega t, \tag{4.61}$$

with

$$x(t) = \sum_{n=1}^{N} V_n \cos\phi_n; \; y(t) = \sum_{n=1}^{N} V_n \sin\phi_n. \tag{4.62}$$

The functions $x(t)$ and $y(t)$ are gaussian random variables with zero mean, since V_n has zero mean, is independent of ϕ_n, and $E\{\cos\phi_n\} = E\{\sin\phi_n\} = 0$. So we can write the probability distribution functions of $x(t)$ and $y(t)$ as

$$f_x(x) = \frac{e^{-x^2/2\sigma^2}}{\sqrt{2\pi\sigma^2}}, \tag{4.63a}$$

$$f_y(y) = \frac{e^{-y^2/2\sigma^2}}{\sqrt{2\pi\sigma^2}}. \tag{4.63b}$$

Now express the quadrature form of $V(t)$ in (4.61) in polar form, as a magnitude and phase:

$$V(t) = r(t)\cos[\omega t + \theta(t)], \tag{4.64}$$

where the magnitude $r(t)$ is defined as

$$r^2(t) = x^2(t) + y^2(t), \tag{4.65}$$

and the phase $\theta(t)$ satisfies

$$\tan\theta(t) = \frac{y(t)}{x(t)}. \tag{4.66}$$

The magnitude $r(t)$ is the envelope of the received signal waveform, as shown in Figure 4.23. To find the statistics of the envelope voltage $r(t)$, we first find the joint pdf of $x(t)$ and $y(t)$, which is simply the product of f_x and f_y since $x(t)$ and $y(t)$ are independent:

$$f_{xy}(x, y) = f_x(x) f_y(y) = \frac{e^{-(x^2+y^2)/2\sigma^2}}{2\pi\sigma^2}. \tag{4.67}$$

Now convert this pdf to the r, θ coordinates using (4.65):

$$f_{r\theta}(r, \theta) = \frac{e^{-r^2/2\sigma^2}}{2\pi\sigma^2}, \quad \text{for } 0 \le r < \infty \quad \text{and} \quad 0 \le \theta < 2\pi. \tag{4.68}$$

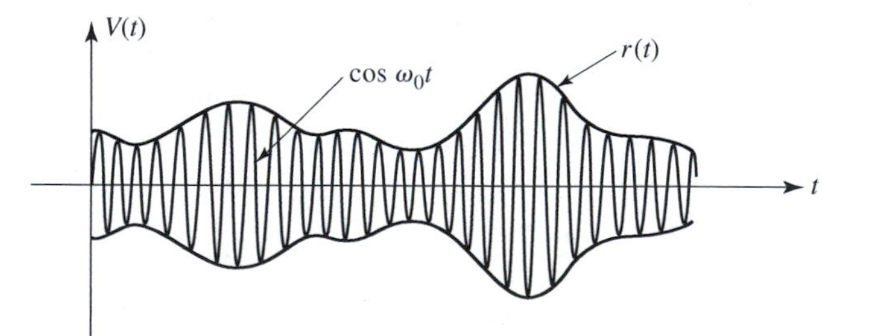

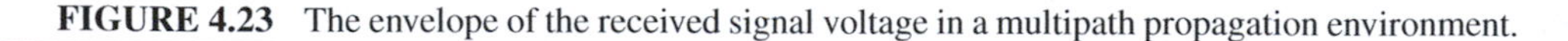

FIGURE 4.23 The envelope of the received signal voltage in a multipath propagation environment.

Finally, the pdf for $r(t)$ can be found by integrating over the θ variable. Since $dx\,dy = r dr d\theta$, the differential element for the θ integration is $r d\theta$:

$$f_r(r) = \int_{\theta=0}^{2\pi} f_{r\theta}(r,\theta) r\, d\theta = \frac{re^{-r^2/2\sigma^2}}{\sigma^2}, \quad \text{for } 0 \le r < \infty. \tag{4.69}$$

This is the Rayleigh probability distribution function, and it gives the statistics of the received signal envelope in the presence of flat fading. We will use this result in Chapter 9 to determine the effect of fading on the bit error rates of digital modulation schemes.

REFERENCES

[1] C. A. Balanis, **Antenna Theory: Analysis and Design**, 2nd edition, Wiley, New York, 1997.

[2] W. L. Stutzman and G. A. Thiele, **Antenna Theory and Design**, 2nd edition, Wiley, New York, 1998.

[3] R. E. Collin, **Antennas and Radiowave Propagation**, McGraw-Hill, New York, 1985.

[4] D. M. Pozar, **Microwave Engineering**, 2nd edition, Wiley, New York, 1998.

[5] L. J. Ippolito, R. D. Kaul, and R. G. Wallace, **Propagation Effects Handbook for Satellite Systems Design**, 3rd edition, NASA Publication 1082(03), June 1983.

[6] K. Fujimoto and J. R. James, Eds., **Mobile Antenna Systems Handbook**, Artech House, Dedham, MA, 1994.

[7] D. M. Pozar and D. H. Schaubert, Eds., **Microstrip Antennas: The Analysis and Design of Microstrip Antennas and Arrays**, IEEE Press, New York, 1995.

[8] T. S. Rappaport, **Wireless Communications: Principles and Practice**, Prentice Hall, Englewood Cliffs, NJ, 1996.

[9] W. C. Jakes, **Microwave Mobile Communications**, IEEE Press, Englewood Cliffs, NJ, 1974.

[10] A. Mehrotra, **Cellular Radio Performance Engineering**, Artech House, Dedham, MA, 1994.

PROBLEMS

4.1 A base station antenna operating at 860 MHz consists of an array of three parallel half-wave dipoles spaced 0.45λ apart. Find the far-field distance of the antenna.

4.2 An antenna has a radiation pattern function given by $F_\theta(\theta, \phi) = A \sin\theta \sin\phi$. Find the main beam position, the 3 dB beamwidth, and the directivity (in dB) for this antenna.

4.3 A small loop antenna at the origin of the x-y plane has a far-field pattern given by

$$E_\theta(r,\theta,\phi) = 0$$

$$E_\phi(r,\theta,\phi) = V_0 \sin\theta \frac{e^{-jk_0 r}}{r} \text{ v/m}.$$

Find the main beam position, the 3 dB beamwidth, and the directivity.

4.4 A monopole antenna on a large ground plane has a far-field pattern function given by $F_\theta(\theta,\phi) = A\sin\theta$, for $0 < \theta < \pi/2$. The radiated field is zero for $\pi/2 < \theta < \pi$. Find the directivity (in dB) of this antenna.

4.5 A DBS reflector antenna operating at 12.4 GHz has a diameter of 18″. If the aperture efficiency is 65%, find the directivity.

4.6 A reflector antenna used for a cellular base station backhaul radio link operates at 38 GHz, with a gain of 32 dB. If the aperture efficiency is 60%, find the beamwidth of the antenna if the main beam is assumed to have equal beamwidths in both planes.

4.7 Find an expression for the effective aperture area of the electrically small dipole antenna of Example 4.2. If the dipole operates at 900 MHz and is 3 cm long with a diameter of 0.5 mm, compare the physical cross-sectional area of the dipole with its effective aperture area.

4.8 At a distance of 300 m from an antenna operating at 5.8 GHz, the radiated power density in the main beam is measured to be 7.5×10^{-3} W/m^2. If the input power to the antenna is known to be 85 W, find the gain of the antenna.

4.9 A transmitting antenna has an input current of 0.04 A (peak) and radiates a far-zone electric field given by

$$E_\theta(r, \theta, \phi) = \frac{e^{-jk_0 r}}{r} \begin{cases} 21.8 & \text{for} \quad 0 < \theta < 12^\circ \\ 0 & \text{for} \quad 12^\circ < \theta < 180^\circ \end{cases} \text{ v/m (peak).}$$

Find the gain of the antenna, in dB, and the radiation resistance, if the efficiency of the antenna is known to be 80%.

4.10 A wireless local area network operating at 2.4 GHz has an average transmit EIRP of 22 dBm. The receiver is located 500 m away, with an antenna gain of 3 dB. If the receiver appears as a 50 Ω load to the receive antenna, find the rms receiver input voltage.

4.11 A cellular base station is to be connected to its Mobile Telephone Switching Office (MTSO) located 5 km away. Two possibilities are to be evaluated: (1) a radio link operating at 28 GHz, with $G_t = G_r = 25$ dB; and (2) a wired link using coaxial line having an attenuation of 0.05 dB/m, with four 30 dB repeater amplifiers along the line. If the minimum required received power level for both cases is the same, which option will require the smallest transmit power?

4.12 A 28 GHz common-carrier radio link uses a tower-mounted reflector antenna with a gain of 32 dB. If the transmitter power is 3 W, find the minimum distance within the main beam of the antenna for which the U.S.-recommended safe power density limit of 10 mW/cm^2 is not exceeded. How does this distance change for a position within the sidelobe region of the antenna, if we assume a worst-case sidelobe level of 15 dB below the main beam? Are these distances in the far field of the antenna (assuming a round reflector, and an aperture efficiency of 60%)?

4.13 Consider a 900 MHz radio transmitter driving a half-wave dipole, with an input impedance of $Z_A = 72 + j40$ Ω, and a radiation efficiency of 80%. What is the value of the generator impedance required to provide maximum power transfer between the transmitter and the antenna, and the required rms generator voltage if it is desired to radiate a total power of 100 mW?

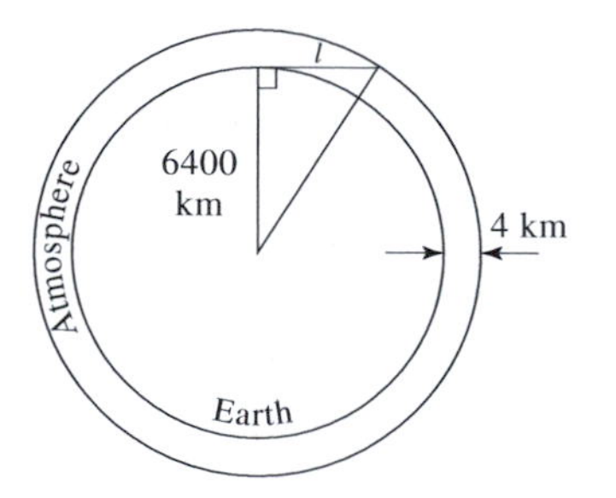

4.14 The atmosphere does not have a definite thickness, since it gradually thins with altitude, with a consequent decrease in attenuation. But if we use a simplified "orange peel" model, and assume that the atmosphere can be approximated by a uniform layer of fixed thickness, we can estimate the background noise temperature seen through the atmosphere. Thus, let the thickness of the atmosphere be 4000 m and find the maximum distance ℓ to the edge of the atmosphere along the horizon, as shown in the figure below (the radius of the earth is 6400 km). Now assume an average atmospheric attenuation of 0.005 dB/km, with a background noise temperature beyond the atmosphere of 4 K, and find the noise temperature seen on earth by treating the cascade of the background noise with the attenuation of the atmosphere. Do this for an ideal antenna pointing toward the zenith, and toward the horizon.

4.15 A key premise in many popular science fiction stories and movies is the idea that radio and TV signals from earth can travel through space and be received by listeners in another star system. Show that this is a fallacy by calculating the maximum distance from earth where a signal could be received with a SNR = 0 dB, in the presence of a 4 K interstellar background noise temperature. To be specific, assume that an "I Love Lucy" rerun is being broadcast on TV channel 4 (67 MHz), with a 4 MHz bandwidth, a transmitter power of 1000 W, transmit and receive antenna gains of 4 dB, and a perfectly noiseless receiver. Relate this distance to the nearest planet in our solar system. How much would this distance decrease if an SNR of 30 dB is required at the receiver? (30 dB is a typical value for good reception of an analog video signal.)

4.16 If a microwave signal at 2 GHz is to be transmitted to the nearest star (Alpha Centuri), what is the required transmitter power for a received SNR of 0 dB? Assume large dish antennas with $G =$ 60 dB for transmit and receive, a 4 K background noise temperature, and a receiver bandwidth of 1 kHz.

4.17 Consider the GPS receiver system shown below. The guaranteed minimum $L1$ (1575 MHz) carrier power received by an antenna on Earth having a gain of 0 dBi is $S_i = -160$ dBW. A GPS receiver is usually specified as requiring a minimum carrier-to-noise ratio, relative to a 1 Hz bandwidth, of C/N (Hz). If the receiver antenna actually has a gain G_A, and a noise temperature T_A, derive an expression for the maximum allowable amplifier noise figure F, assuming an amplifier gain G, and a connecting line loss, L. Evaluate this expression for $C/N = 32$ dB-Hz, $G_A = 5$ dB, $T_A = 300$ K, $G = 10$ dB, and $L = 25$ dB.

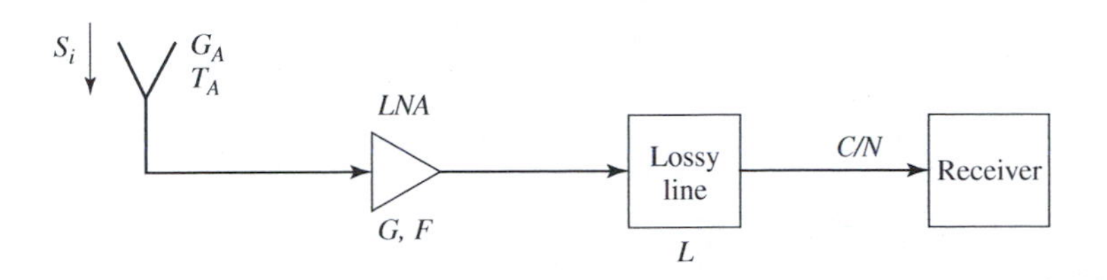

4.18 Consider the replacement of a DBS dish antenna with a microstrip array antenna. A microstrip array offers an aesthetically pleasing flat profile, but suffers from relatively high dissipative loss in its feed network, which leads to a high noise temperature. If the background noise temperature is $T_B = 50$ K, with an antenna gain of 33.5 dB and a receiver LNB noise figure of 1.1 dB, find the overall G/T for the microstrip array antenna and the LNB, if the array has a total loss of 2.5 dB. Assume the antenna is at a physical temperature of 290 K.

4.19 A high gain antenna array operating at 2.4 GHz is pointed toward a region of the sky for which the background temperature can be assumed to be at a uniform temperature of 5 K. A noise temperature of 105 K is measured for the antenna temperature. If the physical temperature of the antenna is 290 K, what is its radiation efficiency?

4.20 The AMPS cellular telephone system operates with a mobile receiver frequency of 882 MHz. If the base station transmits with an EIRP of 100 W, and the mobile receiver has an antenna with a gain of 2 dB and a noise temperature of 200 K, find the maximum operating range if the minimum SNR at the output of the receiver is required to be 18 dB. The channel bandwidth is 30 kHz, and the receiver noise figure is 6 dB. Assume the Friis formula applies to this idealized problem.

4.21 The directivity of a square microstrip array of $N \times N$ elements with $\lambda/2$ spacing is given by $D = \pi N^2$. If a corporate microstrip feed network is used, the power loss increases approximately with size as $L = e^{\alpha N \lambda}$, where α is the attenuation of the feed lines in neper/λ. For large N, the exponential increase in loss leads to a reduction in gain as size increases. Thus, maximum values of array gain and G/T exist for a given value of attenuation. Derive an expression for the gain of the array, and find the optimum value of N resulting in maximum gain. Next, derive an expression for the G/T of the antenna, assuming a uniform background noise temperature T_b, and a physical temperature T_0 for the array. Find the optimum value of N that maximizes G/T for $T_b = 50$ K, and for $T_b = 290$ K. Evaluate the optimum gain and G/T for $\alpha = 0.016$ neper/λ.

4.22 Derive the expression for the radiation resistance of a small loop antenna given in (4.42). The constant $V_0 = k_0^2 \eta_0 b^2 I_0 / 4$.

4.23 An electrically small dipole operates at 900 MHz. If the dipole is made from copper wire with a length L and radius a, compute and plot the radiation efficiency for $0.01 \leq L/\lambda \leq 0.1$, with $a = 10^{-3}\lambda$ and $a = 10^{-5}\lambda$.

4.24 Repeat Problem 4.23 for a small loop antenna, with a circumference in the range $0.01 \leq 2\pi b/\lambda \leq 0.1$.

4.25 Consider the effect of a ground reflection at 900 MHz, with a transmit antenna height of 50 m, and a range distance of 2000 m to the receive antenna. Plot the received signal voltage magnitude, normalized to the receive voltage under free-space conditions, versus receive antenna height h_2, for $0 \leq h_2 \leq 80$ m. Assume vertical polarization, and a ground reflection coefficient of $\Gamma = -1$.

4.26 For a Rayleigh probability distribution function given by $f_r(r) = \frac{r}{\sigma^2} e^{-r^2/2\sigma^2}$, for $r > 0$, find the cumulative distribution function. Next, find the rms value of the Rayleigh pdf. Use these results to plot the percentage of time that a signal amplitude is at least R_0 (dB) below the rms value for a Rayleigh fading channel, for 0 dB $\leq R_0 \leq$ 20 dB.

Chapter Five

Filters

Filters are two-port networks used to control the frequency response in an RF or microwave system by allowing transmission at frequencies within the *passband* of the filter, and attenuation within the *stopband* of the filter. Common filter responses include *low-pass*, *high-pass*, *bandpass*, and *bandstop* (or *bandreject*). Filters are indispensable components in wireless systems, used in receivers for rejecting signals outside the operating band, attenuating undesired mixer products, and for setting the IF bandwidth of the receiver. In transmitters, filters are used to control the spurious responses of upconverting mixers, to select the desired sidebands, and to limit the bandwidth of the radiated signal.

Because of the importance of filters in radio and other applications, a large amount of material on the theory and design of filters is available in the literature. Our purpose here is to give a brief introduction to filter design theory using the insertion loss method, and to describe some of the practical filter designs that are commonly used in modern wireless systems. More complete treatments of filter theory and practice can be found in references [1]–[3].

Many techniques have been proposed for the design and analysis of filter circuits, but the *insertion loss method* is generally preferred for the flexibility and accuracy that it provides. The insertion loss method is based on network synthesis techniques, and can be used to design filters having a specific type of frequency response. The technique begins with the design of a low-pass filter prototype that is normalized in terms of impedance and cutoff frequency. Impedance and frequency scaling and transformations are then used to convert the normalized design to the one having the desired frequency response, cutoff frequency, and impedance level. Additional transformations, such as *Richard's transformation*, *impedance/admittance inverters*, and the *Kuroda identities*, can be used to facilitate filter implementation in terms of practical components such as transmission lines sections, stubs, and resonant elements. Analysis and design examples of low-pass, bandpass, and high-pass filters will be presented.

5.1 FILTER DESIGN BY THE INSERTION LOSS METHOD

The ideal filter would have zero insertion loss in the passband, infinite attenuation in the stopband, and a linear phase response (to avoid signal distortion) in the passband. Of course, such filters do not exist in practice, so compromises must be made; herein lies the art of filter design.

The insertion loss method allows a high degree of control over the passband and stopband amplitude and phase characteristics, with a systematic way to synthesize a desired response. The necessary design trade-offs can be evaluated to best meet the application requirements. If, for example, a minimum insertion loss is most important, a binomial frequency response can be used; a Chebyshev response would satisfy a requirement for the sharpest cutoff. If it is possible to sacrifice the attenuation rate, a better phase response can be obtained by using a linear phase filter design. And in all cases, the insertion loss method allows filter performance to be improved in a straightforward manner, at the expense of a higher order, or more complex, filter.

Characterization by Power Loss Ratio

In the insertion loss method a filter response is defined by its insertion loss, or *power loss ratio*, P_{LR}:

$$P_{\mathrm{LR}} = \frac{\text{Power available from source}}{\text{Power delivered to load}} = \frac{P_{\mathrm{inc}}}{P_{\mathrm{load}}} = \frac{1}{1 - |\Gamma(\omega)|^2}, \tag{5.1}$$

where $\Gamma(\omega)$ is the reflection coefficient seen looking into the filter. Observe that this quantity is the reciprocal of $|S_{12}|^2$ if both load and source are matched. The insertion loss (IL) in dB is

$$\mathrm{IL} = 10 \log P_{\mathrm{LR}}. \tag{5.2}$$

Because of the causal properties of passive networks, $|\Gamma(\omega)|^2$ is an even function of ω (see Problem 5.1). Therefore we can write $|\Gamma(\omega)|^2$ as a polynomial in ω^2:

$$|\Gamma(\omega)|^2 = \frac{M(\omega^2)}{M(\omega^2) + N(\omega^2)}, \tag{5.3}$$

where M and N are real polynomials in ω^2. Substituting this form in (5.1) gives

$$P_{\mathrm{LR}} = 1 + \frac{M(\omega^2)}{N(\omega^2)}. \tag{5.4}$$

Thus, for a filter to be physically realizable its power loss ratio must be of the form of (5.4). Notice that specifying the power loss ratio simultaneously constrains the reflection coefficient, $\Gamma(\omega)$. We now discuss several practical filter responses.

Maximally flat. This characteristic is also called the binomial or Butterworth response, and is optimum in the sense that it provides the flattest possible passband response for a given filter complexity, or order. For a low-pass filter, the maximally flat response is defined by

$$P_{\mathrm{LR}} = 1 + k^2 \left(\frac{\omega}{\omega_c}\right)^{2N}, \tag{5.5}$$

where N is the order of the filter, and ω_c is the cutoff frequency. The passband extends from $\omega = 0$ to $\omega = \omega_c$; at the band edge the power loss ratio is $1 + k^2$. If we choose this as the -3 dB point, as is common in practice, then we have $k = 1$. For $\omega > \omega_c$, the attenuation increases monotonically with frequency, as shown in Figure 5.1. For $\omega \gg \omega_c$,

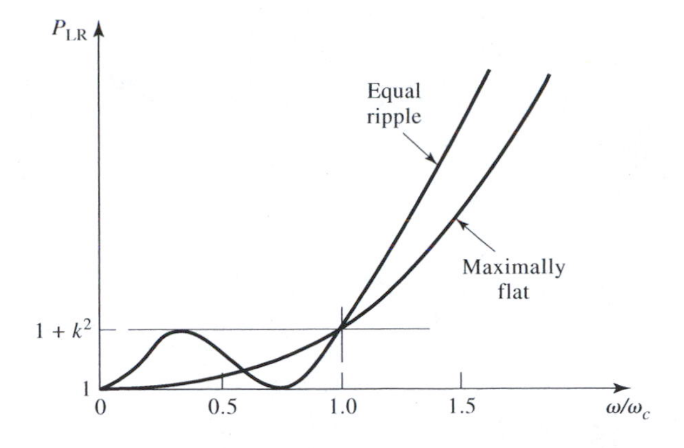

FIGURE 5.1 Maximally flat and equal-ripple low-pass filter responses ($N = 3$).

$P_{LR} \cong k^2(\omega/\omega_c)^{2N}$, which shows that the insertion loss increases at the rate of $20N$ dB per decade increase in frequency. The term *maximally flat* arises from the fact that it can be shown that the first $(2N - 1)$ derivatives of (5.5) are zero at $\omega = 0$.

Equal-ripple. If a Chebyshev polynomial is used to specify the insertion loss of an N-order low-pass filter as

$$P_{LR} = 1 + k^2 T_N^2\left(\frac{\omega}{\omega_c}\right), \tag{5.6}$$

where $T_N(x)$ is a Chebyshev polynomial of order N (see Appendix E), then a sharper cutoff characteristic will result, although the passband response will have ripples of amplitude $1 + k^2$, as shown in Figure 5.1, since $T_N(x)$ oscillates between ± 1 for $|x| \le 1$. Thus, k^2 determines the passband ripple level. For large x, $T_N(x) \cong (2x)^N/2$, so for $\omega \gg \omega_c$ the insertion loss becomes asymptotic to

$$P_{LR} \cong \frac{k^2}{4}\left(\frac{2\omega}{\omega_c}\right)^{2N},$$

which also increases at the rate of $20N$ dB/decade. But the insertion loss for the Chebyshev case is $(2^{2N})/4$ greater than the binomial response, at any given frequency where $\omega \gg \omega_c$.

Linear Phase. The above filters specify the amplitude response, but in some applications (such as multiplexing filters in frequency-division multiplexed communications system) it is important to have a linear phase response in the passband to avoid signal distortion. It turns out that a sharp-cutoff response is generally incompatible with a good phase response, so the phase response of the filter must be deliberately synthesized, usually resulting in an inferior amplitude cutoff characteristic. A linear phase characteristic can be achieved with the following phase response:

$$\phi(\omega) = A\omega\left[1 + p\left(\frac{\omega}{\omega_c}\right)^{2N}\right], \tag{5.7}$$

where $\phi(\omega)$ is the phase of the voltage transfer function of the filter, and p is a constant. A related quantity is the group delay, defined as

$$\tau = \frac{d\phi}{d\omega} = A\left[1 + p(2N + 1)\left(\frac{\omega}{\omega_c}\right)^{2N}\right], \tag{5.8}$$

which shows that the group delay for a linear phase filter is a maximally flat function.

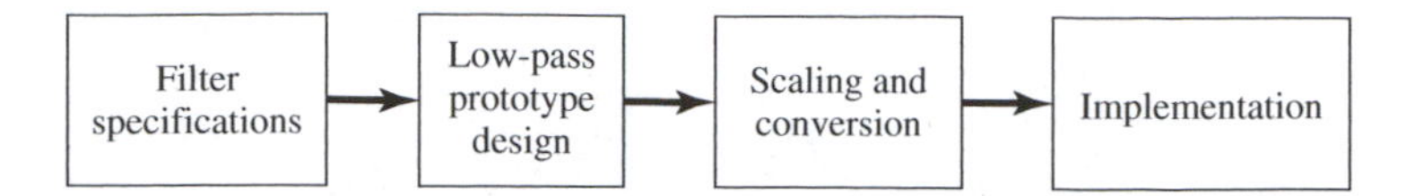

FIGURE 5.2 The process of filter design by the insertion loss method.

More general filter specifications can be obtained, but the above cases are the most common. We will next discuss the design of low-pass filter prototypes which are normalized in terms of impedance and frequency; this type of normalization simplifies the design of filters for arbitrary frequency, impedance, and type (low-pass, high-pass, bandpass, or bandstop). The low-pass prototypes are then scaled to the desired frequency and impedance, and the lumped-element components replaced with distributed circuit elements for implementation at microwave frequencies. This design process is illustrated in Figure 5.2.

Maximally Flat Low-Pass Filter Prototype

Consider the two-element low-pass filter prototype circuit shown in Figure 5.3; we will derive the normalized element values, L and C, for a maximally flat response. We assume a source impedance of 1 Ω, and a cutoff frequency of $\omega_c = 1$. From (5.5), the desired power loss ratio will be, for $N = 2$,

$$P_{\text{LR}} = 1 + \omega^4. \tag{5.9}$$

The input impedance of this filter is

$$Z_{\text{in}} = j\omega L + \frac{R(1 - j\omega RC)}{1 + \omega^2 R^2 C^2}. \tag{5.10}$$

Since the input reflection coefficient is

$$\Gamma = \frac{Z_{\text{in}} - 1}{Z_{\text{in}} + 1},$$

the power loss ratio can be written as

$$P_{\text{LR}} = \frac{1}{1 - |\Gamma|^2} = \frac{1}{1 - [(Z_{\text{in}} - 1)/(Z_{\text{in}} + 1)][(Z_{\text{in}}^* - 1)/(Z_{\text{in}}^* + 1)]} = \frac{|Z_{\text{in}} + 1|^2}{2(Z_{\text{in}} + Z_{\text{in}}^*)}.$$

Now,

$$Z_{\text{in}} + Z_{\text{in}}^* = \frac{2R}{1 + \omega^2 R^2 C^2}$$

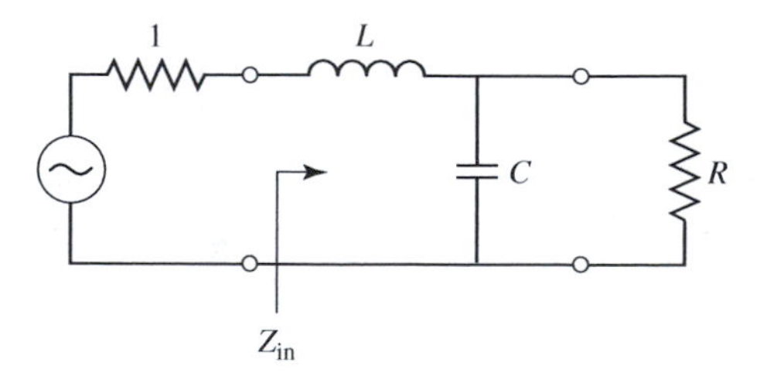

FIGURE 5.3 Low-pass filter prototype for $N = 2$.

and

$$|Z_{\text{in}} + 1|^2 = \left(\frac{R}{1+\omega^2R^2C^2} + 1\right)^2 + \left(\omega L - \frac{\omega CR^2}{1+\omega^2R^2C^2}\right)^2,$$

so (5.10) becomes

$$\begin{aligned} P_{\text{LR}} &= \frac{1+\omega^2R^2C^2}{4R}\left[\left(\frac{R}{1+\omega^2R^2C^2} + 1\right)^2 + \left(\omega L - \frac{\omega CR^2}{1+\omega^2R^2C^2}\right)^2\right] \\ &= \frac{1}{4R}(R^2 + 2R + 1 + R^2C^2\omega^2 + \omega^2L^2 + \omega^4L^2C^2R^2 - 2\omega^2LCR^2) \\ &= 1 + \frac{1}{4R}[(1-R)^2 + (R^2C^2 + L^2 - 2LCR^2)\omega^2 + L^2C^2R^2\omega^4] \end{aligned} \tag{5.11}$$

Notice that this expression is a polynomial in ω^2, as it should be, according to the above discussion. Comparing to the desired response of (5.9) shows that $R = 1$, since $P_{\text{LR}} = 1$ for $\omega = 0$. In addition, the coefficient of ω^2 must vanish, so

$$C^2 + L^2 - 2LC = (C - L)^2 = 0,$$

or $L = C$. Then for the coefficient of ω^4 to be unity, we must have

$$\frac{1}{4}L^2C^2 = \frac{1}{4}L^4 = 1,$$

or $L = C = \sqrt{2}$.

In principle, this procedure can be extended to find the element values for filters with an arbitrary number of elements, N, but clearly this is not practical for large N. For a normalized low-pass design, however, the element values for the ladder-type circuits of Figure 5.4 can be tabulated [1]. Table 5.1 gives such element values for maximally flat low-pass filter prototypes for $N = 1$ to 10. (Notice that the values for $N = 1$ agree with the above analytical solution.). This data is used with either of the ladder circuits of Figure 5.4 in the following way. The element values are numbered from g_0 at the generator impedance to g_{N+1} at the load impedance, for a filter having N reactive elements. The elements alternate

TABLE 5.1 Element Values for Maximally Flat Low-Pass Filter Prototypes ($g_0 = 1, \omega_c = 1, N = 1$ to 10)

N	g_1	g_2	g_3	g_4	g_5	g_6	g_7	g_8	g_9	g_{10}	g_{11}
1	2.0000	1.0000									
2	1.4142	1.4142	1.0000								
3	1.0000	2.0000	1.0000	1.0000							
4	0.7654	1.8478	1.8478	0.7654	1.0000						
5	0.6180	1.6180	2.0000	1.6180	0.6180	1.0000					
6	0.5176	1.4142	1.9318	1.9318	1.4142	0.5176	1.0000				
7	0.4450	1.2470	1.8019	2.0000	1.8019	1.2470	0.4450	1.0000			
8	0.3902	1.1111	1.6629	1.9615	1.9615	1.6629	1.1111	0.3902	1.0000		
9	0.3473	1.0000	1.5321	1.8794	2.0000	1.8794	1.5321	1.0000	0.3473	1.0000	
10	0.3129	0.9080	1.4142	1.7820	1.9754	1.9754	1.7820	1.4142	0.9080	0.3129	1.0000

Source: Reprinted from G. L. Matthaei, L. Young, and E. M. T. Jones, *Microwave Filters, Impedance-Matching Networks, and Coupling Structures* (Dedham, MA: Artech House, 1980), with permission.

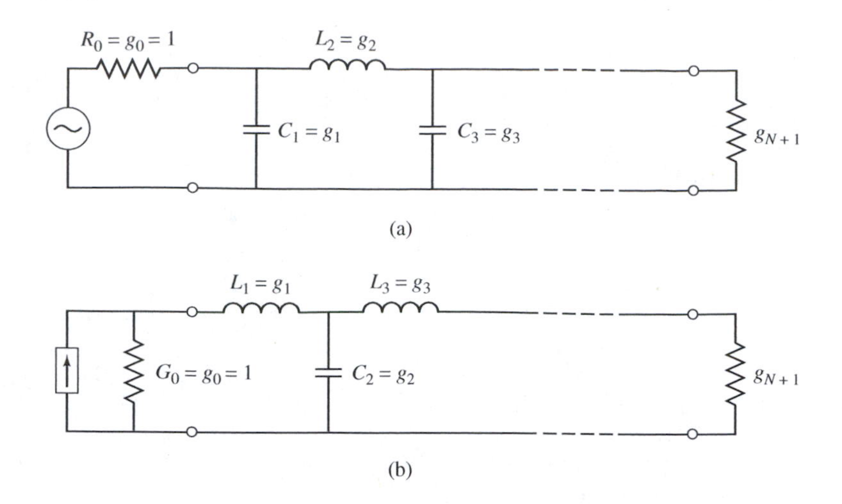

FIGURE 5.4 Ladder circuits for low-pass filter prototypes and their element definitions. (a) Prototype beginning with a shunt element. (b) Prototype beginning with a series element.

between series and shunt connections, and so g_k has the following definition:

$$g_0 = \begin{cases} \text{generator resistance (network of Figures 5.4a)} \\ \text{generator conductance (network of Figures 5.4b)} \end{cases}$$

$$\underset{(k=1 \text{ to } N)}{g_k} = \begin{cases} \text{inductance for series inductors} \\ \text{capacitor for shunt capacitors} \end{cases}$$

$$g_{N+1} = \begin{cases} \text{load resistance if } g_N \text{ is a shunt capacitor} \\ \text{load conductance if } g_N \text{ is a series inductor} \end{cases}$$

Then the circuits of Figure 5.4 can be considered as the duals of each other, and both will give the same response.

Finally, as a matter of practical design procedure, it will be necessary to determine the size, or order, of the filter. This is usually dictated by a specification on the insertion loss at some frequency within the stopband of the filter. Figure 5.5 shows the attenuation characteristics for various N, versus normalized frequency. If a filter with $N > 10$ is required, a good result can usually be obtained by cascading two designs of lower order. Alternatively, (5.5) can be used directly to find N, given a desired level of attenuation at a particular frequency.

EXAMPLE 5.1 LOW-PASS FILTER DESIGN

A maximally flat low-pass filter is to be designed with a cutoff frequency of 8 GHz and a minimum attenuation of 20 dB at 11 GHz. How many filter elements are required?

Solution

We have $\omega = 2\pi = 11$ GHz and $\omega_c/2\pi = 8$ GHz, so we can compute the value for the horizontal axis of Figure 5.5 as

$$\left|\frac{\omega}{\omega_c}\right| - 1 = \frac{11}{8} - 1 = 0.375.$$

Then from Figure 5.5 we see that an attenuation of 20 dB at this frequency requires

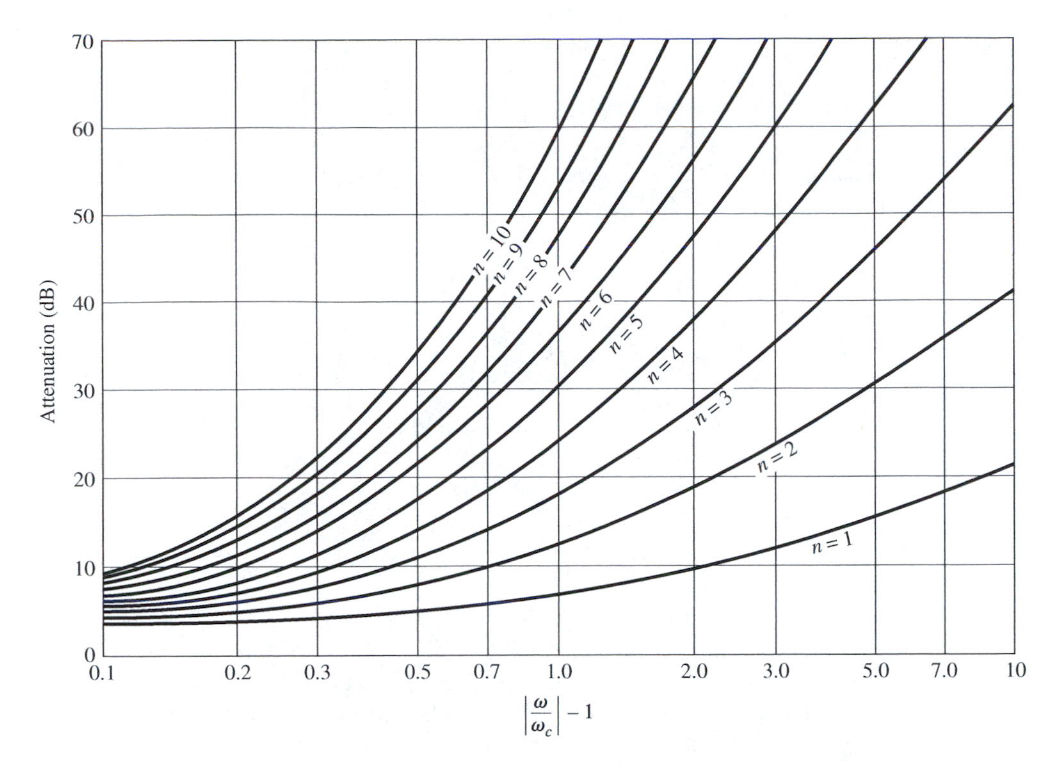

FIGURE 5.5 Attenuation versus normalized frequency for maximally flat filter prototypes. Adapted from G. L. Matthaei, L. Young, and E. M. T. Jones, *Microwave Filters, Impedance-Matching Networks, and Coupling Structures* (Dedham, MA: Artech House, 1980), with permission.

that $N \geq 8$. Further design details for this filter will be discussed in the following section. ○

Equal-Ripple Low-Pass Filter Prototype

For an equal-ripple low-pass filter with a cutoff frequency $\omega_c = 1$, the power loss ratio from (5.6) is

$$P_{\mathrm{LR}} = 1 + k^2 T_N^2(\omega), \tag{5.12}$$

where $1 + k^2$ is the ripple level in the passband. Since the Chebyshev polynomials have the property that

$$T_N(0) = \begin{cases} 0 & \text{for } N \text{ odd} \\ 1 & \text{for } N \text{ even} \end{cases},$$

(5.12) shows that the filter will have a unity power loss ratio at $\omega = 0$ for N odd, but a power loss ratio of $1 + k^2$ at $\omega = 0$ for N even. Thus, there are two cases to consider, depending on N.

For the two-element filter of Figure 5.3, the power loss ratio is given in terms of the component values in (5.11). From Appendix E, the Chebyshev polynomial of order 2 is given by $T_2(x) = 2x^2 - 1$, so equating (5.12) to (5.11) gives

$$1 + k^2(4\omega^4 - 4\omega^2 + 1) = 1 + \frac{1}{4R}[(1-R)^2 + (R^2C^2 + L^2 - 2LCR^2)\omega^2 + L^2C^2R^2\omega^4], \tag{5.13}$$

which can be solved for R, L, and C if the ripple level (as determined by k^2) is known. Thus, at $\omega = 0$ we have that

$$k^2 = \frac{(1-R)^2}{4R},$$

or

$$R = 1 + 2k^2 \pm 2k\sqrt{1+k^2} \qquad \text{(for } N \text{ even)}. \tag{5.14}$$

Equating coefficients of ω^2 and ω^4 yields the additional relations

$$4k^2 = \frac{1}{4R}L^2C^2R^2,$$
$$-4k^2 = \frac{1}{4R}(R^2C^2 + L^2 - 2LCR^2),$$

which can be used to find L and C. Note that (5.14) gives a value for R that is not unity, so there will be an impedance mismatch if the load actually has a unity (normalized) impedance; this can be corrected with a quarter-wave transformer, or by using an additional filter element to make N odd. For odd N, it can be shown that $R = 1$. (This is because there is a unity power loss ratio at $\omega = 0$ for N odd.)

Tables exist for designing equal-ripple low-pass filters with a normalized source impedance and cutoff frequency ($\omega_c' = 1$) [1], and can be applied to either of the ladder circuits of Figure 5.4. This design data depends on the specified passband ripple level; Table 5.2 lists element values for normalized low-pass Chebyshev filter prototypes having 0.5 dB or 3.0 dB ripple, for $N = 1$ to 10. Notice that the load impedance $g_{N+1} \neq 1$ for even N. If the stopband attenuation is specified, the curves in Figures 5.6a,b can be used to determine the necessary value of N for these ripple values.

Linear Phase Low-Pass Filter Prototype

Filters having a maximally flat time delay, or a linear phase response, can be designed in the same way, but things are somewhat more complicated because the phase of the voltage transfer function is not as simply expressed as is its amplitude. Design values have been derived for such filters [1], however, again for the ladder circuits of Figure 5.4, and are given in Table 5.3 for a normalized source impedance and cutoff frequency ($\omega_c' = 1$). The resulting group delay in the passband will be $\tau_d = 1/\omega_c' = 1$.

5.2 FILTER SCALING AND TRANSFORMATION

In this section we describe the process of converting a normalized low-pass filter prototype to a filter circuit having a prescribed impedance level, cutoff frequency, and frequency response.

Impedance Scaling

In the prototype design, the source and load resistances are unity (except for the case of equal-ripple filters with N even, which have nonunity load resistance). A source resistance

TABLE 5.2 Element Values for Equal Ripple Low-Pass Filter Prototypes ($g_0 = 1, \omega_c = 1, N = 1$ to 10)

					0.5 dB Ripple						
N	g_1	g_2	g_3	g_4	g_5	g_6	g_7	g_8	g_9	g_{10}	g_{11}
1	0.6986	1.0000									
2	1.4029	0.7071	1.9841								
3	1.5963	1.0967	1.5963	1.0000							
4	1.6703	1.1926	2.3661	0.8419	1.9841						
5	1.7058	1.2296	2.5408	1.2296	1.7058	1.0000					
6	1.7254	1.2479	2.6064	1.3137	2.4758	0.8696	1.9841				
7	1.7372	1.2583	2.6381	1.3444	2.6381	1.2583	1.7372	1.0000			
8	1.7451	1.2647	2.6564	1.3590	2.6964	1.3389	2.5093	0.8796	1.9841		
9	1.7504	1.2690	2.6678	1.3673	2.7239	1.3673	2.6678	1.2690	1.7504	1.0000	
10	1.7543	1.2721	2.6754	1.3725	2.7392	1.3806	2.7231	1.3485	2.5239	0.8842	1.9841
					3.0 dB Ripple						
N	g_1	g_2	g_3	g_4	g_5	g_6	g_7	g_8	g_9	g_{10}	g_{11}
1	1.9953	1.0000									
2	3.1013	0.5339	5.8095								
3	3.3487	0.7117	3.3487	1.0000							
4	3.4389	0.7483	4.3471	0.5920	5.8095						
5	3.4817	0.7618	4.5381	0.7618	3.4817	1.0000					
6	3.5045	0.7685	4.6061	0.7929	4.4641	0.6033	5.8095				
7	3.5182	0.7723	4.6386	0.8039	4.6386	0.7723	3.5182	1.0000			
8	3.5277	0.7745	4.6575	0.8089	4.6990	0.8018	4.4990	0.6073	5.8095		
9	3.5340	0.7760	4.6692	0.8118	4.7272	0.8118	4.6692	0.7760	3.5340	1.0000	
10	3.5384	0.7771	4.6768	0.8136	4.7425	0.8164	4.7260	0.8051	4.5142	0.6091	5.8095

Source: Reprinted from G. L. Matthaei, L. Young, and E. M. T. Jones, *Microwave Filters, Impedance-Matching Networks, and Coupling Structures* (Dedham, MA: Artech House, 1980), with permission.

of R_0 can be obtained by multiplying the impedances of the prototype design by R_0. Then, if we let primes denote impedance scaled quantities, we have the new filter component values given by

$$L' = R_0 L, \tag{5.15a}$$

$$C' = \frac{C}{R_0}, \tag{5.15b}$$

$$R'_s = R_0, \tag{5.15c}$$

$$R'_L = R_0 R_L, \tag{5.15d}$$

where L, C, and R_L are the component values for the original prototype.

Frequency Scaling for Low-Pass Filters

To scale the cutoff frequency of a low-pass filter prototype from unity to ω_c requires that we scale the frequency dependence of the filter by the factor $1/\omega_c$, which is accomplished

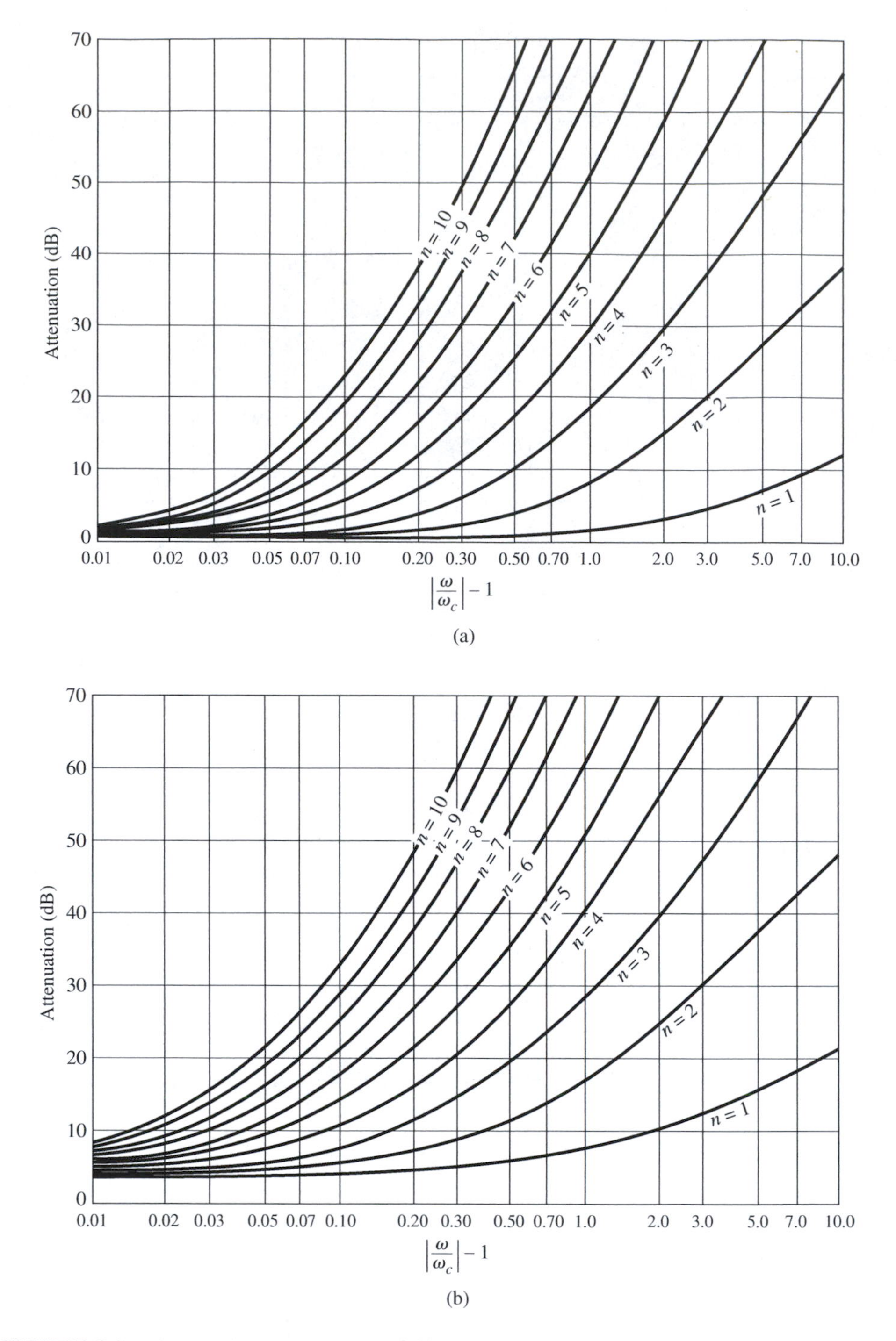

FIGURE 5.6 Attenuation versus normalized frequency for equal-ripple filter prototypes. (a) 0.5 dB ripple level. (b) 3.0 dB ripple level. Adapted from G. L. Matthaei, L. Young, and E. M. T. Jones, *Microwave Filters, Impedance-Matching Networks, and Coupling Structures* (Dedham, MA: Artech House, 1980), with permission.

TABLE 5.3 Element Values for Maximally Flat Time Delay Low-Pass Filter Prototypes ($g_0 = 1, \omega_c = 1, N = 1$ to 10)

N	g_1	g_2	g_3	g_4	g_5	g_6	g_7	g_8	g_9	g_{10}	g_{11}
1	2.0000	1.0000									
2	1.5774	0.4226	1.0000								
3	1.2550	0.5528	0.1922	1.0000							
4	1.0598	0.5116	0.3181	0.1104	1.0000						
5	0.9303	0.4577	0.3312	0.2090	0.0718	1.0000					
6	0.8377	0.4116	0.3158	0.2364	0.1480	0.0505	1.0000				
7	0.7677	0.3744	0.2944	0.2378	0.1778	0.1104	0.0375	1.0000			
8	0.7125	0.3446	0.2735	0.2297	0.1867	0.1387	0.0855	0.0289	1.0000		
9	0.6678	0.3203	0.2547	0.2184	0.1859	0.1506	0.1111	0.0682	0.0230	1.0000	
10	0.6305	0.3002	0.2384	0.2066	0.1808	0.1539	0.1240	0.0911	0.0557	0.0187	1.0000

Source: Reprinted from G. L. Matthaei, L. Young, and E. M. T. Jones, *Microwave Filters, Impedance-Matching Networks, and Coupling Structures* (Dedham, MA: Artech House, 1980), with permission.

by replacing ω by ω/ω_c:

$$\omega \leftarrow \frac{\omega}{\omega_c}. \tag{5.16}$$

Then the new power loss ratio will be

$$P'_{\text{LR}}(\omega) = P_{\text{LR}}\left(\frac{\omega}{\omega_c}\right),$$

where ω_c is the new cutoff frequency; cutoff occurs when $\omega/\omega_c = 1$, or $\omega = \omega_c$. This transformation can be viewed as a stretching, or expansion, of the original passband, as illustrated in Figure 5.7a,b.

The new element values are determined by applying the substitution of (5.16) to the series reactances, $j\omega L_k$, and shunt susceptances, $j\omega C_k$, of the prototype filter, Thus,

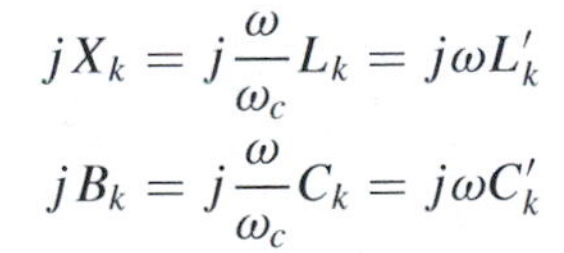

$$jX_k = j\frac{\omega}{\omega_c}L_k = j\omega L'_k$$

$$jB_k = j\frac{\omega}{\omega_c}C_k = j\omega C'_k$$

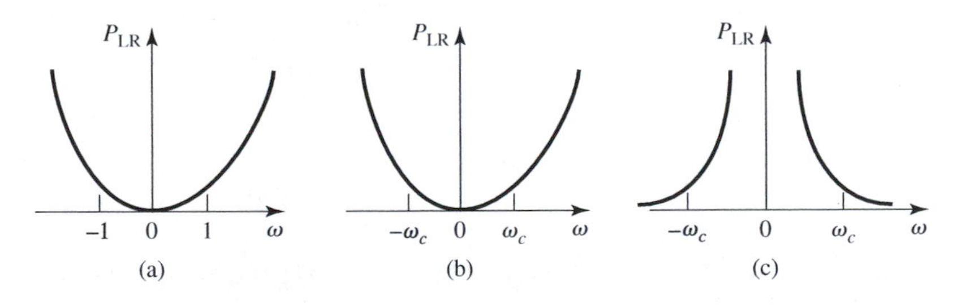

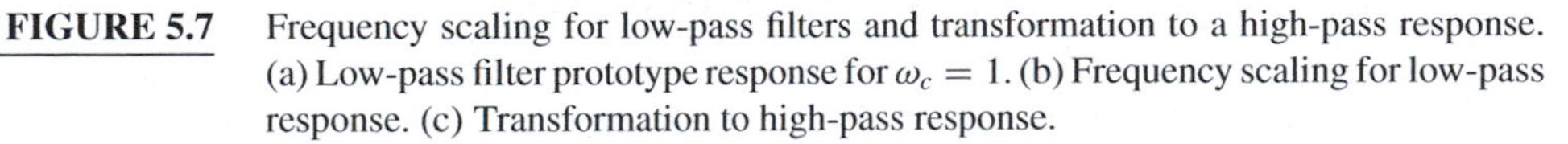

FIGURE 5.7 Frequency scaling for low-pass filters and transformation to a high-pass response. (a) Low-pass filter prototype response for $\omega_c = 1$. (b) Frequency scaling for low-pass response. (c) Transformation to high-pass response.

which shows that the new element values are given by

$$L'_k = \frac{L_k}{\omega_c}, \tag{5.17a}$$

$$C'_k = \frac{C_k}{\omega_c}. \tag{5.17b}$$

When both impedance and frequency scaling are required, the results of (5.15) can be combined with (5.17) to give

$$L'_k = \frac{R_0 L_k}{\omega_c}, \tag{5.18a}$$

$$C'_k = \frac{C_k}{R_0 \omega_c}. \tag{5.18b}$$

Low-pass to High-pass Transformation

The frequency substitution where,

$$\omega \leftarrow -\frac{\omega_c}{\omega}, \tag{5.19}$$

can be used to convert a low-pass response to a high-pass response, as shown in Figure 5.7c. This substitution maps $\omega = 0$ to $\omega = \pm\infty$, and maps $\omega = \pm\infty$ to $\omega = 0$. Cutoff occurs when $\omega = \pm\omega_c$. The negative sign in (5.19) is needed to convert inductors (and capacitors) to realizable capacitors (and inductors). Applying (5.19) to the series reactances, $j\omega L_k$, and the shunt susceptances, $j\omega C_k$, of the prototype filter gives

$$jX_k = -j\frac{\omega_c}{\omega}L_k = \frac{1}{j\omega C'_k},$$

$$jB_k = -j\frac{\omega_c}{\omega}C_k = \frac{1}{j\omega L'_k},$$

which shows that series inductors L_k must be replaced with capacitors C'_k, and shunt capacitors C_k must be replaced with inductors with L'_k. The new component values are given by

$$C'_k = \frac{1}{\omega_c L_k}, \tag{5.20a}$$

$$L'_k = \frac{1}{\omega_c C_k}. \tag{5.20b}$$

Impedance scaling can be included by using (5.15) to give

$$C'_k = \frac{1}{R_0 \omega_c L_k}, \tag{5.21a}$$

$$L'_k = \frac{R_0}{\omega_c C_k}. \tag{5.21b}$$

EXAMPLE 5.2 LOW-PASS FILTER DESIGN COMPARISON

Design a maximally flat low-pass filter with a cutoff frequency of 2 GHz, impedance of 50 Ω, and at least 15 dB attenuation at 3 GHz. Compute and plot the amplitude response and group delay for $f = 0$ to 4 GHz, and compare these results with those for an equal-ripple (3.0 dB ripple) and a linear phase filter having the same order.

Solution

First find the required order of the maximally flat filter to satisfy the insertion loss specification at 3 GHz. We have that $|\omega/\omega_c| - 1 = 0.5$; from Figure 5.5 we see that $N = 5$ will be sufficient. Then Table 5.1 gives the low-pass prototype element values as

$$g_1 = 0.618 = C_1$$
$$g_2 = 1.618 = L_2$$
$$g_3 = 2.000 = C_3$$
$$g_4 = 1.618 = L_4$$
$$g_5 = 0.618 = C_5$$

Then (5.18) can be used to obtain the scaled element values:

$$C_1' = \frac{C_1}{R_0\omega_c} = \frac{0.618}{(50)(2\pi)(2 \times 10^9)} = 0.984 \text{ pF},$$

$$L_2' = \frac{R_0 L_2}{\omega_c} = \frac{(50)(1.618)}{(2\pi)(2 \times 10^9)} = 6.438 \text{ nH},$$

$$C_3' = \frac{C_3}{R_0\omega_c} = \frac{2.000}{(50)(2\pi)(2 \times 10^9)} = 3.183 \text{ pF},$$

$$L_4' = \frac{R_0 L_4}{\omega_c} = \frac{(50)(1.618)}{(2\pi)(2 \times 10^9)} = 6.438 \text{ nH},$$

$$C_5' = \frac{C_5}{R_0\omega_c} = \frac{0.618}{(50)(2\pi)(2 \times 10^9)} = 0.984 \text{ pF}.$$

The final filter circuit is shown in Figure 5.8; the ladder circuit of Figure 5.4a was used, but that of Figure 5.4b could have been used just as well.

The component values for the equal-ripple filter and the linear phase filter, for $N = 5$, can be determined from Tables 5.2 and 5.3. The amplitude and group delay results for these three filters are shown in Figure 5.9. These results clearly show the trade-offs involved with the three types of filters. The equal-ripple response has the sharpest cutoff, but the worst group delay characteristics. The maximally flat response has a flatter attenuation characteristic in the passband, but a slightly lower cutoff rate. The linear phase filter has the worst cutoff rate, but very good group delay characteristic. ○

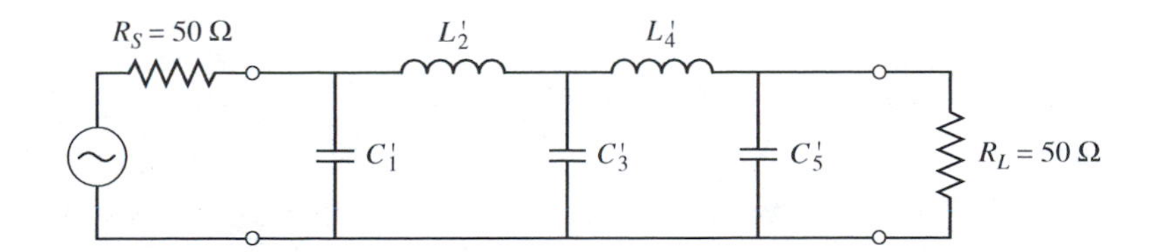

FIGURE 5.8 Low-pass maximally flat filter circuit for Example 5.2.

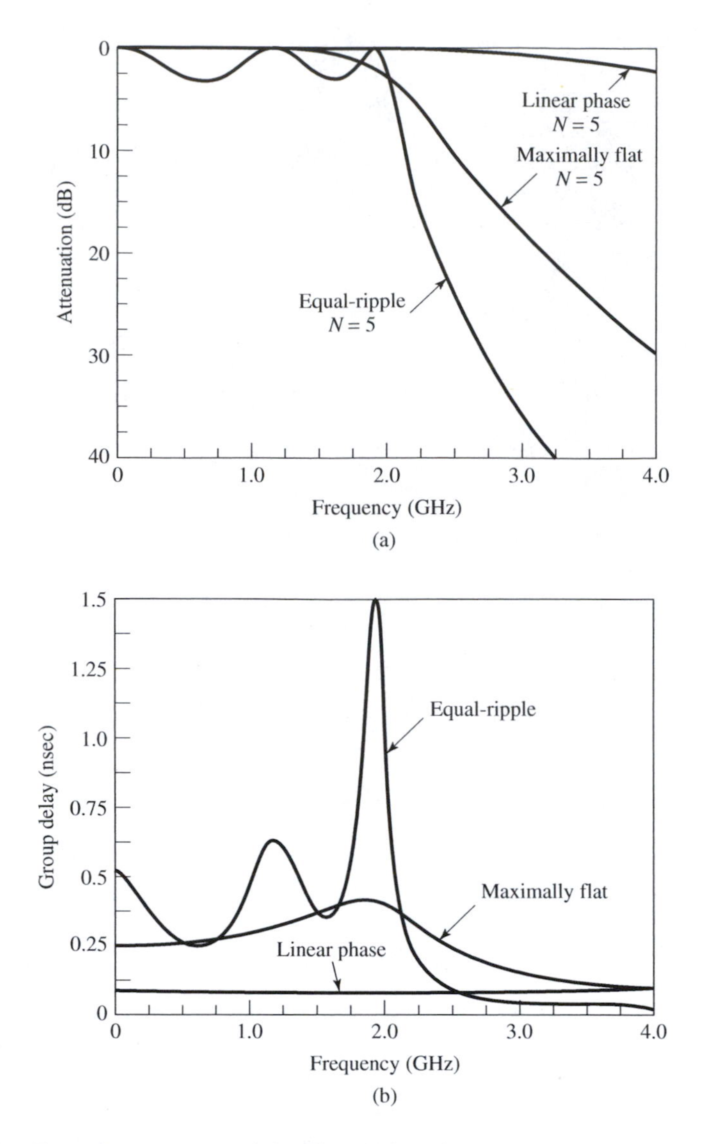

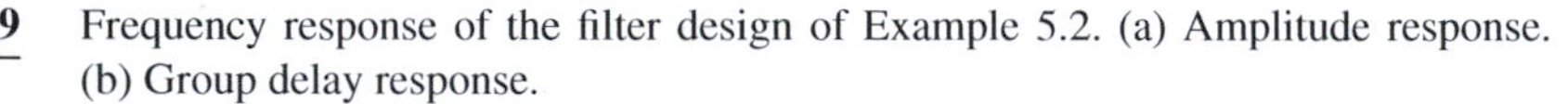
FIGURE 5.9 Frequency response of the filter design of Example 5.2. (a) Amplitude response. (b) Group delay response.

Bandpass and Bandstop Transformation

Low-pass prototype filter designs can also be transformed to produce the bandpass or bandstop response illustrated in Figure 5.10. If ω_1 and ω_2 denote the edges of the passband, then a bandpass response can be obtained using the following frequency substitution:

$$\omega \leftarrow \frac{\omega_0}{\omega_2 - \omega_1}\left(\frac{\omega}{\omega_0} - \frac{\omega_0}{\omega}\right) = \frac{1}{\Delta}\left(\frac{\omega}{\omega_0} - \frac{\omega_0}{\omega}\right), \tag{5.22}$$

where

$$\Delta = \frac{\omega_2 - \omega_1}{\omega_0}, \tag{5.23}$$

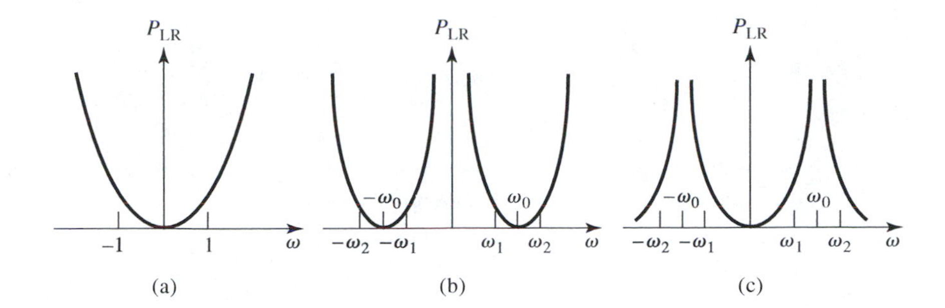

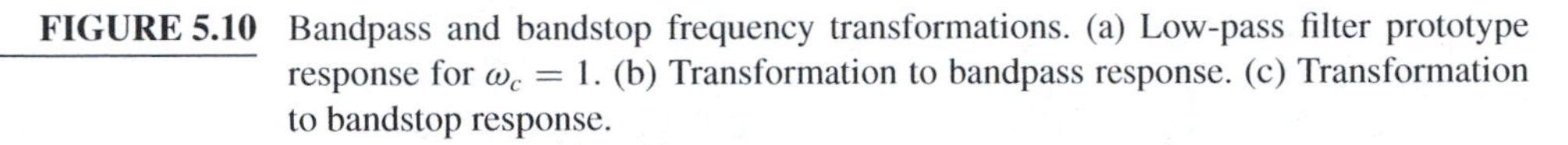

FIGURE 5.10 Bandpass and bandstop frequency transformations. (a) Low-pass filter prototype response for $\omega_c = 1$. (b) Transformation to bandpass response. (c) Transformation to bandstop response.

is the fractional bandwidth of the passband. The center frequency, ω_0, could be chosen as the arithmetic mean of ω_1 and ω_2, but the equations are simpler if it is chosen as their geometric mean:

$$\omega_0 = \sqrt{\omega_1 \omega_2}. \tag{5.24}$$

Then the transformation of (5.22) will map the bandpass characteristics of Figure 5.10b to the normalized low-pass response of Figure 5.10a, as follows:
When $\omega = \omega_0$,

$$\frac{1}{\Delta}\left(\frac{\omega}{\omega_0} - \frac{\omega_0}{\omega}\right) = 0;$$

When $\omega = \omega_1$,

$$\frac{1}{\Delta}\left(\frac{\omega}{\omega_0} - \frac{\omega_0}{\omega}\right) = \frac{1}{\Delta}\left(\frac{\omega_1^2 - \omega_0^2}{\omega_0 \omega_1}\right) = -1;$$

When $\omega = \omega_2$,

$$\frac{1}{\Delta}\left(\frac{\omega}{\omega_0} - \frac{\omega_0}{\omega}\right) = \frac{1}{\Delta}\left(\frac{\omega_2^2 - \omega_0^2}{\omega_0 \omega_2}\right) = 1.$$

The new filter elements are determined by using (5.22) in the expression for the series reactance and shunt susceptances. Thus,

$$jX_k = \frac{j}{\Delta}\left(\frac{\omega}{\omega_0} - \frac{\omega_0}{\omega}\right)L_k = j\frac{\omega L_k}{\Delta \omega_0} - j\frac{\omega_0 L_k}{\Delta \omega} = j\omega L_k' - j\frac{1}{\omega C_k'},$$

which shows that a series inductor, L_k, in the low-pass prototype is transformed to a series LC circuit with element values given by

$$L_k' = \frac{L_k}{\Delta \omega_0}, \tag{5.25a}$$

$$C_k' = \frac{\Delta}{\omega_0 L_k}. \tag{5.25b}$$

Similarly,

$$jB_k = \frac{j}{\Delta}\left(\frac{\omega}{\omega_0} - \frac{\omega_0}{\omega}\right)C_k = j\frac{\omega C_k}{\Delta\omega_0} - j\frac{\omega_0 C_k}{\Delta\omega} = j\omega C'_k - j\frac{1}{\omega L'_k},$$

which shows that a shunt capacitor, C_k, in the low-pass prototype is transformed to a shunt LC circuit with element values given by

$$L'_k = \frac{\Delta}{\omega_0 C_k}, \tag{5.25c}$$

$$C'_k = \frac{C_k}{\Delta\omega_0}. \tag{5.25d}$$

The low-pass filter elements are thus converted to series resonant circuits (having a low impedance at resonance) in the series arms, and to parallel resonant circuits (having a high impedance at resonance) in the shunt arms. Notice that both series and parallel resonator elements have a resonant frequency of ω_0.

The inverse transformation can be used to obtain a bandstop response. Thus,

$$\omega \leftarrow \Delta\left(\frac{\omega}{\omega_0} - \frac{\omega_0}{\omega}\right)^{-1}, \tag{5.26}$$

where Δ and ω_0 have the same definitions as in (5.23) and (5.24). Then series inductors of the low-pass prototype are converted to parallel LC circuits having element values given by

$$L'_k = \frac{\Delta L_k}{\omega_0}, \tag{5.27a}$$

$$C'_k = \frac{1}{\omega_0 \Delta L_k}. \tag{5.27b}$$

The shunt capacitor of the low-pass prototype is converted to series LC circuits having

TABLE 5.4 Summary of Prototype Filter Transformations

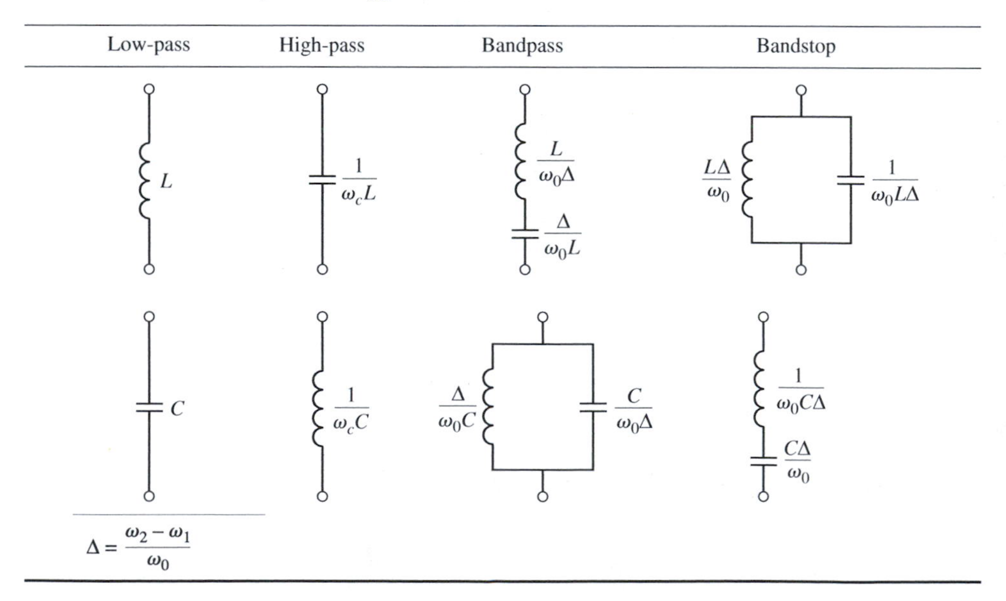

element values given by

$$L'_k = \frac{1}{\omega_0 \Delta C_k}, \tag{5.27c}$$

$$C'_k = \frac{\Delta C_k}{\omega_0}. \tag{5.27d}$$

The element transformations from a low-pass prototype to a high-pass, bandpass, or bandstop filter are summarized in Table 5.4. These results do not include impedance scaling, which can be made using the results in (5.15).

EXAMPLE 5.3 BANDPASS FILTER DESIGN

Design a bandpass filter having a 0.5 dB equal-ripple response, with $N = 3$. The center frequency is 1 GHz, the fractional bandwidth is 10%, and the impedance is 50 Ω.

Solution

From Table 5.2 the element values for the low-pass prototype circuit of Figure 5.4b are given as

$$\begin{aligned} g_1 &= 1.5963 = L_1 \\ g_2 &= 1.0967 = C_2 \\ g_3 &= 1.5963 = L_3 \\ g_4 &= 1.000 \;\;= R_L \end{aligned}$$

Then (5.15) and (5.25) give the impedance-scaled and frequency-transformed element values for the circuit of Figure 5.11 as

$$L'_1 = \frac{L_1 Z_0}{\omega_0 \Delta} = 127.0 \text{ nH},$$

$$C'_1 = \frac{\Delta}{\omega_0 L_1 Z_0} = 0.199 \text{ pF},$$

$$L'_2 = \frac{\Delta Z_0}{\omega_0 C_2} = 0.726 \text{ nH},$$

$$C'_2 = \frac{C_2}{\omega_0 \Delta Z_0} = 34.91 \text{ pF},$$

$$L'_3 = \frac{L_3 Z_0}{\omega_0 \Delta} = 127.0 \text{ nH},$$

$$C'_3 = \frac{\Delta}{\omega_0 L_3 Z_0} = 0.199 \text{ pF}.$$

The resulting amplitude response is shown in Figure 5.12. ○

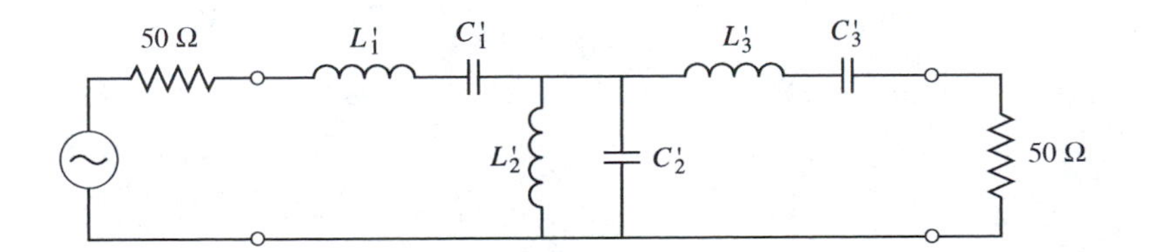

FIGURE 5.11 Bandpass filter circuit for Example 5.3.

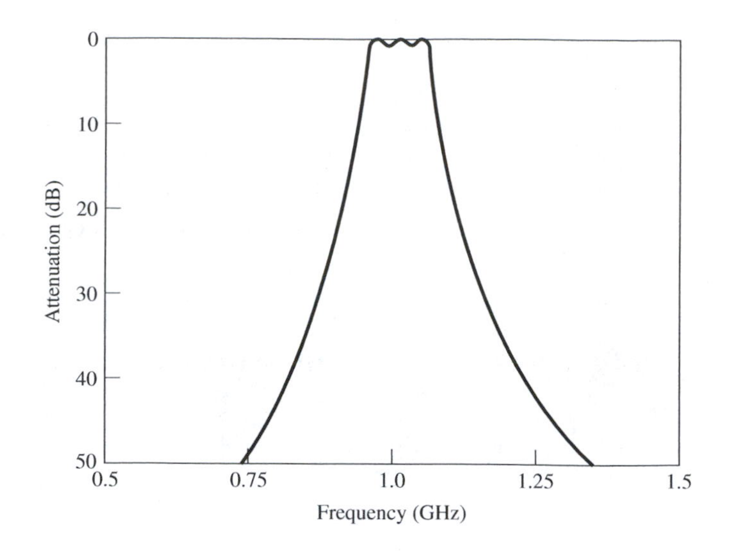

FIGURE 5.12 Amplitude response for the bandpass filter of Example 5.3.

5.3 LOW-PASS AND HIGH-PASS FILTERS USING TRANSMISSION LINE STUBS

The lumped-element filters discussed in the previous sections generally work well at low frequencies, but two problems arise at higher RF and microwave frequencies. First, lumped elements such as inductors and capacitors are generally available only for a limited range of values, and are difficult to implement at high frequencies. Instead, distributed components, such open- or short-circuited transmission line stubs, can be used as reactive elements. In addition, at microwave frequencies the electrical distance between filter components is not negligible. *Richard's transformation* can be used to convert lumped elements to transmission line stubs, while *Kuroda's identities* can be used to separate filter elements by using transmission line sections. Because such additional transmission line sections do not affect the filter response, this type of design is called *redundant* filter synthesis. It is possible to design microwave filters that take advantage of these sections to improve the filter response, but such *nonredundant* synthesis does not have a lumped-element counterpart [4].

Richard's Transformation

The transformation,

$$\Omega = \tan \beta\ell = \tan\left(\frac{\omega\ell}{v_p}\right), \tag{5.28}$$

maps the ω plane to the Ω plane, which repeats with a period of $\omega\ell/v_p = 2\pi$. This transformation was introduced by P. Richard [5] to synthesize an LC network using open- and short-circuited transmission lines. Thus, if we replace the frequency variable ω with Ω, the reactance of an inductor can be written as

$$jX_L = j\Omega L = jL\tan\beta\ell, \tag{5.29a}$$

and the susceptance of a capacitor can be written as

$$jB_c = j\Omega C = jC\tan\beta\ell. \tag{5.29b}$$

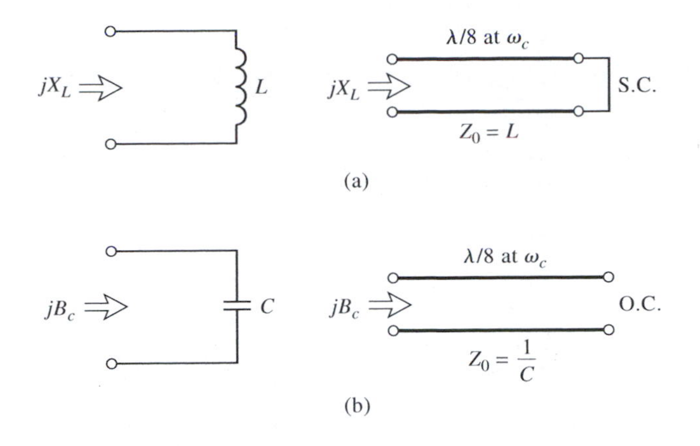

FIGURE 5.13 Richard's transformation. (a) For an inductor transformed to a short-circuited stub. (b) For a capacitor transformed to an open-circuited stub.

These results indicate that an inductor can be replaced with a short-circuited stub of length $\beta\ell$ and characteristic impedance L, while a capacitor can be replaced with an open-circuited stub of length $\beta\ell$ and characteristic impedance $1/C$. A unity (normalized) filter impedance is assumed here.

Cutoff occurs at unity frequency for a low-pass filter prototype; to obtain the same cutoff frequency for the Richard's-transformed filter, (5.28) shows that

$$\Omega = 1 = \tan \beta\ell,$$

which gives a stub length of $\ell = \lambda/8$, where λ is the wavelength of the line at the cutoff frequency, ω_c. At the frequency $\omega_0 = 2\omega_c$, the lines will be $\lambda/4$ long, and an attenuation pole will occur. At frequencies away from ω_c, the impedance of the stubs will no longer match the original lumped-element impedances, and the filter response will differ from the desired prototype response. Also, the response will be periodic in frequency, repeating every $4\omega_c$.

In principle, then, the inductors and capacitors of a lumped-element filter design can be replaced with short-circuited and open-circuited stubs, as illustrated in Figure 5.13. Since the lengths of all stubs are the same ($\lambda/8$ at ω_c), these lines are called *commensurate lines*.

Kuroda's Identities

The four Kuroda identities use redundant transmission line sections to achieve a more practical microwave filter implementation by performing any of the following operations:

- Physically separate transmission line stubs
- Transform series stubs into shunt stubs, or vice versa
- Change impractical characteristic impedances into more realizable ones

The additional transmission lines are called *unit elements*, and are $\lambda/8$ long at ω_c; the unit elements are thus commensurate with the stubs obtained by Richard's transform from the prototype design.

The four Kuroda identities are illustrated in Table 5.5, where each box represents a unit element, or transmission line, of the indicated characteristic impedance and length ($\lambda/8$ at ω_c). The inductors and capacitors represent short-circuit and open-circuit stubs, respectively. We will prove the equivalence of the first case, and then show how to use these identities in Example 5.4.

TABLE 5.5 The Four Kuroda Identities

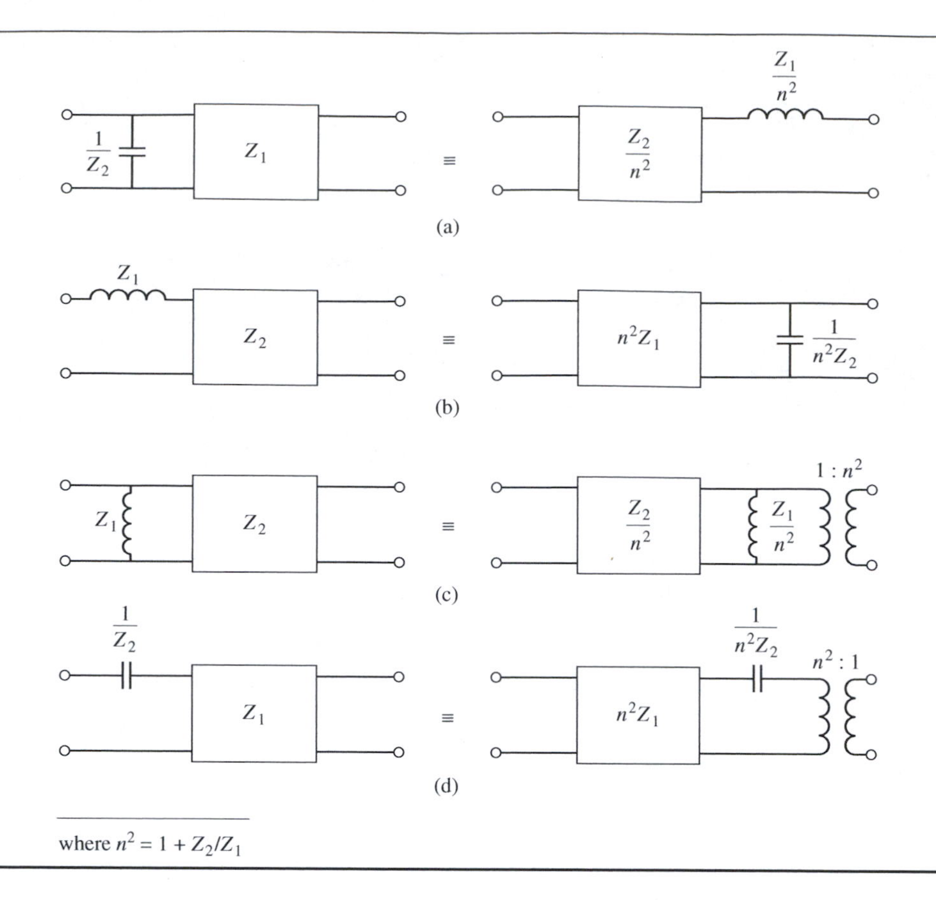

where $n^2 = 1 + Z_2/Z_1$

The two circuits of identity (a) in Table 5.5 can be redrawn as shown in Figure 5.14a,b, where a load resistance R_L is used to terminate both circuits. We will show that these two circuits are identical by showing that the input impedances are equal for all frequencies. For the circuit of Figure 5.14a, the transmission line impedance formula of (2.26) gives the admittance seen looking into the transmission line section as

$$Y' = \frac{Z_1 + jR_L \tan\beta\ell}{Z_1(R_L + jZ_1 \tan\beta\ell)} = \frac{Z_1 + jR_L\Omega}{Z_1(R_L + jZ_1\Omega)}, \tag{5.30}$$

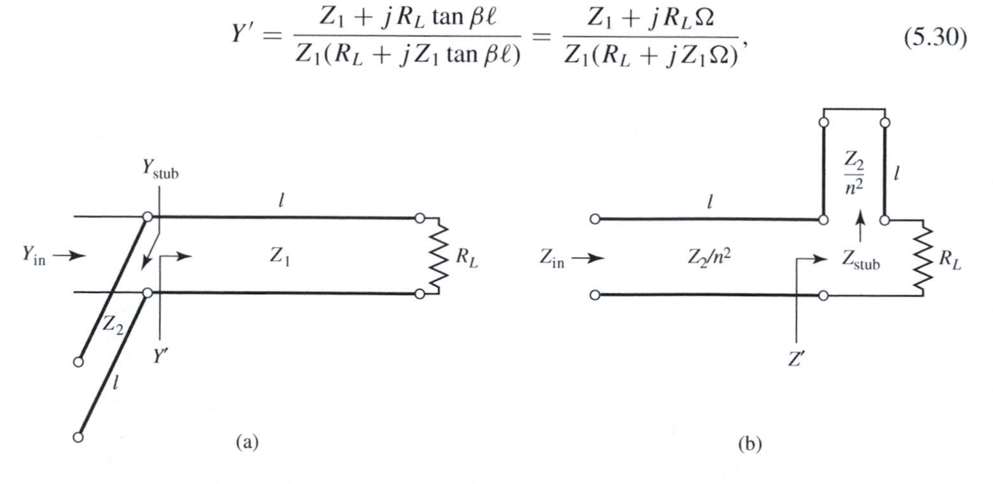

FIGURE 5.14 Equivalent circuits illustrating Kuroda identity (a) in Table 5.5. (a) Original circuit with load R_L. (b) Transformed circuit with load R_L.

where $\Omega = \tan \beta\ell$, as before. The input admittance of the open-circuited stub is, from (2.30),

$$Y_{\text{stub}} = \frac{j}{Z_2} \tan \beta\ell = \frac{j\Omega}{Z_2}. \tag{5.31}$$

Thus the overall admittance seen at the input to the circuit is

$$Y_{\text{in}} = Y_{\text{stub}} + Y' = \frac{j\Omega}{Z_2} + \frac{Z_1 + jR_L\Omega}{Z_1(R_L + jZ_1\Omega)} = \frac{(Z_2 - Z_1\Omega^2) + jn^2R_L\Omega}{Z_2(R_L + jZ_1\Omega)}, \tag{5.32}$$

where $n^2 = 1 + Z_2/Z_1$, as defined in Table 5.5.

Carrying out the corresponding analysis of the circuit of Figure 5.14b gives a stub impedance of

$$Z_{\text{stub}} = j\frac{Z_1}{n^2} \tan \beta\ell = \frac{jZ_1}{n^2}\Omega, \tag{5.33}$$

and an effective load impedance on the transmission line section of

$$Z' = Z_{\text{stub}} + R_L = R_L + \frac{jZ_1}{n^2}\Omega. \tag{5.34}$$

Using the transmission line impedance formula gives the overall input impedance as

$$Z_{\text{in}} = \frac{Z_2}{n^2}\frac{Z' + j\frac{Z_2}{n^2}\Omega}{\frac{Z_2}{n^2} + jZ'\Omega} = \frac{Z_2}{n^2}\frac{R_L + j\frac{Z_1}{n^2}\Omega + j\frac{Z_2}{n^2}\Omega}{\frac{Z_2}{n^2} + jR_L\Omega - \frac{Z_1}{n^2}\Omega^2} = Z_2\frac{R_L + jZ_1\Omega}{(Z_2 - Z_1\Omega^2) + jn^2R_L\Omega}. \tag{5.35}$$

Comparison with (5.32) shows that $Z_{\text{in}} = 1/Y_{\text{in}}$. Since this equality applies for all frequencies (Ω), the two circuits of Figure 5.14 are identical, and can be used interchangeably. Similar derivations can be used to establish the validity of the remaining Kuroda identities.

EXAMPLE 5.4 LOW-PASS FILTER DESIGN USING STUBS

Design a low-pass filter for fabrication using microstrip lines. The cutoff frequency is 4 GHz, and the impedance is 50 Ω. Use a third-order design, with a 3 dB equal-ripple passband characteristic.

Solution

From Table 5.2, the normalized low-pass prototype element values are

$$\begin{aligned} g_1 &= 3.3487 = L_1 \\ g_2 &= 0.7117 = C_2 \\ g_3 &= 3.3487 = L_3 \\ g_4 &= 1.0000 = R_L, \end{aligned}$$

with the lumped-element circuit shown in Figure 5.15a.

The next step is to use Richard's transformations to convert series inductors to series short-circuited stubs, and shunt capacitors to shunt open-circuited stubs, as shown in Figure 5.15b. According to (5.29), the characteristic impedance of a series stub (inductor) is L, and the characteristic impedance of a shunt stub (capacitor) is $1/C$. For commensurate line synthesis, all stubs are $\lambda/8$ long at $\omega = \omega_c$. It is usually most convenient to work with normalized quantities until the last step in the design.

The series stubs shown in Figure 5.15b would be difficult to implement in microstrip form, because series connections cannot be made to other microstrip lines. Thus we will use a Kuroda identity to convert these to shunt stubs. First we must add unit elements at each end of the filter circuit, as shown in Figure 5.15c. These redundant elements do not affect filter performance since they are matched to the source and load impedances ($Z_0 = 1$). Then we can apply Kuroda identity (b) from Table 5.5 to both ends of the filter. In both cases we have that

$$n^2 = 1 + \frac{Z_2}{Z_1} = 1 + \frac{1}{3.3487} = 1.299.$$

The transformed circuit is shown in Figure 5.15d, where all stubs are now in shunt with the transmission lines.

Finally, we impedance and frequency scale the circuit, which simply involves multiplying the normalized characteristic impedances by 50 Ω, and choosing the line and stub lengths to be $\lambda/8$ at 4 GHz. The final circuit is shown in Figure 5.15e, with the microstrip layout in Figure 5.15f.

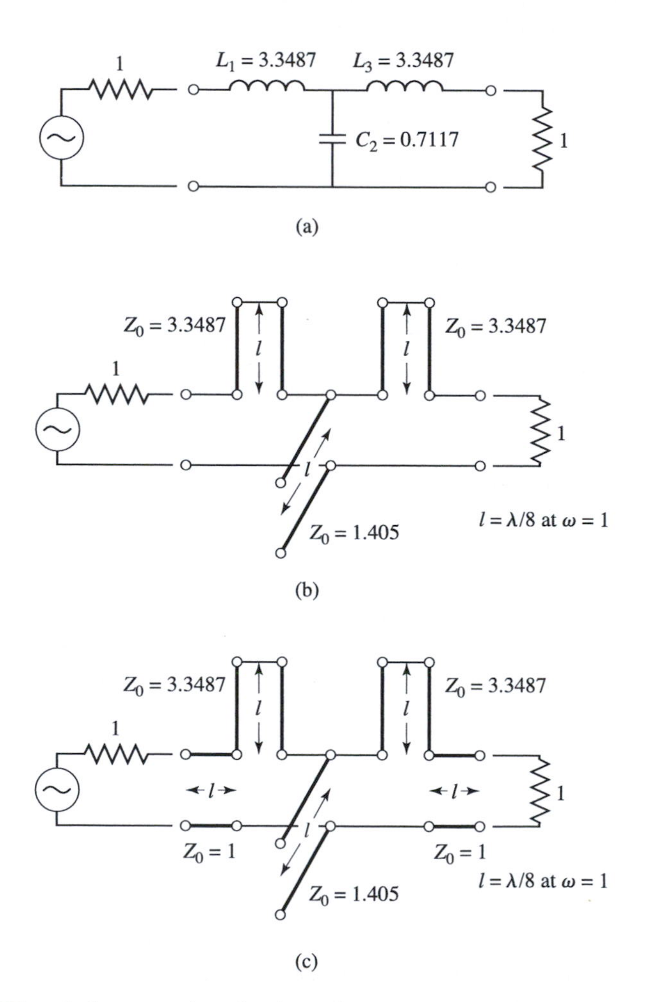

FIGURE 5.15 Filter design procedure for Example 5.4. (a) Lumped element low-pass filter prototype. (b) Using Richard's transformations to convert inductors and capacitors to series and shunt stubs. (c) Adding unit elements at ends of filter. (d) Applying the second Kuroda identity. (e) After impedance and frequency scaling. (f) Microstrip fabrication of final filter.

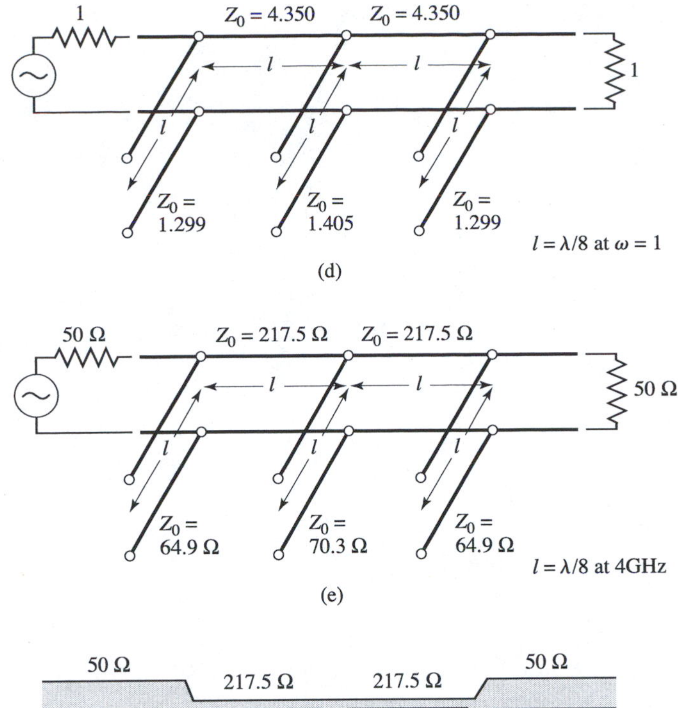

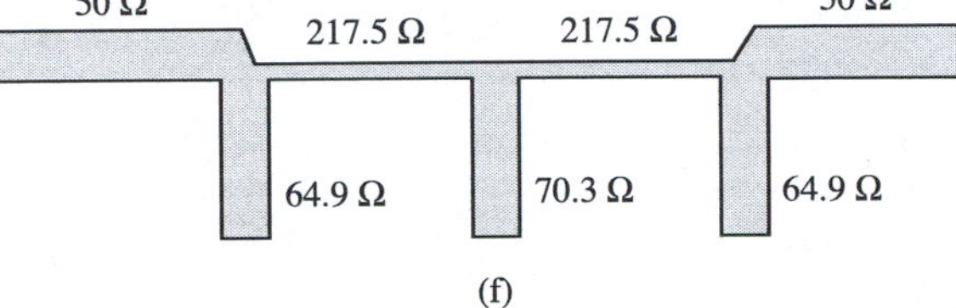

FIGURE 5.15 *(Continued)*.

The calculated amplitude response of this design is plotted in Figure 5.16, along with the response of the lumped-element version of the filter (scaled from Figure 5.15a). Note that the passband characteristics are very similar up to 4 GHz, but the distributed-element filter has a much sharper cutoff response. Also notice that the distributed-element filter has a response that repeats every 16 GHz, as a result of the periodic nature of Richard's transformation. ○

Similar procedures can be used for bandstop filters, but the Kuroda identities are not useful for high-pass or bandpass filters.

5.4 STEPPED-IMPEDANCE LOW-PASS FILTERS

A relatively easy way to implement low-pass filters in microstrip or stripline form is to use alternating sections of very high and very low characteristic lines. Such filters are usually referred to a *stepped-impedance*, or *hi-Z*, *low-Z*, filters, and are popular because they are easy to design and take up less space than a similar low-pass filter using stubs. Because of the approximations involved, however, their electrical performance is often not as good as that of stub filters, so the use of such filters is usually limited to applications where a sharp cutoff is not required, such as for rejection of out-of-band mixer products.

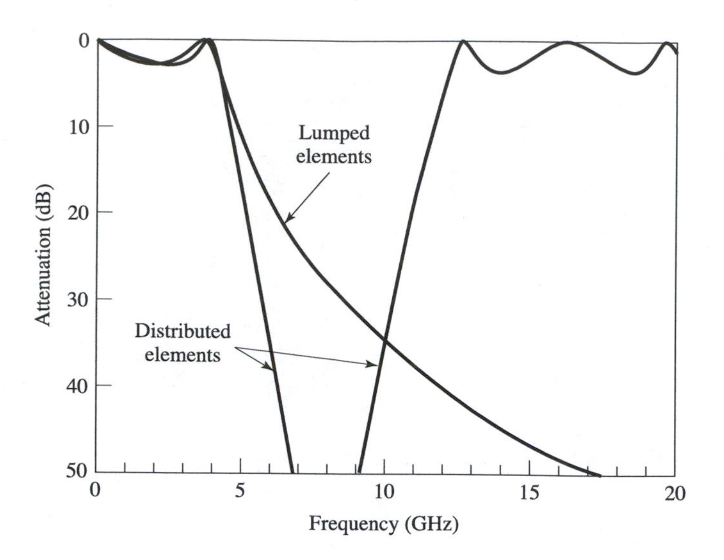

FIGURE 5.16 Amplitude response of lumped-element and distributed-element low-pass filter of Example 5.4.

Approximate Equivalent Circuits for Short Transmission Line Sections

We begin by finding the approximate equivalent circuits for a short length of transmission line having either a very large or a very small characteristic impedance. The open-circuit impedance matrix elements for a transmission line of length ℓ and characteristic impedance Z_0 can easily be found as follows:

$$Z_{11} = Z_{22} = \left.\frac{V_1}{I_1}\right|_{I_2=0} = -jZ_0 \cot \beta\ell, \tag{5.36a}$$

$$Z_{12} = Z_{21} = \left.\frac{V_2}{I_1}\right|_{I_2=0} = -jZ_0 \csc \beta\ell. \tag{5.36b}$$

The series elements of a T-equivalent circuit for the transmission line section are then given as $Z_{11} - Z_{12}$ for the series arms, and Z_{12} for the shunt arm. The series arm impedances simplify as follows:

$$Z_{11} - Z_{12} = -jZ_0\left[\frac{\cos \beta\ell - 1}{\sin \beta\ell}\right] = jZ_0 \tan\left(\frac{\beta\ell}{2}\right). \tag{5.37}$$

Now if the length of the line is small, so that $\beta\ell < \pi/2$, the series elements will have a positive reactance (inductors), while the shunt element has a negative reactance (capacitor). We thus have the equivalent circuit shown in Figure 5.17a, where

$$\frac{X}{2} = Z_0 \tan\left(\frac{\beta\ell}{2}\right), \tag{5.38a}$$

$$B = \frac{1}{Z_0} \sin \beta\ell. \tag{5.38b}$$

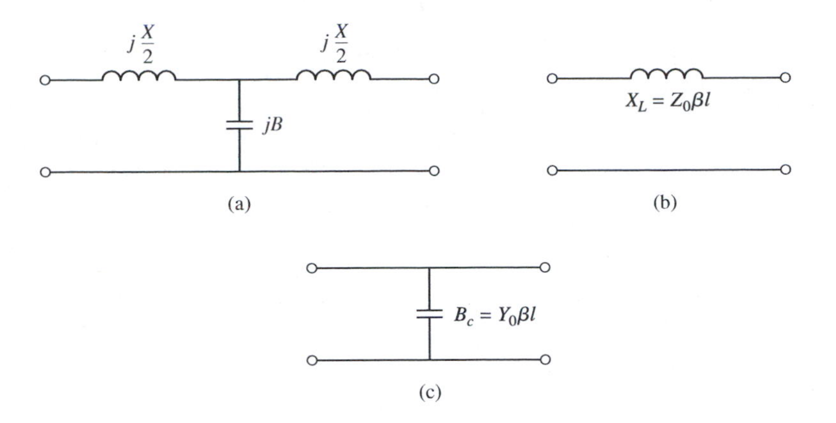

FIGURE 5.17 Approximate equivalent circuits for short sections of transmission lines. (a) T-equivalent circuit for a transmission line section having $\beta\ell \ll \pi/2$. (b) Equivalent circuit for small $\beta\ell$ and large Z_0. (c) Equivalent circuit for small $\beta\ell$ and small Z_0.

Now assume a short length of line (say $\beta\ell < \pi/4$), and a large characteristic impedance. Then (5.38) approximately reduces to

$$X \cong Z_0\beta\ell, \tag{5.39a}$$

$$B \cong 0, \tag{5.39b}$$

which implies the equivalent circuit of Figure 5.17b (a series inductor). Alternatively, for a short length of line and a small characteristic impedance, (5.38) approximately reduces to

$$X \cong 0, \tag{5.40a}$$

$$B \cong Y_0\beta\ell, \tag{5.40b}$$

which implies the equivalent circuit of Figure 5.17c (a shunt capacitor). Based on these results, we see that the series inductors of a low-pass prototype filter can be replaced with high-impedance transmission line sections ($Z_0 = Z_h$), and the shunt capacitors can be replaced with low-impedance transmission line sections ($Z_0 = Z_\ell$). The ratio Z_h/Z_ℓ should be as high as possible, so the actual values of Z_h and Z_ℓ are usually set to the highest and lowest characteristic impedances that can be practically fabricated (the thinnest and widest lines, respectively). The lengths of the lines can then be determined from (5.39) and (5.40); to get the best response near cutoff, these lengths should be evaluated at $\omega = \omega_c$. Combining the results of (5.39) and (5.40) with the impedance scaling equations of (5.15) allows the electrical lengths of the inductor sections to be calculated as

$$\beta\ell = \frac{LR_0}{Z_h} \quad \text{(inductor)}, \tag{5.41a}$$

and the electrical length of the capacitor sections as

$$\beta\ell = \frac{CZ_\ell}{R_0} \quad \text{(capacitor)}, \tag{5.41b}$$

where R_0 is the filter impedance and L and C are the normalized element values (the g_ks) of the low-pass prototype filter.

EXAMPLE 5.5 STEPPED-IMPEDANCE FILTER DESIGN

Design a stepped-impedance low-pass filter having a maximally flat response and a cutoff frequency of 2.5 GHz. It is necessary to have at least 20 dB attenuation at 4.0 GHz. The filter impedance is 50 Ω; the highest practical line impedances is 150 Ω, and the lowest is 10 Ω.

Solution

We first determine the required order of the filter based on the out-of-band attenuation specification. To use Figure 5.5, we calculate

$$\frac{\omega}{\omega_c} - 1 = \frac{4.0}{2.5} - 1 = 0.6,$$

then the figure indicates that $N = 6$ should give the desired attenuation at 4.0 GHz. Table 5.1 gives the low-pass prototype values as

$$\begin{aligned} g_1 &= 0.517 = C_1 \\ g_2 &= 1.414 = L_2 \\ g_3 &= 1.932 = C_3 \\ g_4 &= 1.932 = L_4 \\ g_5 &= 1.414 = C_5 \\ g_6 &= 0.517 = L_6. \end{aligned}$$

The low-pass prototype filter circuit is shown in Figure 5.18a.

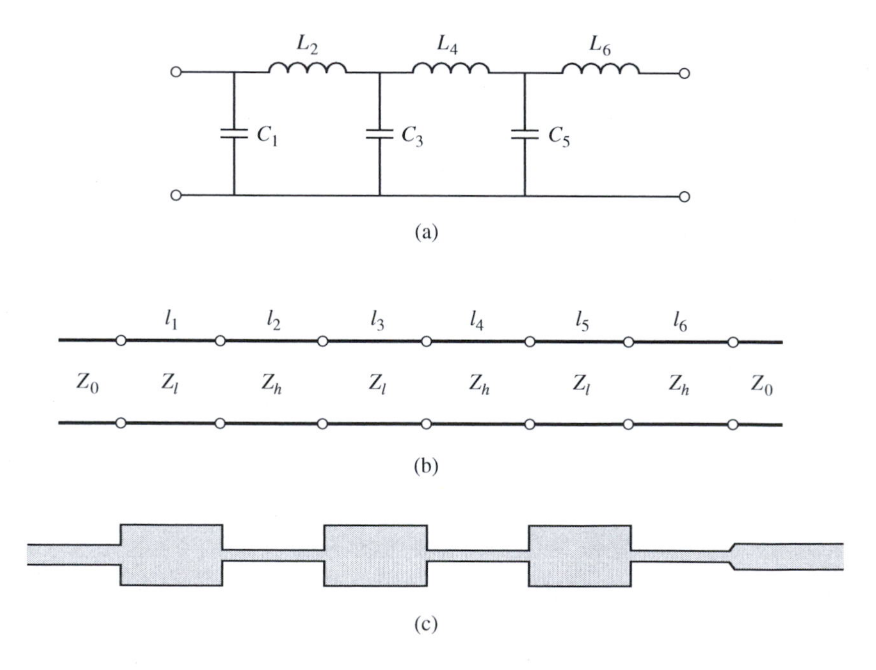

FIGURE 5.18 Filter design for Example 5.5. (a) Low-pass filter prototype circuit. (b) Stepped-impedance implementation. (c) Microstrip layout of final filter.

Next, we use (5.41) to find the electrical lengths of the hi-Z, low-Z transmission line sections to replace the series inductors and shunt capacitors:

$$\beta\ell_1 = g_1 \frac{Z_\ell}{R_0} = 5.9^\circ,$$

$$\beta\ell_2 = g_2 \frac{R_0}{Z_h} = 27.0^\circ,$$

$$\beta\ell_3 = g_3 \frac{Z_\ell}{R_0} = 22.1^\circ,$$

$$\beta\ell_4 = g_4 \frac{R_0}{Z_h} = 36.9^\circ,$$

$$\beta\ell_5 = g_5 \frac{Z_\ell}{R_0} = 16.2^\circ,$$

$$\beta\ell_6 = g_6 \frac{R_0}{Z_h} = 9.9^\circ.$$

The final filter circuit is shown in Figure 5.18b, where $Z_\ell = 10\,\Omega$ and $Z_h = 150\,\Omega$. Note that $\beta\ell < \pi/4$ in all cases. A layout of the filter in microstrip is shown in Figure 5.18c.

Figure 5.19 shows the calculated amplitude response, compared with the response of the corresponding lumped-element filter (scaled from Figure 5.18a). The passband characteristics are very similar, but the lumped-element circuit gives more attenuation at higher frequencies. This is because the stepped-impedance filter elements depart significantly from the lumped-element values at the higher frequencies. The stepped-impedance filter may have other passbands at higher frequencies, but the response will not be perfectly periodic because the sections are not commensurate in length. ○

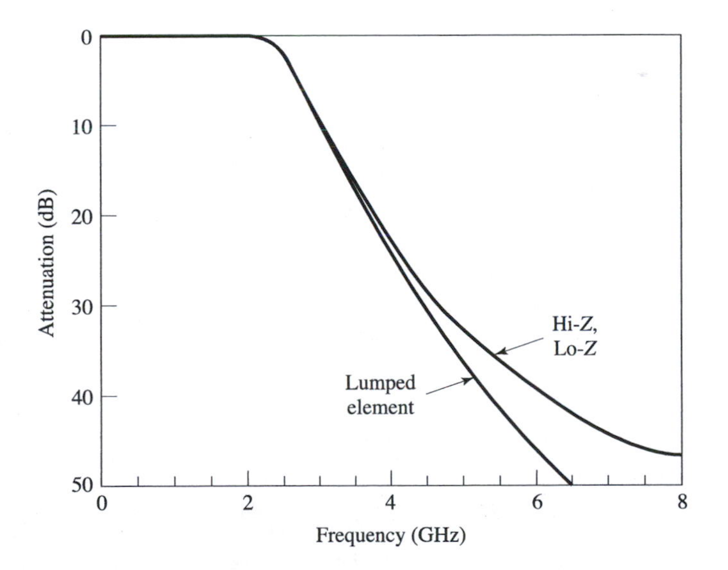

FIGURE 5.19 Amplitude response of the stepped-impedance low-pass filter of Example 5.5, compared with the corresponding lumped-element design.

5.5 BANDPASS FILTERS USING TRANSMISSION LINE RESONATORS

Bandpass filters perform a variety of critical functions in wireless systems, being used to reject out-of-band and image signals in the front end of a receiver, to attenuate undesired mixer products in transmitters and receivers, and to set the IF bandwidth of the receiver system. Because of their importance, a large number of different types of bandpass filters have been developed (see [1]–[3]), but we can only treat some of the basic principles of operation and designs here. We begin with a discussion of impedance and admittance inverters, which form the basis of design for many different types of bandpass filters. Then we present analysis and design details for two types of bandpass filters using quarter-wave resonators. These types of filters are among the most commonly used in practical wireless systems.

Impedance and Admittance Inverters

As seen in Section 5.2, bandpass filter prototypes require shunt elements consisting of parallel LC resonators and series elements consisting of series LC resonators. Such an arrangement is very difficult to implement using transmission line sections, for which it is preferable to have either all shunt, or all series, elements. While the Kuroda identities are useful for transforming capacitors or inductors to either series or shunt transmission line stubs, they are not useful for transforming LC resonators. For this purpose, impedance (K) and admittance (J) inverters can be used. Such techniques are especially useful for bandpass and bandstop filters having narrow (<10%) bandwidths.

The conceptual operation of impedance and admittance inverters is illustrated in Figure 5.20a. An impedance inverter converts a load impedance to its inverse, while an

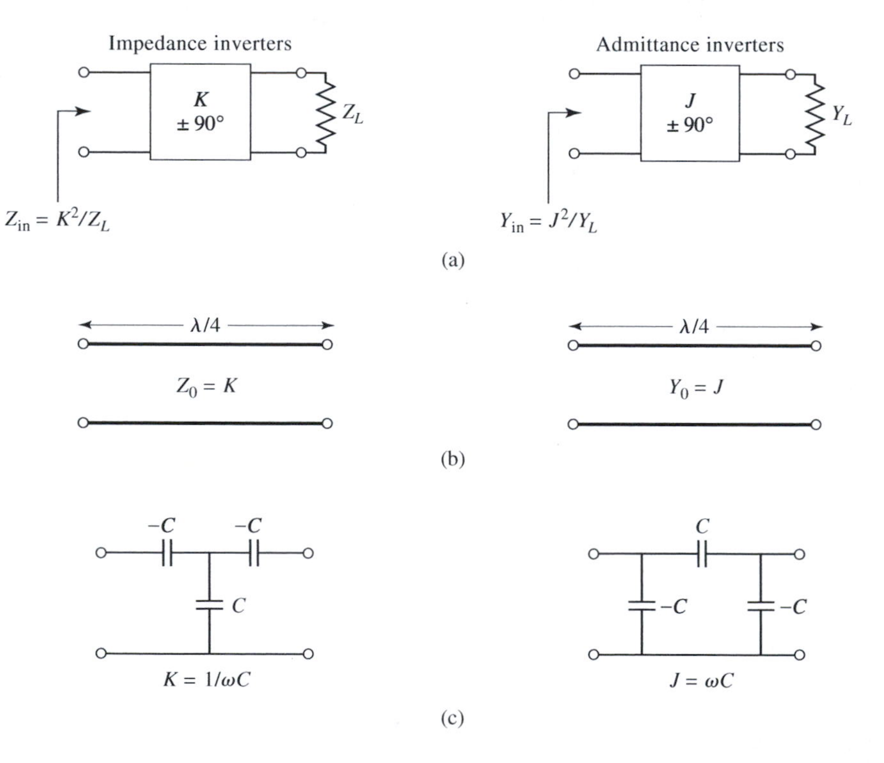

FIGURE 5.20 Impedance and admittance inverters. (a) Operation of impedance and admittance inverters. (b) Implementation as quarter-wave transformers. (c) Implementation using T and π capacitor circuits.

admittance inverter converts a load admittance to its inverse:

$$Z_{\text{in}} = \frac{K^2}{Z_L}, \tag{5.42a}$$

$$Y_{\text{in}} = \frac{J^2}{Y_L}, \tag{5.42b}$$

where K is the impedance inverter constant, and J is the admittance inverter constant. The utility of impedance and admittance inverters is that they can be used to transform between series-connected and shunt-connected elements. Thus, a series LC resonator can be transformed to a parallel LC resonator, or vice versa. The procedure for doing this will be illustrated for particular filter types in the following sections.

In its simplest form, a K or J inverter can be constructed using a quarter-wave transformer of the appropriate characteristic impedance, as shown in Figure 5.20b. It is clear from the relation for the input impedance of a quarter-wave, $Z_{\text{in}} = Z_0^2/Z_L$, that the results of (5.42) follow by setting Z_0 to the inverter constant K or J. Several other types of circuits can be derived for use as impedance and admittance inverters [1]–[3]. One of these is shown in Figure 5.20c, consisting of either a T network (K inverter), or a π network (J inverter), of capacitors. Note that the capacitor value is related to the inverter constant as shown in the figure, and that some of the capacitors have negative values. The procedure for using this type of inverter is shown next.

Bandpass Filters Using Quarter-Wave Coupled Quarter-Wave Resonators

Since quarter-wave short-circuited transmission line stubs look like parallel resonant circuits [2], they can be used as the shunt parallel LC resonators for bandpass filters. Quarter-wavelength connecting lines between the stubs will act as admittance inverters, effectively converting alternate shunt stubs to series resonators. Such an arrangement is shown in Figure 5.21; both the stubs and the connecting lines are $\lambda/4$ long at the center frequency of the passband, ω_0. The characteristic impedance of the connecting lines is Z_0, the impedance of the filter.

For a narrow passband bandwidth (small Δ), the response of such a filter using N stubs is essentially the same as that of a lumped element bandpass filter of order N. The circuit topology of this filter is convenient in that only shunt stubs are used, but a disadvantage in practice is that the required characteristic impedances of the stub lines are often unrealistically low. A similar design employing open-circuited stubs can be used for bandstop filters [1]–[2].

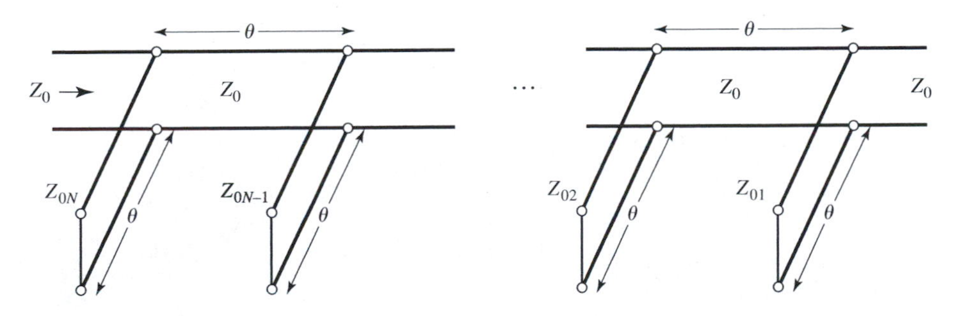

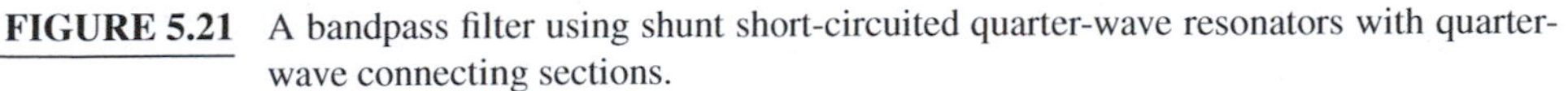

FIGURE 5.21 A bandpass filter using shunt short-circuited quarter-wave resonators with quarter-wave connecting sections.

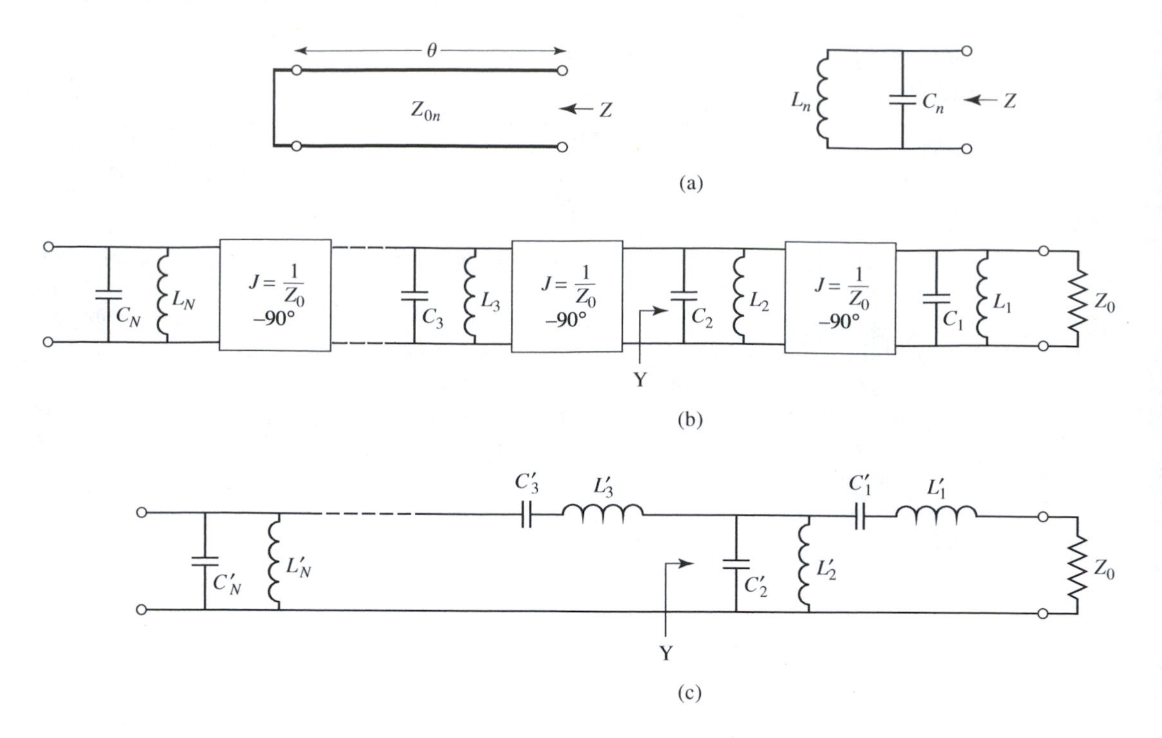

FIGURE 5.22 Equivalent circuit for the bandpass filter of Figure 5.21. (a) Equivalent circuit for a short-circuited stub for θ near $\pi/2$. (b) Equivalent filter circuit after replacing stubs with parallel LC resonators and quarter-wave connecting lines with admittance inverters. (c) Equivalent lumped-element bandpass filter.

Consider an N-order bandpass filter of the form shown in Figure 5.21. We will derive design equations for the stub characteristic impedances, Z_{0n}, in terms of the element values of a low-pass prototype having the desired response. This can be accomplished by using equivalent circuits for the resonant stubs and connecting lines, and equating the response to that of a lumped element bandpass filter. Note that a given LC resonator has two degrees of freedom: L and C, or equivalently, ω_0 and the slope of the admittance at resonance. For a stub resonator the corresponding degrees of freedom are the resonant length and characteristic impedance of the transmission line.

As shown in Figure 5.22a, the equivalent circuit of a short-circuited transmission line stub can be approximated as a parallel LC resonator when its length is near 90°. The input admittance of a short-circuited transmission line of characteristic impedance Z_{0n} is

$$Y = \frac{-j}{Z_{0n}} \cot\theta, \tag{5.43}$$

where $\theta = \pi/2$ for $\omega = \omega_0$. If we let $\omega = \omega_0 + \Delta\omega$, where $\Delta\omega \ll \omega_0$, then $\theta = \frac{\pi}{2}(1 + \frac{\Delta\omega}{\omega_0})$, which allows the admittance of (5.43) to be approximated as

$$Y = \frac{-j}{Z_{0n}} \cot\left(\frac{\pi}{2} + \frac{\pi\,\Delta\omega}{2\omega_0}\right) = \frac{j}{Z_{0n}} \tan\frac{\pi\,\Delta\omega}{2\omega_0} \cong \frac{j\pi\,\Delta\omega}{2Z_{0n}\omega_0}, \tag{5.44}$$

for frequencies in the vicinity of the center frequency, ω_0. The admittance near resonance of the parallel LC network of Figure 5.22a can be approximated as

$$Y = j\omega C_n + \frac{1}{j\omega L_n} = j\sqrt{\frac{C_n}{L_n}}\left(\omega\sqrt{C_n L_n} - \frac{1}{\omega\sqrt{C_n L_n}}\right)$$
$$= j\sqrt{\frac{C_n}{L_n}}\left(\frac{\omega}{\omega_0} - \frac{\omega_0}{\omega}\right) \cong 2jC_n\Delta\omega, \tag{5.45}$$

where $C_n L_n = 1/\omega_0^2$. Equating (5.44) and (5.45) gives the characteristic impedance of the transmission line stub in terms of the resonator parameters as

$$Z_{0n} = \frac{\pi\omega_0 L_n}{4} = \frac{\pi}{4\omega_0 C_n}. \tag{5.46}$$

Next, we consider the quarter-wave sections of line between the stubs as ideal admittance inverters, with $J = 1/Z_0$. Then the bandpass filter of Figure 5.21 can be represented by the equivalent circuit shown in Figure 5.22b, which further can be shown to be equivalent to the lumped-element circuit of Figure 5.22c by basic circuit analysis. Thus, with reference to the terminated (with Z_0) circuit of Figure 5.22b, the admittance, Y, seen looking toward the L_2C_2 resonator is

$$Y = j\omega C_2 + \frac{1}{j\omega L_2} + \frac{1}{Z_0^2}\left[j\omega C_1 + \frac{1}{j\omega L_1} + \frac{1}{Z_0}\right]^{-1}$$
$$= j\sqrt{\frac{C_2}{L_2}}\left(\frac{\omega}{\omega_0} - \frac{\omega_0}{\omega}\right) + \frac{1}{Z_0^2}\left[j\sqrt{\frac{C_1}{L_1}}\left(\frac{\omega}{\omega_0} - \frac{\omega_0}{\omega}\right) + \frac{1}{Z_0}\right]^{-1} \tag{5.47}$$

where use has been made of the fact that $L_1C_1 = L_2C_2 = 1/\omega_0^2$. The admittance at the corresponding point in the equivalent circuit of Figure 5.22c (also terminated in Z_0) is found as

$$Y = j\omega C_2' + \frac{1}{j\omega L_2'}\left[j\omega L_1' + \frac{1}{j\omega C_1'} + Z_0\right]^{-1}$$
$$= j\sqrt{\frac{C_2'}{L_2'}}\left(\frac{\omega}{\omega_0} - \frac{\omega_0}{\omega}\right) + \left[j\sqrt{\frac{L_1'}{C_1'}}\left(\frac{\omega}{\omega_0} - \frac{\omega_0}{\omega}\right) + Z_0\right]^{-1} \tag{5.48}$$

These two results are exactly equivalent for all frequencies if the following conditions are satisfied:

$$\sqrt{\frac{C_2}{L_2}} = \sqrt{\frac{C_2'}{L_2'}}, \tag{5.49a}$$

and

$$Z_0^2\sqrt{\frac{C_1}{L_1}} = \sqrt{\frac{L_1'}{C_1'}}. \tag{5.49b}$$

Using the fact that $L_1'C_1' = L_2'C_2' = 1/\omega_0^2$ allows these two equations to be solved for L_1 and L_2:

$$L_1 = \frac{Z_0^2}{\omega_0^2 L_1'}, \tag{5.50a}$$
$$L_2 = L_2'. \tag{5.50b}$$

Then using (5.46) and the impedance-scaled bandpass filter elements from Table 5.4 gives the required characteristic impedances for the first two stubs as

$$Z_{01} = \frac{\pi \omega_0 L_1}{4} = \frac{\pi Z_0^2}{4\omega_0 L_1'} = \frac{\pi Z_0 \Delta}{4g_1}, \tag{5.51a}$$

$$Z_{02} = \frac{\pi \omega_0 L_2}{4} = \frac{\pi \omega_0 L_2'}{4} = \frac{\pi Z_0 \Delta}{4g_2}. \tag{5.51b}$$

By extension, it can be shown that the general result for the characteristic impedance of the nth stub in a filter of order N is given by

$$Z_{0n} = \frac{\pi Z_0 \Delta}{4g_n}. \tag{5.52}$$

These results apply only to filters having input and output impedances of Z_0, and so cannot be used for equal-ripple designs with N even.

EXAMPLE 5.6 BANDPASS FILTER DESIGN USING QUARTER-WAVE COUPLED RESONATORS

Design a third-order bandpass filter with a 0.5 dB equal-ripple response using quarter-wave coupled quarter-wave short-circuited stub resonators. The center frequency is 2.5 GHz, and the bandwidth is 15%. The impedance is 50 Ω. What is the resulting attenuation at 2.0 GHz?

Solution

We first calculate the attenuation at 2.0 GHz. Using (5.22) to convert 2.0 GHz to normalized low-pass form gives

$$\omega \leftarrow \frac{1}{\Delta}\left(\frac{\omega}{\omega_0} - \frac{\omega_0}{\omega}\right) = \frac{1}{0.15}\left(\frac{2.0}{2.5} - \frac{2.5}{2.0}\right) = -3.00.$$

Then, to use Figure 5.6a, the value on the horizontal axis is

$$\left|\frac{\omega}{\omega_c}\right| - 1 = |-3.00| - 1 = 2.00,$$

from which we find the attenuation as 30 dB.

From Table 5.2 we find the required g_n's for 0.5 dB ripple and $N = 3$. Then (5.52) gives the necessary characteristic impedances:

n	g_n	Z_{0n} (Ω)
1	1.5963	3.69
2	1.0967	5.37
3	1.5963	3.69

All stubs and connecting lines are $\lambda/4$ long at 2.5 GHz. The calculated response of the filter is shown in Figure 5.23. Note that 30 dB attenuation is achieved at 2 GHz, as expected. Also note that the characteristic impedances for the stubs are very low, making practical implementation of this type of filter very difficult. This difficulty is avoided with the capacitively coupled resonator filter discussed in the following section. ○

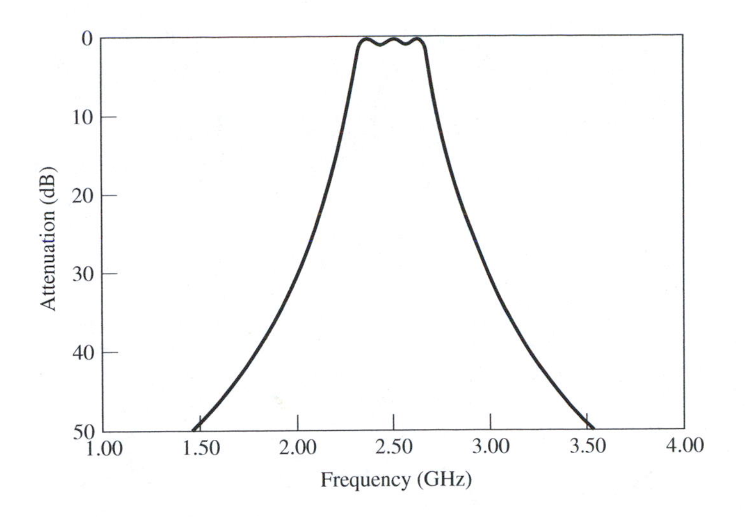

FIGURE 5.23 Amplitude response of the quarter-wave coupled quarter-wave resonator bandpass filter of Example 5.6.

Bandpass Filters Using Capacitively Coupled Quarter-Wave Resonators

A related type of bandpass filter is shown in Figure 5.24, where short-circuited shunt resonators are capacitively coupled with series capacitors. An Nth order filter will use N stubs, which are slightly shorter than $\lambda/4$ at the filter center frequency. The short-circuited stub resonators can be made from sections of coaxial line using ceramic materials having very high dielectric constant and low loss, resulting in a very compact design even at UHF frequencies [6]. Such filters are often referred to as *ceramic resonator* filters, and are presently the most common type of RF bandpass filter used in portable wireless systems. Virtually every modern cellular/PCS telephone, wireless LAN, and GPS receiver employs between two and four of these filters.

Operation and design of this filter can be understood by beginning with the general bandpass filter circuit of Figure 5.25a, where shunt LC resonators alternate with admittance inverters. As in the case of the previous bandpass filter, the function of the admittance inverters is to convert alternate shunt resonators to series resonators; the extra inverters at the ends serve to scale the impedance level of the filter to a realistic level. Using an analysis similar to that used for the previous bandpass filter, the admittance inverter constants can

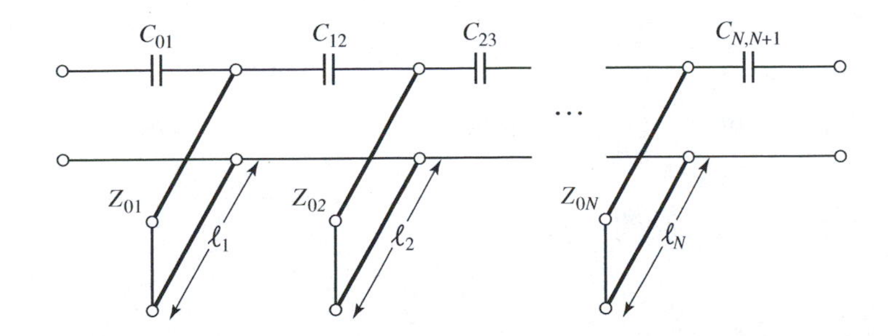

FIGURE 5.24 A bandpass filter using capacitively coupled shunt short-circuited quarter-wave resonators.

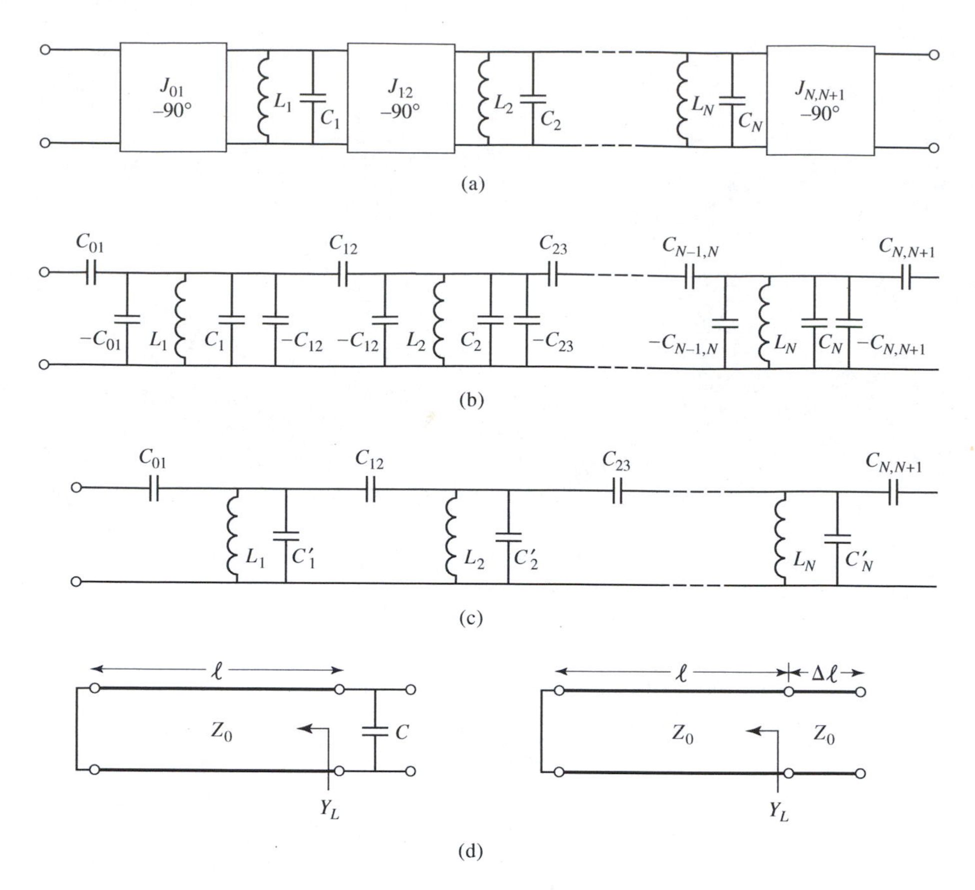

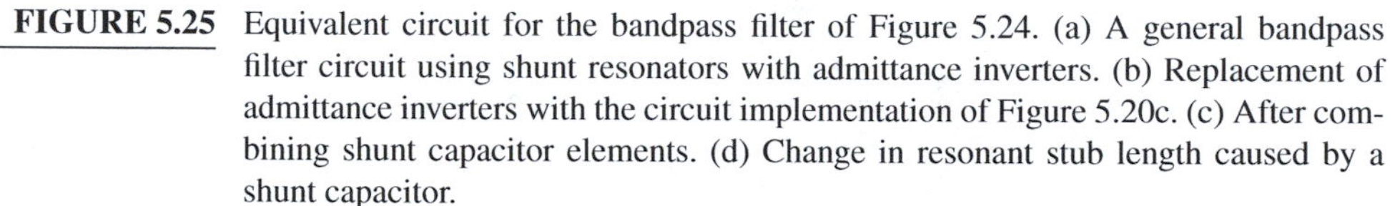

FIGURE 5.25 Equivalent circuit for the bandpass filter of Figure 5.24. (a) A general bandpass filter circuit using shunt resonators with admittance inverters. (b) Replacement of admittance inverters with the circuit implementation of Figure 5.20c. (c) After combining shunt capacitor elements. (d) Change in resonant stub length caused by a shunt capacitor.

be derived as [1]

$$Z_0 J_{01} = \sqrt{\frac{\pi \Delta}{4 g_1}}, \tag{5.53a}$$

$$Z_0 J_{n,n+1} = \frac{\pi \Delta}{4\sqrt{g_n g_{n+1}}}, \tag{5.53b}$$

$$Z_0 J_{N,N+1} = \sqrt{\frac{\pi \Delta}{4 g_N g_{N+1}}}. \tag{5.53c}$$

Similarly, the coupling capacitor values can be found as

$$C_{01} = \frac{J_{01}}{\omega_0 \sqrt{1 - (Z_0 J_{01})^2}}, \tag{5.54a}$$

$$C_{n,n+1} = \frac{J_{n,n+1}}{\omega_0}, \tag{5.54b}$$

$$C_{N,N+1} = \frac{J_{N,N+1}}{\omega_0 \sqrt{1 - (Z_0 J_{N,N+1})^2}}. \tag{5.54c}$$

Note that the end capacitors are treated differently than the internal elements.

Now replace the admittance inverters of Figure 5.25a with the equivalent π-network of Figure 5.20c, to produce the equivalent lumped-element circuit shown in Figure 5.25b. Note that the shunt capacitors of the admittance inverter circuits are negative, but these elements combine in parallel with the larger capacitor of the LC resonator to yield a positive capacitance value. The resulting circuit is shown in Figure 5.25c, where the effective resonator capacitor values are given by

$$C_n' = C_n + \Delta C_n = C_n - C_{n-1,n} - C_{n,n+1}, \tag{5.55}$$

where $\Delta C_n = -C_{n-1,n} - C_{n,n+1}$ represents the change in the resonator capacitance caused by the parallel addition of the inverter elements.

Finally, the shunt LC resonators of Figure 5.25c are replaced with short-circuited transmission stubs, as shown in Figure 5.24. Note that the resonant frequency of the stub resonators is no longer ω_0, since the resonator capacitor values have been modified by the ΔC_n's. This implies that the length of the resonator is less than $\lambda/4$ long at ω_0, the filter center frequency. The transformation of the stub length to account for the change in capacitance is illustrated in Figure 5.25d. A short-circuited length of line with a shunt capacitor at its input has an input admittance of

$$Y = Y_L + j\omega_0 C, \tag{5.56a}$$

where $Y_L = \frac{-j}{Z_0}\cot\beta\ell$. If the capacitor is replaced with a short length, $\Delta\ell$, of transmission line, the input admittance would be

$$Y = \frac{1}{Z_0}\frac{Y_L + j\frac{1}{Z_0}\tan\beta\Delta\ell}{\frac{1}{Z_0} + jY_L\tan\beta\Delta\ell} \cong Y_L + j\frac{\beta\Delta\ell}{Z_0}. \tag{5.56b}$$

The last approximation follows for $\beta\Delta\ell \ll 1$, which is true in practice for filters of this type. Comparing (5.56b) with (5.56a) gives the change in stub length in terms of the capacitor value:

$$\Delta\ell = \frac{Z_0\omega_0 C}{\beta} = \left(\frac{Z_0\omega_0 C}{2\pi}\right)\lambda. \tag{5.57}$$

Note that if $C < 0$, then $\Delta\ell < 0$, indicating a shortening of the stub length. Thus the overall stub length is given by

$$\ell_n = \frac{\lambda}{4} + \left(\frac{Z_0\omega_0\Delta C_n}{2\pi}\right)\lambda, \tag{5.58}$$

where ΔC_n is defined in (5.55). The characteristic impedance of the stub resonators is Z_0.

Dielectric material properties play a critical role in the performance of dielectric resonator filters. Materials with high dielectric constants are required in order to provide miniaturization at the frequencies typically used for wireless applications. Losses must be low to provide resonators with high Q, leading to low passband insertion loss and maximum attenuation in the stopbands. And the dielectric constant must be stable with changes in temperature to avoid drifting of the filter passband over normal operating conditions. Most materials that are commonly used in dielectric resonator filters are ceramics such as Barium tetratitanate, Zinc/Strontium titanate, and various titanium oxide compounds. For example, a Zinc/Strontium titanate ceramic material has a dielectric constant of 36, with a Q of 10,000 at 4 GHz, and a dielectric constant temperature coefficient of -7 ppm/C°.

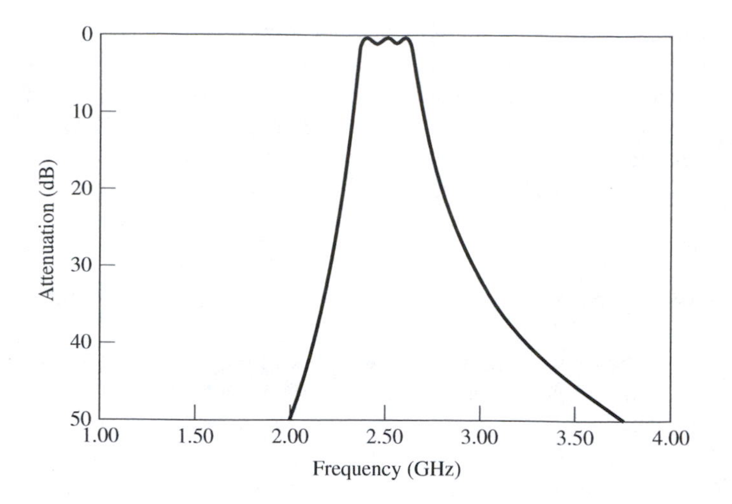

FIGURE 5.26 Amplitude response of the capacitively coupled quarter-wave resonator bandpass filter of Example 5.7.

EXAMPLE 5.7 BANDPASS FILTER DESIGN USING CAPACITIVELY COUPLED RESONATORS

Design a third-order bandpass filter with a 0.5 dB equal-ripple response using capacitively coupled quarter-wave short-circuited stub resonators. The center frequency is 2.5 GHz, and the bandwidth is 10%. The impedance is 50 Ω. What is the resulting attenuation at 3.0 GHz?

Solution

We first calculate the attenuation at 3.0 GHz. Using (5.22) to convert 3.0 GHz to normalized low-pass form gives

$$\omega \leftarrow \frac{1}{\Delta}\left(\frac{\omega}{\omega_0} - \frac{\omega_0}{\omega}\right) = \frac{1}{0.1}\left(\frac{3.0}{2.5} - \frac{2.5}{3.0}\right) = 3.667.$$

Then, to use Figure 5.6a, the value on the horizontal axis is

$$\left|\frac{\omega}{\omega_c}\right| - 1 = |-3.667| - 1 = 2.667,$$

from which we find the attenuation as 35 dB.

Next we calculate the admittance inverter constants and coupling capacitor values using (5.53) and (5.54):

n	g_n	$Z_0 J_{n-1,n}$	$C_{n-1,n}$ (pF)
1	1.5963	$Z_0 J_{01} = 0.2218$	$C_{01} = 0.2896$
2	1.0967	$Z_0 J_{12} = 0.0594$	$C_{12} = 0.0756$
3	1.5963	$Z_0 J_{23} = 0.0594$	$C_{23} = 0.0756$
4	1.0000	$Z_0 J_{34} = 0.2218$	$C_{34} = 0.2896$

Then (5.55), (5.57), and (5.58) are used to find the required resonator lengths:

n	ΔC_n (pF)	$\Delta \ell_n(\lambda)$	ℓ
1	−0.3652	−0.04565	73.6°
2	−0.1512	−0.0189	83.2°
3	−0.3652	−0.04565	73.6°

Note that the resonator lengths are slightly less than $90°(\lambda/4)$. The calculated amplitude response of this design is shown in Figure 5.26. The stopband rolloff at high frequencies is less than at lower frequencies, and the attenuation at 3 GHz is seen to be about 30 dB, while our calculated value for a canonical lumped-element bandpass filter was 35 dB. ○

REFERENCES

[1] G. L. Matthaei, L. Young, and E. M. T. Jones, **Microwave Filters, Impedance Matching Networks, and Coupling Structures**, Artech House, Dedham, MA, 1980.

[2] D. M. Pozar, **Microwave Engineering**, 2nd edition, Wiley, New York, 1998.

[3] J. A. G. Malherbe, **Microwave Transmission Line Filters**, Artech House, Dedham, MA, 1979.

[4] W. A. Davis, **Microwave Semiconductor Circuit Design**, Van Nostrand Reinhold, New York, 1984.

[5] P. I. Richard, "Resistor-Transmission Line Circuits," *Proc. of the IRE*, vol. 36, pp. 217–220, February 1948.

[6] M. Sagawa, M. Makimoto, and S. Yamashita, "A Design Method of Bandpass Filters Using Dielectric-Filled Coaxial Resonators," *IEEE Transactions on Microwave Theory and Techniques*, vol. MTT-33, pp. 152–157, February 1985.

PROBLEMS

5.1 Use the causality properties of $V(\omega)$, $I(\omega)$, and $Z(\omega)$ to show that $|\Gamma(\omega)|^2$ is an even function of ω.

5.2 Solve the design equations of Section 5.1 for the elements of an $N = 1$ equal-ripple filter if the ripple specification is 1 dB.

5.3 Design a low-pass maximally flat filter having a passband of 0 to 3 GHz, and an attenuation of 20 dB at 5 GHz. The characteristic impedance is 75 Ω.

5.4 Design a five-section high-pass filter with a 3 dB equal-ripple response, a cutoff frequency of 1 GHz, and an impedance of 50 Ω. What is the resulting attenuation at 0.6 GHz?

5.5 Design a four-section bandpass filter having a maximally flat group delay response. The bandwidth should be 5%, with a center frequency of 2 GHz. The impedance is 50 Ω.

5.6 Design a three-section bandstop filter with a 0.5 dB equal-ripple response, a bandwidth of 10% centered at 3 GHz, and an impedance of 75 Ω. What is the resulting attenuation at 3.1 GHz?

5.7 Design a low-pass fourth-order maximally flat filter using only series stubs. The cutoff frequency is 2.5 GHz, and the impedance is 50 Ω.

5.8 Verify the second Kuroda identity of Table 5.5 by finding the ABCD matrices for both circuits.

5.9 Design a low-pass fourth-order maximally flat filter using only shunt stubs. The cutoff frequency is 2.5 GHz, and the impedance is 50 Ω.

5.10 Design a band-stop fourth-order maximally flat filter using only shunt stubs. The cutoff frequency is 2.5 GHz, the bandwidth is 50%, and the impedance is 50 Ω.

5.11 Derive the open-circuit impedance matrix elements given in (5.36) for a two-port network consisting of a transmission line of length ℓ and characteristic impedance Z_0.

5.12 Design a stepped-impedance low-pass filter having a cutoff frequency of 4.0 GHz and a fifth-order, 0.5 dB equal ripple response. Assume $R_0 = 100\ \Omega$, $Z_\ell = 15\ \Omega$, and $Z_h = 200\ \Omega$.

5.13 Demonstrate that the circuits of Figure 5.20c act as ideal impedance and admittance inverters when terminated with a load, Z_L or Y_L.

5.14 A bandpass filter is to be used in a PCS receiver operating in the 824–849 MHz band, and must provide at least 30 dB isolation at the lowest end of the transmit frequency band (869–894 MHz). Design a 0.5 dB equal-ripple bandpass filter meeting these specifications using quarter-wave line coupled quarter-wave resonators. Assume an impedance of 50 Ω.

5.15 Repeat Problem 5.14 using capacitively coupled quarter-wave resonators.

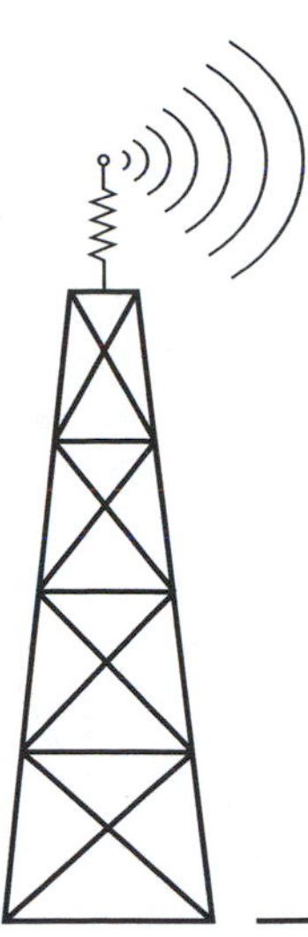

Chapter Six

Amplifiers

Amplification is a critical function in wireless receivers and transmitters. Virtually all microwave and RF amplifiers today use three-terminal solid-state devices such as gallium arsenide *field effect transistors* (FETs), silicon (Si) or silicon germanium (SiGe) *bipolar transistors, heterojunction bipolar transistors* (HBTs), and *high electron mobility transistors* (HEMTs) [1]–[5]. Microwave transistor amplifiers are rugged, low-cost, reliable, and can be easily integrated in both hybrid and monolithic integrated circuits with mixers, oscillators, switches, and related components. They can presently be used at frequencies up to 100 GHz in a wide variety of applications requiring low noise figure, broad bandwidth, and medium power capacity. While microwave tube amplifiers are still sometimes required for very high power and/or very high frequency applications, continuing improvement in the performance of microwave transistors is steadily reducing the need for microwave tubes.

We begin this chapter with a brief overview of microwave FET and bipolar transistors, their small-signal equivalent circuits, and some biasing considerations. Since our emphasis will be on circuit design using transistors, as opposed to the physics of the device itself, we will treat transistors primarily in terms of their terminal characteristics, using either S parameters or an equivalent circuit model. Next we develop some general results for the gain and stability of a two-port network in terms of its S parameters, and apply this theory to the design of single-stage transistor amplifiers in Section 6.4. Section 6.5 discusses noise considerations and the design of low-noise transistor amplifiers. We conclude with a brief treatment of power amplifier design. An understanding of the topics of S parameters and stub tuning, as discussed in Chapter 2, are required for this chapter; references [1]–[3] are suggested for further background on this material.

6.1 FET AND BIPOLAR TRANSISTOR MODELS

Microwave and RF transistors are used as amplifiers, oscillators, switches, phase shifters, mixers, and active filters. Most of these applications use either silicon bipolar transistors or GaAs field effect transistors. Silicon bipolar device technology is very mature and inexpensive compared to GaAs transistor technology. Bipolar transistors are capable of higher gain and power capacity at lower frequencies, but GaAs FETs generally have better noise figures and can operate at much higher frequencies. Present silicon bipolar transistors are limited to applications below about 10 GHz, but recent developments such as silicon-germanium devices and heterojunction bipolar transistors allow operation at much higher frequencies. GaAs FETs can be used at frequencies in excess of 100 GHz. Table 6.1 compares the gain and noise figure versus frequency for some typical microwave transistors [2]:

In this section we give a brief discussion of the basic construction of GaAs FETs and silicon bipolar transistors, along with small-signal equivalent circuit models for these devices, and DC biasing considerations. The design of amplifiers and oscillators relies primarily on the terminal characteristics of the transistor, and these can be expressed either in terms of the two-port S parameters of the device, or in terms of the component values of an equivalent circuit. We will use the S parameter method for most of our design work, as this is a procedure that is both accurate and convenient, although it does have the drawback of requiring knowledge of the transistor S parameters (usually through measurement) over the frequency band of interest. This is usually not a serious problem unless a very wide frequency range is being considered, since the S parameters of microwave transistors typically change fairly slowly with frequency. In contrast, the use of a good transistor equivalent circuit model involves only a few circuit parameters which are generally stable over a wide frequency range. An equivalent circuit model can also provide a closer linkage between the operation of the device and its physical parameters.

Field Effect Transistors

Field effect transistors can be used at frequencies well into the millimeter wave range with high gain and low noise figure, making them the device of choice for hybrid and monolithic integrated circuits at frequencies above 5–10 GHz [3]. Figure 6.1 shows the construction of a typical GaAs FET. The desirable gain and noise features of the GaAs FET are a result of the higher electron mobility of GaAs compared to silicon, and the absence of shot noise. In operation, electrons are drawn from the source to the drain by the positive V_{ds} supply voltage. An input signal voltage on the gate then modulates the flow of these majority carriers, producing voltage amplification. The maximum frequency of operation

TABLE 6.1 Comparison of Gain and Noise Figure of Microwave Transistors (gain and noise figure in dB)

Frequency	GaAs FET		GaAs HEMT		Silicon Bipolar		GaAs HBT	
GHz	Gain	F_{min}	Gain	F_{min}	Gain	F_{min}	Gain	F_{min}
4	20	0.5	—	—	15	2.5	—	—
8	16	0.7	—	—	9	4.5	—	—
12	12	1.0	22	0.5	6	8.0	20	4.0
18	8	1.2	16	0.9	—	—	16	—
36	—	—	12	1.7	—	—	10	—
60	—	—	8	2.6	—	—	7	—

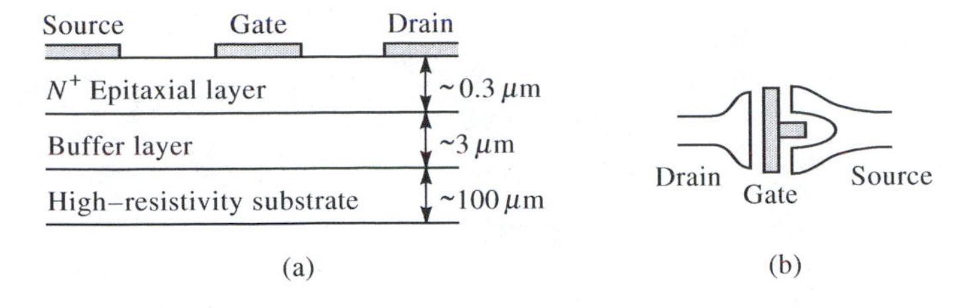

FIGURE 6.1 (a) Cross section of a GaAs FET. (b) Top view, showing drain, gate, and source contacts.

is limited by the gate length; presently manufactured FETs have gate lengths on the order of 0.3 to 0.6 μm, with corresponding upper frequency limits of 100–50 GHz.

A small-signal equivalent circuit for a microwave GaAs FET is shown in Figure 6.2, for a common-source configuration. Typical component values for this circuit model are:

R_i (series gate resistance) $= 7\ \Omega$

R_{ds} (drain-to-source resistance) $= 400\ \Omega$

C_{gs} (gate-to-source capacitance) $= 0.3$ pF

C_{ds} (drain-to-source capacitance) $= 0.12$ pF

C_{gd} (gate-to-drain capacitance) $= 0.01$ pF

g_m (transconductance) $= 40$ mS

This model does not include package parasitics, which typically introduce small series resistances and inductances at the three terminals due to ohmic contacts and bonding leads. The dependent current generator $g_m V_c$ depends on the voltage across the gate-to-source capacitor C_{gs}, leading to a value of $|S_{21}| > 1$ under normal operating conditions (where port 1 is at the gate, and port 2 is at the drain). The reverse signal path, given by S_{12}, is due solely to the capacitance C_{gd}. As can be seen from the above data, this is typically a very small capacitor which can often be ignored in practice. In this case, $S_{12} = 0$, and the device is said to be *unilateral*.

The equivalent circuit model of Figure 6.2 can be used to determine the upper frequency of operation for the transistor. The *short-circuit current gain*, G_i^{sc}, is defined as the ratio of drain to gate current when the output is short-circuited. For the unilateral case, where C_{gd} is assumed to be zero, this can be derived as

$$G_i^{sc} = \left|\frac{I_d}{I_g}\right| = \left|\frac{g_m V_c}{I_g}\right| = \frac{g_m}{\omega C_{gs}}. \tag{6.1}$$

The upper frequency limit, f_T, is the frequency where the short-circuit current gain is unity;

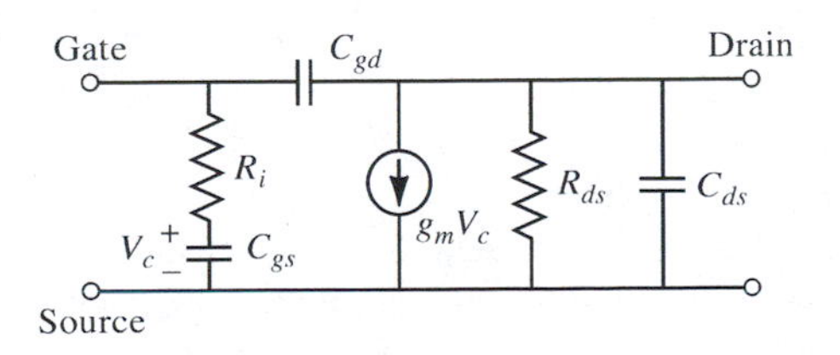

FIGURE 6.2 Small-signal equivalent circuit for a GaAs FET in a common-source configuration.

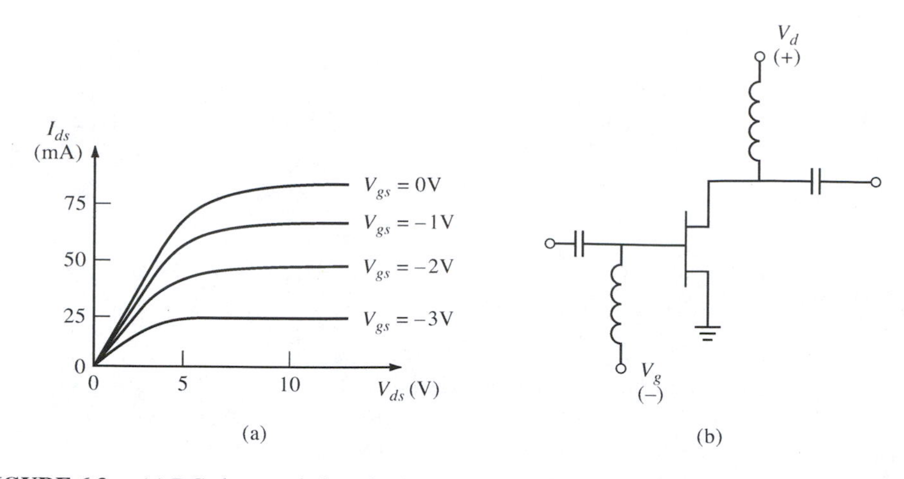

FIGURE 6.3 (a) DC characteristics of a GaAs FET. (b) Biasing and decoupling ciruit for a GaAs FET.

thus we have that

$$f_T = \frac{g_m}{2\pi C_{gs}}. \tag{6.2}$$

For proper operation, the transistor must be DC biased at an appropriate operating point. This depends on the application (low-noise, high-gain, or high-power), the class of the amplifier (class A, class AB, class B), and the type of transistor (bipolar, FET, HBT, HEMT). Figure 6.3a shows a typical family of DC I_{ds} versus V_{ds} curves for a GaAs FET. For low-noise design, the drain current is generally chosen to be about 15% of I_{dss} (the saturated drain-to-source current). High power circuits generally use higher values of drain current. DC bias voltage must be applied to the gate and drain, without disturbing the RF signal paths. This can be done as shown in Figure 6.3b, which shows the biasing and decoupling circuitry for a dual polarity supply. The RF chokes provide a very low DC resistance for biasing, and a very high impedance at RF frequencies to prevent the microwave signal from being shorted by the bias supply. Similarly, the input and output decoupling capacitors block DC from the input and output lines, while allowing passage of microwave signals. There are many other types of bias circuits that provide compensation for temperature and device variations, and that can work with single-polarity power supplies.

Bipolar Transistors

Bipolar transistors are usually of the *npn* type, and are often preferred over GaAs FETs at frequencies below 2 to 4 GHz because of higher gain and lower cost. Bipolar transistors are subject to shot noise as well as thermal noise effects, so their noise figure is not as good as that of FETs. Figure 6.4 shows the construction of a typical silicon bipolar transistor. In contrast to the FET, the bipolar transistor is current driven, with the base current modulating the collector current. The upper frequency limit of the bipolar transistor is controlled primarily by the base length, which is on the order of 0.1 μm.

A small-signal equivalent circuit model for a microwave bipolar transistor is shown in Figure 6.5, for a common emitter configuration. Typical values for the components of the

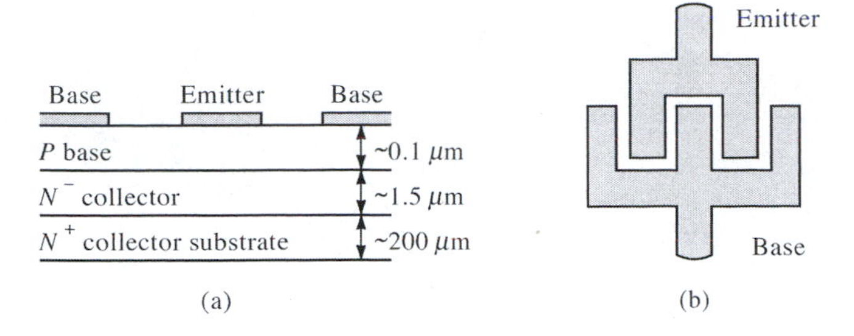

FIGURE 6.4 (a) Cross section of a microwave silicon biolar transistor. (b) Top view, showing base and emitter contacts.

equivalent circuit are:

R_b (base resistance) $= 7\ \Omega$
R_π (equivalent π resistance) $= 110\ \Omega$
C_π (equivalent π capacitance) $= 18$ pF
C_c (collector capacitance) $= 18$ pF
g_m (transconductance) $= 900$ mS

Observe that the transconductance is much higher than that of the GaAs FET, leading to higher power gain at lower frequencies. The larger capacitances in the bipolar transistor model serve to reduce the gain at higher frequencies. The model in Figure 6.5 is popular because of its similarity to the FET equivalent circuit, but more sophisticated equivalent circuits may be advantageous for use over wide frequency ranges [2]. In addition, this model does not include parasitic resistances and inductances due to the base and emitter leads.

The equivalent circuit of Figure 6.5 can be used to estimate the upper frequency limit, f_T, where the short-circuit current gain is unity. The result is similar to that found above for the FET:

$$f_T = \frac{g_m}{2\pi C_\pi}. \tag{6.3}$$

Figure 6.6a shows typical DC operating characteristics for a bipolar transistor. As with the FET, the biasing point for a bipolar transistor depends on the application and type of transistor, with low collector currents generally giving the best noise figure, and higher collector currents giving the best power gain. Figure 6.6b shows a typical bias and decoupling circuit for a bipolar transistor that requires only a single polarity supply.

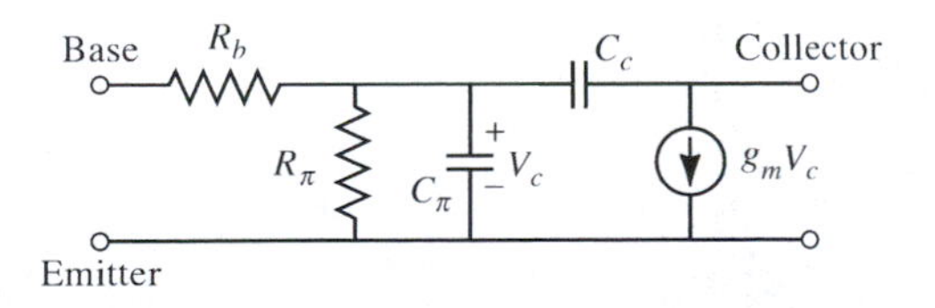

FIGURE 6.5 Simplified hybrid-π equivalent circuit for a microwave bipolar transistor in a common-emitter configuration.

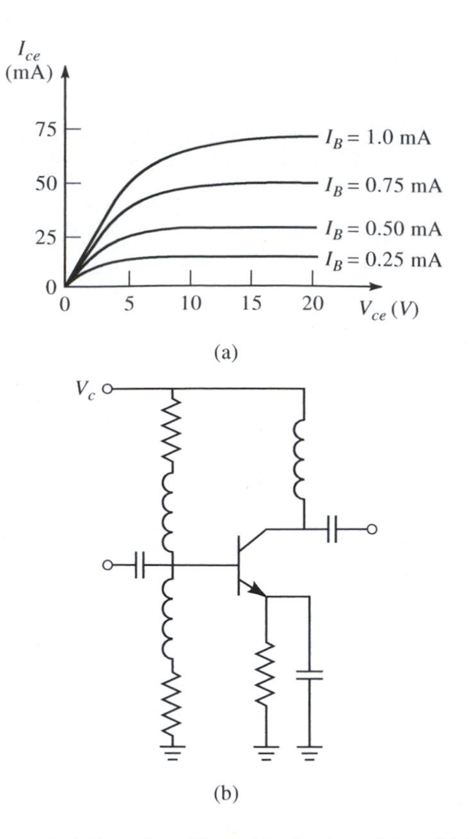

FIGURE 6.6 (a) DC characteristics of a silicon bipolar transistor. (b) Biasing and decoupling circuit for a bipolar transistor.

6.2 TWO-PORT POWER GAINS

In this section we develop expressions for several different types of gain for a general two-port circuit in terms of the S parameters of the network. These results will be used in later sections for the design of transistor amplifiers.

Definitions of Two-Port Power Gains

Consider an arbitrary two-port network with scattering matrix $[S]$, connected to source and load impedances Z_S and Z_L, respectively, as shown in Figure 6.7. We will define and derive expressions for three types of power gain in terms of the S parameters of the two-port

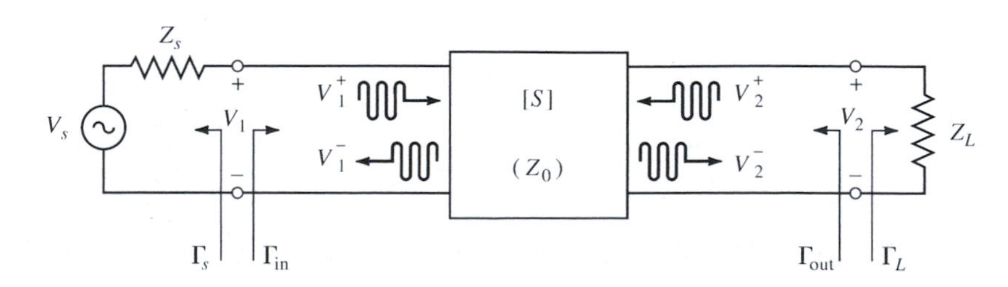

FIGURE 6.7 A two-port network with general source and load impedances.

network and the reflection coefficients, Γ_S and Γ_L, of the source and load.

- *Power gain* $= G = P_L/P_{in}$ is the ratio of power dissipated in the load Z_L to the power delivered to the input of the two-port network. This gain is independent of Z_S, although some active circuits are strongly dependent on Z_S.
- *Available gain* $= G_A = P_{avn}/P_{avs}$ is the ratio of the power available from the two-port network to the power available from the source. This assumes conjugate matching of both the source and the load, and depends on Z_s but not Z_L.
- *Transducer power gain* $= G_T = P_L/P_{avs}$ is the ratio of the power delivered to the load to the power available from the source. This depends on both Z_S and Z_L.

These definitions differ primarily in the way the source and load are matched to the two-port device; if the input and output are both conjugately matched to the two port, then the gain is maximized and $G = G_A = G_T$.

With reference to Figure 6.7, the reflection coefficient seen looking toward the load is

$$\Gamma_L = \frac{Z_L - Z_0}{Z_L + Z_0}, \tag{6.4a}$$

while the reflection coefficient seen looking toward the source is

$$\Gamma_S = \frac{Z_S - Z_0}{Z_S + Z_0}, \tag{6.4b}$$

where Z_0 is the characteristic impedance reference for the S parameters of the two-port network.

In general, the input impedance of the terminated two-port network will be mismatched with a reflection coefficient given by Γ_{in}, which can be determined using a signal flow-graph [1], or by the following analysis. From the definition of S parameters (see Section 2.3) and the fact that $V_2^+ = \Gamma_L V_2^-$, we have

$$V_1^- = S_{11}V_1^+ + S_{12}V_2^+ = S_{11}V_1^+ + S_{12}\Gamma_L V_2^-, \tag{6.5a}$$

$$V_2^- = S_{21}V_1^+ + S_{22}V_2^+ = S_{21}V_1^+ + S_{22}\Gamma_L V_2^-. \tag{6.5b}$$

Eliminating V_2^- from (6.5a) and solving for V_1^-/V_1^+ gives

$$\Gamma_{in} = \frac{V_1^-}{V_1^+} = S_{11} + \frac{S_{12}S_{21}\Gamma_L}{1 - S_{22}\Gamma_L} = \frac{Z_{in} - Z_0}{Z_{in} + Z_0}, \tag{6.6a}$$

where Z_{in} is the impedance seen looking into port 1 of the terminated network. Similarly, the reflection coefficient seen looking into port 2 of the network when port 1 is terminated by Z_S is

$$\Gamma_{out} = \frac{V_2^-}{V_2^+} = S_{22} + \frac{S_{12}S_{21}\Gamma_S}{1 - S_{11}\Gamma_S}. \tag{6.6b}$$

By voltage division,

$$V_1 = V_s \frac{Z_{in}}{Z_S + Z_{in}} = V_1^+ + V_1^- = V_1^+(1 + \Gamma_{in}).$$

Solving (6.6a) for Z_{in} gives

$$Z_{in} = Z_0 \frac{1 + \Gamma_{in}}{1 - \Gamma_{in}},$$

and using this result in the previous equation and solving for V_1^+ in terms of V_S gives

$$V_1^+ = \frac{V_S}{2} \frac{(1 - \Gamma_S)}{(1 - \Gamma_S \Gamma_{\text{in}})}. \tag{6.7}$$

If peak values are assumed for all voltages, the average power delivered to the network is

$$P_{\text{in}} = \frac{1}{2Z_0} |V_1^+|^2 (1 - |\Gamma_{\text{in}}|^2) = \frac{|V_S|^2}{8Z_0} \frac{|1 - \Gamma_S|^2}{|1 - \Gamma_S \Gamma_{\text{in}}|^2} (1 - |\Gamma_{\text{in}}|^2), \tag{6.8}$$

where (6.7) was used. The power delivered to the load is

$$P_L = \frac{|V_2^-|^2}{2Z_0} (1 - |\Gamma_L|^2). \tag{6.9}$$

Solving for V_2^- from (6.5b), substituting into (6.9), and using (6.7) gives

$$P_L = \frac{|V_1^+|^2}{2Z_0} \frac{|S_{21}|^2 (1 - |\Gamma_L|^2)}{|1 - S_{22}\Gamma_L|^2} = \frac{|V_S|^2}{8Z_0} \frac{|S_{21}|^2 (1 - |\Gamma_L|^2) |1 - \Gamma_S|^2}{|1 - S_{22}\Gamma_L|^2 |1 - \Gamma_S \Gamma_{\text{in}}|^2}. \tag{6.10}$$

The power gain can then be expressed as

$$G = \frac{P_L}{P_{\text{in}}} = \frac{|S_{21}|^2 (1 - |\Gamma_L|^2)}{(1 - |\Gamma_{\text{in}}|^2) |1 - S_{22}\Gamma_L|^2}. \tag{6.11}$$

The power available from the source, P_{avs}, is the maximum power that can be delivered to the network. This occurs when the input impedance of the terminated network is conjugately matched to the source impedance [1]. Thus, from (6.8),

$$P_{\text{avs}} = P_{\text{in}}\Big|_{\Gamma_{\text{in}} = \Gamma_S^*} = \frac{|V_s|^2}{8Z_0} \frac{|1 - \Gamma_s|^2}{(1 - |\Gamma_s|^2)}. \tag{6.12}$$

Similarly, the power available from the network, P_{avn}, is the maximum power that can be delivered to the load. Thus, from (6.10),

$$P_{\text{avn}} = P_L\Big|_{\Gamma_L = \Gamma_{\text{out}}^*} = \frac{|V_S|^2}{8Z_0} \frac{|S_{21}|^2 (1 - |\Gamma_{\text{out}}|^2) |1 - \Gamma_S|^2}{|1 - S_{22}\Gamma_{\text{out}}^*|^2 |1 - \Gamma_S \Gamma_{\text{in}}|^2}\Bigg|_{\Gamma_L = \Gamma_{\text{out}}^*}. \tag{6.13}$$

In (6.13), Γ_{in} must be evaluated for $\Gamma_L = \Gamma_{\text{out}}^*$. From (6.6a), it can be shown that

$$|1 - \Gamma_S \Gamma_{\text{in}}|^2\Big|_{\Gamma_L = \Gamma_{\text{out}}^*} = \frac{|1 - S_{11}\Gamma_S|^2 (1 - |\Gamma_{\text{out}}|^2)}{|1 - S_{22}\Gamma_{\text{out}}^*|^2},$$

which allows (6.13) to be reduced to

$$P_{\text{avn}} = \frac{|V_S|^2}{8Z_0} \frac{|S_{21}|^2 |1 - \Gamma_S|^2}{|1 - S_{11}\Gamma_S|^2 (1 - |\Gamma_{\text{out}}|^2)}. \tag{6.14}$$

Observe that P_{avs} and P_{avn} have been expressed in terms of the source voltage, V_S, which is independent of the input or load impedances. There would be confusion if these quantities were expressed in terms of V_1^+, since V_1^+ is different for each of the calculations of P_L, P_{avs}, and P_{avn}.

Using (6.14) and (6.12), the available power gain is then

$$G_A = \frac{P_{\text{avn}}}{P_{\text{avs}}} = \frac{|S_{21}|^2 (1 - |\Gamma_S|^2)}{|1 - S_{11}\Gamma_S|^2 (1 - |\Gamma_{\text{out}}|^2)}. \tag{6.15}$$

From (6.10) and (6.12), the transducer power gain is

$$G_T = \frac{P_L}{P_{avs}} = \frac{|S_{21}|^2(1 - |\Gamma_S|^2)(1 - |\Gamma_L|^2)}{|1 - \Gamma_S \Gamma_{in}|^2 |1 - S_{22}\Gamma_L|^2}. \tag{6.16}$$

Special Cases

A special case of the transducer power gain occurs when both the input and output are matched for zero reflection (in contrast to conjugate matching). Then $\Gamma_L = \Gamma_S = 0$, and (6.16) reduces to

$$G_T = |S_{21}|^2. \tag{6.17}$$

Another special case is the *unilateral transducer power gain*, G_{TU}, where $S_{12} = 0$ (or is negligibly small). This nonreciprocal characteristic is common to many practical amplifier circuits. From (6.6a), $\Gamma_{in} = S_{11}$ when $S_{12} = 0$, so (6.16) gives the unilateral transducer gain as

$$G_{TU} = \frac{|S_{21}|^2(1 - |\Gamma_S|^2)(1 - |\Gamma_L|^2)}{|1 - S_{11}\Gamma_S|^2 |1 - S_{22}\Gamma_L|^2} \tag{6.18}$$

EXAMPLE 6.1 COMPARISON OF POWER GAIN DEFINITIONS

A microwave transistor has the following S parameters at 10 GHz, with a 50 Ω reference impedance:

$$S_{11} = 0.45\angle 150^\circ$$
$$S_{12} = 0.01\angle -10^\circ$$
$$S_{21} = 2.05\angle 10^\circ$$
$$S_{22} = 0.40\angle -150^\circ$$

The source impedance is $Z_S = 20\ \Omega$ and the load impedance is $Z_L = 30\ \Omega$. Compute the power gain, the available gain, and the transducer power gain.

Solution

From (6.4a,b) the reflection coefficients at the source and load are

$$\Gamma_S = \frac{Z_S - Z_0}{Z_S + Z_0} = \frac{20 - 50}{20 + 50} = -0.429,$$

$$\Gamma_L = \frac{Z_L - Z_0}{Z_L + Z_0} = \frac{30 - 50}{30 + 50} = -0.250.$$

From (6.6a,b) the reflection coefficients seen looking at the input and output of the terminated network are

$$\Gamma_{in} = S_{11} + \frac{S_{12}S_{21}\Gamma_L}{1 - S_{22}\Gamma_L}$$
$$= 0.45\angle 150^\circ + \frac{(0.01\angle -10^\circ)(2.05\angle 10^\circ)(-0.250)}{1 - (0.40\angle -150^\circ)(-0.250)} = 0.455\angle 150^\circ,$$

$$\Gamma_{out} = S_{22} + \frac{S_{12}S_{21}\Gamma_S}{1 - S_{11}\Gamma_S}$$
$$= 0.40\angle -150^\circ + \frac{(0.01\angle -10^\circ)(2.05\angle 10^\circ)(-0.429)}{1 - (0.45\angle 150^\circ)(-0.429)} = 0.408\angle -151^\circ.$$

Then from (6.1) the power gain is

$$G = \frac{|S_{21}|^2(1 - |\Gamma_L|^2)}{(1 - |\Gamma_{in}|^2)|1 - S_{22}\Gamma_L|^2}$$

$$= \frac{(2.05)^2[1 - (0.250)^2]}{|1 - (0.40\angle{-150°})(-0.250)|^2[1 - (0.455)^2]} = 5.94.$$

From (6.15) the available power gain is

$$G_A = \frac{|S_{21}|^2(1 - |\Gamma_S|^2)}{|1 - S_{11}\Gamma_S|^2(1 - |\Gamma_{out}|^2)}$$

$$= \frac{(2.05)^2[1 - (0.429)^2]}{|1 - (0.45\angle 150°)(-0.429)|^2[1 - (0.408)^2]} = 5.85.$$

From (6.16) the transducer power gain is

$$G_T = \frac{|S_{21}|^2(1 - |\Gamma_S|^2)(1 - |\Gamma_L|^2)}{|1 - \Gamma_S\Gamma_{in}|^2|1 - S_{22}\Gamma_L|^2}$$

$$= \frac{(2.05)^2[1 - (0.429)^2][1 - (0.250)^2]}{|1 - (0.40\angle{-150°})(-0.250)|^2|1 - (-0.429)(0.455\angle 150°)|^2} = 5.49$$

○

Further Discussion of Two-Port Power Gains

A single-stage microwave transistor amplifier can be modeled by the circuit of Figure 6.8, where a matching network is used on both sides of the transistor to transform the input and output impedance Z_0 to the source and load impedances Z_S and Z_L. The most useful gain definition for amplifier design is the transducer power gain of (6.16), which accounts for both source and load mismatch. Thus, from (6.16), we can define separate effective gain factors for the input (source) matching network, the transistor itself, and the output (load) matching network as follows:

$$G_S = \frac{1 - |\Gamma_S|^2}{|1 - \Gamma_{in}\Gamma_S|^2}, \tag{6.19a}$$

$$G_0 = |S_{21}|^2, \tag{6.19b}$$

$$G_L = \frac{1 - |\Gamma_L|^2}{|1 - S_{22}\Gamma_L|^2}. \tag{6.19c}$$

Then the overall transducer gain is $G_T = G_S G_0 G_L$. The effective gains from G_S and G_L are due to the impedance matching of the transistor to the impedances Z_S and Z_L.

If the transistor is unilateral, so that $S_{12} = 0$ or is small enough to be ignored, then (6.6) reduces to $\Gamma_{in} = S_{11}$, $\Gamma_{out} = S_{22}$, and the unilateral transducer gain reduces to $G_{TU} =$

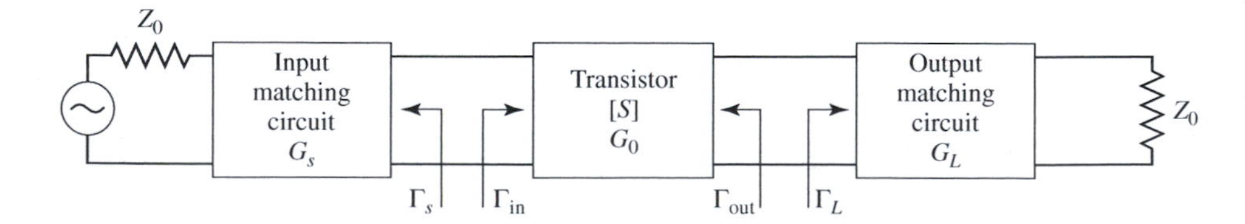

FIGURE 6.8 The general transistor amplifier circuit.

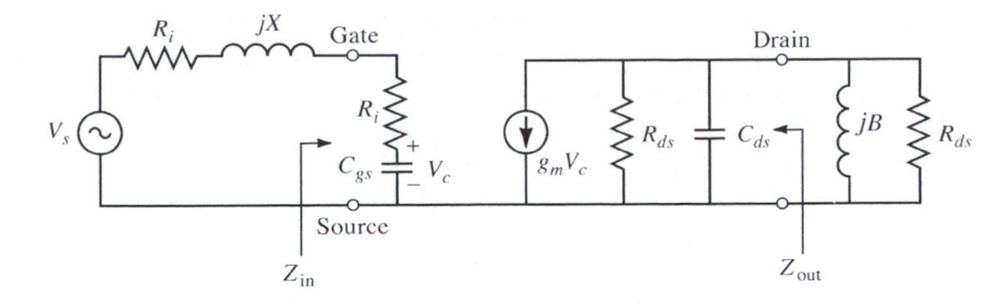

FIGURE 6.9 Unilateral FET equivalent circuit, and source and load terminations for the calculation of unilateral transducer power gain.

$G_S G_0 G_L$, where

$$G_S = \frac{1 - |\Gamma_S|^2}{|1 - S_{11}\Gamma_S|^2}, \tag{6.20a}$$

$$G_0 = |S_{21}|^2, \tag{6.20b}$$

$$G_L = \frac{1 - |\Gamma_L|^2}{|1 - S_{22}\Gamma_L|^2}. \tag{6.20c}$$

These results have been derived using the S parameters of the transistor, but it is possible to obtain alternative expressions for gain in terms of the equivalent circuit parameters of the transistor. As an example, consider the evaluation of the unilateral transducer gain for a conjugately matched GaAs FET using the equivalent circuit of Figure 6.2 (with $C_{gd} = 0$). To conjugately match the transistor, we choose source and load impedances as shown in Figure 6.9. Setting the series source inductive reactance $X = 1/(\omega C_{gs})$ will make $Z_{\text{in}} = Z_S^*$, and setting the shunt load inductive susceptance $B = -\omega C_{ds}$ will make $Z_{\text{out}} = Z_L^*$; this effectively eliminates the reactive elements from the FET equivalent circuit. Then by voltage division $V_c = V_S/(2j\omega R_i C_{gs})$, and the gain can be easily evaluated as

$$G_{\text{TU}} = \frac{P_L}{P_{\text{avs}}} = \frac{\frac{1}{8}|g_m V_c|^2 R_{ds}}{\frac{1}{8}|V_s|^2/R_i} = \frac{g_m^2 R_{ds}}{4\omega^2 R_i C_{gs}^2} = \frac{R_{ds}}{4R_i}\left(\frac{f_T}{f}\right)^2. \tag{6.21}$$

where the last step has been written in terms of the cutoff frequency, f_T, from (6.2). This shows the interesting result that the gain of a conjugately matched FET amplifier drops off as $1/f^2$, or 6 dB per octave.

6.3 STABILITY

We now discuss the stability of a transistor amplifier circuit. In the circuit of Figure 6.8, oscillation is possible if either the input or output port impedance has a negative real part; this would then imply that $|\Gamma_{\text{in}}| > 1$ or $|\Gamma_{\text{out}}| > 1$. Because Γ_{in} and Γ_{out} depend on the source and load matching networks, the stability of the amplifier depends on Γ_S and Γ_L as presented by the matching networks. Thus, we define two types of stability:

- *Unconditional stability*: The network is unconditionally stable if $|\Gamma_{\text{in}}| < 1$ and $|\Gamma_{\text{out}}| < 1$ for all passive source and load impedances (i.e., $|\Gamma_S| < 1$ and $|\Gamma_L| < 1$).
- *Conditional stability*: The network is conditionally stable if $|\Gamma_{\text{in}}| < 1$ and $|\Gamma_{\text{out}}| < 1$ only for a certain range of passive source and load impedances. This case is also referred to as *potentially unstable*.

Note that the stability condition of a network is frequency dependent, since the input and output matching networks are generally frequency dependent. Thus it is possible for an amplifier to be stable at its design frequency, but unstable at other frequencies. Careful amplifier design should consider this possibility. In addition, while the following discussion is rigorously valid for the circuit of Figure 6.8, there are two situations where these results may not apply: if the network is nonlinear (then S parameters do not apply), or if there is feedback in the circuit.

Stability Circles

Applying the above requirements for unconditional stability to (6.6) gives the following conditions that must be satisfied by Γ_S and Γ_L if the amplifier is to be unconditionally stable:

$$|\Gamma_{\text{in}}| = \left| S_{11} + \frac{S_{12}S_{21}\Gamma_L}{1 - S_{22}\Gamma_L} \right| < 1, \tag{6.22a}$$

$$|\Gamma_{\text{out}}| = \left| S_{22} + \frac{S_{12}S_{21}\Gamma_S}{1 - S_{11}\Gamma_S} \right| < 1. \tag{6.22b}$$

If the device is unilateral ($S_{12} = 0$), these conditions reduce to the simple results that $|S_{11}| < 1$ and $|S_{22}| < 1$ are sufficient for unconditional stability. Otherwise, the inequalities of (6.22) define a range of values for Γ_S and Γ_L where the amplifier will be stable. Finding this range for Γ_S and Γ_L can be facilitated by using a Smith chart, since the solutions to (6.22) form input and output *stability circles*. Stability circles are defined as the loci in the Γ_L (or Γ_S) plane for which $|\Gamma_{\text{in}}| = 1$ (or $|\Gamma_{\text{out}}| = 1$). The stability circles thus define the boundaries between stable and potentially unstable regions of values of Γ_S and Γ_L. Γ_S and Γ_L must lie on the Smith chart, since $|\Gamma_S| < 1$ and $|\Gamma_L| < 1$ for passive matching networks and loads.

We can derive the equation for the output stability circle as follows. First use (6.22a) to express the condition that $|\Gamma_{\text{in}}| = 1$ as

$$\left| S_{11} + \frac{S_{12}S_{21}\Gamma_L}{1 - S_{22}\Gamma_L} \right| = 1, \tag{6.23}$$

or

$$|S_{11}(1 - S_{22}\Gamma_L) + S_{12}S_{21}\Gamma_L| = |1 - S_{22}\Gamma_L|.$$

Now define Δ as the determinant of the scattering matrix:

$$\Delta = S_{11}S_{22} - S_{12}S_{21}. \tag{6.24}$$

Then we can write the above result as

$$|S_{11} - \Delta\Gamma_L| = |1 - S_{22}\Gamma_L|. \tag{6.25}$$

Now square both sides of (6.25) and simplify to obtain

$$|S_{11}|^2 + |\Delta|^2|\Gamma_L|^2 - (\Delta\Gamma_L S_{11}^* + \Delta^*\Gamma_L^* S_{11}) = 1 + |S_{22}|^2|\Gamma_L|^2 - (S_{22}^*\Gamma_L^* + S_{22}\Gamma_L)$$

$$(|S_{22}|^2 - |\Delta|^2)\Gamma_L\Gamma_L^* - (S_{22} - \Delta S_{11}^*)\Gamma_L - (S_{22}^* - \Delta^* S_{11})\Gamma_L^* = |S_{11}|^2 - 1$$

$$\Gamma_L\Gamma_L^* - \frac{(S_{22} - \Delta S_{11}^*)\Gamma_L + (S_{22}^* - \Delta^* S_{11})\Gamma_L^*}{|S_{22}|^2 - |\Delta|^2} = \frac{|S_{11}|^2 - 1}{|S_{22}|^2 - |\Delta|^2}. \tag{6.26}$$

Next, complete the square by adding $|S_{22} - \Delta S_{11}^*|^2/(|S_{22}|^2 - |\Delta|^2)^2$ to both sides:

$$\left| \Gamma_L - \frac{(S_{22} - \Delta S_{11}^*)^*}{|S_{22}|^2 - |\Delta|^2} \right|^2 = \frac{|S_{11}|^2 - 1}{|S_{22}|^2 - |\Delta|^2} + \frac{|S_{22} - \Delta S_{11}^*|^2}{(|S_{22}|^2 - |\Delta|^2)^2}$$

or

$$\left|\Gamma_L - \frac{(S_{22} - \Delta S_{11}^*)^*}{|S_{22}|^2 - |\Delta|^2}\right| = \left|\frac{S_{12}S_{21}}{|S_{22}|^2 - |\Delta|^2}\right|. \tag{6.27}$$

In the complex Γ plane an equation of the form $|\Gamma - C| = R$ represents a circle with center at C (a complex number), and a radius R (a real number). Thus (6.27) defines the output (load) stability circle with a center C_L and radius R_L, where

$$C_L = \frac{(S_{22} - \Delta S_{11}^*)^*}{|S_{22}|^2 - |\Delta|^2} \quad \text{(center)}, \tag{6.28a}$$

$$R_L = \left|\frac{S_{12}S_{21}}{|S_{22}|^2 - |\Delta|^2}\right| \quad \text{(radius)}. \tag{6.28b}$$

Similar results can be derived for the input (source) stability circle by interchanging S_{11} and S_{22}:

$$C_S = \frac{(S_{11} - \Delta S_{22}^*)^*}{|S_{11}|^2 - |\Delta|^2} \quad \text{(center)}, \tag{6.29a}$$

$$R_S = \left|\frac{S_{12}S_{21}}{|S_{11}|^2 - |\Delta|^2}\right| \quad \text{(radius)}. \tag{6.29b}$$

Given the S parameters of the transistor, we can plot the input and output stability circles to define where $|\Gamma_{in}| = 1$ and $|\Gamma_{out}| = 1$. Then on one side of the input stability circle we will have $|\Gamma_{out}| < 1$, while on the other side we will have $|\Gamma_{out}| > 1$. Similarly, we will have $|\Gamma_{in}| < 1$ on one side of the output stability circle, and $|\Gamma_{in}| > 1$ on the other side. So we need to determine which areas on the Smith chart represent the stable region, for which $|\Gamma_{in}| < 1$ and $|\Gamma_{out}| < 1$.

Consider the output stability circles plotted in the Γ_L plane for $|S_{11}| < 1$ and $|S_{11}| > 1$, as shown in Figure 6.10. If we set $Z_L = Z_0$, then $\Gamma_L = 0$ and (6.22a) shows that $|\Gamma_{in}| = |S_{11}|$. Now if $|S_{11}| < 1$, then $|\Gamma_{in}| < 1$, so the point where $\Gamma_L = 0$ must be in a stable region. This means that the center of the Smith chart (where $\Gamma_L = 0$) is in the stable region, so all of the Smith chart (for which $|\Gamma_L| < 1$) that is outside the stability circle defines the stable range for Γ_L. This region is shaded in Figure 6.10a. Alternatively, if we set $Z_L = Z_0$ but

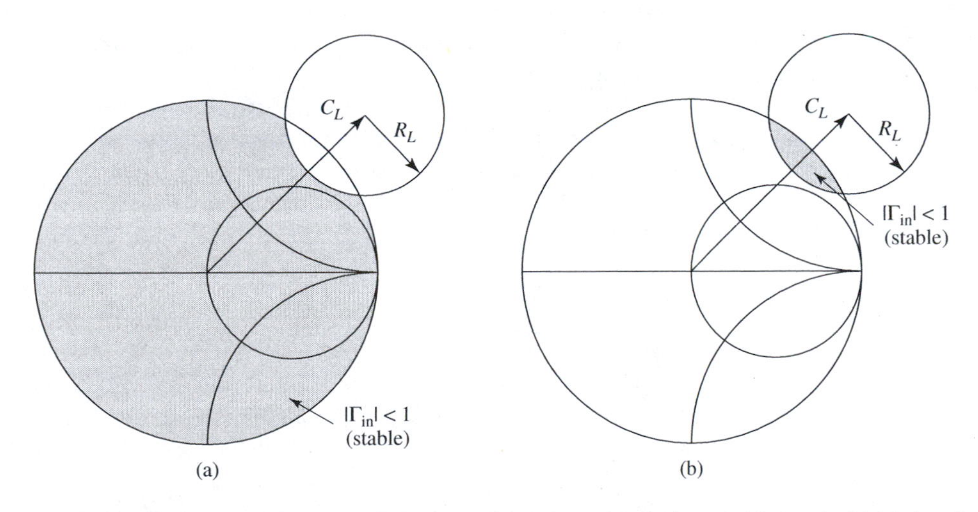

FIGURE 6.10 Output stability circles for a conditionally stable device. (a) $|S_{11}| < 1$. (b) $|S_{11}| > 1$.

have $|S_{11}| > 1$, then $|\Gamma_{\text{in}}| > 1$ for $\Gamma_L = 0$, and the center of the Smith chart must be in an unstable region. In this case the stable region is the inside region of the stability circle that intersects the Smith chart, as illustrated in Figure 6.10b. Similar results apply to the input stability circle.

If the device is unconditionally stable, the stability circles must be completely outside (or totally enclose) the Smith chart. We can state this result mathematically as

$$||C_L| - R_L| > 1, \qquad \text{for } |S_{11}| < 1, \tag{6.30a}$$

$$||C_S| - R_S| > 1, \qquad \text{for } |S_{22}| < 1. \tag{6.30b}$$

If $|S_{11}| > 1$ or $|S_{22}| > 1$, the amplifier cannot be unconditionally stable because we can always have a source or load impedance of Z_0 leading to $\Gamma_S = 0$ or $\Gamma_L = 0$, thus causing $|\Gamma_{\text{in}}| > 1$ or $|\Gamma_{\text{out}}| > 1$. If the device is only conditionally stable, operating points for Γ_S and Γ_L must be chosen in stable regions, and it is good practice to check the stability at several frequencies near the design frequency. If it is possible to accept a design with less than maximum gain, a transistor can usually be made to be unconditionally stable by using resistive loading [3].

Tests for Unconditional Stability

The stability circles discussed above can be used to determine regions for Γ_S and Γ_L where the amplifier circuit will be conditionally stable, but simpler tests can be used to determine unconditional stability. One of these is the *K-Δ test*, where it can be shown that a device will be unconditionally stable if *Rollet's condition*, defined as

$$K = \frac{1 - |S_{11}|^2 - |S_{22}|^2 + |\Delta|^2}{2|S_{12}S_{21}|} > 1, \tag{6.31}$$

along with the auxiliary condition that

$$|\Delta| = |S_{11}S_{22} - S_{12}S_{21}| < 1, \tag{6.32}$$

are simultaneously satisfied. These two conditions are necessary and sufficient for unconditional stability, and are easily evaluated. If the device S parameters do not satisfy the K-Δ test, the device is not unconditionally stable, and stability circles must be used to determine if there are values of Γ_S and Γ_L for which the device will be conditionally stable. Also, recall from the previous paragraph that we must have $|S_{11}| < 1$ and $|S_{22}| < 1$ if the device is to be unconditionally stable.

While the K-Δ test of (6.31)–(6.32) is a mathematically rigorous condition for unconditional stability, it cannot be used to compare the relative stability of two or more devices since it involves constraints on two separate parameters. Recently, however, a new criterion has been proposed [6] that combines the S parameters in a test involving only a single parameter, μ, defined as

$$\mu = \frac{1 - |S_{11}|^2}{|S_{22} - \Delta S_{11}^*| + |S_{12}S_{21}|} > 1. \tag{6.33}$$

Thus, if $\mu > 1$, the device is unconditionally stable. In addition, larger values of μ imply greater stability.

We can derive the μ-test of (6.33) by starting with the expression from (6.6b) for Γ_{out}:

$$\Gamma_{\text{out}} = S_{22} + \frac{S_{12}S_{21}\Gamma_S}{1 - S_{11}\Gamma_S} = \frac{S_{22} - \Delta\Gamma_S}{1 - S_{11}\Gamma_S}, \tag{6.34}$$

where Δ is the determinant of the S matrix defined in (6.24). Unconditional stability implies that $|\Gamma_{out}| < 1$ for any passive source termination, Γ_S. The reflection coefficient for a passive source impedance must lie within the unit circle on a Smith chart, and the outer boundary of this circle can be written as $\Gamma_S = e^{j\phi}$. The expression given in (6.34) maps this circle into another circle in the Γ_{out} plane. We can show this by substituting $\Gamma_S = e^{j\phi}$ into (6.34) and solving for $e^{j\phi}$:

$$e^{j\phi} = \frac{S_{22} - \Gamma_{out}}{\Delta - S_{11}\Gamma_{out}}.$$

Taking the magnitude of both sides gives

$$\left|\frac{S_{22} - \Gamma_{out}}{\Delta - S_{11}\Gamma_{out}}\right| = 1.$$

Squaring both sides and expanding gives

$$|\Gamma_{out}|^2(1 - |S_{11}|^2) + \Gamma_{out}(\Delta^* S_{11} - S_{22}^*) + \Gamma_{out}^*(\Delta S_{11}^* - S_{22}) = |\Delta|^2 - |S_{22}|^2.$$

Next, divide by $1 - |S_{11}|^2$ to obtain

$$|\Gamma_{out}|^2 + \frac{(\Delta^* S_{11} - S_{22}^*)\Gamma_{out} + (\Delta S_{11}^* - S_{22})\Gamma_{out}^*}{1 - |S_{11}|^2} = \frac{|\Delta|^2 - |S_{22}|^2}{1 - |S_{11}|^2}.$$

Now complete the square by adding $\frac{|\Delta^* S_{11} - S_{22}^*|^2}{(1-|S_{11}|^2)^2}$ to both sides:

$$\left|\Gamma_{out} + \frac{\Delta S_{11}^* - S_{22}}{1 - |S_{11}|^2}\right|^2 = \frac{|\Delta|^2 - |S_{22}|^2}{1 - |S_{11}|^2} + \frac{|\Delta^* S_{11} - S_{22}^*|^2}{(1 - |S_{11}|^2)^2} = \frac{|S_{12}S_{21}|^2}{(1 - |S_{11}|^2)^2}. \tag{6.35}$$

This equation is of the form $|\Gamma_{out} - C| = R$, which represents a circle with center C and radius R in the Γ_{out} plane. Thus the center and radius of the mapped $|\Gamma_s| = 1$ circle are given by

$$C = \frac{S_{22} - \Delta S_{11}^*}{1 - |S_{11}|^2}, \tag{6.36a}$$

$$R = \frac{|S_{12}S_{21}|}{1 - |S_{11}|^2}. \tag{6.36b}$$

If points within this circular region are to satisfy $|\Gamma_{out}| < 1$, then we must have that

$$|C| + R < 1. \tag{6.37}$$

Substituting (6.36) into (6.37) gives

$$|S_{22} - \Delta S_{11}^*| + |S_{12}S_{21}| < 1 - |S_{11}|^2,$$

which after rearranging yields the μ-test of (6.33):

$$\frac{1 - |S_{11}|^2}{|S_{22} - \Delta S_{11}^*| + |S_{12}S_{21}|} > 1$$

The K-Δ test of (6.31)–(6.32) can be derived from a similar starting point, as in reference [1], or more simply from the μ-test of (6.33). Rearranging (6.33) and squaring gives

$$|S_{22} - \Delta S_{11}^*|^2 < (1 - |S_{11}|^2 - |S_{12}S_{21}|)^2. \tag{6.38}$$

It can be verified by direct expansion that

$$|S_{22} - \Delta S_{11}^*|^2 = |S_{12}S_{21}|^2 + (1 - |S_{11}|^2)(|S_{22}|^2 - |\Delta|^2),$$

so (6.38) expands to

$$|S_{12}S_{21}|^2 + (1 - |S_{11}|^2)(|S_{22}|^2 - |\Delta|^2) < (1 - |S_{11}|^2)(1 - |S_{11}|^2 - 2|S_{12}S_{21}|) + |S_{12}S_{21}|^2.$$

Simplifying gives

$$|S_{22}|^2 - |\Delta|^2 < 1 - |S_{11}|^2 - 2|S_{12}S_{21}|,$$

which yields the Rollet condition of (6.31) after rearranging:

$$\frac{1 - |S_{11}|^2 - |S_{22}|^2 + |\Delta|^2}{2|S_{12}S_{21}|} = K > 1.$$

In addition to (6.31), the K-Δ test also requires an auxiliary condition to guarantee unconditional stability. Although we derived Rollet's condition from the necessary and sufficient result of the μ-test, the squaring step used in (6.38) introduces an ambiguity in the sign of the right-hand side, thus requiring an additional condition. This can be derived by requiring that the right-hand side of (6.38) be positive before squaring. Thus,

$$|S_{12}S_{21}| < 1 - |S_{11}|^2.$$

Because similar conditions can be derived for the input side of the circuit, we can interchange S_{11} and S_{22} to obtain the analogous condition that

$$|S_{12}S_{21}| < 1 - |S_{22}|^2.$$

Adding these two inequalities gives

$$2|S_{12}S_{21}| < 2 - |S_{11}|^2 - |S_{22}|^2.$$

From the triangle inequality we know that

$$|\Delta| = |S_{11}S_{22} - S_{12}S_{21}| \leq |S_{11}S_{22}| + |S_{12}S_{21}|,$$

so we have that

$$|\Delta| < |S_{11}||S_{22}| + 1 - \frac{1}{2}|S_{11}|^2 - \frac{1}{2}|S_{22}|^2 < 1 - \frac{1}{2}(|S_{11}|^2 - |S_{22}|^2) < 1,$$

which is identical to (6.32).

EXAMPLE 6.2 TRANSISTOR STABILITY

The S parameters for the HP HFET-102 GaAs FET at 2 GHz with a bias voltage $V_{gs} = 0$ are given as follows, with $Z_0 = 50\ \Omega$:

$$S_{11} = 0.894\angle{-60.6^\circ}$$
$$S_{12} = 0.020\angle 62.4^\circ$$
$$S_{21} = 3.122\angle 123.6^\circ$$
$$S_{22} = 0.781\angle{-27.6^\circ}$$

Determine the stability of this transistor by using the K-Δ test and the μ-test, and plot the stability circles on a Smith chart.

Solution

From (6.31) and (6.32) we compute K and $|\Delta|$ as

$$|\Delta| = |S_{11}S_{22} - S_{12}S_{21}| = |0.696\angle{-83^\circ}| = 0.696,$$

$$K = \frac{1 - |S_{11}|^2 - |S_{22}|^2 + |\Delta|^2}{2|S_{12}S_{21}|} = 0.607.$$

Thus we have $|\Delta| = 0.696 < 1$, but $K < 1$, so the unconditional stability criteria of (6.31)–(6.32) is not satisfied, and the device is potentially unstable. The stability of this device could also be evaluated using the μ-test, for which (6.33) gives $\mu = 0.86$, again indicating potential instability.

The centers and radii of the stability circles are given by (6.28) and (6.29):

$$C_L = \frac{(S_{22} - \Delta S_{11}^*)^*}{|S_{22}|^2 - |\Delta|^2} = 1.361\angle 47^\circ,$$

$$R_L = \left|\frac{S_{12}S_{21}}{|S_{22}|^2 - |\Delta|^2}\right| = 0.50,$$

$$C_S = \frac{(S_{11} - \Delta S_{22}^*)^*}{|S_{11}|^2 - |\Delta|^2} = 1.132\angle 68^\circ,$$

$$R_S = \left|\frac{S_{12}S_{21}}{|S_{11}|^2 - |\Delta|^2}\right| = 0.199.$$

This data is used to plot the input and output stability circles shown in Figure 6.11. Since $|S_{11}| < 1$ and $|S_{22}| < 1$, the central part of the Smith chart represents the stable operating region for Γ_S and Γ_L. The unstable regions are shaded in the figure. ○

6.4 AMPLIFIER DESIGN USING *S* PARAMETERS

We can now apply the above results to design a single-stage transistor amplifier. The first step in amplifier design is to consider the stability of the device, using either the $K - |\Delta|$ or μ-test to check for unconditional stability, and plotting the stability circles if the device is potentially unstable. Then the input and output matching sections are designed to give a particular value of gain or noise figure. Since G_0 of (6.19b) is fixed for a given transistor, the overall gain of the amplifier will be controlled by the gains, G_S and G_L, of the matching sections. Maximum gain will be realized when these sections provide a conjugate match between the amplifier source or load impedance and the transistor. Because most transistors appear as a significant impedance mismatch (large $|S_{11}|$ and $|S_{22}|$) the resulting frequency response will be narrowband. In the next section we will discuss how to design for less than maximum gain, with a corresponding improvement in bandwidth. A discussion of broadband amplifier design can be found in [1].

Design for Maximum Gain

With reference to Figure 6.8, maximum power transfer from the input matching section to the transistor will occur when the input impedance to the transistor is conjugate matched to the impedance presented by the matching section:

$$\Gamma_{in} = \Gamma_S^*, \tag{6.39a}$$

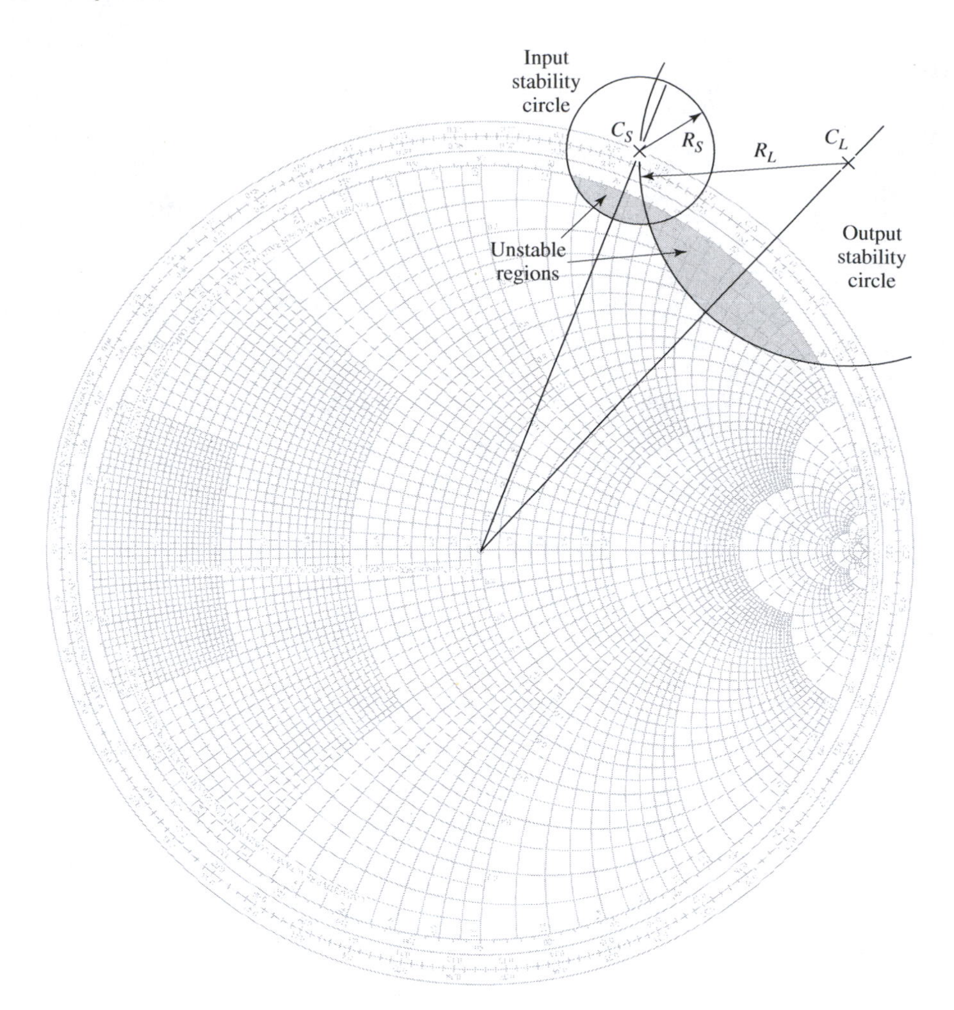

FIGURE 6.11 Stability circles for Example 6.2.

and the maximum power transfer from the transistor to the output matching network will occur when

$$\Gamma_{\text{out}} = \Gamma_L^*. \tag{6.39b}$$

Then, assuming lossless matching sections, these conditions will maximize the overall transducer gain. From (6.16), this maximum gain will be given by

$$G_{T_{\max}} = \frac{1}{1 - |\Gamma_S|^2} |S_{21}|^2 \frac{1 - |\Gamma_L|^2}{|1 - S_{22}\Gamma_L|^2}. \tag{6.40}$$

In the general case with a bilateral transistor ($|S_{12}| \neq 0$), Γ_{in} is affected by Γ_{out}, and vice versa, so that the input and output sections must be matched simultaneously. Using (6.39) in (6.16) gives the necessary equations:

$$\Gamma_S^* = S_{11} + \frac{S_{12}S_{21}\Gamma_L}{1 - S_{22}\Gamma_L}, \tag{6.41a}$$

$$\Gamma_L^* = S_{22} + \frac{S_{12}S_{21}\Gamma_S}{1 - S_{11}\Gamma_S}. \tag{6.41b}$$

We can solve for Γ_S by first rewriting these equations as follows:

$$\Gamma_S = S_{11}^* + \frac{S_{12}^* S_{21}^*}{1/\Gamma_L^* - S_{22}^*},$$

$$\Gamma_L^* = \frac{S_{22} - \Delta \Gamma_S}{1 - S_{11}\Gamma_S},$$

where $\Delta = S_{11}S_{22} - S_{12}S_{21}$, as before. Substituting this expression for Γ_L^* into the expression for Γ_S and expanding gives

$$\begin{aligned}&\Gamma_S(1 - |S_{22}|^2) + \Gamma_S^2(\Delta S_{22}^* - S_{11}) \\ &\quad = \Gamma_S(\Delta S_{11}^* S_{22}^* - |S_{11}|^2 - \Delta S_{12}^* S_{21}^*) + S_{11}^*(1 - |S_{22}|^2) + S_{12}^* S_{21}^* S_{22}\end{aligned}$$

Using the result that $\Delta(S_{11}^* S_{22}^* - S_{12}^* S_{21}^*) = |\Delta|^2$ allows this to be rewritten as a quadratic equation for Γ_S:

$$(S_{11} - \Delta S_{22}^*)\Gamma_S^2 + (|\Delta|^2 - |S_{11}|^2 + |S_{22}|^2 - 1)\Gamma_S + (S_{11}^* - \Delta^* S_{22}) = 0. \tag{6.42}$$

The solution is then

$$\Gamma_S = \frac{B_1 \pm \sqrt{B_1^2 - 4|C_1|^2}}{2C_1}. \tag{6.43a}$$

Similarly, the solution for Γ_L can be written as

$$\Gamma_L = \frac{B_2 \pm \sqrt{B_2^2 - 4|C_2|^2}}{2C_2}. \tag{6.43b}$$

The variables B_1, C_1, B_2, C_2 are defined as

$$B_1 = 1 + |S_{11}|^2 - |S_{22}|^2 - |\Delta^2|, \tag{6.44a}$$

$$B_2 = 1 + |S_{22}|^2 - |S_{11}|^2 - |\Delta^2|, \tag{6.44b}$$

$$C_1 = S_{11} - \Delta S_{22}^*, \tag{6.44c}$$

$$C_2 = S_{22} - \Delta S_{11}^*. \tag{6.44d}$$

Solutions to (6.43) are only possible if the quantity within the square root is positive, and it can be shown that this is equivalent to requiring $K > 1$. Thus unconditionally stable devices can always be conjugately matched for maximum gain, and potentially unstable devices can be conjugately matched if $K > 1$ and $|\Delta| < 1$. The results are much simpler for the unilateral case. When $S_{12} = 0$, (6.41) shows that $\Gamma_S = S_{11}^*$ and $\Gamma_L = S_{22}^*$, and then the maximum transducer gain of (6.40) reduces to

$$G_{TU_{\max}} = \frac{1}{1 - |S_{11}|^2}|S_{21}|^2 \frac{1}{1 - |S_{22}|^2}. \tag{6.45}$$

Maximum Stable Gain

The maximum transducer power gain given by (6.40) occurs when the source and load are conjugately matched to the transistor, as given by the conditions of (6.39). If the transistor is unconditionally stable, so that $K > 1$, the maximum transducer power gain of (6.40) can be simply rewritten as follows:

$$G_{T_{\max}} = \frac{|S_{21}|}{|S_{12}|}(K - \sqrt{K^2 - 1}). \tag{6.46}$$

This result can be obtained by substituting (6.43) and (6.44) for Γ_S and Γ_L into (6.40) and simplifying. The maximum transducer power gain is also sometimes referred to as the *matched gain*.

The maximum gain does not provide a meaningful result if the device is only conditionally stable, since simultaneous conjugate matching of the source and load are not possible if $K < 1$ (see Problem 6.8). In this case a useful figure of merit is the *maximum stable gain*, defined as the maximum transducer power gain of (6.46) with $K = 1$. Thus,

$$G_{\text{msg}} = \frac{|S_{21}|}{|S_{12}|}. \tag{6.47}$$

The maximum stable gain is easy to compute and offers a convenient way to compare the gain of various devices under stable operating conditions.

EXAMPLE 6.3 CONJUGATELY MATCHED AMPLIFIER DESIGN

Design an amplifier for maximum gain at 4.0 GHz using single-stub matching sections. Calculate and plot the input return loss and the gain from 3 to 5 GHz. Use a GaAs FET with the following S parameters ($Z_0 = 50\ \Omega$):

f (GHz)	S_{11}	S_{21}	S_{12}	S_{22}
3.0	$0.80\angle{-89°}$	$2.86\angle 99°$	$0.03\angle 56°$	$0.76\angle{-41°}$
4.0	$0.72\angle{-116°}$	$2.60\angle 76°$	$0.03\angle 57°$	$0.73\angle{-54°}$
5.0	$0.66\angle{-142°}$	$2.39\angle 54°$	$0.03\angle 62°$	$0.72\angle{-68°}$

Solution

We first check for unconditional stability of the transistor by calculating Δ and K at 4.0 GHz:

$$\Delta = S_{11}S_{22} - S_{12}S_{21} = 0.488\angle{-162°},$$

$$K = \frac{1 - |S_{11}|^2 - |S_{22}|^2 + |\Delta|^2}{2|S_{12}S_{21}|} = 1.195.$$

Since $|\Delta| < 1$ and $K > 1$, the transistor is unconditionally stable at 4.0 GHz. There is therefore no need to plot the stability circles.

For maximum gain, we should design the matching sections for a conjugate match to the transistor. Thus, $\Gamma_S = \Gamma_{\text{in}}^*$ and $\Gamma_L = \Gamma_{\text{out}}^*$, and Γ_S and Γ_L can be determined from (6.43):

$$\Gamma_S = \frac{B_1 \pm \sqrt{B_1^2 - 4|C_1|^2}}{2C_1} = 0.872\angle 123°$$

$$\Gamma_L = \frac{B_2 \pm \sqrt{B_2^2 - 4|C_2|^2}}{2C_2} = 0.876\angle 61°.$$

Then the effective gain factors of (6.19) can be calculated as

$$G_S = \frac{1}{1 - |\Gamma_S|^2} = 4.17 = 6.20\text{ dB},$$

$$G_0 = |S_{21}|^2 = 6.76 = 8.30\text{ dB},$$

$$G_L = \frac{1 - |\Gamma_L|^2}{|1 - S_{22}\Gamma_L|^2} = 1.67 = 2.22\text{ dB}.$$

So the overall transducer gain will be

$$G_{T_{max}} = 6.20 + 8.30 + 2.22 = 16.7 \text{ dB}.$$

The matching networks can easily be determined with a Smith chart using the procedure described in Section 2.4. For the input matching section, we first plot Γ_S, as shown in Figure 6.12a. The impedance, Z_S, represented by this reflection

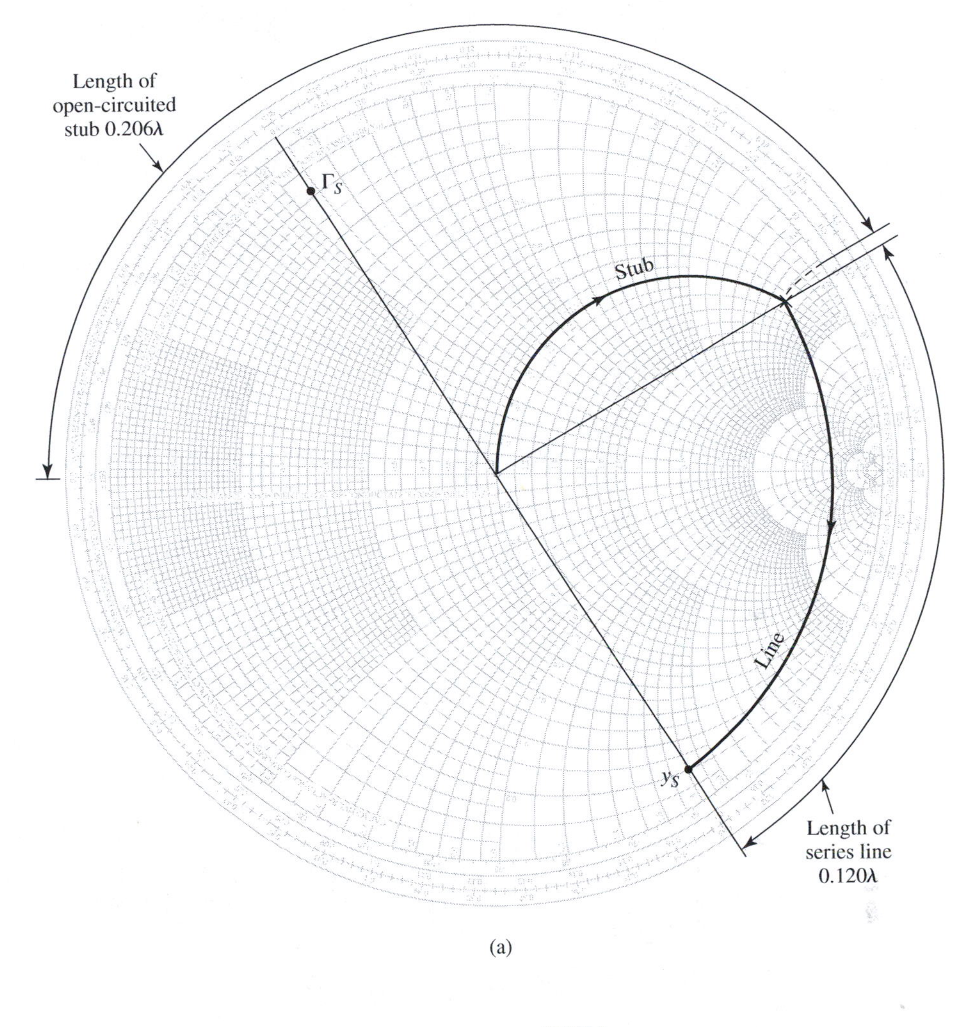

(a)

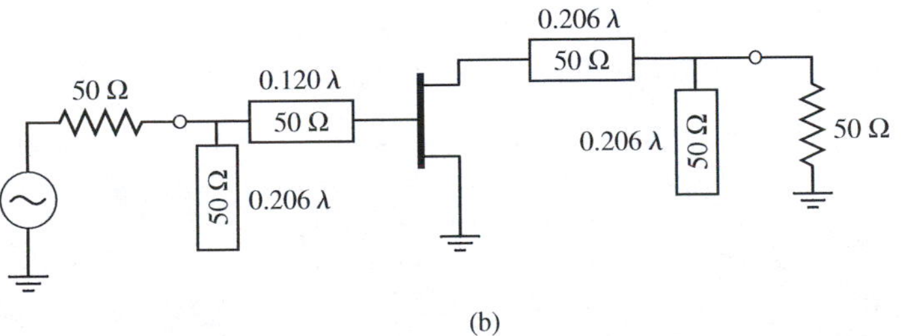

(b)

FIGURE 6.12 Circuit and frequency response for the transistor amplifier of Example 6.3. (a) Smith chart for the design of the input matching network. (b) RF circuit. (c) Frequency response.

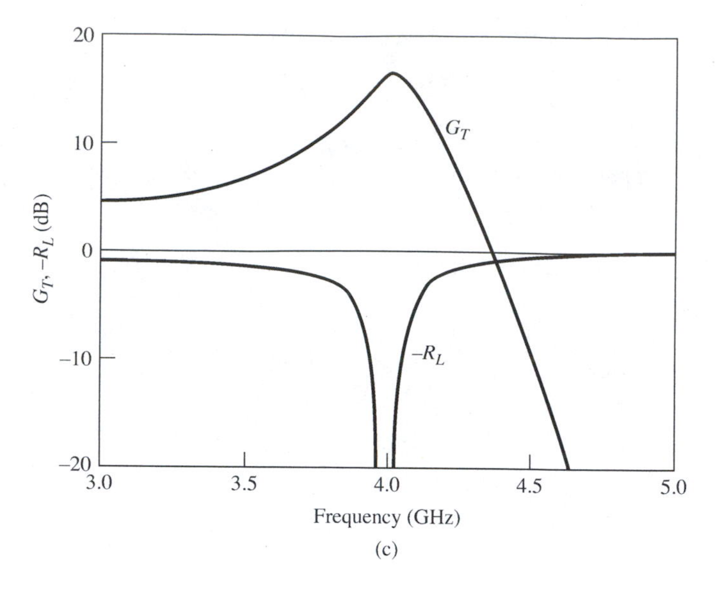

FIGURE 6.12 (*Continued*)

coefficient is the impedance seen looking into the matching section toward the source impedance, Z_0. Thus, the matching section must transform Z_0 to the impedance Z_S. There are several ways of doing this, but we will use an open-circuited shunt stub followed by a length of line. Thus we convert to the normalized admittance y_s and work backward (toward the load on the Smith chart) to find that a line of length 0.120 λ will bring us to the $1 + jb$ circle. Then we see that the required stub admittance is $+j3.5$, for an open-circuited stub length of 0.206 λ. A similar procedure gives a line length of 0.206 λ and a stub length of 0.206 λ for the output matching circuit.

The final amplifier circuit is shown in Figure 6.12b. This circuit only shows the RF components; the amplifier will also require some bias circuitry. The return loss and gain were calculated using a commercial CAD package, interpolating the necessary S parameters from the preceding table. The results are plotted in Figure 6.12c, and show the expected gain of 16.7 dB at 4 GHz, with a very good return loss. The bandwidth where the gain drops by 1 dB is about 2.5%, so this design is relatively narrowband. ○

Constant Gain Circles and Design for Specified Gain

In many cases it is preferable to design for less than the maximum obtainable gain, to improve bandwidth, to obtain a specific value of amplifier gain, or to minimize the effect of device variations. This can be done by designing the input and output matching sections to have less than maximum gains; in other words, impedance mismatches are purposely introduced to reduce the overall gain. The design procedure is facilitated by plotting *constant gain circles* on a Smith chart, to represent loci of Γ_S and Γ_L that give fixed values of gain for the input and output sections (G_S and G_L). To simplify our discussion, we will only treat the case of a unilateral device; the more general case of a bilateral device must sometimes be considered in practice, and is discussed in detail in [2]–[4].

In many practical cases $|S_{12}|$ is small enough to be ignored, and the device can then be assumed to be unilateral. This greatly simplifies the design procedure. The error in the transducer gain caused by approximating $|S_{12}|$ as zero is given by the ratio G_T/G_{TU}. It can be shown that this ratio is bounded by

$$\frac{1}{(1+U)^2} < \frac{G_T}{G_{TU}} < \frac{1}{(1-U)^2}, \tag{6.48}$$

where U is defined as the *unilateral figure of merit*,

$$U = \frac{|S_{11}||S_{12}||S_{21}||S_{22}|}{(1-|S_{11}|^2)(1-S_{22}|^2)}. \tag{6.49}$$

Usually an error of a few tenths of a dB or less will justify the unilateral assumption.

The expressions for G_S and G_L for the unilateral case are given by (6.20a) and (6.20c):

$$G_S = \frac{1-|\Gamma_S|^2}{|1-S_{11}\Gamma_S|^2},$$

$$G_L = \frac{1-|\Gamma_L|^2}{|1-S_{22}\Gamma_L|^2}.$$

These gains are maximized when $\Gamma_S = S_{11}^*$ and $\Gamma_L = S_{22}^*$, resulting in the maximum values given by

$$G_{S_{\max}} = \frac{1}{1-|S_{11}|^2}, \tag{6.50a}$$

$$G_{L_{\max}} = \frac{1}{1-|S_{22}|^2}. \tag{6.50b}$$

Now define the normalized gain factors g_S and g_L as

$$g_S = \frac{G_S}{G_{S_{\max}}} = \frac{1-|\Gamma_S|^2}{|1-S_{11}\Gamma_S|^2}(1-|S_{11}|^2), \tag{6.51a}$$

$$g_L = \frac{G_L}{G_{L_{\max}}} = \frac{1-|\Gamma_L|^2}{|1-S_{22}\Gamma_L|^2}(1-|S_{22}|^2). \tag{6.51b}$$

Then we have that $0 \leq g_S \leq 1$, and $0 \leq g_L \leq 1$.

For fixed values of g_S and g_L, (6.51) represents circles in the Γ_S or Γ_L plane. To show this, consider (6.51a), which can be expanded to give

$$g_S|1-S_{11}\Gamma_S|^2 = (1-|\Gamma_S|^2)(1-|S_{11}|^2),$$

$$(g_S|S_{11}|^2 + 1 - |S_{11}|^2)|\Gamma_S|^2 - g_S\big(S_{11}\Gamma_S + S_{11}^*\Gamma_S^*\big) = 1-|S_{11}|^2 - g_S,$$

$$\Gamma_S\Gamma_S^* - \frac{g_S(S_{11}\Gamma_S + S_{11}^*\Gamma_S^*)}{1-(1-g_S)|S_{11}|^2} = \frac{1-|S_{11}|^2-g_S}{1-(1-g_S)|S_{11}|^2}. \tag{6.52}$$

Now add $g_S^2|S_{11}|^2/[1-(1-g_S)|S_{11}|^2]^2$ to both sides to complete the square:

$$\left|\Gamma_S - \frac{g_S S_{11}^*}{1-(1-g_S)|S_{11}|^2}\right|^2 = \frac{(1-|S_{11}|^2-g_S)[1-(1-g_S)|S_{11}|^2] + g_S^2|S_{11}|^2}{[1-(1-g_S)|S_{11}|^2]^2}.$$

Simplifying gives

$$\left|\Gamma_S - \frac{g_S S_{11}^*}{1-(1-g_S)|S_{11}|^2}\right| = \frac{\sqrt{1-g_S}(1-|S_{11}|^2)}{1-(1-g_S)|S_{11}|^2}, \tag{6.53}$$

which is the equation of a circle with its center and radius given by

$$C_S = \frac{g_S S_{11}^*}{1 - (1 - g_S)|S_{11}|^2}, \tag{6.54a}$$

$$R_S = \frac{\sqrt{1 - g_S}(1 - |S_{11}|^2)}{1 - (1 - g_S)|S_{11}|^2}. \tag{6.54b}$$

The results for the constant gain circles of the output section can be shown to be

$$C_L = \frac{g_L S_{22}^*}{1 - (1 - g_L)|S_{22}|^2}, \tag{6.55a}$$

$$R_L = \frac{\sqrt{1 - g_L}(1 - |S_{22}|^2)}{1 - (1 - g_L)|S_{22}|^2}. \tag{6.55b}$$

The centers of each family of circles lie along straight lines given by the angle of S_{11}^* or S_{22}^*. Note that when g_S (or g_L) $= 1$ (maximum gain), the radius R_S (or R_L) $= 0$, and the center reduces to S_{11}^* (or S_{22}^*), as expected. Also, it can be shown that the 0 dB gain circles ($G_S = 1$ or $G_L = 1$) will always pass through the center of the Smith chart. These results can be used to plot a family of circles of constant gain for the input and output sections. Then Γ_S and Γ_L can be chosen along these circles to provide the desired gains. The choices for Γ_S and Γ_L are not unique, but it makes sense to choose points close to the center of the Smith chart to minimize the mismatch, and thus maximize the bandwidth. Alternatively, as we will see in the next section, the input network mismatch can be chosen to provide a low-noise design.

EXAMPLE 6.4 AMPLIFIER DESIGN FOR SPECIFIED GAIN

Design an amplifier to have a gain of 11 dB at 4.0 GHz. Plot the constant gain circles for $G_S = 2$ dB and 3 dB, and $G_L = 0$ dB and 1 dB. Calculate and plot the input return loss and overall amplifier gain from 3 to 5 GHz. Use an FET with the following S parameters ($Z_0 = 50\ \Omega$):

f (GHz)	S_{11}	S_{21}	S_{12}	S_{22}
3	$0.80\angle{-90^\circ}$	$2.8\angle{100^\circ}$	0	$0.66\angle{-50^\circ}$
4	$0.75\angle{-120^\circ}$	$2.5\angle{80^\circ}$	0	$0.60\angle{-70^\circ}$
5	$0.71\angle{-140^\circ}$	$2.3\angle{60^\circ}$	0	$0.58\angle{-85^\circ}$

Solution

Since $S_{12} = 0$ and $|S_{11}| < 1$ and $|S_{22}| < 1$, the transistor is unilateral and unconditionally stable. From (6.50) we calculate the maximum matching section gains as

$$G_{S_{\max}} = \frac{1}{1 - |S_{11}|^2} = 2.29 = 3.6 \text{ dB},$$

$$G_{L_{\max}} = \frac{1}{1 - |S_{22}|^2} = 1.56 = 1.9 \text{ dB}.$$

The gain of the mismatched transistor is

$$G_0 = |S_{21}|^2 = 6.25 = 8.0 \text{ dB},$$

so the maximum unilateral transducer gain is

$$G_{TU_{\max}} = 3.6 + 1.9 + 8.0 = 13.5 \text{ dB}.$$

Thus we have 2.5 dB more gain than is required by the specifications. We use (6.51), (6.54), and (6.55) to calculate the following data for the constant gain circles:

$$G_S = 3 \text{ dB} \quad g_S = 0.875 \quad C_S = 0.706\angle 120^\circ \quad R_S = 0.166$$
$$G_S = 2 \text{ dB} \quad g_S = 0.691 \quad C_S = 0.627\angle 120^\circ \quad R_S = 0.294$$
$$G_L = 1 \text{ dB} \quad g_L = 0.806 \quad C_L = 0.520\angle 70^\circ \quad R_L = 0.303$$
$$G_L = 0 \text{ dB} \quad g_L = 0.640 \quad C_L = 0.440\angle 70^\circ \quad R_L = 0.440$$

The constant gain circles are shown in Figure 6.13a. We can choose $G_S = 2$ dB and $G_L = 1$ dB, for an overall amplifier gain of 11 dB. Then we select Γ_S and Γ_L along these circles as shown, to minimize the distance from the center of the chart (this places Γ_S and Γ_L along the radial lines at 120° and 70°, respectively). Thus,

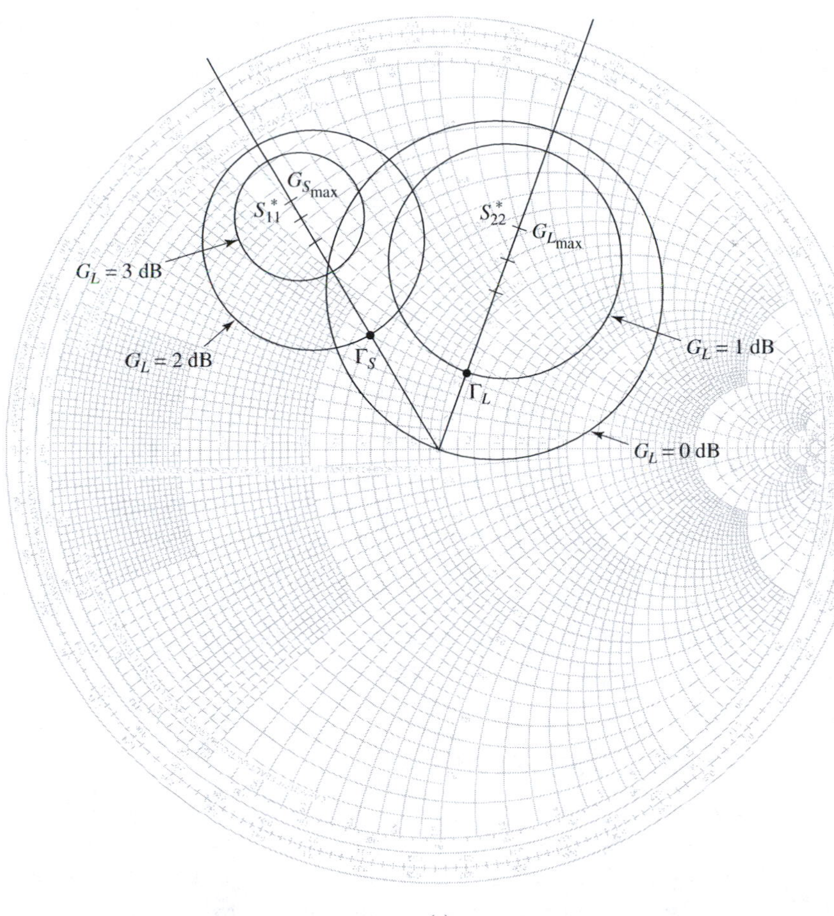

(a)

FIGURE 6.13 Circuit and frequency response for the transistor amplifier of Example 6.4. (a) Constant gain circles. (b) RF circuit. (c) Transducer gain and return loss versus frequency.

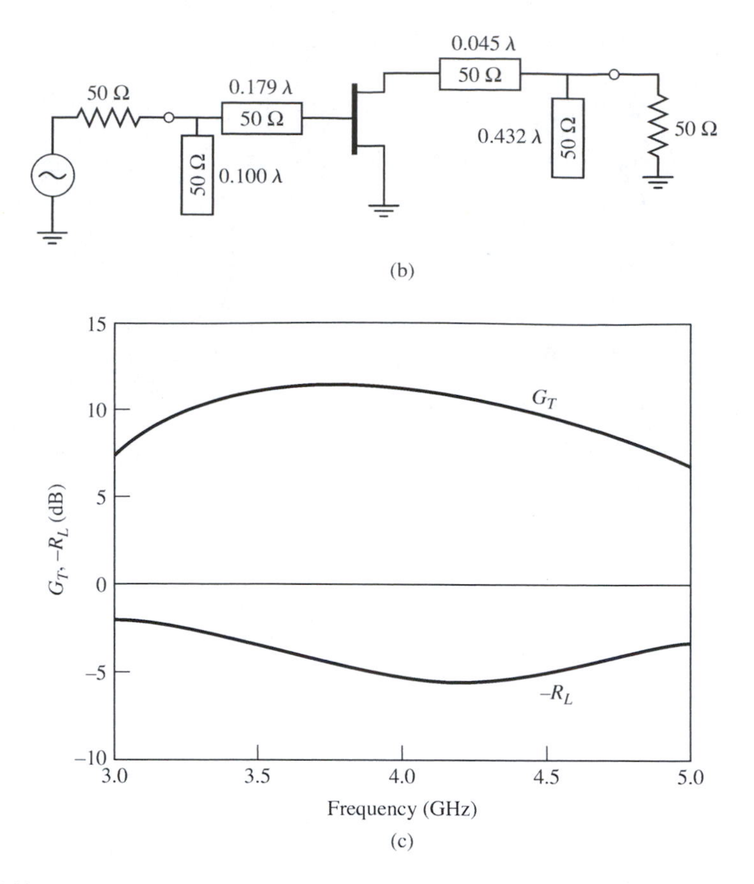

FIGURE 6.13 (*Continued*)

$\Gamma_S = 0.33\angle 120°$ and $\Gamma_L = 0.22\angle 70°$, and the matching networks can be designed using shunt stubs as in Example 6.3.

The final amplifier circuit is shown in Figure 6.13b. The frequency response was calculated using CAD software, with interpolation of the given S parameter data. The results are shown in Figure 6.13c, where it is seen that the desired gain of 11 dB is achieved at 4.0 GHz. The bandwidth over which the gain varies by ± 1 dB or less is about 25%, which is considerably better than the gain bandwidth of the maximum gain design of Example 6.3. The return loss, however, is not very good, being only about 5 dB at the design frequency. This is due to the deliberate mismatch introduced into the matching sections to achieve the specified gain. ○

6.5 LOW-NOISE AMPLIFIER DESIGN

Besides stability and gain, another important design consideration for an RF or microwave amplifier is its noise figure. In receiver applications especially, it is often required to have a preamplifier with as low a noise figure as possible since, as we saw in Section 2.5, the first stage of a receiver front end usually has the dominant effect on the noise performance of the overall system. Generally it is not possible to obtain both minimum noise figure and maximum gain for an amplifier, so some sort of compromise must be made. This can be done by using constant gain circles and *circles of constant noise figure* to select a usable trade-off between noise figure and gain. Here we will derive the equations for constant noise figure circles, and show how they are used in transistor amplifier design.

As derived in [5], the noise figure of a two-port amplifier can be expressed as

$$F = F_{\min} + \frac{R_N}{G_S}|Y_S - Y_{\text{opt}}|^2, \tag{6.56}$$

where the following definitions apply:

$Y_S = G_S + jB_S =$ source admittance presented to transistor.

$Y_{\text{opt}} =$ optimum source admittance that results in minimum noise figure.

$F_{\min} =$ minimum noise figure of transistor, obtained when $Y_S = Y_{\text{opt}}$.

$R_N =$ equivalent noise resistance of transistor.

$G_S =$ real part of source admittance.

Instead of the admittances Y_S and Y_{opt}, we can use the reflection coefficients Γ_S and Γ_{opt}, where

$$Y_S = \frac{1}{Z_0}\frac{1-\Gamma_S}{1+\Gamma_S}, \tag{6.57a}$$

$$Y_{\text{opt}} = \frac{1}{Z_0}\frac{1-\Gamma_{\text{opt}}}{1+\Gamma_{\text{opt}}}. \tag{6.57b}$$

Γ_S is the source reflection coefficient defined in Figure 6.8. The quantities $F_{\min}$, Γ_{opt}, and R_N are characteristics of the particular transistor being used, and are called the *noise parameters* of the device; they may be given by the manufacturer, or measured.

Using (6.57), the quantity $|Y_S - Y_{\text{opt}}|^2$ can be expressed in terms of Γ_S and Γ_{opt}:

$$|Y_S - Y_{\text{opt}}|^2 = \frac{4}{Z_0^2}\frac{|\Gamma_S - \Gamma_{\text{opt}}|^2}{|1+\Gamma_S|^2|1+\Gamma_{\text{opt}}|^2}. \tag{6.58}$$

Also,

$$G_S = \text{Re}\{Y_S\} = \frac{1}{2Z_0}\left(\frac{1-\Gamma_S}{1+\Gamma_S} + \frac{1-\Gamma_S^*}{1+\Gamma_S^*}\right) = \frac{1}{Z_0}\frac{1-|\Gamma_S|^2}{|1+\Gamma_S|^2}. \tag{6.59}$$

Using these results in (6.56) gives the noise figure as

$$F = F_{\min} + \frac{4R_N}{Z_0}\frac{|\Gamma_S - \Gamma_{\text{opt}}|^2}{(1-|\Gamma_S|^2)|1+\Gamma_{\text{opt}}|^2}. \tag{6.60}$$

For a fixed noise figure, F, we can show that this result defines a circle in the Γ_S plane. First, define the *noise figure parameter*, N, as

$$N = \frac{|\Gamma_S - \Gamma_{\text{opt}}|^2}{1-|\Gamma_S|^2} = \frac{F - F_{\min}}{4R_N/Z_0}|1+\Gamma_{\text{opt}}|^2, \tag{6.61}$$

which is a constant, for a given noise figure and set of noise parameters. Then rewrite (6.61) as

$$(\Gamma_S - \Gamma_{\text{opt}})(\Gamma_S^* - \Gamma_{\text{opt}}^*) = N(1-|\Gamma_S|^2),$$

$$\Gamma_S\Gamma_S^* - (\Gamma_S\Gamma_{\text{opt}}^* + \Gamma_S^*\Gamma_{\text{opt}}) + \Gamma_{\text{opt}}\Gamma_{\text{opt}}^* = N - N|\Gamma_S|^2,$$

$$\Gamma_S\Gamma_S^* - \frac{(\Gamma_S\Gamma_{\text{opt}}^* + \Gamma_S^*\Gamma_{\text{opt}})}{N+1} = \frac{N - |\Gamma_{\text{opt}}|^2}{N+1}.$$

Now add $|\Gamma_{opt}|^2/(N+1)^2$ to both sides to complete the square to obtain

$$\left|\Gamma_S - \frac{\Gamma_{opt}}{N+1}\right| = \frac{\sqrt{N(N+1-|\Gamma_{opt}|^2)}}{N+1}. \tag{6.62}$$

This expression defines circles of constant noise figure with centers at

$$C_F = \frac{\Gamma_{opt}}{N+1}, \tag{6.63a}$$

and radii of

$$R_F = \frac{\sqrt{N\left(N+1-|\Gamma_{opt}|^2\right)}}{N+1}. \tag{6.63b}$$

EXAMPLE 6.5 LOW-NOISE AMPLIFIER DESIGN

A GaAs FET is biased for minimum noise figure and has the following S parameters and noise parameters at 4 GHz ($Z_0 = 50\ \Omega$): $S_{11} = 0.60\angle{-60^\circ}$, $S_{21} = 1.9\angle 81^\circ$, $S_{12} = 0.05\angle 26^\circ$, $S_{22} = 0.5\angle{-60^\circ}$; $F_{min} = 1.6$ dB, $\Gamma_{opt} = 0.62\angle 100^\circ$, $R_N = 20\ \Omega$. For design purposes, assume the device is unilateral, and calculate the maximum error in G_T resulting from this assumption. Then design an amplifier having a 2.0 dB noise figure with the maximum gain that is possible with this noise figure.

Solution

We first compute the unilateral figure of merit from (6.49):

$$U = \frac{|S_{11}||S_{12}||S_{21}||S_{22}|}{(1-|S_{11}|^2)(1-|S_{22}|^2)} = 0.059.$$

Then from (6.48) the ratio G_T/G_{TU} is bounded as

$$\frac{1}{(1+U)^2} < \frac{G_T}{G_{TU}} < \frac{1}{(1-U)^2},$$

or

$$0.891 < \frac{G_T}{G_{TU}} < 1.130.$$

In dB,

$$-0.50 < G_T - G_{TU} < 0.53 \text{ dB},$$

where G_T and G_{TU} are now in dB. Thus, we should expect less than about ± 0.5 dB error in gain due to the approximation of a unilateral device.

Next, we use (6.61) and (6.63) to compute the center and radius of the 2 dB noise figure circle:

$$N = \frac{F - F_{min}}{4R_N/Z_0}|1+\Gamma_{opt}|^2 = \frac{1.58 - 1.445}{4(20/50)}|1 + 0.62\angle 100^\circ|^2 = 0.0986,$$

$$C_F = \frac{\Gamma_{opt}}{N+1} = 0.56\angle 100^\circ,$$

$$R_F = \frac{\sqrt{N(N+1-\Gamma_{opt}|^2)}}{N+1} = 0.24.$$

This noise figure circle is plotted in Figure 6.14a. Minimum noise figure ($F_{min} = 1.6$ dB) occurs for $\Gamma_S = \Gamma_{opt} = 0.62\angle 100^\circ$. Next, we calculate data for

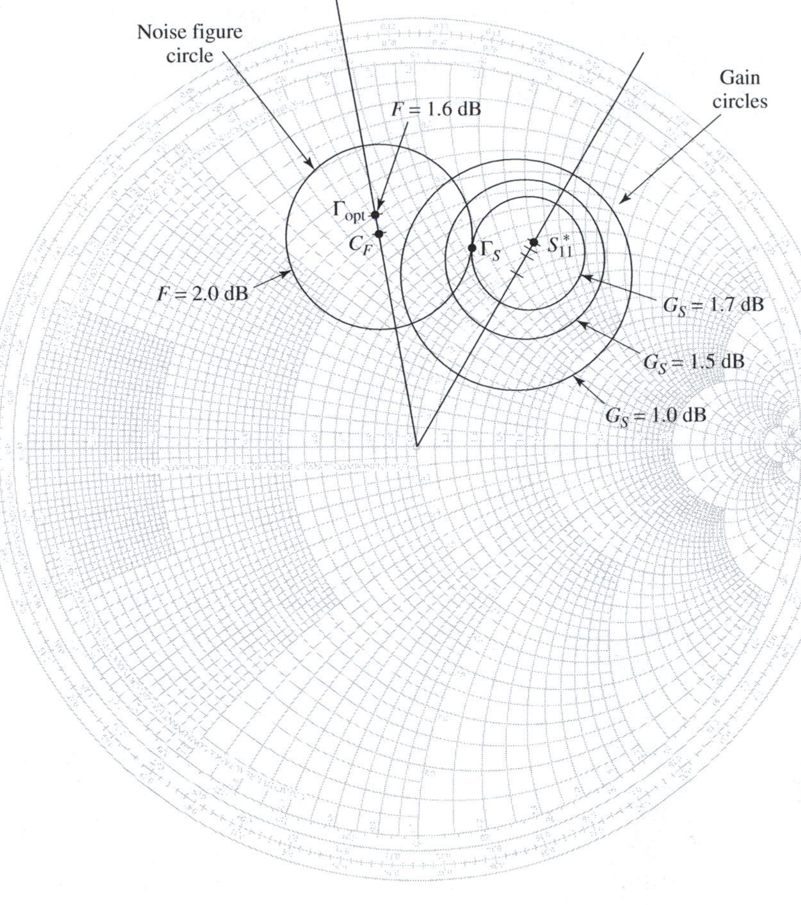

(a)

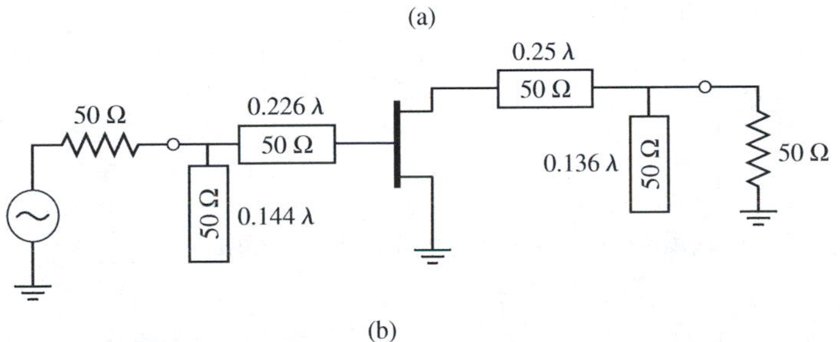

(b)

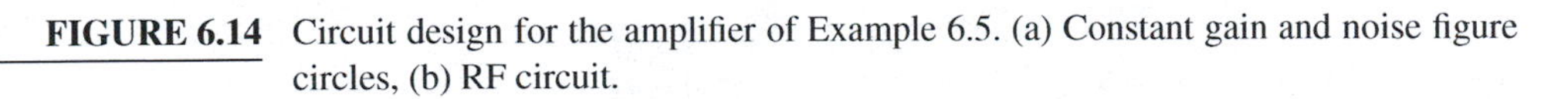

FIGURE 6.14 Circuit design for the amplifier of Example 6.5. (a) Constant gain and noise figure circles, (b) RF circuit.

several input section constant gain circles. From (6.54) we compute the following data:

G_S (dB)	g_S	C_S	R_S
1.0	0.805	$0.52\angle 60°$	0.300
1.5	0.904	$0.56\angle 60°$	0.205
1.7	0.946	$0.58\angle 60°$	0.150

These circles are also plotted in Figure 6.14a. We see that the $G_S = 1.7$ dB gain circle just intersects the $F = 2$ dB noise figure circle, and that any higher gain will result in a worse noise figure. From the Smith chart the optimum solution is then $\Gamma_S = 0.53\angle 75°$, yielding $G_S = 1.7$ dB and $F = 2.0$ dB.

For the output section we chose $\Gamma_L = S_{22}^* = 0.5\angle 60°$ for a maximum G_L of

$$G_L = \frac{1}{1 - |S_{22}|^2} = 1.33 = 1.25 \text{ dB}.$$

The transistor gain is

$$G_0 = |S_{21}|^2 = 3.61 = 5.58 \text{ dB}.$$

so the overall transducer gain will be

$$G_{TU} = G_S + G_0 + G_L = 8.53 \text{ dB}.$$

The complete AC circuit for the amplifier, using open-circuited shunt stubs in the matching sections, is shown in Figure 6.14b. A computer analysis of the circuit (with $S_{12} \neq 0$) gave a gain of 8.36 dB. ○

6.6 POWER AMPLIFIERS

Power amplifiers are used in the final stages of wireless transmitters to increase the radiated power level. Typical output powers may be on the order of 0.3 to 0.6 W for a handheld cellular or PCS phone, or in the range of 10–100 W for base station transmitters. Important considerations for RF and microwave power amplifiers are efficiency, gain, and intermodulation effects. Single transistors can provide output powers of 10 to 100 W at UHF frequencies, while devices at higher frequencies are generally limited to output powers of 0.5 to 1 W. Various power combining techniques can be used in conjunction with multiple transistors if higher output powers are required [2]–[4].

So far we have considered only *small-signal amplifiers*, where the input signal power is small enough that the transistor can be assumed to operate as a linear device. The S parameters of linear devices are well defined and do not depend on the input power level or output load impedance, a fact that greatly simplifies the design of fixed-gain and low-noise amplifiers. For high input powers (in the range of the 1 dB compression point or third-order intercept point, for example), transistors do not behave linearly. In this case the impedances seen at the input and output of the transistor will depend on the input power level, and this greatly complicates the design of power amplifiers.

Characteristics of Power Amplifiers and Amplifier Classes

The power amplifier is usually the primary consumer of DC power in most handheld wireless devices, so amplifier efficiency is a very important consideration. One measure of amplifier efficiency is the ratio of RF output power to DC input power:

$$\eta = \frac{P_{\text{out}}}{P_{\text{DC}}}. \tag{6.64}$$

One drawback of this definition is that it does not account for the RF power delivered at the input to the amplifier. Since most power amplifiers have relatively low gains, the efficiency of (6.64) tends to overate the actual efficiency. A better measure that includes the effect of input power is the *power added efficiency*, defined as

$$\eta_{\text{PAE}} = PAE = \frac{P_{\text{out}} - P_{\text{in}}}{P_{\text{DC}}} = \left(1 - \frac{1}{G}\right)\frac{P_{\text{out}}}{P_{\text{DC}}} = \left(1 - \frac{1}{G}\right)\eta, \tag{6.65}$$

where G is the power gain of the amplifier. Silicon transistor amplifiers in the 800–900 MHz band have power added efficiencies on the order of 80%, but efficiency drops quickly with increasing frequency. Power amplifiers are often designed to provide the best efficiency, even if this means that the resulting gain is less than the maximum possible.

Another useful parameter for power amplifiers is the *compressed gain*, G_1, defined as the gain of the amplifier at the 1 dB compression point. Thus, if G_0 is the small-signal (linear) power gain, we have

$$G_1\ (\mathrm{dB}) = G_0\ (\mathrm{dB}) - 1. \tag{6.66}$$

As we have seen in Chapter 3, nonlinearities can lead to the generation of spurious frequencies and intermodulation distortion. This is a serious issue in wireless transmitters, especially in a multicarrier system, where spurious signals may appear in adjacent channels. Linearity is also critical for nonconstant envelope modulations, such as amplitude shift keying and higher quadrature amplitude modulation methods.

Class A amplifiers are inherently linear circuits, where the transistor is biased to conduct over the entire range of the input signal cycle. Because of this, class A amplifiers have a theoretical maximum efficiency of 50%. Most small-signal and low-noise amplifiers operate as class A circuits. In contrast, the transistor in a class B amplifier is biased to conduct only during one-half of the input signal cycle. Usually two complementary transistors are operated in a class B push-pull amplifier to provide amplification over the entire cycle. The theoretical efficiency of a class B amplifier is 78%. Class C amplifiers are operated with the transistor near cutoff for more than half of the input signal cycle, and generally use a resonant circuit in the output stage to recover the fundamental. Class C amplifiers can achieve efficiencies near 100%, but can only be used with constant envelope modulations. Higher classes, such as class D, E, F, and S, use the transistor as a switch to pump a highly resonant tank circuit, and achieve very high efficiencies. The majority of wireless transmitters operating at UHF frequencies or above rely on class A, AB, or B power amplifiers because of the need for low distortion products.

Large-Signal Characterization of Transistors

A transistor behaves linearly for signal powers well below the 1 dB compression point (P_1), and so the small-signal S-parameters should not depend on either the input power level or the output termination impedance. But for power levels comparable to or greater than P_1, where the nonlinearity of the transistor becomes apparent, the measured S parameters will depend on input power level and the output termination impedance (as well as frequency, bias conditions, and temperature). Thus large-signal S parameters are not uniquely defined and do not satisfy linearity, and cannot be used in place of small-signal parameters. (For device stability calculations, however, small-signal S parameters can generally be used with good results.)

A more useful way to characterize transistors under large-signal operating conditions is to measure the gain and output power as a function of source and load impedances. One way of doing this is to determine the large-signal source and load reflection coefficients, Γ_{SP} and Γ_{LP}, that maximize power gain for a particular output power (often chosen as P_1), and versus frequency. Table 6.2 shows typical large-signal source and load reflection coefficients for a typical NPN silicon bipolar power transistor, along with the small-signal S parameters.

Another way of characterizing the large-signal behavior of a transistor is to plot contours of constant power output on a Smith chart as a function of the load reflection coefficient, Γ_{LP}, with the transistor conjugately matched at its input. These are called *load-pull contours*, and can be obtained using an automated measurement set-up with computer-controlled

TABLE 6.2 Small-Signal S Parameters and Large-Signal Reflection Coefficients (Silicon Bipolar Power Transistor)

f (MHz)	S_{11}	S_{12}	S_{21}	S_{22}	Γ_{SP}	Γ_{LP}	G_p (dB)
800	$0.76\angle 176^\circ$	$4.10\angle 76^\circ$	$0.065\angle 49^\circ$	$0.35\angle -163^\circ$	$0.856\angle -167^\circ$	$0.455\angle 129^\circ$	13.5
900	$0.76\angle 172^\circ$	$3.42\angle 72^\circ$	$0.073\angle 52^\circ$	$0.35\angle -167^\circ$	$0.747\angle -177^\circ$	$0.478\angle 161^\circ$	12.0
1000	$0.76\angle 169^\circ$	$3.08\angle 69^\circ$	$0.079\angle 53^\circ$	$0.36\angle -169^\circ$	$0.797\angle -187^\circ$	$0.491\angle 185^\circ$	10.0

electromechanical stub tuners. A typical set of load-pull contours is shown in Figure 6.15. Load-pull contours are similar in function to the constant gain contours of Section 6.4, but are not perfect circles due to the nonlinearities of the device.

Nonlinear equivalent circuit models can also be developed and used to predict the large-signal performance of FETs and BJTs [7]. The dominant nonlinear parameters for a microwave FET are C_{gs}, g_m, C_{gd}, and R_{ds}. An important consideration in modeling large-signal transistors is the fact that most parameters are dependent on device temperature, which of course increases with output power. Equivalent circuit models can be very useful when combined with computer-aided design software.

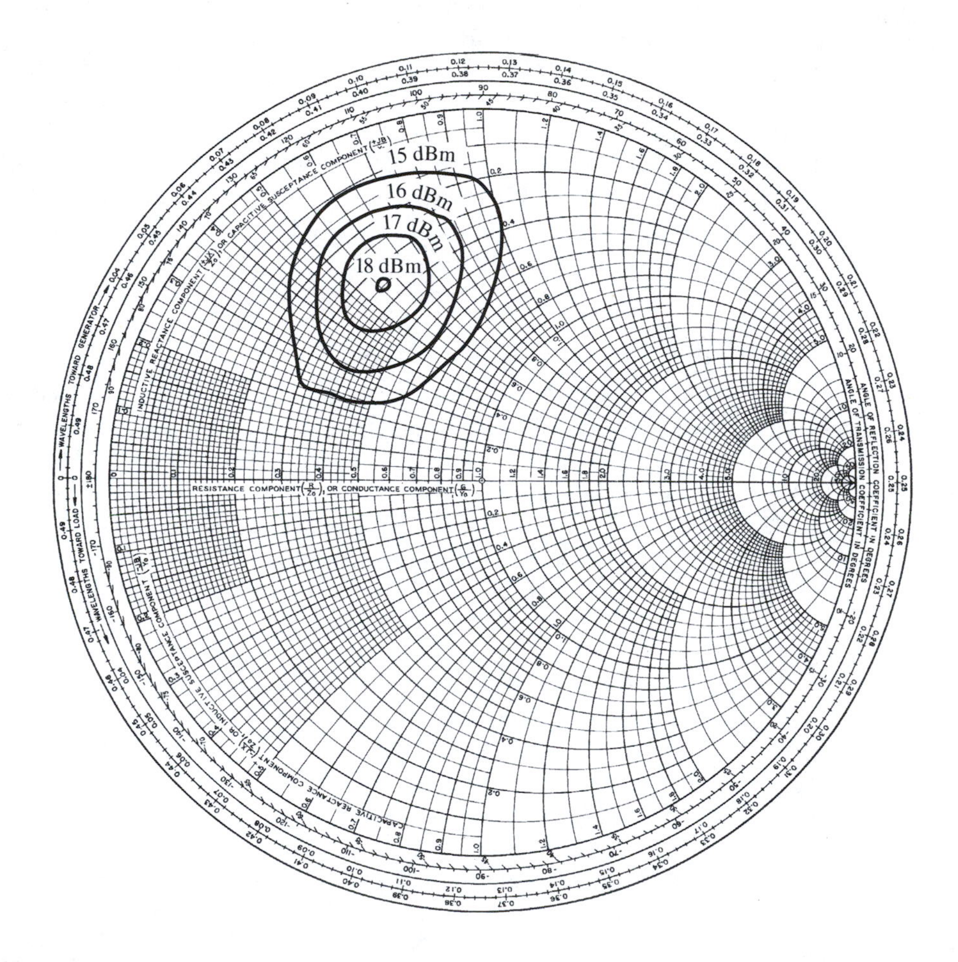

FIGURE 6.15 Constant output power contours versus load impedance for a typical power FET.

Design of Class A Power Amplifiers

In this section we will discuss the use of large-signal parameters for the design of class A amplifiers. Since class A amplifiers are ideally linear, it is sometimes possible to use small-scale S parameters for design, but better results are usually obtained if large-signal parameters are available. As with small-signal amplifier design, the first step is to check the stability of the device. Since instabilities begin at low signal levels, small-signal S parameters can be used for this purpose. Stability is especially important for power amplifiers, as high-power oscillations can easily damage active devices and related circuitry.

The transistor should be chosen on the basis of frequency range and power output, ideally with about 20% more power capacity than is required by the design. Silicon bipolar transistors have higher power outputs than GaAs FETs at frequencies up to a few GHz, and are generally cheaper. Good thermal contact of the transistor package to a heat sink is essential for any amplifier with more than a few tenths of a watt power output. Input matching networks are generally designed for maximum power transfer (conjugate matching), while output matching networks are designed for maximum output power (as derived from Γ_{LP}). The optimum values of source and load reflection coefficients are different from those obtained from small-signal S parameters via (6.43). Low-loss matching elements are important for good efficiency, particularly in the output stage, where currents are highest. Internally matched chip transistors are sometimes available, and have the advantage of reducing the effect of parasitic package reactances, thus improving efficiency and bandwidth.

EXAMPLE 6.6 DESIGN OF A CLASS A POWER AMPLIFIER

Design a power amplifier at 900 MHz using a Motorola MRF858S NPN Silicon bipolar transistor with an output power of 3 W. Design input and output impedance matching sections for the amplifier, find the required input power, and compute the power added efficiency. Use the given S parameters to compute the source and load reflection coefficients for conjugate matching, and compare to the actual large-signal values for Γ_{LP} and Γ_{SP}.

The small-signal S parameters of the MRF858S transistor at 900 MHz are: $S_{11} = 0.940\angle 164°$, $S_{12} = 0.031\angle 59°$, $S_{21} = 1.222\angle 43°$, $S_{22} = 0.570\angle -165°$. For a emitter-collector voltage $V_{CE} = 24$ V and a collector current of $I_C = 0.5$ A, the output power at the 1 dB compression point is 3.6 W, and the power gain is 12 dB. The source and load impedances are $Z_{\text{in}} = 1.2 + j3.5\ \Omega$, and $Z_{\text{out}} = 9.0 + j14.5\ \Omega$.

Solution

We begin by establishing the stability of the device. Using the small-signal S parameters in (6.31) and (6.32) gives

$$\begin{aligned}|\Delta| &= |S_{11}S_{22} - S_{12}S_{21}| \\ &= |(0.940\angle 164°)(0.570\angle -165°) - (0.031\angle 59°)(1.222\angle 43°)| \\ &= 0.546\end{aligned}$$

$$K = \frac{1 - |S_{11}|^2 - |S_{22}|^2 + |\Delta|^2}{2|S_{12}S_{21}|} = \frac{1 - (0.940)^2 - (0.570)^2 + (0.546)^2}{2(0.031)(1.222)} = 1.177,$$

showing that the device is unconditionally stable.

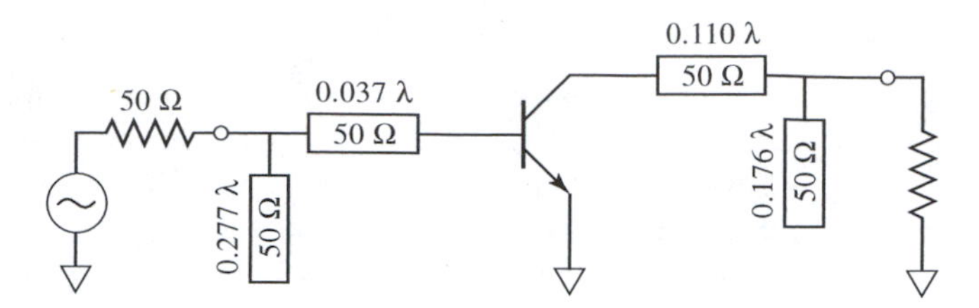

FIGURE 6.16 RF circuit for the amplifier of Example 6.6.

Converting the large-signal input and output impedances to reflection coefficients gives

$$\Gamma_{\text{in}} = 0.953\angle 172^\circ,$$
$$\Gamma_{\text{out}} = 0.716\angle -147^\circ.$$

Using the small-signal S parameters in (6.43) to find the source and load reflection coefficients for conjugate matching gives

$$\Gamma_S = \frac{B_1 \pm \sqrt{B_1^2 - 4|C_1|^2}}{2C_1} = 0.963\angle -166^\circ,$$
$$\Gamma_L = \frac{B_2 \pm \sqrt{B_2^2 - 4|C_2|^2}}{2C_2} = 0.712\angle 134^\circ.$$

Note that these values approximately satisfy the relationships of (6.39), that $\Gamma_S = \Gamma_{\text{in}}^*$ and $\Gamma_L = \Gamma_{\text{out}}^*$, but not exactly, due to the fact that the S parameters used to calculate Γ_S and Γ_L do not apply for large power levels. Thus we should use the given large-signal reflection coefficients and let

$$\Gamma_S = \Gamma_{\text{in}}^* = 0.953\angle -172^\circ,$$
$$\Gamma_L = \Gamma_{\text{out}}^* = 0.716\angle 147^\circ.$$

Then the input and output matching networks can be designed as described in Example 6.3. The complete AC amplifier circuit is shown in Figure 6.16.

For an output power of 3 W, the required input drive power is

$$P_{\text{in}}\,(\text{dBm}) = P_{\text{out}}\,(\text{dBm}) - G_p\,(\text{dB}) = 10\log(3000) - 12 = 22.8\,\text{dBm} = 189\,\text{mW}.$$

Then the power added efficiency of the amplifier can be found from (6.65) to be

$$\eta_{\text{PAE}} = \frac{P_{\text{out}} - P_{\text{in}}}{P_{\text{DC}}} = \frac{3.0 - 0.189}{(24)(0.5)} = 23.4\% \qquad \bigcirc$$

REFERENCES

[1] D. M. Pozar, **Microwave Engineering**, 2nd edition, Wiley, New York, 1998.

[2] G. D. Vendelin, A. M. Pavio, and U. L. Rohde, **Microwave Circuit Design Using Linear and Nonlinear Techniques**, Wiley, New York, 1990.

[3] G. Gonzalez, **Microwave Transistor Amplifiers: Analysis and Design**, 2nd edition Prentice-Hall, NJ, 1997.

[4] I. Bahl and P. Bhartia, **Microwave Solid-State Circuit Design**, Wiley Interscience, New York, 1988.

[5] C. Gentile, **Microwave Amplifiers and Oscillators**, McGraw-Hill, New York, 1987.

[6] M. L. Edwards and J. H. Sinksy, "A New Criteria for Linear 2-Port Stability Using a Single Geometrically Derived Parameter," *IEEE Trans. Microwave Theory and Techniques*, vol. MTT-40, pp. 2803–2811, December 1992.

[7] W. R. Curtice and M. Ettenberg, "A Nonlinear GaAs FET model for Use in the Design of Output Circuits for Power Amplifiers," *IEEE Trans. Microwave Theory and Techniques*, vol. MTT-33, pp. 1383–1394, December 1985.

PROBLEMS

6.1 Use the equivalent circuit of Figure 6.5 to derive the expression for the short-circuit current gain of a bipolar transistor. Assume a unilateral device, where $C_c = 0$.

6.2 Derive expressions for the y (admittance) parameters of an FET using the unilateral equivalent circuit model. Evaluate these parameters at 5 GHz for the following FET characteristics: $R_i = 7\ \Omega$, $R_{ds} = 400\ \Omega$, $C_{gs} = 0.3$ pF, $C_{ds} = 0.12$ pF, $C_{gd} = 0$, $g_m = 30$ mS. Convert the y parameters to S parameters for a 50 Ω ohm system impedance, and find the unilateral transducer gain assuming conjugately matched source and load impedances. Compare with the value computed using (6.21).

6.3 Consider the RF network shown below, consisting of a 50 Ω source, a 3 dB matched attenuator, and a 50 Ω load. Compute the available gain, the transducer power gain, and the actual power gain. How do these gains change if the load is changed to 25 Ω?

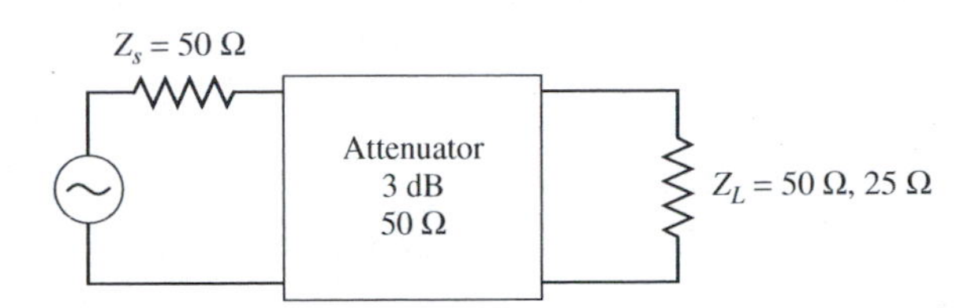

6.4 An RF transistor has the following S parameters at 1.8 GHz: $S_{11} = 0.34\angle{-170°}$, $S_{21} = 4.3\angle{80°}$, $S_{12} = 0.06\angle{70°}$, $S_{22} = 0.45\angle{-25°}$. Determine the stability of the device, and plot the stability circles if the device is potentially unstable.

6.5 Repeat Problem 6.4 for the following transistor S parameters: $S_{11} = 0.8\angle{-90°}$, $S_{21} = 5.1\angle{80°}$, $S_{12} = 0.3\angle{70°}$, $S_{22} = 0.62\angle{-40°}$.

6.6 Use the μ-parameter test to determine which of the following devices are unconditionally stable, and of those, which has the greatest stability.

Device	S_{11}	S_{12}	S_{21}	S_{22}
A	$0.34\angle{-170°}$	$0.06\angle{70°}$	$4.3\angle{80°}$	$0.45\angle{-25°}$
B	$0.75\angle{-60°}$	$0.2\angle{70°}$	$5.0\angle{90°}$	$0.51\angle{60°}$
C	$0.65\angle{-140°}$	$0.04\angle{60°}$	$2.4\angle{50°}$	$0.70\angle{-65°}$

6.7 Show that for a unilateral device, where $S_{12} = 0$, the μ-parameter test of (6.38) implies that $|S_{11}| < 1$ and $|S_{22}| < 1$ for unconditional stability.

6.8 Prove that the condition for a positive discriminant in (6.43a), that is, $B_1^2 > 4|C_1|^2$, is equivalent to the condition that $K^2 > 1$.

6.9 Design a transistor amplifier for maximum gain at 5.0 GHz using a GaAs FET with the following S parameters ($Z_0 = 50\ \Omega$): $S_{11} = 0.65\angle{-140°}$, $S_{21} = 2.4\angle{50°}$, $S_{12} = 0.04\angle{60°}$, $S_{22} = 0.70\angle{-65°}$. Use open-circuited stubs for input and output matching.

6.10 Design an amplifier with maximum G_{TU} using a transistor with the following S parameters at 1.8 GHz ($Z_0 = 50\ \Omega$): $S_{11} = 0.61\angle{-170°}$, $S_{21} = 2.24\angle{32°}$, $S_{12} = 0$, $S_{22} = 0.72\angle{-83°}$.

6.11 Design an amplifier to have a gain of 10 dB at 2.4 GHz, using a transistor with the following S parameters ($Z_0 = 50\ \Omega$): $S_{11} = 0.61\angle{-170°}$, $S_{21} = 2.24\angle 32°$, $S_{12} = 0$, $S_{22} = 0.72\angle{-83°}$. Plot (and use) constant gain circles for $G_S = 1$ dB and $G_L = 2$ dB. Use matching sections with open-circuited shunt stubs.

6.12 Compute the unilateral figure of merit for the transistor of Problem 6.4. What is the maximum error in the transducer gain if an amplifier is designed under the assumption that the device is unilateral?

6.13 Show that the 0 dB gain circle for $G_S (G_S = 1)$, defined by (6.54), will pass through the center of the Smith chart.

6.14 A GaAs FET has the following scattering and noise parameters at 8 GHz ($Z_0 = 50\ \Omega$): $S_{11} = 0.7\angle{-110°}$, $S_{21} = 3.5\angle 60°$, $S_{12} = 0.02\angle 60°$, $S_{22} = 0.8\angle{-70°}$, $F_{\text{min}} = 2.5$ dB, $\Gamma_{\text{opt}} = 0.7\angle 120°$, $R_N = 15\ \Omega$. Design an amplifier with minimum noise figure, and maximum possible gain. Use open-circuited shunt stubs in the matching sections.

6.15 A GaAs FET has the following scattering and noise parameters at 6 GHz ($Z_0 = 50\ \Omega$): $S_{11} = 0.6\angle{-60°}$, $S_{21} = 2.0\angle 81°$, $S_{12} = 0$, $S_{22} = 0.7\angle{-60°}$, $F_{\text{min}} = 2.0$ dB, $\Gamma_{\text{opt}} = 0.62\angle 100°$, $R_N = 20\ \Omega$. Design an amplifier to have a gain of 6 dB, and the minimum noise figure possible with this gain. Use open-circuited shunt stubs in the mathcing sections.

6.16 Repeat Problem 6.15, but design the amplifier for a noise figure of 2.5 dB, and the maximum gain that can be achieved with this noise figure.

6.17 Use the transistor data given in Table 6.2 to design a power amplifier at 1 GHz with a power output of 1 W. Design the input and output matching circuits using the given large-signal reflection coefficients. Compute the required input power level.

Chapter Seven

Mixers

A *mixer* is a three-port device that uses a nonlinear or time-varying element to achieve frequency conversion. As introduced in Chapter 1, an ideal mixer produces an output consisting of the sum and difference frequencies of its two input signals. Operation of practical RF and microwave mixers are usually based on the nonlinearity provided by either a diode or a transistor. As we saw in Chapter 3, a nonlinear component can generate a wide variety of harmonics and other products of input frequencies, so filtering must be used to select the desired frequency components. Modern wireless transmitters and receivers typically use several mixers and filters to perform the function of frequency conversion between baseband signal frequencies and RF carrier frequencies.

We begin this chapter with an overview of the process of frequency conversion and mixer characteristics, which include losses, noise effects, and intermodulation distortion. We then discuss diode mixer circuits and present several models to treat the nonlinear aspect of the diode element. Next, we discuss the basic FET mixer, and conclude with circuits of more specialized mixers.

7.1 MIXER CHARACTERISTICS

Here we describe the fundamental operation of frequency conversion in a mixer, the important role of the image frequency, and basic mixer characteristics.

Frequency Conversion

The symbol and functional diagram for a mixer are shown in Figure 7.1. The mixer symbol is intended to imply that the output is proportional to the product of the two input signals. We will see that this is an idealized view of mixer operation, which in actuality

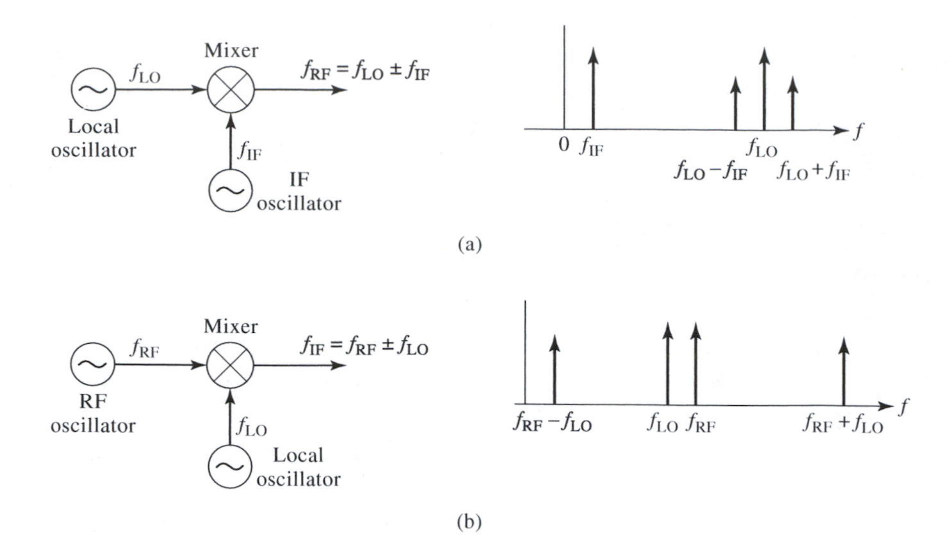

FIGURE 7.1 Frequency conversion using a mixer. (a) Up-conversion. (b) Down-conversion.

produces a large variety of harmonics and other undesired products of the input signals due to the nonlinearity of the mixer.

Figure 7.1a illustrates the operation of *frequency up-conversion*, as occurs in a transmitter. A local oscillator (LO) signal at the relatively high frequency f_{LO} is connected to one of the input ports of the mixer. The LO signal can be represented as

$$v_{LO}(t) = \cos 2\pi f_{LO} t. \tag{7.1}$$

A lower frequency baseband or intermediate frequency (IF) signal is applied to the other mixer input. This signal typically contains the information or data to be transmitted, and can be expressed for our purposes as

$$v_{IF}(t) = \cos 2\pi f_{IF} t. \tag{7.2}$$

The output of the idealized mixer is given by the product of the LO and IF signals:

$$\begin{aligned} v_{RF}(t) &= K v_{LO}(t) v_{IF}(t) = K \cos 2\pi f_{LO} t \cos 2\pi f_{IF} t \\ &= \frac{K}{2}[\cos 2\pi (f_{LO} - f_{IF})t + \cos 2\pi (f_{LO} + f_{IF})t] \end{aligned} \tag{7.3}$$

where K is a constant accounting for the voltage conversion loss of the mixer. The RF output is seen to consist of the sum and differences of the input signal frequencies:

$$f_{RF} = f_{LO} \pm f_{IF} \tag{7.4}$$

The spectrum of the input and output signals are shown in Figure 7.1a, where we see that the mixer has the effect of modulating the LO signal with the IF signal. The sum and difference frequencies at $f_{LO} \pm f_{IF}$ are called the *sidebands* of the carrier frequency f_{LO}, with $f_{LO} + f_{IF}$ being the *upper sideband* (USB), and $f_{LO} - f_{IF}$ being the *lower sideband* (LSB). A *double-sideband* (DSB) signal contains both upper and lower sidebands, as in (7.3), while a *single-sideband* (SSB) signal can be produced by filtering or by using a single-sideband mixer.

Conversely, Figure 7.1b shows the process of *frequency down-conversion*, as used in a receiver. In this case an RF input signal of the form

$$v_{RF}(t) = \cos 2\pi f_{RF} t, \tag{7.5}$$

is applied to the input of the mixer, along with the local oscillator signal of (7.1). The output of the mixer is

$$\begin{aligned} v_{IF}(t) &= K v_{RF}(t) v_{LO}(t) = K \cos 2\pi f_{RF} t \cos 2\pi f_{LO} t \\ &= \frac{K}{2}[\cos 2\pi (f_{RF} - f_{LO})t + \cos 2\pi (f_{RF} + f_{LO})t] \end{aligned} \tag{7.6}$$

Thus the mixer output consists of the sum and difference of the input signal frequencies. The spectrum for these signals is shown in Figure 7.1b. In practice, the RF and LO frequencies are relatively close together, so the sum frequency is approximately twice the RF frequency, while the difference is much smaller than f_{RF}. The desired IF output in a receiver is the difference frequency, $f_{RF} - f_{LO}$, which is easily selected by low-pass filtering:

$$f_{IF} = f_{RF} - f_{LO} \tag{7.7}$$

Note that the preceding discussion only considers the sum and difference outputs as generated by multiplication of the input signals, whereas in a realistic mixer many more products will be generated due to the more involved nonlinearity of the diode or transistor. These products are usually undesirable, and removed by filtering.

Image Frequency

In a receiver the RF input signal at frequency f_{RF} is typically delivered from the antenna, which may receive RF signals over a relatively wide band of frequencies. For a receiver with a local oscillator frequency f_{LO} and intermediate frequency f_{IF}, (7.7) gives the RF input frequency that will be down-converted to the IF frequency as

$$f_{RF} = f_{LO} + f_{IF}, \tag{7.8a}$$

since the insertion of (7.8a) into (7.7) yields f_{IF} (after low-pass filtering). Now consider the RF input frequency given by

$$f_{IM} = f_{LO} - f_{IF}. \tag{7.8b}$$

Insertion of (7.8b) into (7.7) yields $-f_{IF}$ (after low-pass filtering). Mathematically, this frequency is identical to f_{IF} because of the fact that the Fourier spectrum of any real signal is symmetric about zero frequency, and thus contains negative frequencies as well as positive. The RF frequency defined in (7.8b) is called the *image response*. The image response is important in receiver design because a received RF signal at the image frequency of (7.8b) is indistinguishable at the IF stage from the desired RF signal of frequency (7.8a), unless steps are taken in the RF stages of the receiver to preselect signals only within the desired RF frequency band. This issue will be further discussed in Chapter 10 on receiver design.

The choice of which RF frequency in (7.8) is the desired and which is the image response is arbitrary, depending on whether the LO frequency is above or below the desired RF frequency. Another way of viewing this difference is to note that f_{IF} in (7.8) may be negative. Observe that the desired and image frequencies of (7.8a) and (7.8b) are separated by $2f_{IF}$.

Another implication of (7.7) and the fact that f_{IF} may be negative is that there are two local oscillator frequencies that can be used for a given RF and IF frequency:

$$f_{LO} = f_{RF} \pm f_{IF}, \tag{7.9}$$

since taking the difference frequency of f_{RF} with these two LO frequencies gives $\pm f_{IF}$. These two frequencies correspond to the upper and lower sidebands when a mixer is operated as an up-converter. In practice, most receivers use a local oscillator set at the upper sideband, $f_{LO} = f_{RF} + f_{IF}$, because this requires a smaller LO tuning ratio when the receiver must select RF signals over a given band.

EXAMPLE 7.1 IMAGE FREQUENCY

The IS-54 digital cellular telephone system uses a receive frequency band of 869–894 MHz, with a first IF frequency of 87 MHz, and a channel bandwidth of 30 kHz. What are the two possible ranges for the LO frequency? If the upper LO frequency range is used, determine the image frequency range. Does the image frequency fall within the receive passband?

Solution

By (7.9), the two possible LO frequency ranges are

$$f_{LO} = f_{RF} \pm f_{IF} = (869 \text{ to } 894) \pm 87 = \begin{cases} 956 \text{ to } 981 \text{ MHz} \\ 782 \text{ to } 807 \text{ MHz} \end{cases}$$

Using the 956–981 MHz LO, (7.7) gives the IF frequency as

$$f_{IF} = f_{RF} - f_{LO} = (869 \text{ to } 894) - (956 \text{ to } 981) = -87 \text{ MHz},$$

so from (7.8b) the RF image frequency range is

$$f_{IM} = f_{LO} - f_{IF} = (956 \text{ to } 981) + 87 = 1043 \text{ to } 1068 \text{ MHz},$$

which is well outside the receive passband. ○

Conversion Loss

Mixer design requires impedance matching at three ports, complicated by the fact that several frequencies and their harmonics are involved. Ideally, each mixer port would be matched at its particular frequency (RF, LO, or IF), and undesired frequency products would be absorbed with resistive loads, or blocked with reactive terminations. Resistive loads increase mixer losses, however, and reactive loads can be very frequency sensitive. In addition, there are inherent losses in the frequency conversion process because of the generation of undesired harmonics and other frequency products.

An important figure of merit for a mixer is therefore the *conversion loss*, which is defined as the ratio of available RF input power to the available IF output power, expressed in dB:

$$L_c = 10 \log \frac{\text{available RF input power}}{\text{available IF output power}} \geq 0 \text{ dB} \tag{7.10}$$

Conversion loss accounts for resistive losses in a mixer as well as loss in the frequency conversion process from RF to IF ports. Conversion loss applies to both up-conversion and down-conversion, even though the context of the above definition is for the latter case. Since the RF stages of receivers operate at much lower power levels than do transmitters, minimum conversion loss is more critical for receivers because of the importance of minimizing losses in the RF stages to minimize receiver noise figure.

Practical diode mixers typically have conversion losses between 4 and 7 dB in the 1–10 GHz range. Transistor mixers have lower conversion loss, and may even have *conversion gain* of a few dB. One factor that strongly affects conversion loss is the local oscillator power level; minimum conversion loss often occurs for LO powers between 0 and 10 dBm. This power level is large enough that the accurate characterization of mixer performance often requires nonlinear analysis.

Noise Figure

Noise is generated in mixers by the diode or transistor elements, and by thermal sources due to resistive losses. Noise figures of practical mixers range from 1 dB to 5 dB, with diode mixers generally achieving lower noise figures than transistor mixers.

The noise figure of a mixer depends on whether its input is a single sideband signal or a double sideband signal. This is because the mixer will down-convert noise at both sideband frequencies (since these have the same IF), but the power of a SSB signal is one-half that of a DSB signal (for the same amplitude). To derive the relation between the noise figure for these two cases, first consider a DSB input signal of the form

$$v_{\mathrm{DSB}}(t) = A[\cos(\omega_{\mathrm{LO}} - \omega_{\mathrm{IF}})t + \cos(\omega_{\mathrm{LO}} + \omega_{\mathrm{IF}})t]. \tag{7.11}$$

Upon mixing with an LO signal $\cos\omega_{\mathrm{LO}}t$ and low-pass filtering, the down-converted IF signal will be

$$v_{\mathrm{IF}}(t) = \frac{AK}{2}\cos(\omega_{\mathrm{IF}}t) + \frac{AK}{2}\cos(-\omega_{\mathrm{IF}}t) = AK\cos\omega_{\mathrm{IF}}t, \tag{7.12}$$

where K is a constant accounting for the conversion loss for each sideband. The power of the DSB input signal of (7.11) is

$$S_i = \frac{A^2}{2} + \frac{A^2}{2} = A^2,$$

and the power of the output IF signal is

$$S_o = \frac{A^2K^2}{2}.$$

For noise figure, the input noise power is defined as $N_i = kT_0B$, where $T_0 = 290$ K and B is the IF bandwidth. The output noise power is equal to the input noise plus N_{added}, the noise power added by the mixer, divided by the conversion loss (assuming a reference at the mixer input):

$$N_o = \frac{(kT_0B + N_{\mathrm{added}})}{L_c}.$$

Then using the definition of noise figure gives the DSB noise figure of the mixer as

$$F_{\mathrm{DSB}} = \frac{S_iN_o}{S_oN_i} = \frac{2}{K^2L_c}\left(1 + \frac{N_{\mathrm{added}}}{kT_0B}\right). \tag{7.13}$$

The corresponding analysis for the SSB case begins with a SSB input signal of the form

$$v_{\mathrm{SSB}}(t) = A\cos(\omega_{\mathrm{LO}} - \omega_{\mathrm{IF}})t. \tag{7.14}$$

Upon mixing with the LO signal $\cos\omega_{\mathrm{LO}}t$ and low-pass filtering, the down-converted IF signal will be

$$v_{\mathrm{IF}}(t) = \frac{AK}{2}\cos(\omega_{\mathrm{IF}}t). \tag{7.15}$$

The power of the SSB input signal of (7.14) is

$$S_i = \frac{A^2}{2},$$

and the power of the output IF signal is

$$S_o = \frac{A^2 K^2}{8}.$$

The input and output noise powers are the same as for the DSB case, so the noise figure for an SSB input signal is

$$F_{\text{SSB}} = \frac{S_i N_o}{S_o N_i} = \frac{4}{K^2 L_c}\left(1 + \frac{N_{\text{added}}}{k T_0 B}\right). \tag{7.16}$$

Comparison with (7.13) shows that the noise figure of the SSB case is twice that of the DSB case:

$$F_{\text{SSB}} = 2F_{\text{DSB}}. \tag{7.17}$$

Intermodulation Distortion

Since mixers involve nonlinearity, they will produce intermodulation products. Typical values of P_3 for mixers range from 15 dBm to 30 dBm.

Isolation

Another important characteristic of a mixer is the isolation between the RF and LO ports. Ideally, the LO and RF ports would be decoupled, but internal impedance mismatches and limitations of coupler performance often result in some LO power being coupled out of the RF port. This is a potential problem for receivers that drive the RF port directly from the antenna, because LO power coupled through the mixer to the RF port will be radiated by the antenna. Because such signals will likely interfere with other services or users, the FCC sets stringent limits on the power radiated by receivers. This problem can be largely alleviated by using a bandpass filter between the antenna and mixer, or by using an RF amplifier ahead of the mixer. Isolation between the LO and RF ports is highly dependent on the type of coupler used for diplexing these two inputs, but typical values range from 20 dB to 40 dB.

7.2 DIODE MIXERS

In Section 7.1 we treated the mixer from the idealized view that its output was proportional to the product of its input signals, which has the effect of producing sum and difference frequencies when the inputs are sinusoidal. Here we present a treatment of more realistic mixers, and show that the output does indeed contain a term proportional to the product of the inputs, but many higher order terms as well. In this section we discuss mixers using diodes as the nonlinear element; mixers using FETs will be discussed in the following section.

The nonlinear V-I characteristics of a diode make it useful for rectifiers, detectors, and mixers [1], but analysis is difficult because of this nonlinearity. We will discuss several approaches to this problem. The first is to assume that the signal power presented to the diode is small enough that a Taylor series approximation can be made for the diode current in terms of the diode voltage. This is known as the *small-signal approximation*, and leads to useful qualitative results for many rectifier, detector, and mixer circuits [1]. The small-signal approximation generally does not give very accurate results in the practical case when a mixer is driven with a relatively high LO power. For this reason we also present a *large-signal* model for the diode mixer, using a fully nonlinear analysis. Finally, we present an alternative model for a mixer, based on an ideal switching circuit.

Small-Signal Diode Characteristics

The I-V response of a diode can be written as

$$I(V) = I_s(e^{\alpha V} - 1), \tag{7.18}$$

where V is the voltage across the diode, I is the current through the diode, and I_s is the reverse saturation current [1]–[3]. The constant $\alpha = q/nkT$, where q is the charge of an electron, k is Boltzmann's constant, T is the temperature in Kelvin, and n is the diode ideality factor. For typical RF diodes, I_s is between 10^{-6} and 10^{-15} A, and α is approximately 1/(28 mV) for $T = 290$ K. The ideality factor, n, depends on the structure of the diode, and can vary from 1.2 for Schottky barrier diodes to about 2.0 for point-contact silicon diodes. The I-V response of (7.18) is shown in Figure 7.2.

Now consider the total diode voltage and current to consist of small AC signals, $v(t)$ and $i(t)$, superimposed on a DC bias I_0 and V_0:

$$I = I_0 + i(t), \tag{7.19a}$$

$$V = V_0 + v(t), \tag{7.19b}$$

where $I_0 = I(V_0)$ is the DC bias current. If we assume that $i(t)$ and $v(t)$ represent only small excursions about the constant bias terms, we can represent the total current as a Taylor series in terms of the applied AC signal voltage:

$$I(V) = I_0 + G_d v(t) + \frac{1}{2} G'_d v^2(t) + \cdots, \tag{7.20}$$

where G_d is the dynamic conductance of the diode, defined as

$$G_d = \left.\frac{dI}{dV}\right|_{V=V_0} = \alpha I_s e^{\alpha V_0} = \alpha(I_0 + I_s), \tag{7.21a}$$

and G'_d is the derivative of the dynamic conductance:

$$G'_d = \left.\frac{dG_d}{dV}\right|_{V=V_0} = \left.\frac{d^2 I}{dV^2}\right|_{V=V_0} = \alpha^2 I_s e^{\alpha V_0} = \alpha G_d. \tag{7.21b}$$

The Taylor series of (7.20) constitutes the small-signal approximation for a diode. The first two terms, representing DC bias current and the linear diode response, are of little

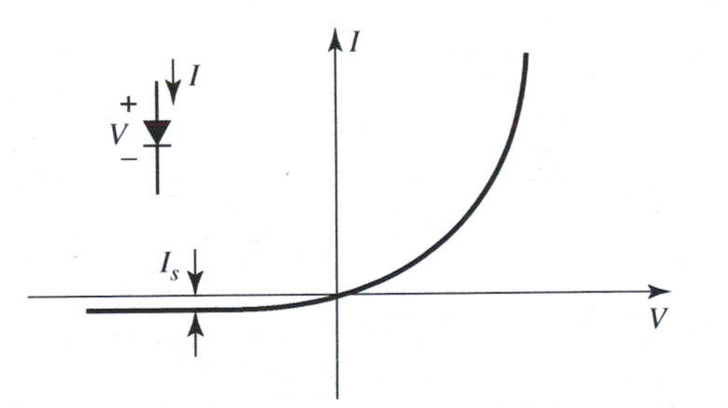

FIGURE 7.2 V-I characteristic of a diode.

interest to us because no frequency conversion occurs through these terms. The third term, containing v^2, represents the *square-law* response of the diode, and is responsible for the dominant frequency conversion terms. We will now apply the small-signal approximation to a basic diode mixer circuit.

Single-Ended Mixer

A basic diode mixer circuit is shown in Figure 7.3a. This type of mixer is called a *single-ended mixer* because it uses a single diode element. The RF and LO inputs are combined in a *diplexer*, which superimposes the two input voltages to drive the diode. The diplexing function is easily implemented using an RF coupler or hybrid junction to provide combining as well as isolation between the two inputs. The diode may be biased with a DC bias voltage, which must be decoupled from the RF signal paths. This is done by using DC blocking capacitors on either side of the diode, and an RF choke between the diode and the bias voltage source. The AC output of the diode is passed through a low-pass filter to provide the desired IF output voltage. This description is for application as a down-converter, but the same mixer can be used for up-conversion since each port may be used interchangeably as an input or output port.

The AC equivalent circuit of the mixer is shown in Figure 7.3b, where the RF and LO input voltages are represented as two series-connected voltage sources. Let the RF input voltage be a cosine wave of frequency ω_{RF}:

$$v_{RF}(t) = V_{RF} \cos \omega_{RF} t, \tag{7.22}$$

and let the LO input voltage be a cosine wave of frequency ω_{LO}:

$$v_{LO}(t) = V_{LO} \cos \omega_{LO} t. \tag{7.23}$$

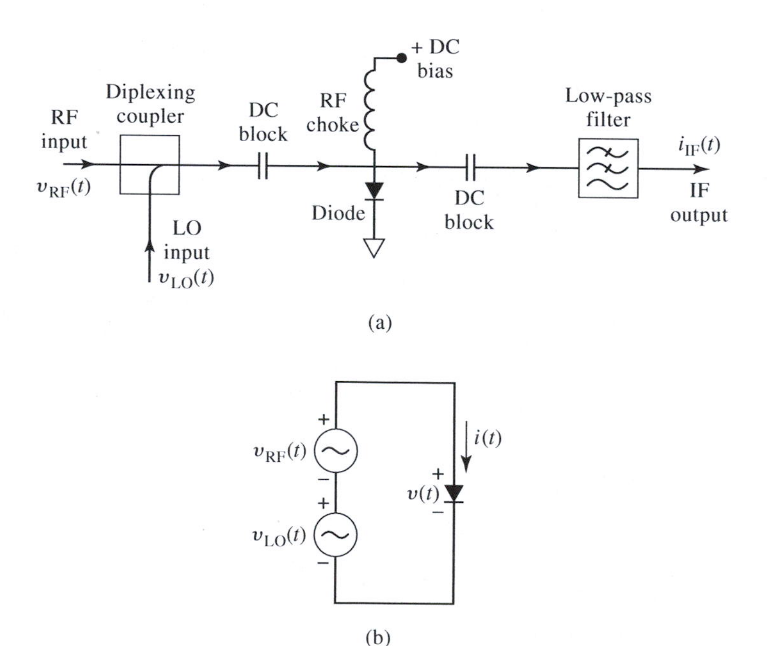

FIGURE 7.3 (a) Circuit for a single-ended mixer. (b) Idealized equivalent circuit of the single-ended mixer.

Using the small-signal approximation of (7.20) gives the total diode current as

$$i(t) = I_0 + G_d[v_{\mathrm{RF}}(t) + v_{\mathrm{LO}}(t)] + \frac{G'_d}{2}[v_{\mathrm{RF}}(t) + v_{\mathrm{LO}}(t)]^2 + \cdots \tag{7.24}$$

The first term in (7.24) is the DC bias current, which will be blocked from the IF output by the DC blocking capacitors. The second term is a replication of the RF and LO input signals, which will be filtered out by the low-pass IF filter. This leaves the third term, which can be rewritten using trigonometric identities as

$$\begin{aligned}
i(t) &= \frac{G'_d}{2}[V_{\mathrm{RF}}\cos\omega_{\mathrm{RF}}t + V_{\mathrm{LO}}\cos\omega_{\mathrm{LO}}t]^2 \\
&= \frac{G'_d}{2}\left[V_{\mathrm{RF}}^2\cos^2\omega_{\mathrm{RF}}t + 2V_{\mathrm{RF}}V_{\mathrm{LO}}\cos\omega_{\mathrm{RF}}t\cos\omega_{\mathrm{LO}}t + V_{\mathrm{LO}}^2\cos^2\omega_{\mathrm{LO}}t\right] \\
&= \frac{G'_d}{4}\left[V_{\mathrm{RF}}^2(1+\cos 2\omega_{\mathrm{RF}}t) + V_{\mathrm{LO}}^2(1+\cos 2\omega_{\mathrm{LO}}t)\right. \\
&\quad \left. + 2V_{\mathrm{RF}}V_{\mathrm{LO}}\cos(\omega_{\mathrm{RF}} - \omega_{\mathrm{LO}})t + 2V_{\mathrm{RF}}V_{\mathrm{LO}}\cos(\omega_{\mathrm{RF}} + \omega_{\mathrm{LO}})t\right]
\end{aligned}$$

This result is seen to contain several new signal components, only one of which produces the desired IF difference product. The two DC terms again will be blocked by the blocking capacitors, and the $2\omega_{\mathrm{RF}}$, $2\omega_{\mathrm{LO}}$, and $\omega_{\mathrm{RF}} + \omega_{\mathrm{LO}}$ terms will be blocked by the low-pass filter. This leaves the IF output current as

$$i_{\mathrm{IF}}(t) = \frac{G'_d}{2}V_{\mathrm{RF}}V_{\mathrm{LO}}\cos\omega_{\mathrm{IF}}t, \tag{7.25}$$

where $\omega_{\mathrm{IF}} = \omega_{\mathrm{RF}} - \omega_{\mathrm{LO}}$ is the IF frequency. The spectrum of the down- converting single-ended mixer is thus identical to that of the idealized mixer shown in Figure 7.1b.

Large-Signal Model

While the small-signal analysis of a mixer demonstrates the key process of frequency conversion, it is not accurate enough to provide a realistic result for conversion loss. This is primarily because the power supplied to the mixer LO port is usually large enough to violate the small-signal approximation. Here we consider a fully nonlinear analysis of a resistive diode mixer [3]–[4], with the goal of deriving an expression for the conversion loss defined in (7.10). The term "resistive" in this context means that reactances associated with the diode junction and package are ignored, to simplify the analysis. Our results should be useful in understanding the nonlinear operation and losses of the diode mixer, but for actual design purposes modern computer-aided design (CAD) software is preferred [5]. Such software can model the diode nonlinearity, as well as the effects of diode reactances and impedance matching networks.

We again assume a diode I-V characteristic as given by (7.18), with a relatively low-level RF input voltage given by (7.22), and a much larger LO *pump* signal given by (7.23). A DC bias current may also be present, but will not directly enter into our analysis. As we have seen from the small-signal mixer analysis, these two AC input signals generate a multitude of harmonics and other frequency products:

ω_{RF}	RF input signal (low power)
$\omega_{\mathrm{IF}} = \omega_{\mathrm{RF}} - \omega_{\mathrm{LO}}$	IF output signal (low power)
$\omega_{\mathrm{IM}} = \omega_{\mathrm{LO}} - \omega_{\mathrm{IF}}$	image signal (low power)
ω_{LO}	LO input signal (high power)
$n\omega_{\mathrm{LO}}$	harmonics of LO (high power)
$n\omega_{\mathrm{LO}} \pm \omega_{\mathrm{IF}}$	harmonic sidebands of LO (low power)

In a typical mixer, harmonics of the LO and the harmonic sidebands are terminated reactively, and therefore do not lead to much power loss. This leaves three signal frequencies of most importance: ω_{RF}, ω_{IF}, and ω_{IM}. To evaluate conversion loss, we will find the available power of the RF input signal, the power of the IF output signal, and the power lost in the image signal. The image signal is important because it is relatively close in frequency to the RF signal, and thus sees essentially the same load. We will see that approximately half the input power gets converted to the image frequency. Note that the image term at frequency $\omega_{\mathrm{IM}} = \omega_{\mathrm{LO}} - \omega_{\mathrm{RF}} = 2\omega_{\mathrm{LO}} - \omega_{\mathrm{RF}}$ was not explicitly shown in the small-signal expansion of (7.24), since this product is generated by the v^3 term of (7.20).

Under the assumption that the RF input voltage is small, we can write the AC diode current as a Taylor series expansion about the LO voltage as

$$i(v) = I(v_{\mathrm{LO}}) + v\left.\frac{dI}{dV}\right|_{v_{\mathrm{LO}}} + \frac{1}{2}v^2\left.\frac{d^2I}{dV^2}\right|_{v_{\mathrm{LO}}} + \cdots \tag{7.26}$$

This Taylor series is similar in form to (7.20), as used for the small-signal analysis, but with the important difference that the expansion point here is about the LO voltage, whereas (7.20) was expanded about the DC bias point. The first term in (7.26) is due only to the LO input, and does not enter into the calculation of conversion loss. The second term is a function of the RF and LO input voltages, and will provide a good approximation for the three products at frequencies ω_{RF}, ω_{IF}, and ω_{IM}, with a large LO pump signal. The coefficient of the second term has dimensions of conductance, so we can use (7.18) to write the *differential conductance* as

$$g(t) = \left.\frac{dI}{dV}\right|_{v_{\mathrm{LO}}} = \alpha I_s e^{\alpha V}\Big|_{v_{\mathrm{LO}}} = \alpha I_s e^{\alpha V_{\mathrm{LO}}\cos\omega_{\mathrm{LO}}t}. \tag{7.27}$$

Then for small input voltages $v(t)$ we can write the resulting diode current as

$$i(t) = g(t)v(t) \tag{7.28}$$

We see from (7.27) that $g(t)$ is a real number (consistent with our description as a resistive mixer) and is a periodic function of the LO frequency. Thus, $g(t)$ can be expressed as a Fourier cosine series in terms of harmonics of ω_{LO}:

$$g(t) = g_0 + \sum_{n=1}^{\infty} 2g_n \cos n\omega_{\mathrm{LO}}t, \tag{7.29}$$

with Fourier coefficients given by

$$\begin{aligned} g_n &= \frac{1}{T}\int_0^T g(t)\cos n\omega_{\mathrm{LO}}t\,dt = \frac{\alpha\omega_{\mathrm{LO}}I_s}{2\pi}\int_0^{\frac{\omega_{\mathrm{LO}}}{2\pi}} e^{\alpha V_{\mathrm{LO}}\cos\omega_{\mathrm{LO}}t}\cos n\omega_{\mathrm{LO}}t\,dt \\ &= \frac{\alpha I_s}{2\pi}\int_0^{2\pi} e^{\alpha V_{\mathrm{LO}}\cos\theta}\cos n\theta\,d\theta = \alpha I_s I_n(\alpha V_{\mathrm{LO}}) \end{aligned} \tag{7.30}$$

where $I_n(x)$ is the modified Bessel function of order n, defined in Appendix B. Now let the AC diode current consist of three components at the frequencies ω_{RF}, ω_{IF}, and ω_{IM}:

$$i(t) = I_{\mathrm{RF}}\cos\omega_{\mathrm{RF}}t + I_{\mathrm{IF}}\cos\omega_{\mathrm{IF}}t + I_{\mathrm{IM}}\cos\omega_{\mathrm{IM}}t, \tag{7.31}$$

where I_{RF}, I_{IF}, and I_{IM} are the amplitudes of the RF, IF, and image signals to be determined. If the RF voltage of (7.22) is applied to the diode through a source resistance R_g, and the IF and image ports are terminated in load resistances R_{IF} and R_g, respectively, then the voltage

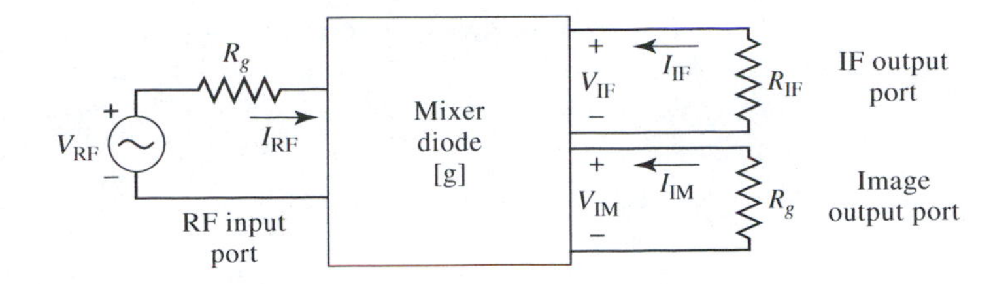

FIGURE 7.4 Equivalent circuit for the large-signal model of the resistive diode mixer.

across the diode can be written as

$$v(t) = V_{RF} \cos \omega_{RF} t - I_{RF} R_g \cos \omega_{RF} t - I_{IF} R_{IF} \cos \omega_{IF} t - I_{IM} R_g \cos \omega_{IM} t. \tag{7.32}$$

The equivalent circuit consists of a three-port network, with one port for each of the frequency components at ω_{RF}, ω_{IF}, and ω_{IM}, as shown in Figure 7.4. We assume the terminations for the RF and image ports are identical, because ω_{RF} is very close to ω_{IM}, while the termination for the IF port may be different.

Using the first three terms of the Fourier series of (7.29) for the diode differential conductance gives

$$g(t) = g_0 + 2g_1 \cos \omega_{LO} t + 2g_2 \cos 2\omega_{LO} t. \tag{7.33}$$

Multiplying the voltage of (7.32) by the conductance in (7.33), and matching like frequency terms with the current of (7.31) gives a system of three equations for the unknown port currents:

$$\begin{bmatrix} I_{RF} \\ I_{IF} \\ I_{IM} \end{bmatrix} = \begin{bmatrix} g_0 & g_1 & g_2 \\ g_1 & g_0 & g_1 \\ g_2 & g_1 & g_0 \end{bmatrix} \begin{bmatrix} V_{RF} - I_{RF} R_g \\ -I_{IF} R_{IF} \\ -I_{IM} R_g \end{bmatrix}, \tag{7.34}$$

where V_{RF} is the source voltage, and the g_n's are defined in (7.30). Note that multiplication of (7.32) by (7.33) creates several frequencies in addition to ω_{RF}, ω_{IF}, and ω_{IM}, but we assume these frequencies to be reactively terminated so that they do not lead to significant power dissipation.

The easiest way to find the available power from the IF port is to first find the Norton equivalent source for the IF port. As shown in Figure 7.5, this consists of a current source equal to I_{SC}, the short-circuit current at the IF port, and G_{IF}, the conductance seen looking into the IF port. This conductance can be found as $G_{IF} = I_{SC}/V_{OC}$, where V_{OC} is the open-circuit voltage of the IF port. The short-circuit IF port current can be found by setting $R_{IF} = 0$ in (7.34) and solving for I_{IF}. After some straightforward algebra, we obtain

$$I_{SC} = -I_{IF}|_{R_{IF}=0} = \frac{g_1 V_{RF}}{1 + g_0 R_g + g_2 R_g}. \tag{7.35}$$

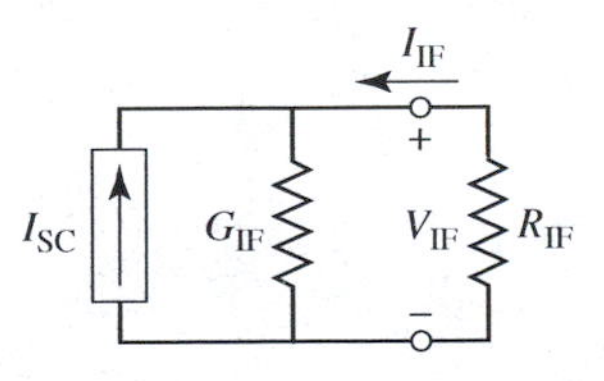

FIGURE 7.5 Norton equivalent circuit for the IF port of the large-signal model of the resistive diode mixer.

The open-circuit IF port voltage is found by setting $I_{\rm IF} = 0$, and solving (7.34) for $V_{\rm IF}$:

$$V_{\rm OC} = V_{\rm IF}|_{I_{\rm IF}=0} = \frac{g_1 V_{\rm RF}}{2g_1^2 R_g - g_0 g_2 R_g - g_0(1 + g_0 R_g)}. \tag{7.36}$$

Then the Norton conductance of the IF port is

$$G_{\rm IF} = \frac{I_{\rm SC}}{V_{\rm OC}} = g_0 - \frac{2g_1^2 R_g}{1 + g_0 R_g + g_2 R_g}. \tag{7.37}$$

The available output power at the IF port is

$$P_{\rm IF-avail} = \frac{|I_{\rm SC}|^2}{4G_{\rm IF}}, \tag{7.38a}$$

and the available input power from the RF source is

$$P_{\rm RF-avail} = \frac{|V_{\rm RF}|^2}{4R_g}. \tag{7.38b}$$

So from (7.10) the conversion loss is (not in dB)

$$L_c = \frac{P_{\rm RF-avail}}{P_{\rm IF-avail}} = \frac{(1 + g_0 R_g + g_2 R_g)\left[g_0(1 + g_0 R_g + g_2 R_g) - 2g_1^2 R_g\right]}{g_1^2 R_g}. \tag{7.39}$$

Note that the conversion loss does not depend on the IF port termination, $R_{\rm IF}$, because of the use of available powers. It does depend on R_g, the RF and image port terminations, so it is possible to minimize the conversion loss by properly selecting R_g. If we let $x = 1/R_g$, $a = g_0 + g_2$, and $b = 2g_1^2/g_0$, then (7.39) can be rewritten as

$$L_c = \frac{2(x + a)(x + a - b)}{bx}. \tag{7.40}$$

Differentiating with respect to x and setting the result to zero gives the optimum value of x as

$$x_{\rm opt} = \sqrt{a(a - b)}, \tag{7.41}$$

for which the minimum value of conversion loss is

$$\begin{aligned} L_{c-\min} &= \frac{2[a + \sqrt{a(a - b)}][a - b + \sqrt{a(a - b)}]}{b\sqrt{a(a - b)}} \\ &= \frac{2[\sqrt{a} + \sqrt{a - b}]^2}{b} = 2\frac{1 + \sqrt{1 - b/a}}{1 - \sqrt{1 - b/a}}. \end{aligned} \tag{7.42}$$

We can evaluate this result by approximating values for g_0, g_1, and g_2. For an LO input power of 10 mW, $V_{\rm LO}$ is about 0.707 V rms, and $\alpha = 1/28$ mV, so $\alpha V_{\rm LO}$, the argument of the modified Bessel functions for g_n given in (7.30), is approximately 25. Thus the modified Bessel functions can be approximated asymptotically using the large-argument formula given in Appendix B, and the g_n's simplified as

$$g_n = \alpha I_s I_n(\alpha V_{\rm LO}) \cong \alpha I_s \frac{e^{\alpha V_{\rm LO}}}{\sqrt{2\pi\alpha V_{\rm LO}}}. \tag{7.43}$$

Thus

$$\frac{b}{a} = \frac{2g_1^2}{g_0(g_0 + g_2)} \cong 1, \tag{7.44}$$

and the minimum conversion loss of (7.42) reduces to $L_c = 2$, or 3 dB. This means that half the RF input power is converted to IF power, and half is converted to power at the image frequency. In principle this result could be improved by terminating the image port with a reactive load, but it is usually difficult in practice to separate the image termination from the RF termination. Also, this result is highly idealized in that it assumes no power loss at higher harmonic frequencies, and it ignores diode reactances.

This same model can be used to derive the SSB noise temperature for the resistive mixer as

$$T_e = \frac{nT}{2}(L_c - 1), \tag{7.45}$$

where n is the diode ideality factor and T is the physical temperature of the diode [3].

Switching Model

The large-signal model suggests that the diode mixer can be viewed as a switch. As the LO voltage cycles between positive and negative values of $\cos \omega_{\text{LO}} t$, the diode becomes conducting or nonconducting, respectively. Thus, the diode conductance (the ratio of diode current to diode voltage) switches between large values and zero at the same rate as the LO voltage. Figure 7.6 shows a typical diode conductance waveform, where $T = 2\pi/\omega_{\text{LO}}$ is the period of the LO waveform.

The conductance waveform of Figure 7.6 can be calculated directly from the diode V-I characteristic of (7.18), or from the Fourier series representation of (7.29). But since a conductance greater than a few Siemens is essentially a short circuit, we can approximate the diode conductance as the square wave shown in Figure 7.7. This square wave has a Fourier transform given by

$$g(t) = \frac{1}{2} + \sum_{n=1}^{\infty} \frac{2}{n\pi} \sin \frac{n\pi}{2} \cos n\omega_{\text{LO}} t, \tag{7.46}$$

which is similar in form to the Fourier series of (7.29).

An equivalent circuit of the diode mixer then consists of the RF input voltage applied across a load resistor in series with an ideal switch, as shown in Figure 7.8. The time-varying

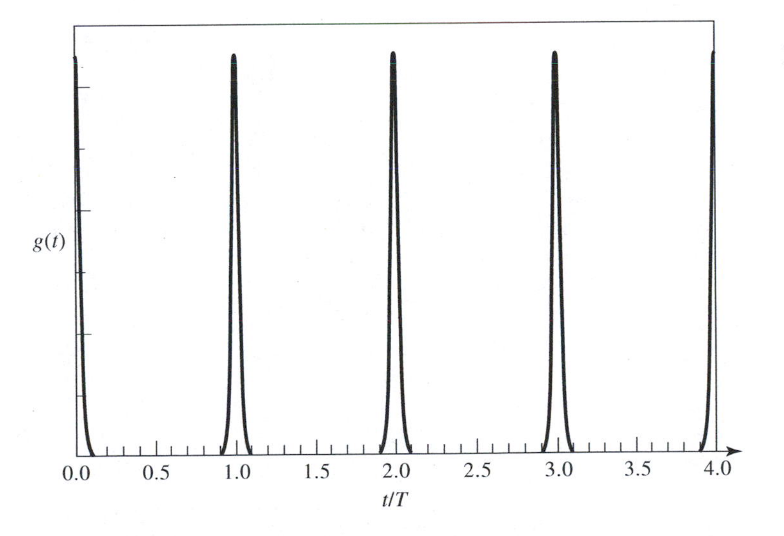

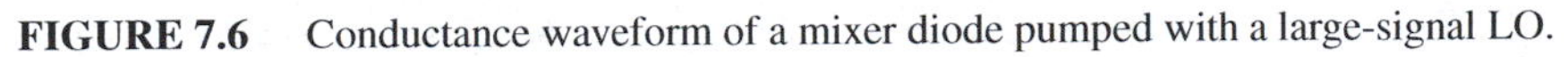

FIGURE 7.6 Conductance waveform of a mixer diode pumped with a large-signal LO.

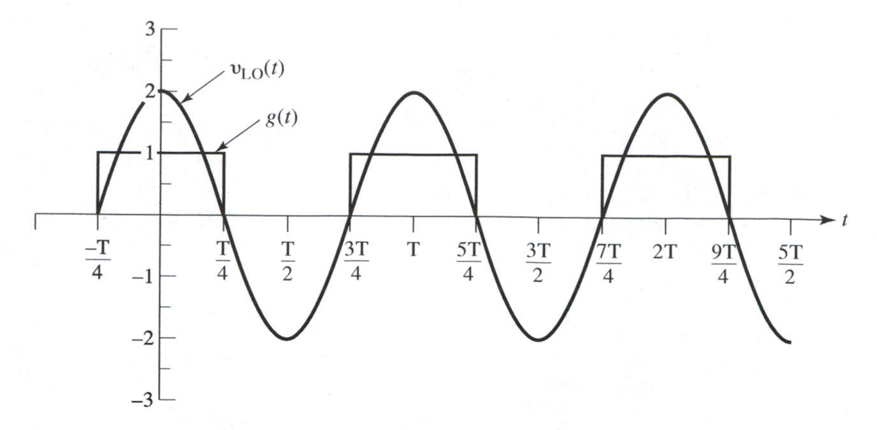

FIGURE 7.7 LO voltage waveform and idealized square-wave diode conductance waveform for the switching model of a diode mixer.

switch conductance is given by (7.46). The diode current can be found by multiplying the RF input voltage of (7.22) by the conductance of (7.46):

$$i(t) = g(t)v_{RF}(t) = V_{RF}\left[\frac{1}{2}\cos\omega_{RF}t + \sum_{n=1}^{\infty}\frac{2}{n\pi}\sin\frac{n\pi}{2}\cos n\omega_{LO}t\cos\omega_{RF}t\right]$$

$$= \frac{1}{2}V_{RF}\left[\cos\omega_{RF}t + \sum_{n=1}^{\infty}\frac{2}{n\pi}\sin\frac{n\pi}{2}[\cos(\omega_{RF}+\omega_{RF})t + \cos(\omega_{RF}-n\omega_{LO})t]\right] \tag{7.47}$$

Filtering all but the lowest-frequency component for the $n = 1$ term of the summation gives the desired IF output as

$$\frac{V_{RF}}{\pi}\cos\omega_{IF}t,$$

with

$$\omega_{IF} = \omega_{RF} - \omega_{LO}.$$

The switching model is useful for mixers of any type, including the FET mixer discussed in the following section. Note that the switching model of a mixer can be considered as a linear, but time-varying, circuit.

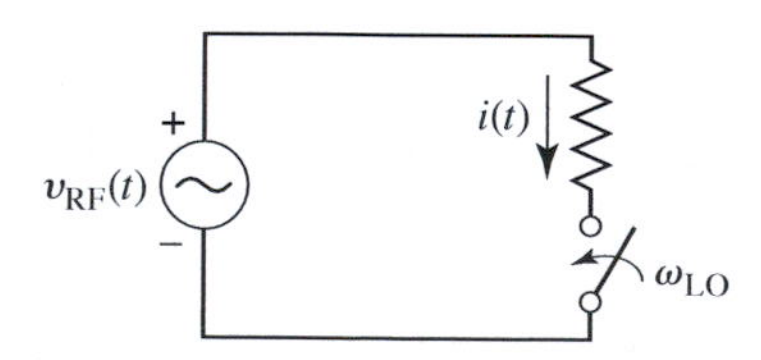

FIGURE 7.8 Equivalent circuit for the switching model of the diode mixer.

7.3 FET MIXERS

Mixers can also be implemented by using the nonlinear properties of transistors. FETs, in particular, offer low noise characteristics and easy integration with other circuitry, such as switches and low-noise amplifiers. Transistor mixers can provide conversion gain, but their noise figure is generally not as good as can be obtained with diode mixers. FET mixers also offer higher dynamic range. The following table compares the characteristics of typical diode and FET mixers.

Mixer Type	Conversion Gain	Noise Figure	1 dB Compression	3rd Order Intercept
Diode	−5 dB	5–7 dB	−6 to −1 dBm	5 dBm
FET	6 dB	7–8 dB	5 to 6 dBm	20 dBm

Because a FET mixer has conversion gain, but usually worse noise figure, the proper comparison with a diode mixer should include the cascade effect of adjacent stages.

In this section we will analyze the single-ended FET mixer, and derive an expression for its conversion gain. We will also discuss a few other popular FET mixer configurations.

Single-Ended FET Mixer

There are several FET parameters that offer nonlinearities that can be used for mixing, but the strongest is the transconductance, g_m, when the FET is operated in a common source configuration with a negative gate bias. Figure 7.9 shows the variation of transconductance with gate bias for a typical FET. When used as an amplifier, the gate bias voltage is near zero, or positive, so the transconductance is near its maximum value, and the transistor operates as a linear device. When the gate bias is near the *pinch-off* region, where the transconductance approaches zero, a small variation of gate voltage can cause a large change in transconductance, leading to a nonlinear response. Thus the LO voltage can be applied to the gate of the FET to pump the transconductance to switch the FET between high and low transconductance states, and provide mixing in much the same manner as the switching model discussed in the previous section.

The circuit for a single-ended FET mixer is shown in Figure 7.10. A diplexing coupler is used to combine the RF and LO signals at the gate of the FET. An impedance matching network is also usually required between the inputs and the FET, which typically presents a very low input impedance. RF chokes are used to bias the gate at a negative voltage near pinch-off, and to provide a positive bias for the drain of the FET. A bypass capacitor at the drain provides a return path for the LO signal, and a low-pass filter provides the final IF output signal.

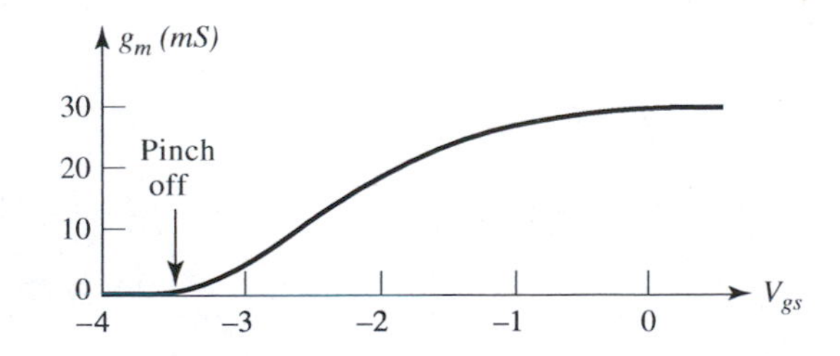

FIGURE 7.9 Variation of FET transconductance versus gate-to-source voltage.

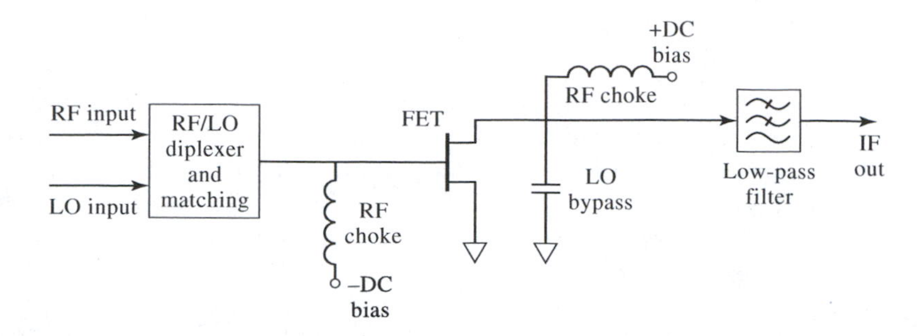

FIGURE 7.10 Circuit for a single-ended FET mixer.

Our analysis of the mixer of Figure 7.10 follows the original work described in reference [6]. The simplified equivalent circuit is shown in Figure 7.11 and is based on the standard unilateral equivalent circuit for a FET. The RF and LO input voltages are given in (7.22) and (7.23). Let $Z_g = R_g + jX_g$ be the Thevenin source impedance for the RF input port, and $Z_L = R_L + jX_L$ be the Thevenin source impedance at the IF output port. These impedances are complex to allow us to conjugately match the input and output ports for maximum power transfer. The LO port has a real generator impedance of Z_0, since we are not concerned with maximum power transfer for the LO signal.

As we did for the large-signal analysis of the diode mixer, we express the LO pumped FET transconductance as a Fourier series in terms of harmonics of the LO signal:

$$g(t) = g_0 + 2\sum_{n=1}^{\infty} g_n \cos n\omega_{\text{LO}}t. \tag{7.48}$$

Because we do not have an explicit formula for the transconductance, we cannot calculate directly the Fourier coefficients of (7.48), but must rely on measurements for these values. As in the case of the switching model, the desired down-conversion result is due to the $n = 1$ term of the Fourier series, so we only need the g_1 coefficient. Measurements typically give a value in the range of 10 mS for g_1.

The conversion gain of the FET mixer can be found as

$$G_c = \frac{P_{\text{IF-avail}}}{P_{\text{RF-avail}}} = \frac{\dfrac{\left|V_D^{\text{IF}}\right|^2 R_L}{|Z_L|^2}}{\dfrac{|V_{\text{RF}}|^2}{4R_g}} = \frac{4R_g R_L}{|Z_L|^2}\left|\frac{V_D^{\text{IF}}}{V_{\text{RF}}}\right|^2, \tag{7.49}$$

where V_D^{IF} is the IF drain voltage, and the impedances Z_g and Z_L are chosen for maximum power transfer at the RF and IF ports. The RF frequency component of the phasor voltage

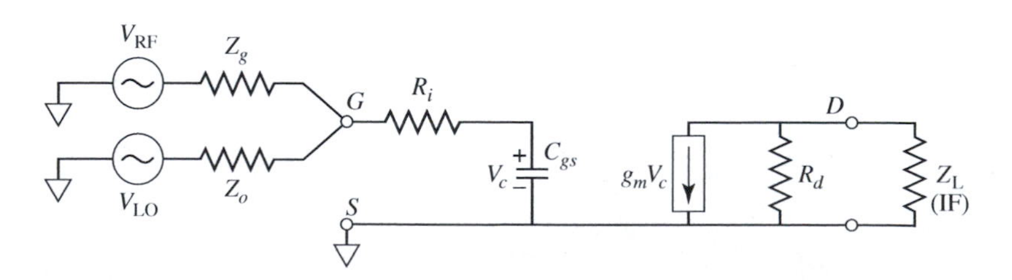

FIGURE 7.11 Equivalent circuit for the FET mixer for Figure 7.10.

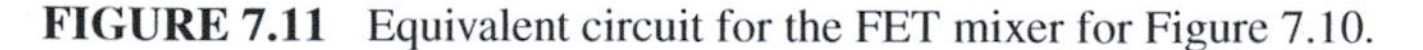

across the gate-to-source capacitance is given in terms of the voltage divider between Z_g, R_i, and C_{gs}:

$$V_c^{\mathrm{RF}} = \frac{V_{\mathrm{RF}}}{j\omega_{\mathrm{RF}} C_{gs}\left[(R_i + Z_g) - \dfrac{j}{\omega_{\mathrm{RF}} C_{gs}}\right]} = \frac{V_{\mathrm{RF}}}{1 + j\omega_{\mathrm{RF}} C_{gs}(R_i + Z_g)}. \tag{7.50}$$

Multiplying the transconductance of (7.48) by $v_c^{\mathrm{RF}}(t) = V_c^{\mathrm{RF}} \cos\omega_{\mathrm{RF}} t$ gives terms of the form

$$g_m(t) v_c^{\mathrm{RF}}(t) = g_0 V_c^{\mathrm{RF}} \cos\omega_{\mathrm{RF}} t + 2g_1 V_c^{\mathrm{RF}} \cos\omega_{\mathrm{RF}} t \cos\omega_{\mathrm{LO}} t + \cdots. \tag{7.51}$$

The down-converted IF frequency component can be extracted from the second term of (7.51) using the usual trigonometric identity:

$$g_m(t) v_c^{\mathrm{RF}}(t)\big|_{\omega_{\mathrm{IF}}} = g_1 V_c^{\mathrm{RF}} \cos\omega_{\mathrm{IF}} t. \tag{7.52}$$

Then the IF component of the drain voltage is, in phasor form,

$$V_D^{\mathrm{IF}} = -g_1 V_c^{\mathrm{RF}} \left(\frac{R_d Z_L}{R_d + Z_L}\right) = \frac{-g_1 V_{\mathrm{RF}}}{1 + j\omega_{\mathrm{RF}} C_{gs}(R_i + Z_g)} \left(\frac{R_d Z_L}{R_d + Z_L}\right), \tag{7.53}$$

where (7.50) has been used. Using this result in (7.49) gives the conversion gain (before conjugate matching) as

$$G_c\big|_{\substack{\text{not}\\ \text{matched}}} = \left(\frac{2g_1 R_d}{\omega_{\mathrm{RF}} C_{gs}}\right)^2 \frac{R_g}{\left[(R_i + R_g)^2 + \left(X_g - \dfrac{1}{\omega_{\mathrm{RF}} C_{gs}}\right)^2\right]} \frac{R_L}{\left[(R_d + R_L)^2 + X_L^2\right]}.$$

We must now conjugately match the RF and IF ports. Thus we let $R_g = R_i$, $X_g = 1/\omega_{\mathrm{RF}} C_{gs}$, $R_L = R_d$, and $X_L = 0$, which reduces the above result to

$$G_c = \frac{g_1^2 R_d}{4\omega_{\mathrm{RF}}^2 C_{gs}^2 R_i}. \tag{7.54}$$

The quantities g_1, R_d, R_i, and C_{gs} are all parameters of the FET. Practical mixer circuits generally use matching circuits to transform the FET impedance to 50 Ω for the RF, LO, and IF ports.

EXAMPLE 7.2 MIXER CONVERSION GAIN

A single-ended FET mixer is to be designed for a wireless local area network receiver operating at 2.4 GHz. The parameters of the FET are: $R_d = 300\ \Omega$, $R_i = 10\ \Omega$, $C_{gs} = 0.3$ pF, and $g_1 = 10$ mS. Calculate the maximum possible conversion gain.

Solution

This is a straightforward application of the formula for conversion gain given in (7.54):

$$G_c = \frac{g_1^2 R_d}{4\omega_{\mathrm{RF}}^2 C_{gs}^2 R_i} = \frac{(10 \times 10^{-3})^2 (300)}{4(2\pi)^2 (2.4 \times 10^9)^2 (10)} = 36.6 = 15.6\ \text{dB}.$$

Note that this value does not include losses due to the necessary impedance matching networks. ○

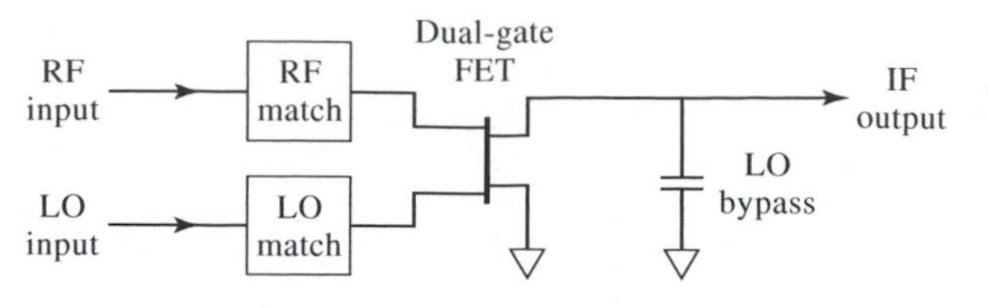

FIGURE 7.12 A dual-gate FET mixer.

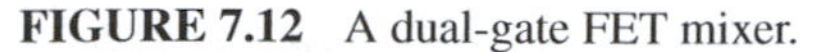

Other FET Mixers

There are several practical variations of mixer circuits that can be implemented using FETs. Figure 7.12 shows a single-ended mixer using a dual-gate FET, where the RF and LO inputs are applied to separate gates of the FET. This provides a high degree of RF-LO isolation, but generally an inferior noise figure relative to the transconductance mixer of Figure 7.10.

Another configuration is shown in Figure 7.13, using two FETs in a differential amplifier configuration. The *balun* (balanced-to-unbalanced) networks on the LO and IF ports provide a transition between a two-wire line that is balanced with respect to ground and a single line that is unbalanced relative to ground. Baluns may be implemented with center-tapped transformers, or with 180° hybrid junctions.

The differential mixer operates as an alternating switch, with the LO turning the top two FETs on and off on alternate cycles of the LO. These FETs are biased slightly above pinch-off, so each FET will be conducting for slightly more than half of each LO cycle. Thus, one of the upper FETs is always conducting, and the lower FET will remain in saturation. The RF and LO ports should each be impedance matched. The IF output circuit must provide a return path to ground for the LO signal.

An extension of the differential FET mixer is the Gilbert cell mixer shown in Figure 7.14. This mixer uses two differential FET mixer stages to form a double balanced mixer. This circuit achieves high RF-LO isolation and a high dynamic range. It also cancels all even-order intermodulation products. This circuit is very popular for wireless integrated circuits.

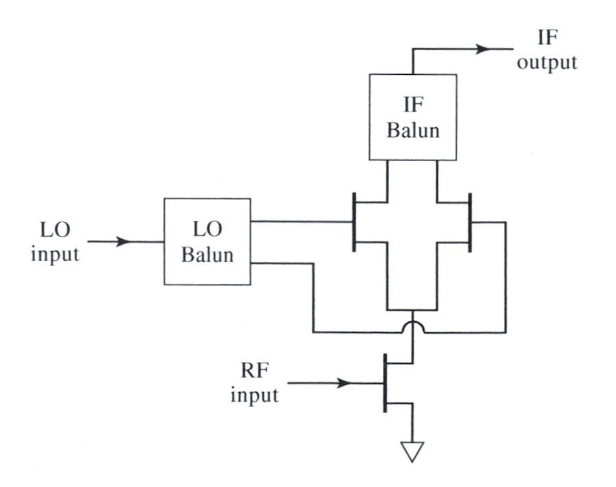

FIGURE 7.13 A differential FET mixer.

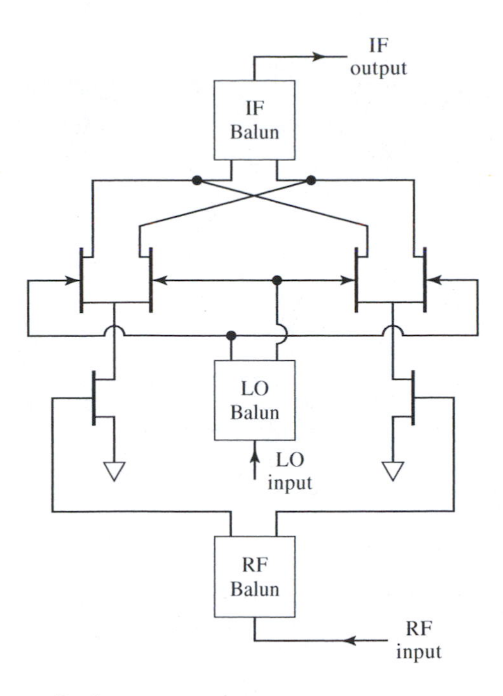

FIGURE 7.14 A Gilbert cell mixer.

7.4 OTHER MIXER CIRCUITS

The single-ended diode and FET mixers discussed above provide frequency conversion, but often have poor RF input matching and RF-LO isolation. This reduces the performance of wireless systems, but fortunately it is possible to improve these characteristics by combining two or more single-ended mixers with hybrid junctions.

Balanced Mixers

RF input matching and RF-LO isolation can be improved through the use of a *balanced mixer*, which consists of two single-ended mixers combined with a hybrid junction. Figure 7.15 shows the basic configuration, with either a 90° hybrid (Figure 7.15a), or a 180° hybrid (Figure 7.15b). As we will see, a balanced mixer using a 90° hybrid junction will ideally lead to a perfect input match at the RF port over a wide frequency range, while the use of a 180° hybrid will ideally lead to perfect RF-LO isolation over a wide frequency range. In addition, both mixers will reject all even-order intermodulation products.

Microwave quadrature or ring hybrids [1] can be used to implement balanced mixers, but at lower frequencies a center-tapped transformer can be used. As shown in Figure 7.16, the secondary of the transformer provides outputs with a 180° phase shift to the two mixer diodes. The LO signal is applied to the center tap of the secondary.

The *double-balanced mixer* of Figure 7.17 uses two hybrid junctions or transformers, and provides good isolation between all three ports, as well as rejection of all even harmonics of the RF and LO signals. This leads to very good conversion loss, but less than ideal input matching at the RF port. The double-balanced mixer also provides a higher third-order intercept point than either a single-ended mixer or a balanced mixer.

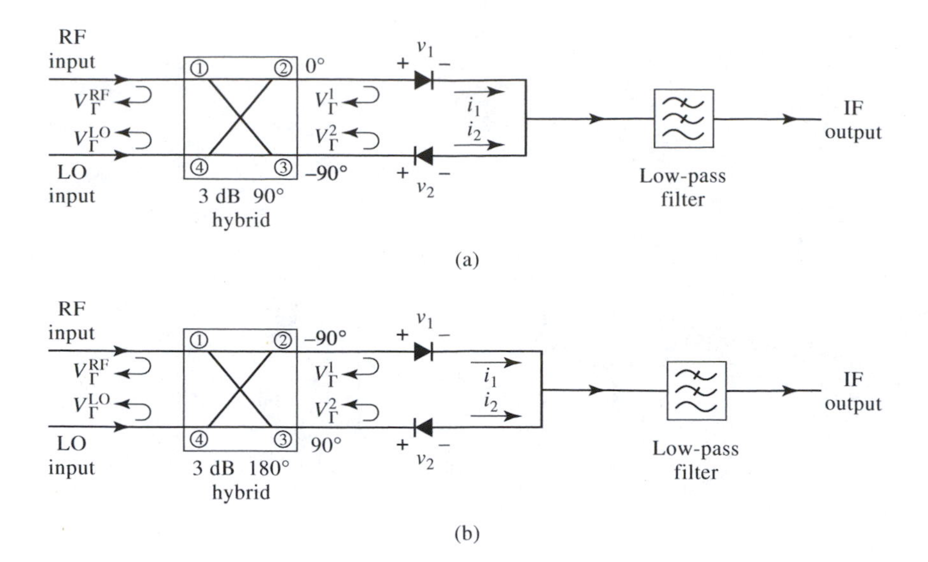

FIGURE 7.15 Balanced mixer circuits. (a) Using a 90° hybrid. (b) Using a 180° hybrid.

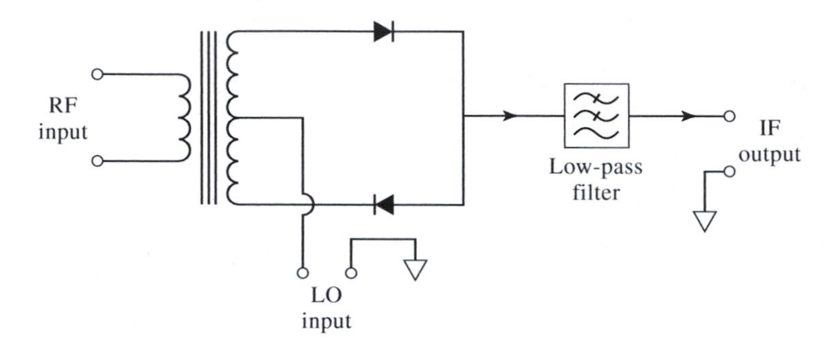

FIGURE 7.16 Balanced mixer using a hybrid transformer.

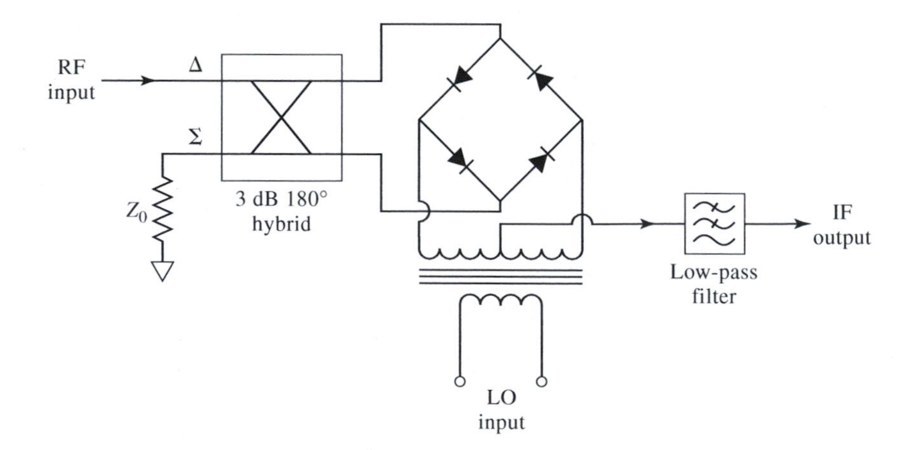

FIGURE 7.17 Double-balanced mixer circuit.

The following table summarizes the characteristics of several types of mixers.

Mixer Type	Number of Diodes	RF Input Match	RF-LO Isolation	Conversion Loss	Third-Order Intercept
Single-ended	1	Poor	Fair	Good	Fair
Balanced (90°)	2	Good	Poor	Good	Fair
Balanced (180°)	2	Fair	Excellent	Good	Fair
Double-balanced	4	Poor	Excellent	Excellent	Excellent
Image reject	2 or 4	Good	Good	Good	Good

Small-Signal Analysis of the Balanced Mixer

We can analyze the performance of a balanced mixer using the small-signal approach that was used in Section 7.2. Here we will concentrate on the balanced mixer with a 90° hybrid, shown in Figure 7.15a, and leave the 180° hybrid case as a problem.

As usual, let the RF and LO voltages be defined as

$$v_{\mathrm{RF}}(t) = V_{\mathrm{RF}} \cos \omega_{\mathrm{RF}} t, \tag{7.55}$$

and

$$v_{\mathrm{LO}}(t) = V_{\mathrm{LO}} \cos \omega_{\mathrm{LO}} t. \tag{7.56}$$

The scattering matrix for the 90° hybrid junction is [1]

$$[S] = \frac{-1}{\sqrt{2}} \begin{bmatrix} 0 & j & 1 & 0 \\ j & 0 & 0 & 1 \\ 1 & 0 & 0 & j \\ 0 & 1 & j & 0 \end{bmatrix}, \tag{7.57}$$

where the ports are numbered as shown in Figure 7.15a. Then the total RF and LO voltages applied to the two diodes can be written as

$$\begin{aligned} v_1(t) &= \frac{1}{\sqrt{2}}[V_{\mathrm{RF}} \cos(\omega_{\mathrm{RF}} t - 90°) + V_{\mathrm{LO}} \cos(\omega_{\mathrm{LO}} t - 180°)] \\ &= \frac{1}{\sqrt{2}}[V_{\mathrm{RF}} \sin \omega_{\mathrm{RF}} t - V_{\mathrm{LO}} \cos \omega_{\mathrm{LO}} t] \end{aligned} \tag{7.58a}$$

$$\begin{aligned} v_2(t) &= \frac{1}{\sqrt{2}}[V_{\mathrm{RF}} \cos(\omega_{\mathrm{RF}} t - 180°) + V_{\mathrm{LO}} \cos(\omega_{\mathrm{LO}} t - 90°)] \\ &= \frac{1}{\sqrt{2}}[-V_{\mathrm{RF}} \cos \omega_{\mathrm{RF}} t + V_{\mathrm{LO}} \sin \omega_{\mathrm{LO}} t] \end{aligned} \tag{7.58b}$$

Using only the quadratic term from the small-signal diode approximation of (7.20) gives the diode currents as

$$i_1(t) = K v_1^2 = \frac{K}{2}\left[V_{\mathrm{RF}}^2 \sin^2 \omega_{\mathrm{RF}} t - 2 V_{\mathrm{RF}} V_{\mathrm{LO}} \sin \omega_{\mathrm{RF}} \cos \omega_{\mathrm{LO}} t + V_{\mathrm{LO}}^2 \cos^2 \omega_{\mathrm{LO}} t\right] \tag{7.59a}$$

$$i_2(t) = -K v_2^2 = \frac{-K}{2}\left[V_{\mathrm{RF}}^2 \cos^2 \omega_{\mathrm{RF}} t - 2 V_{\mathrm{RF}} V_{\mathrm{LO}} \cos \omega_{\mathrm{RF}} \sin \omega_{\mathrm{LO}} t + V_{\mathrm{LO}}^2 \sin^2 \omega_{\mathrm{LO}} t\right] \tag{7.59b}$$

where the negative sign on i_2 accounts for the reversed diode polarity, and K is a constant for the quadratic term of the diode response. Adding these two currents at the input to the low-pass filter gives

$$i_1(t) + i_2(t) = \frac{-K}{2}\left[V_{RF}^2 \cos 2\omega_{RF}t + 2V_{RF}V_{LO} \sin \omega_{IF}t - V_{LO}^2 \cos 2\omega_{LO}t\right],$$

where the usual trigonometric identities have been used, and $\omega_{IF} = \omega_{RF} - \omega_{LO}$ is the IF frequency. Note that the DC components of the diode currents cancel upon combining. After low-pass filtering, the IF output is

$$i_{IF}(t) = -K V_{RF} V_{LO} \sin \omega_{IF}t, \tag{7.60}$$

as desired.

We can also calculate the input match at the RF port, and the coupling between the RF and LO ports. If we assume the diodes are matched, and each exhibits a voltage reflection coefficient Γ at the RF frequency, then the phasor expression for the reflected RF voltages at the diodes will be

$$V_{\Gamma_1} = \Gamma V_1 = \frac{-j\Gamma V_{RF}}{\sqrt{2}}, \tag{7.61a}$$

and

$$V_{\Gamma_2} = \Gamma V_2 = \frac{-\Gamma V_{RF}}{\sqrt{2}}. \tag{7.61b}$$

These reflected voltages appear at ports 2 and 3 of the hybrid, respectively, and combine to form the following outputs at the RF and LO ports:

$$V_{\Gamma}^{RF} = \frac{-jV_{\Gamma_1}}{\sqrt{2}} - \frac{V_{\Gamma_2}}{\sqrt{2}} = -\frac{1}{2}\Gamma V_{RF} + \frac{1}{2}\Gamma V_{RF} = 0 \tag{7.62a}$$

$$V_{\Gamma}^{LO} = \frac{-V_{\Gamma_2}}{\sqrt{2}} - j\frac{V_{\Gamma_1}}{\sqrt{2}} = \frac{1}{2}j\Gamma V_{RF} + \frac{1}{2}j\Gamma V_{RF} = j\Gamma V_{RF} \tag{7.62b}$$

Thus we see that the phase characteristics of the 90° hybrid lead to perfect cancellation of reflections at the RF port. The isolation between the RF and LO ports, however, is dependent on the matching of the diodes, which may be difficult to maintain over a reasonable frequency range.

Image Reject Mixer

We have already discussed the fact that two distinct RF input signals at frequencies $\omega_{RF} = \omega_{LO} \pm \omega_{IF}$ will down-convert to the same IF frequency when mixed with ω_{LO}. These two frequencies are the upper and lower sidebands of a double-sideband signal. The desired response can be arbitrarily selected as either the LSB ($\omega_{LO} - \omega_{IF}$) or the USB ($\omega_{LO} + \omega_{IF}$), assuming a positive IF frequency. The *image reject mixer*, shown in Figure 7.18, can be used to isolate these two responses into separate output signals. The same circuit can also be used for up-conversion, in which case it is usually called a *single-sideband modulator*. In this case, the IF input signal is delivered to either the LSB or the USB port of the IF hybrid, and the associated single sideband signal is produced at the RF port of the mixer.

We can analyze the image reject mixer using the small-signal approximation. Let the RF input signal be expressed as

$$v_{RF}(t) = V_U \cos(\omega_{LO} + \omega_{IF})t + V_L \cos(\omega_{LO} - \omega_{IF})t, \tag{7.63}$$

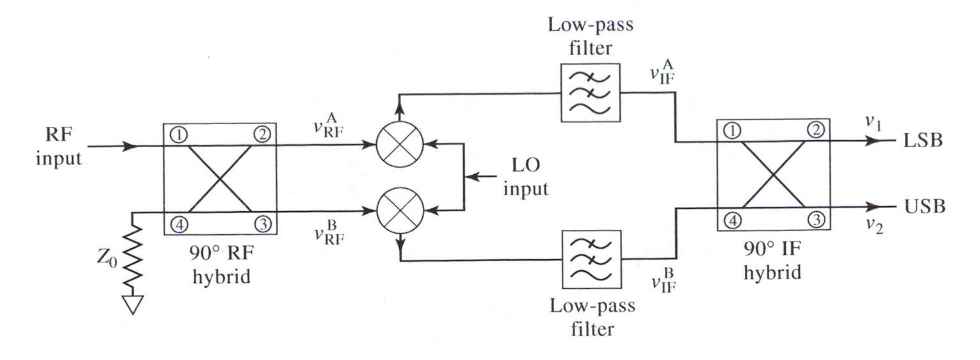

FIGURE 7.18 Circuit for an image reject mixer.

where V_U and V_L represent the amplitudes of the upper and lower sidebands, respectively. Using the S-matrix given in (7.57) for the 90° hybrid gives the RF voltages at the diodes as

$$\begin{aligned} v_A(t) &= \frac{1}{\sqrt{2}}[V_U \cos(\omega_{LO}t + \omega_{IF}t - 90^\circ) + V_L \cos(\omega_{LO}t - \omega_{IF}t - 90^\circ)] \\ &= \frac{1}{\sqrt{2}}[V_U \sin(\omega_{LO} + \omega_{IF})t + V_L \sin(\omega_{LO} - \omega_{IF})t] \end{aligned} \tag{7.64a}$$

$$\begin{aligned} v_B(t) &= \frac{1}{\sqrt{2}}[V_U \cos(\omega_{LO}t + \omega_{IF}t - 180^\circ) + V_L \cos(\omega_{LO}t - \omega_{IF}t - 180^\circ)] \\ &= \frac{-1}{\sqrt{2}}[V_U \cos(\omega_{LO} + \omega_{IF})t + V_L \cos(\omega_{LO} - \omega_{IF})t] \end{aligned} \tag{7.64b}$$

After mixing with the LO signal given in (7.56) and low-pass filtering, the IF inputs to the IF hybrid are

$$v_{IF}^A(t) = \frac{KV_{LO}}{2\sqrt{2}}(V_U - V_L)\sin\omega_{IF}t, \tag{7.65a}$$

$$v_{IF}^B(t) = \frac{-KV_{LO}}{2\sqrt{2}}(V_U + V_L)\cos\omega_{IF}t, \tag{7.65b}$$

where K is the mixer constant for the squared term of the diode response. The phasor representation of the IF signals of (7.65) is

$$V_{IF}^A = \frac{-jKV_{LO}}{2\sqrt{2}}(V_U - V_L), \tag{7.66a}$$

$$V_{IF}^B = \frac{-KV_{LO}}{2\sqrt{2}}(V_U + V_L). \tag{7.66b}$$

Combining these voltages in the IF hybrid gives the following outputs:

$$V_1 = -j\frac{V_{IF}^A}{\sqrt{2}} - \frac{V_{IF}^B}{\sqrt{2}} = \frac{KV_{LO}V_L}{2} \quad \text{(LSB)} \tag{7.67a}$$

$$V_2 = -\frac{V_{IF}^A}{\sqrt{2}} - j\frac{V_{IF}^B}{\sqrt{2}} = \frac{jKV_{LO}V_U}{2} \quad \text{(USB)} \tag{7.67b}$$

which we see are the separate sidebands of the downconverted input signal of (7.63). These outputs can be expressed in time-domain form as

$$v_1(t) = \frac{K V_{\rm LO} V_L}{2} \cos \omega_{\rm IF} t, \tag{7.68a}$$

$$v_2(t) = \frac{-K V_{\rm LO} V_U}{2} \sin \omega_{\rm IF} t, \tag{7.68b}$$

which clearly shows the presence of a 90° phase shift between the two sidebands. Also note that the image rejection mixer does not incur any additional losses beyond the usual conversion losses of the single rejection mixer.

A practical difficulty with image rejection mixers is in fabricating a good hybrid at the relatively low IF frequency. Losses, and hence noise figure, are also usually greater than for a simpler mixer.

REFERENCES

[1] D. M. Pozar, **Microwave Engineering**, 2nd edition. Wiley, New York, 1998.

[2] S. Y. Yngvesson, **Microwave Semiconductor Devices**, Kluwer Academic Publishers, 1991.

[3] K. Chang, **Handbook of Microwave and Optical Components**, vol. 2, Chapter 2, "Mixers and Detectors," by E. L. Kollberg, Wiley InterScience, New York, 1990.

[4] C. T. Torrey and C. A. Whitmer, **Crystal Rectifiers**, MIT Radiation Laboratory Series, vol. 14, McGraw-Hill, New York, 1948.

[5] S. A. Maas, **Microwave Mixers**, 2nd edition, Artech House, Dedham, MA, 1993.

[6] R. A. Pucel, D. Masse, and R. Bera, "Performance of GaAs MESFET Mixers at X Band," *IEEE Trans. Microwave Theory and Techniques*, vol. MTT-24, pp. 351–360, June 1976.

PROBLEMS

7.1 A double-sideband signal of the form $v_{\rm RF}(t) = V_{\rm RF}[\cos(\omega_{\rm LO} - \omega_{\rm IF})t + \cos(\omega_{\rm LO} + \omega_{\rm IF})t]$ is applied to a mixer with an LO voltage given by (7.1). Derive the output of the mixer after low-pass filtering.

7.2 An RF input signal at 600 MHz is down-converted in a mixer to an IF frequency of 80 MHz. What are the two possible LO frequencies, and the corresponding image frequencies?

7.3 Consider a diode mixer with a conversation loss of 5 dB and a noise figure of 4 dB, and a FET mixer with conversion gain of 3 dB and a noise figure of 8 dB. If each of these mixers is followed by an IF amplifier having a gain of 30 dB and a noise figure F_A, as shown below, calculate and plot the overall noise figure for both amplifier-mixer configurations for $F_A = 0$ to 10 dB.

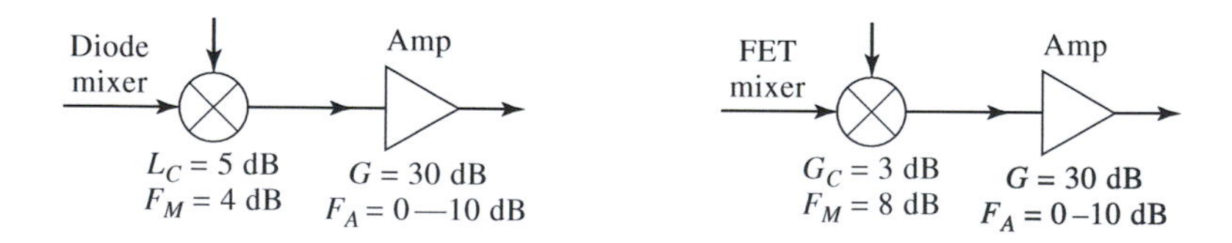

7.4 Let $T_{\rm SSB}$ be the equivalent noise temperature of a mixer receiving a SSB signal, and $T_{\rm DSB}$ be the temperature when it receives a DSB signal. Compute the output noise powers in each case, and show that $T_{\rm SSB} = 2T_{\rm DSB}$, and that therefore $F_{\rm SSB} = 2F_{\rm DSB}$. Assume that the conversion gains for the signal and its image are identical.

7.5 If the noise power N_i = kTB is applied at the RF input port of a mixer having noise figure F (DSB) and conversion loss L_c, what is the available output noise power at the IF port? Assume the mixer is at a physical temperature T_0.

7.6 A diode has an I-V characterisitic given by $i(t) = I_s[e^{30v(t)} - 1]$. Let $v(t) = 0.01 \cos \omega_1 t + 0.01 \cos \omega_2 t$, and expand $i(t)$ in a power series in v, retaining only the v, v^2, v^3 terms. Find the magnitudes of each frequency term.

7.7 Carry out the details of multiplying (7.32) by (7.33) to obtain the set of equations in (7.34).

7.8 Derive the results in (7.35) and (7.36) for the short-circuit current and open-circuit voltage at the IF port of the large-signal mixer model.

7.9 Derive the Fourier series of (7.46) for the square-wave conductance waveform shown in Figure 7.7.

7.10 Analyze a balanced mixer using a 180° hybrid junction. Find the output IF current, and the input reflections at the RF and LO ports. Show that this mixer suppresses even harmonics of the LO. Assume that the RF signal is applied to the sum port of the hybrid, and that the LO signal is applied to the difference port.

7.11 For an image rejection mixer, let the RF hybrid have a dissipative insertion loss of L_R, and the IF hybrid have a dissipative insertion loss of L_I. If the component single-ended mixers each have a conversion loss L_c and noise figure F, derive expressions for the overall conversion loss and noise figure of the image rejection mixer.

7.12 Find the IF output power of the double-sideband signal of (7.63) after it has been down converted using an ideal single-ended mixer. Ignoring dissipative losses, show that this power is the same as the total output power of the signals of (7.68) for the image reject mixer.

Chapter Eight

Transistor Oscillators and Frequency Synthesizers

Oscillators and frequency synthesizers are key components in wireless transmitters and receivers, providing precisely controlled sources for frequency conversion and carrier generation. Simple transistor oscillators usually lack the frequency stability and low-noise performance required for modern wireless systems, so crystal controlled oscillators are often used to provide accurate frequency references. Frequency synthesis methods can then be used to precisely derive higher frequencies from a crystal-controlled source, and allow the generation of the closely-spaced local oscillator frequencies required for multichannel wireless transceivers. Phase-locked loops are often used for this purpose.

Important considerations for oscillators and frequency synthesizers used in wireless systems include the following:

- Tuning range (specified in MHz/V for voltage tuned oscillators)
- Frequency stability (specified in PPM/°C)
- AM and FM (phase) noise (specified in dBc/Hz below carrier, offset from carrier)
- Harmonics (specified in dBc below carrier)

Typical frequency stability requirements for wireless systems range from 2 PPM/°C to 0.5 PPM/°C, while phase noise requirements typically range from −80 dBc/Hz to −110 dBc/Hz at a 10 kHz offset from the carrier.

We begin with a general analysis of radio frequency transistor oscillator design, which includes the well-known Hartley and Colpitts oscillators, as well as crystal-controlled oscillators and voltage-controlled oscillators. Next we consider oscillators for use at microwave frequencies, which differ from their lower frequency counterparts primarily due to different transistor characteristics and the ability to make practical use of negative-resistance devices and high-Q microwave components such as transmission line and dielectric resonators. Then we consider analog and digital methods of frequency synthesis, and focus on the very important phase-locked loop circuit.

8.1 RADIO FREQUENCY OSCILLATORS

In the most general sense, an oscillator is a nonlinear circuit that converts DC power to an AC waveform. Most oscillators used in wireless systems provide sinusoidal outputs, thereby minimizing undesired harmonics and noise sidebands. The basic conceptual operation of a sinusoidal oscillator can be described with the linear feedback circuit shown in Figure 8.1. An amplifier with voltage gain A has an output voltage V_o. This voltage passes through a feedback network with a frequency dependent transfer function $H(\omega)$, and is added to the input V_i of the circuit. Thus the output voltage can be expressed as

$$V_o(\omega) = AV_i(\omega) + H(\omega)AV_o(\omega), \tag{8.1}$$

which can be solved to yield the output voltage in terms of the input voltage as

$$V_o(\omega) = \frac{A}{1 - AH(\omega)} V_i(\omega). \tag{8.2}$$

If the denominator of (8.2) becomes zero at a particular frequency, it is possible to achieve a nonzero output voltage for a zero input voltage, thus forming an oscillator. This is known as the *Nyquist criterion*, or the *Barkhausen criterion*. In contrast to the design of an amplifier, where we design to achieve maximum stability, oscillator design depends on an unstable circuit.

The oscillator circuit of Figure 8.1 is useful conceptually, but provides little helpful information for the design of practical transistor oscillators. Thus we consider next a general analysis of transistor oscillator circuits.

General Analysis

There are a large number of possible RF oscillator circuits using bipolar or field-effect transistors in either common emitter/source, base/gate, or collector/drain configurations. Various types of feedback networks lead to the well-known Hartley, Colpitts, Clapp, and Pierce oscillator circuits [1]–[3]. All of these variations can be represented by the general oscillator circuit shown in Figure 8.2.

The equivalent circuit on the right-hand side of Figure 8.2 is used to model either a bipolar or a field-effect transistor. As discussed in Chapter 6, we have assumed here a unilateral transistor, which is usually a good approximation in practice. We can simplify the analysis by assuming real input and output admittances of the transistor, defined as G_i and G_o, respectively, with a transistor transconductance g_m. The feedback network on the left side of the circuit is formed from three admittances in a bridged-T configuration. These components are usually reactive elements (capacitors or inductors) in order to provide a frequency selective transfer function with high Q. A common emitter/source configuration can be obtained by setting $V_2 = 0$, while common base/gate or common collector/drain configurations can be modeled by setting either $V_1 = 0$ or $V_4 = 0$, respectively. As shown, the

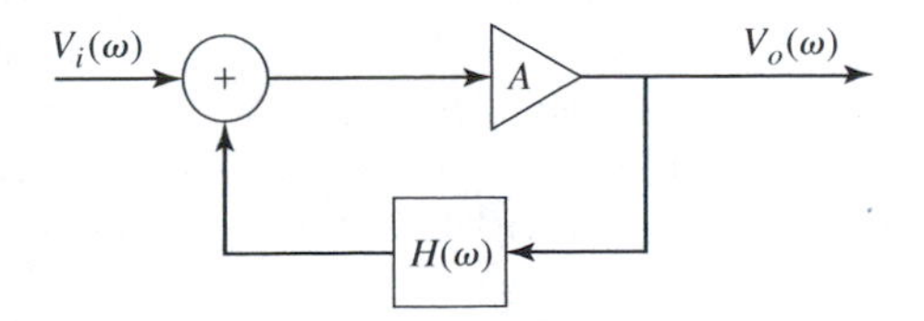

FIGURE 8.1 Block diagram of a sinusoidal oscillator using an amplifier with a frequency-dependent feedback path.

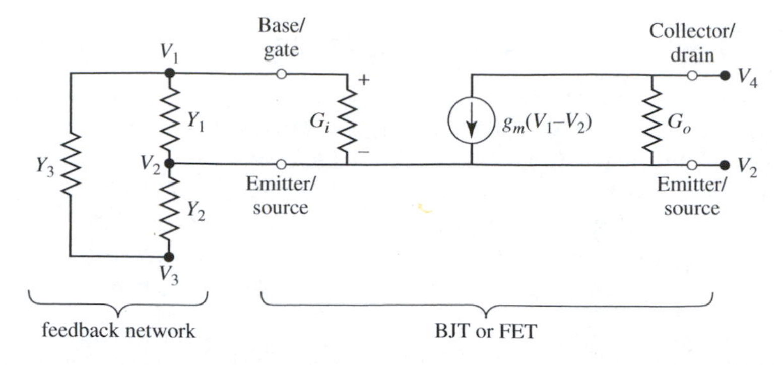

FIGURE 8.2 General circuit for a transistor oscillator. The transistor may be either a bipolar junction transistor or a field effect transistor. This circuit can used for common emitter/source, base/gate, or collector/drain configurations by grounding either V_2, V_1, or V_4, respectively. Feedback is provided by connecting node V_3 to V_4.

circuit of Figure 8.2 does not include a feedback path—this can be achieved by connecting node V_3 to node V_4.

Writing Kirchoff's law for the four voltage nodes of the circuit of Figure 8.2 gives the following matrix equation:

$$\begin{bmatrix} (Y_1 + Y_3 + G_i) & -(Y_1 + G_i) & -Y_3 & 0 \\ -(Y_1 + G_i + g_m) & (Y_1 + Y_2 + G_i + G_o + g_m) & -Y_2 & -G_o \\ -Y_3 & -Y_2 & (Y_2 + Y_3) & 0 \\ g_m & -(G_o + g_m) & 0 & G_o \end{bmatrix} \begin{bmatrix} V_1 \\ V_2 \\ V_3 \\ V_4 \end{bmatrix} = 0 \tag{8.3}$$

Recall from circuit analysis that if the ith node of the circuit is grounded, so that $V_i = 0$, the matrix of (8.3) will be modified by eliminating the ith row and column, reducing the order of the matrix by one. Additionally, if two nodes are connected together, the matrix is modified by adding the corresponding rows and columns.

Oscillators Using a Common Emitter BJT

As a specific example, consider an oscillator using a bipolar junction transistor in a common emitter configuration. In this case we have $V_2 = 0$, with feedback provided from the collector, so that $V_3 = V_4$. In addition, the output admittance of the transistor is negligible, so we set $G_o = 0$. These conditions serve to reduce the matrix of (8.3) to the following:

$$\begin{bmatrix} (Y_1 + Y_3 + G_i) & -Y_3 \\ (g_m - Y_3) & (Y_2 + Y_3) \end{bmatrix} \begin{bmatrix} V_1 \\ V \end{bmatrix} = 0, \tag{8.4}$$

where $V = V_3 = V_4$. If the circuit is to operate as an oscillator, then (8.4) must be satisfied for nonzero values of V_1 and V, so the determinant of the matrix must be zero. If the feedback network consists only of lossless capacitors and inductors, then Y_1, Y_2, and Y_3 must be imaginary, so we let $Y_1 = jB_1$, $Y_2 = jB_2$, and $Y_3 = jB_3$. Also, recall that the transconductance g_m and transistor input conductance are G_i are real. Then the determinant of (8.4) simplifies to

$$\begin{vmatrix} G_i + j(B_1 + B_3) & -jB_3 \\ g_m - jB_3 & j(B_2 + B_3) \end{vmatrix} = 0 \tag{8.5}$$

Separately equating the real and imaginary parts of the determinant to zero gives two equations:

$$\frac{1}{B_1} + \frac{1}{B_2} + \frac{1}{B_3} = 0 \tag{8.6a}$$

and

$$\frac{1}{B_3} + \left(1 + \frac{g_m}{G_i}\right)\frac{1}{B_2} = 0. \tag{8.6b}$$

If we convert susceptances to reactances, and let $X_1 = 1/B_1$, $X_2 = 1/B_2$, and $X_3 = 1/B_3$, then (8.6a) can be written as

$$X_1 + X_2 + X_3 = 0. \tag{8.7a}$$

Using (8.6a) to eliminate B_3 from (8.6b) reduces that equation to the following:

$$X_1 = \frac{g_m}{G_i} X_2. \tag{8.7b}$$

Since g_m and G_i are positive, (8.7b) implies that X_1 and X_2 have the same sign, and therefore are either both capacitors, or both inductors. Equation (8.7a) then shows that X_3 must be opposite in sign from X_1 and X_2, and therefore the opposite type of component. This conclusion leads to two of the most commonly used oscillator circuits.

If X_1 and X_2 are capacitors and X_3 is an inductor, we have a *Colpitts* oscillator. Let $X_1 = -1/\omega_0 C_1$, $X_2 = -1/\omega_0 C_2$, and $X_3 = \omega_0 L_3$. Then (8.7a) becomes

$$\frac{-1}{\omega_0}\left(\frac{1}{C_1} + \frac{1}{C_2}\right) + \omega_0 L_3 = 0,$$

which can be solved for the frequency of oscillation, ω_0, as

$$\omega_0 = \sqrt{\frac{1}{L_3}\left(\frac{C_1 + C_2}{C_1 C_2}\right)}. \tag{8.8}$$

Using these same substitutions in (8.7b) gives a necessary condition for oscillation of the Colpitts circuit as

$$\frac{C_2}{C_1} = \frac{g_m}{G_i}. \tag{8.9}$$

The resulting common-emitter Colpitts oscillator circuit is shown in Figure 8.3a.

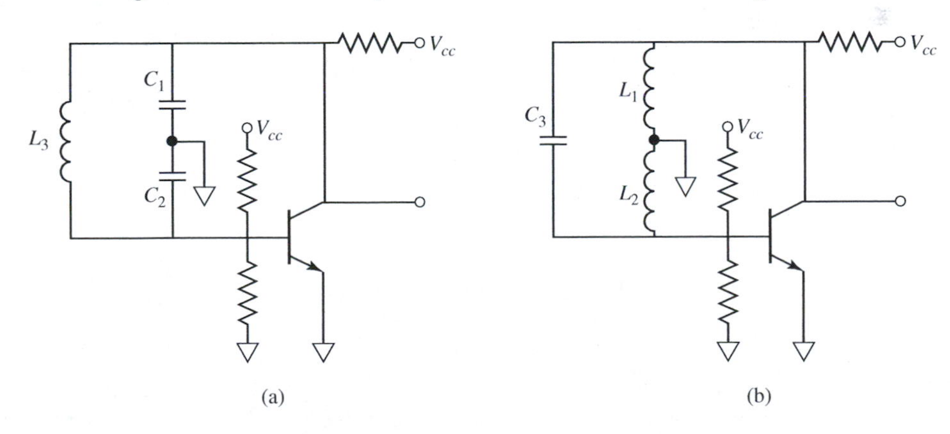

FIGURE 8.3 Transistor oscillator circuits using a common-emitter BJT. (a) Colpitts oscillator. (b) Hartley oscillator.

Alternatively, if we choose X_1 and X_2 to be inductors, and X_3 to be a capacitor, then we have a *Hartley* oscillator. Let $X_1 = \omega_0 L_1$, $X_2 = \omega_0 L_2$, and $X_3 = -1/\omega_0 C_3$. Then (8.7a) becomes

$$\omega_0(L_1 + L_2) - \frac{1}{\omega_0 C_3} = 0,$$

which can be solved for ω_0 to give

$$\omega_0 = \sqrt{\frac{1}{C_3(L_1 + L_2)}}. \tag{8.10}$$

These same substitutions used in (8.7b) gives a necessary condition for oscillation of the Hartley circuit as

$$\frac{L_1}{L_2} = \frac{g_m}{G_i}. \tag{8.11}$$

The resulting common-emitter Hartley oscillator circuit is shown in Figure 8.3b.

Oscillators Using a Common Gate FET

Next consider an oscillator using an FET in a common gate configuration. In this case $V_1 = 0$, and again $V_3 = V_4$ provides the feedback path. For an FET the input admittance can be neglected, so we set $G_i = 0$. Then the matrix of (8.3) reduces to

$$\begin{bmatrix} (Y_1 + Y_2 + g_m + G_o) & -(Y_2 + G_o) \\ -(G_o + g_m + Y_2) & (Y_2 + Y_3 + G_o) \end{bmatrix} \begin{bmatrix} V_2 \\ V \end{bmatrix} = 0, \tag{8.12}$$

where $V = V_3 = V_4$.

Again we assume the feedback network is composed of lossless reactive elements, so that Y_1, Y_2, and Y_3 can be replaced with their susceptances. Setting the determinant of (8.12) to zero then gives

$$\begin{vmatrix} (g_m + G_o) + j(B_1 + B_2) & -G_o - jB_2 \\ -(G_o + g_m) - jB_2 & G_o + j(B_2 + B_3) \end{vmatrix} = 0. \tag{8.13}$$

Equating the real and imaginary parts to zero gives two equations:

$$\frac{1}{B_1} + \frac{1}{B_2} + \frac{1}{B_3} = 0, \tag{8.14a}$$

and

$$\frac{G_o}{B_3} + \frac{g_m}{B_1} + \frac{G_o}{B_1} = 0. \tag{8.14b}$$

As before, let X_1, X_2, and X_3 be the reciprocals of the corresponding susceptances. Then (8.14a) can be rewritten as

$$X_1 + X_2 + X_3 = 0. \tag{8.15a}$$

Using (8.14a) to eliminate B_3 from (8.14b) reduces that equation to

$$\frac{X_2}{X_1} = \frac{g_m}{G_o}. \tag{8.15b}$$

Since g_m and G_o are positive, (8.15b) shows that X_1 and X_2 must have the same sign, while (8.15a) indicates that X_3 must have the opposite sign. If X_1 and X_2 are chosen to be negative, then these elements will be capacitive and X_3 will be inductive. This corresponds to a Colpitts oscillator. Since (8.15a) is identical to (8.7a), its solution gives the result for the resonant frequency for the common gate Colpitts oscillator as

$$\omega_0 = \sqrt{\frac{1}{L_3}\left(\frac{C_1 + C_2}{C_1 C_2}\right)}, \tag{8.16}$$

which is identical to the result obtained in (8.8) for the common emitter Colpitts oscillator. This is because the resonant frequency is determined by the feedback network, which is identical in both cases. The further condition for oscillation given by (8.15b) reduces to

$$\frac{C_1}{C_2} = \frac{g_m}{G_o}. \tag{8.17}$$

If we choose X_1 and X_2 to be positive (inductive), then X_3 will be capacitive, and we have a Hartley oscillator. The resonant frequency of the common gate Hartley oscillator is given by

$$\omega_0 = \sqrt{\frac{1}{C_3(L_1 + L_2)}}, \tag{8.18}$$

which is identical to the result of (8.10) for the common emitter Hartley oscillator. Equation (8.15b) reduces to

$$\frac{L_2}{L_1} = \frac{g_m}{G_o}. \tag{8.19}$$

The circuits for common gate Colpitts and Hartley oscillators are similar to the circuits shown in Figure 8.3, if the BJT is replaced with an FET device.

Practical Considerations

It must be emphasized that the above analysis is based on very idealized assumptions, and in practice successful oscillator design requires attention to factors such as the reactances associated with the input and output transistor ports, the variation of transistor properties with temperature, transistor bias and decoupling circuitry, and the effect of inductor losses. For these purposes computer aided design software can be very helpful [3].

The above analysis can be extended to account for more realistic feedback network inductors having series resistance, which invariably occurs in practice. For example, consider the case of a common emitter BJT Colpitts oscillator, with the impedance of the inductor given by $Z_3 = 1/Y_3 = R + j\omega L_3$. Substituting into (8.4) and setting the real and imaginary parts of the determinant to zero gives the following result for resonant frequency:

$$\omega_0 = \sqrt{\frac{1}{L_3}\left(\frac{1}{C_1} + \frac{1}{C_2} + \frac{G_i R}{C_1}\right)} = \sqrt{\frac{1}{L_3}\left(\frac{1}{C_1'} + \frac{1}{C_2}\right)}. \tag{8.20}$$

This equation is similar to the result of (8.8) for the lossless inductor, except that C_1' is defined as

$$C_1' = \frac{C_1}{1 + RG_i}. \tag{8.21}$$

The corresponding condition for oscillation is

$$\frac{R}{G_i} = \frac{1 + g_m/G_i}{\omega_0^2 C_1 C_2} - \frac{L_3}{C_1}. \tag{8.22}$$

This result sets the maximum value of the series resistance R; the left side of (8.22) should generally be chosen to be less than the right hand side to ensure oscillation.

EXAMPLE 8.1 COLPITTS OSCILLATOR DESIGN

Design a 50 MHz Colpitts oscillator using a transistor in a common emitter configuration with $\beta = g_m/G_i = 30$, and a transistor input resistance of $R_i = 1/G_i = 1200\ \Omega$. Use an inductor with $L_3 = 0.10\ \mu$H, with a Q of 100. What is the minimum Q of the inductor for which oscillation will be sustained?

Solution

From (8.20) the series combination of C_1' and C_2 is found to be

$$\frac{C_1' C_2}{C_1' + C_2} = \frac{1}{\omega_0^2 L_3} = \frac{1}{(2\pi)^2(50 \times 10^6)^2(0.1 \times 10^{-6})} = 100 \text{ pF}.$$

This value can be obtained in several ways, but here we will choose $C_1' = C_2 = 200$ pF.

From circuit analysis [4] we know that the Q of an inductor is related to its series resistance by $Q = \omega L/R$, so the series resistance of the 0.1 μH inductor is

$$R = \frac{\omega_0 L_3}{Q} = \frac{(2\pi)(50 \times 10^6)(0.1 \times 10^{-6})}{100} = 0.31\ \Omega$$

Then (8.21) gives C_1 as

$$C_1 = C_1'(1 + RG_i) = (200 \text{ pF})\left(1 + \frac{0.31}{1200}\right) = 200 \text{ pF},$$

which we see is essentially unchanged from the value found by neglecting the inductor loss. Using (8.22) with the above values gives

$$\frac{R}{G_i} = \frac{1+\beta}{\omega_0^2 C_1 C_2} - \frac{L_3}{C_1}$$

$$(0.31)(1200) < \frac{1+30}{(2\pi)^2(50 \times 10^6)^2(200 \times 10^{-12})^2} - \frac{0.1 \times 10^{-6}}{200 \times 10^{-12}}$$

$$372. < 7852. - 500. = 7352,$$

which indicates that the condition for oscillation will be satisfied. This condition can be used to find the minimum inductor Q by first solving for the maximum value of series resistance R:

$$R_{max} = \frac{1}{R_i}\left(\frac{1+\beta}{\omega_0^2 C_1 C_2} - \frac{L_3}{C_1}\right) = \frac{7352.}{1200} = 6.13\ \Omega$$

So the minimum Q is

$$Q_{min} = \frac{\omega_0 L_3}{R_{max}} = \frac{(2\pi)(50 \times 10^6)(0.1 \times 10^{-6})}{6.13} = 5.1$$

○

Crystal Oscillators

As we have seen from the above analysis, the resonant frequency of an oscillator is determined from the condition that a 180° phase shift occurs between the input and output of the transistor. If the resonant feedback circuit has a high Q, so that there is a very rapid change in the phase shift with frequency, the oscillator will have good frequency stability.

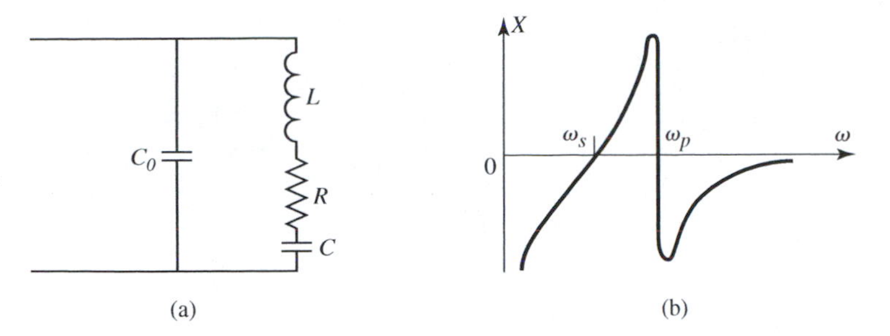

FIGURE 8.4 (a) Equivalent circuit of a crystal. (b) Input reactance of a crystal resonator.

Quartz crystals are useful for this purpose, especially at frequencies below a few hundred MHz, where LC resonators seldom have Qs greater than a few hundred. Quartz crystals may have unloaded Qs as high as 100,000 and temperature drift less than 0.001%/C°. Crystal-controlled oscillators therefore find extensive use as stable frequency sources in wireless systems. Further stability can be obtained by controlling the temperature of the quartz crystal.

A quartz crystal resonator consists of a small slab of quartz mounted between two metallic plates. Mechanical oscillations can be excited in the crystal through the piezoelectric effect. The equivalent circuit of a quartz crystal near its lowest resonant mode is shown in Figure 8.4a. This circuit has series and parallel resonant frequencies, ω_s and ω_p, given by

$$\omega_s = \frac{1}{\sqrt{LC}}, \tag{8.23a}$$

$$\omega_p = \frac{1}{\sqrt{L\left(\frac{C_0 C}{C_0 + C}\right)}}. \tag{8.23b}$$

The reactance of the circuit of Figure 8.4a is plotted in Figure 8.4b, where we see that the reactance is inductive in the frequency range between the series and parallel resonances. This is the usual operating point of the crystal, so that the crystal may be used in place of the inductor in a Colpitts or Pierce oscillator. A typical crystal oscillator circuit is shown in Figure 8.5.

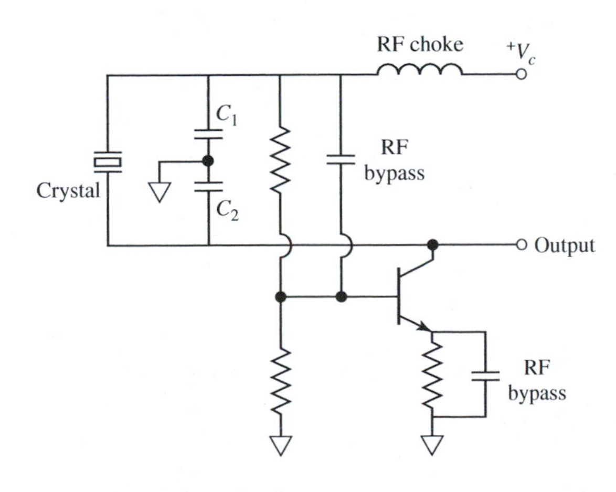

FIGURE 8.5 Pierce crystal oscillator circuit.

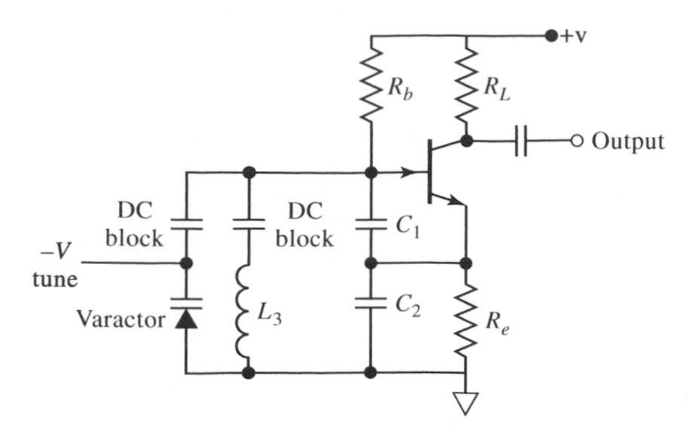

FIGURE 8.6 A varactor-tuned voltage-controlled transistor oscillator circuit.

Voltage-Controlled Oscillators

In many wireless applications it is necessary to vary the frequency of the local oscillator. This requirement occurs in AM and FM broadcast receivers, and in multichannel telecommunications systems such as cellular telephones and wireless local area networks. Because the resonant frequency of an oscillator is controlled by an LC network, changing the frequency of an oscillator requires changing either the inductance or capacitance, and it is usually preferred to do this electronically. While it is possible to use tunable ferromagnetic inductors for this purpose, it is usually easier and cheaper to use voltage-controlled capacitors, such as *varactors*. A varactor is a diode whose junction capacitance may be controlled by changing the DC reverse bias applied to the diode. The resulting configuration is called a *voltage-controlled oscillator* (VCO). A typical varactor may have a junction capacitance that varies from 5 pF to 30 pF as the bias voltage varies from 20 to 1 V. Since resonant frequency varies as $1/\sqrt{C}$, a linear variation of frequency with tuning voltage, v, requires that the junction capacitance vary as $1/v^2$; hyperabrupt junction varactors have characteristics that approximate this behavior fairly closely.

Varactors can be used in a variety of configurations to provide voltage tuning of an oscillator. Generally a varactor is used in either series or parallel with a capacitor in the feedback network to provide a fine-tuning range about the quiescent resonant frequency. In addition, DC blocking capacitors and/or RF chokes must be used to provide a reverse bias voltage without detuning or shorting the RF circuit. A typical varactor-tuned VCO circuit is shown in Figure 8.6. This design uses a varactor in shunt across the capacitors of a Colpitts oscillator; this type of circuit is known as a *Clapp* oscillator.

8.2 MICROWAVE OSCILLATORS

Microwave circuit design often involves qualitative differences from the techniques used at lower RF frequencies, because of differences in transistor characteristics, circuit layout methods, and test equipment. At microwave frequencies S parameter methods are the preferred choice, primarily because it is very difficult to measure voltages or currents directly at these frequencies, while incident and reflected signals can be measured reliably and accurately. In this section we discuss some of the basic principles of microwave oscillator design, including negative resistance oscillators, FET oscillators, and dielectric resonator oscillators. This material is drawn largely from [4].

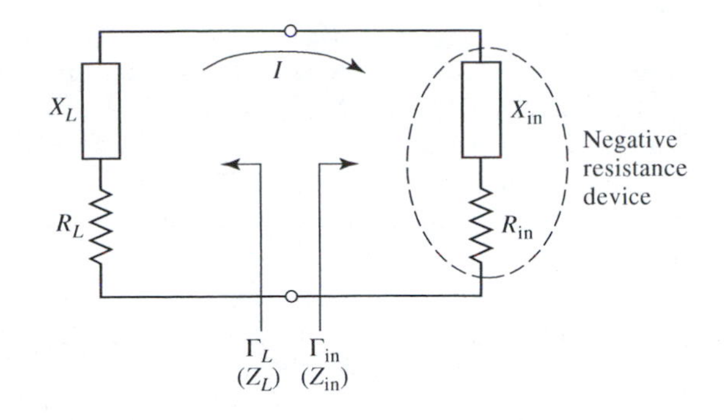

FIGURE 8.7 Circuit for a one-port negative-resistance oscillator.

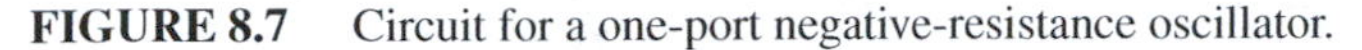

Negative Resistance Oscillators

Here we discuss some of the basic principles of the operation and design of one-port negative resistance oscillators; much of this material will also apply to two-port (transistor) oscillators. One-port negative resistance oscillators include circuits that use IMPATT or Gunn diodes, where the active device can be biased to produce an impedance having a negative real part.

Figure 8.7 shows a canonical RF circuit for a one-port negative resistance oscillator, where $Z_{in} = R_{in} + jX_{in}$ is the input impedance of the active device. In general, this impedance is current (or voltage) dependent, as well as frequency dependent, which we can indicate by writing $Z_{in}(I, j\omega) = R_{in}(I, j\omega) + jX_{in}(I, j\omega)$. The device is terminated with a passive load impedance, $Z_L = R_L + jX_L$. Applying Kirchoff's voltage law gives

$$(Z_L + Z_{in})I = 0. \tag{8.24}$$

If oscillation is occurring, so that the RF current I is nonzero, then the following conditions must be satisfied:

$$R_L + R_{in} = 0, \tag{8.25a}$$

$$X_L + X_{in} = 0. \tag{8.25b}$$

(These conditions are analogous to setting the real and imaginary parts of the determinantal equations of (8.5) or (8.13) to zero in the case of the transistor oscillator circuits considered in Section 8.1.)

Since the load is passive, $R_L > 0$ and (8.25a) indicates that $R_{in} < 0$. Thus, while a positive resistance implies energy dissipation, a negative resistance implies an energy source. The condition of (8.25b) controls the frequency of oscillation. The requirement of (8.24), that $Z_L = -Z_{in}$ for steady-state oscillation, implies that the reflection coefficients Γ_L and Γ_{in} are related as

$$\Gamma_L = \frac{Z_L - Z_0}{Z_L + Z_0} = \frac{-Z_{in} - Z_0}{-Z_{in} + Z_0} = \frac{Z_{in} + Z_0}{Z_{in} - Z_0} = \frac{1}{\Gamma_{in}}. \tag{8.26}$$

The process of oscillation depends on the nonlinear behavior of Z_{in}, as follows. Initially, it is necessary for the overall circuit to be unstable at a certain frequency, that is, $R_{in}(I, j\omega) + R_L < 0$. Then any transient excitation, such as noise or the turning on of the power supply, will cause an oscillation to build up at the frequency, ω. As I increases, $R_{in}(I, j\omega)$ must become less negative until the current I_0 is reached such that $R_{in}(I_0, j\omega_0) + R_L = 0$, and

$X_{in}(I_0, j\omega_0) + X_L(j\omega_0) = 0$. Then the oscillator will be running in a stable state. The final frequency, ω_0, generally differs from the start-up frequency because X_{in} is current dependent, so that $X_{in}(I, j\omega) \neq X_{in}(I_0, j\omega_0)$.

Thus we see that the conditions of (8.25) are not enough to guarantee a stable state of oscillation. In particular, stability requires that any perturbation in current or frequency will be damped out, allowing the oscillator to return to its original state. This condition can be quantified by considering the effect of a small change, δI, in the current and a small change, δs, in the complex frequency $s = \alpha + j\omega$. If we let $Z_T(I, s) = Z_{in}(I, s) + Z_L(s)$, then we can write a Taylor series for $Z_T(I, s)$ about the operating point I_0, ω_0 as

$$Z_T(I, s) = Z_T(I_0, s_0) + \left.\frac{\partial Z_T}{\partial s}\right|_{s_0, I_0} \delta s + \left.\frac{\partial Z_T}{\partial I}\right|_{s_0, I_0} \delta I = 0, \tag{8.27}$$

since $Z_T(I, s)$ must still equal zero if oscillation is occurring. In (8.27), $s_0 = j\omega_0$ is the complex frequency at thde original operating point. Now we use the fact that $Z_T(I_0, s_0) = 0$, and that $\frac{\partial Z_T}{\partial s} = -j\frac{\partial Z_T}{\partial \omega}$, to solve (8.27) for $\delta s = \delta\alpha + j\delta\omega$:

$$\delta s = \delta\alpha + j\omega = \left.\frac{-\partial Z_T/\partial I}{\partial Z_T/\partial s}\right|_{s_0, I_0} \delta I = \frac{-j(\partial Z_T/\partial I)(\partial Z_T^*/\partial \omega)}{|\partial Z_T/\partial \omega|^2}\delta I. \tag{8.28}$$

Now if the transient caused by δI and $\delta\omega$ is to decay, we must have $\delta\alpha < 0$ when $\delta I > 0$. Equation (8.28) then implies that

$$\text{Im}\left(\frac{\partial Z_T}{\partial I}\frac{\partial Z_T^*}{\partial \omega}\right) < 0,$$

or

$$\frac{\partial R_T}{\partial I}\frac{\partial X_T}{\partial \omega} - \frac{\partial X_T}{\partial I}\frac{\partial R_T}{\partial \omega} > 0. \tag{8.29}$$

For a passive (series) load impedance, $\partial R_L/\partial I = \partial X_L/\partial I = \partial R_L/\partial \omega = 0$, so (8.29) reduces to

$$\frac{\partial R_{in}}{\partial I}\frac{\partial}{\partial \omega}(X_L + X_{in}) - \frac{\partial X_{in}}{\partial I}\frac{\partial R_{in}}{\partial \omega} > 0. \tag{8.30}$$

As discussed above, we usually have $\partial R_{in}/\partial I > 0$. So (8.30) can be satisfied if $\partial(X_L + X_{in})/\partial\omega \gg 0$, which implies that a high-Q circuit will result in maximum oscillator stability. Cavity, crystal, and dielectric resonators are therefore often preferred over LC resonators for practical oscillator design.

As in the case of RF transistor oscillator design, the above analysis is very idealized, and does not consider all the factors that must often be considered in practice. These include the selection of a device operating point, frequency pulling due to changes in the output load impedance, large-signal effects, and noise characteristics. Such topics are left to more advanced texts [3].

EXAMPLE 8.2 NEGATIVE RESISTANCE OSCILLATOR DESIGN

A one-port oscillator uses a negative-resistance diode having $\Gamma_{in} = 1.25\angle 40°$, with $Z_0 = 50\ \Omega$, at its desired operating point, for $f = 6$ GHz. Design a load matching network for a 50 Ω load impedance.

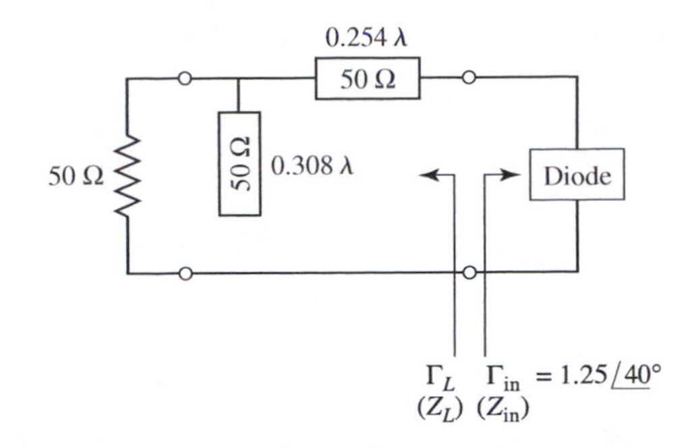

FIGURE 8.8 Load matching circuit for the one-port oscillator of Example 8.2.

Solution

From either a Smith chart, or by direct calculation, we find the input impedance as

$$Z_{\text{in}} = -44 + j123\ \Omega.$$

Then by (8.25) the load impedance must be

$$Z_L = 44 - j123\ \Omega.$$

A shunt stub and a series line section can be used to convert 50 Ω to Z_L, as shown in the circuit of Figure 8.8. ○

Transistor Oscillators

In a transistor oscillator a negative resistance one-port network is effectively created by terminating a potentially unstable transistor with an impedance designed to drive the device in an unstable region. The circuit model is shown in Figure 8.9; output power can be tapped from either side of the transistor. In the case of an amplifier, we prefer a device with a high degree of stability—ideally, an unconditionally stable device. For an oscillator, however, we require a device with a high degree of instability. Typically, common source or common gate FET configurations are used (or common emitter or common base in the case of bipolar transistors), often with positive feedback to enhance the instability of the device. After the transistor configuration is selected, the output stability circle can be drawn in the Γ_T plane, and Γ_T selected to produce a large value of negative resistance at the input to the transistor. Then the load impedance Z_L can be chosen to match Z_{in}. Because such a design uses the small-signal S parameters, and because R_{in} will become less negative as the oscillator power builds up, it is necessary to choose R_L so that $R_L + R_{\text{in}} < 0$. Otherwise, oscillation

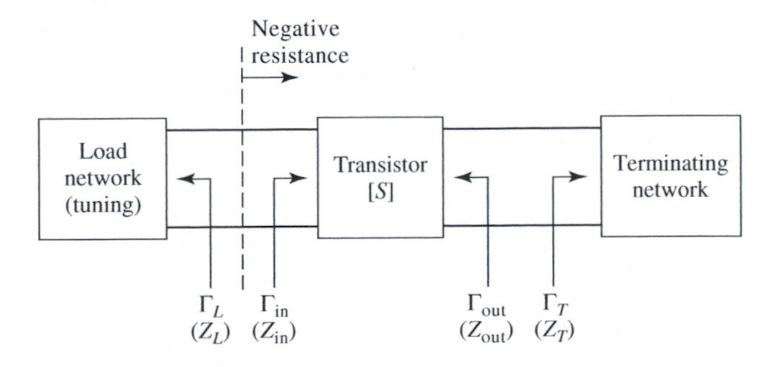

FIGURE 8.9 Circuit for a two-port transistor oscillator.

will cease when the increasing power increases R_{in} to the point where $R_L + R_{in} > 0$. In practice, a value of

$$R_L = \frac{-R_{in}}{3} \tag{8.31a}$$

is typically used. The reactive part of Z_L is chosen to resonate the circuit at the operating frequency

$$X_L = -X_{in}. \tag{8.31b}$$

When oscillation occurs between the load network and the transistor, oscillation will simultaneously occur at the output port, which we can show as follows. For steady-state oscillation at the input port, we must have $\Gamma_L \Gamma_{in} = 1$, as derived in (8.26). Then from (6.6a) (after replacing Γ_L with Γ_T) we have

$$\frac{1}{\Gamma_L} = \Gamma_{in} = S_{11} + \frac{S_{12}S_{21}\Gamma_T}{1 - S_{22}\Gamma_T} = \frac{S_{11} - \Delta\Gamma_T}{1 - S_{22}\Gamma_T}, \tag{8.32}$$

where $\Delta = S_{11}S_{22} - S_{21}S_{12}$. Solving for Γ_T gives

$$\Gamma_T = \frac{1 - S_{11}\Gamma_L}{S_{22} - \Delta\Gamma_L}. \tag{8.33}$$

Then from (6.6b) (after replacing Γ_s with Γ_L) we have that

$$\Gamma_{out} = S_{22} + \frac{S_{12}S_{21}\Gamma_L}{1 - S_{11}\Gamma_L} = \frac{S_{22} - \Delta\Gamma_L}{1 - S_{11}\Gamma_L}, \tag{8.34}$$

which shows that $\Gamma_T \Gamma_{out} = 1$, and hence $Z_T = -Z_{out}$. Thus, the condition for oscillation of the terminating network is satisfied. Note that the appropriate S parameters to use in the above development are generally the large signal parameters of the transistor, if available.

EXAMPLE 8.3 TRANSISTOR OSCILLATOR DESIGN

Design a transistor oscillator at 4 GHz using a GaAs FET in a common gate configuration, with a 5 nH inductor in series with the gate to increase the instability. Choose a terminating network to match a 50 Ω load, and an appropriate tuning network.

The S parameters of the transistor in a common source configuration are, with $Z_0 = 50\,\Omega$: $S_{11} = 0.72\angle{-116°}$, $S_{21} = 2.60\angle{76°}$, $S_{12} = 0.03\angle{57°}$, $S_{22} = 0.73\angle{-54°}$.

Solution

The first step is to convert the common source S parameters to the S parameters that apply to the transistor in a common gate configuration with a series inductor. (See Figure 8.10a.) This is most easily done using a microwave CAD package. The new S parameters are

$$S'_{11} = 2.18\angle{-35°},$$
$$S'_{21} = 2.75\angle{96°},$$
$$S'_{12} = 1.26\angle{18°},$$
$$S'_{22} = 0.52\angle{155°}.$$

Note that S'_{11} is significantly larger than $|S_{11}|$, which suggests that the configuration of Figure 8.10a is more unstable than the common source configuration.

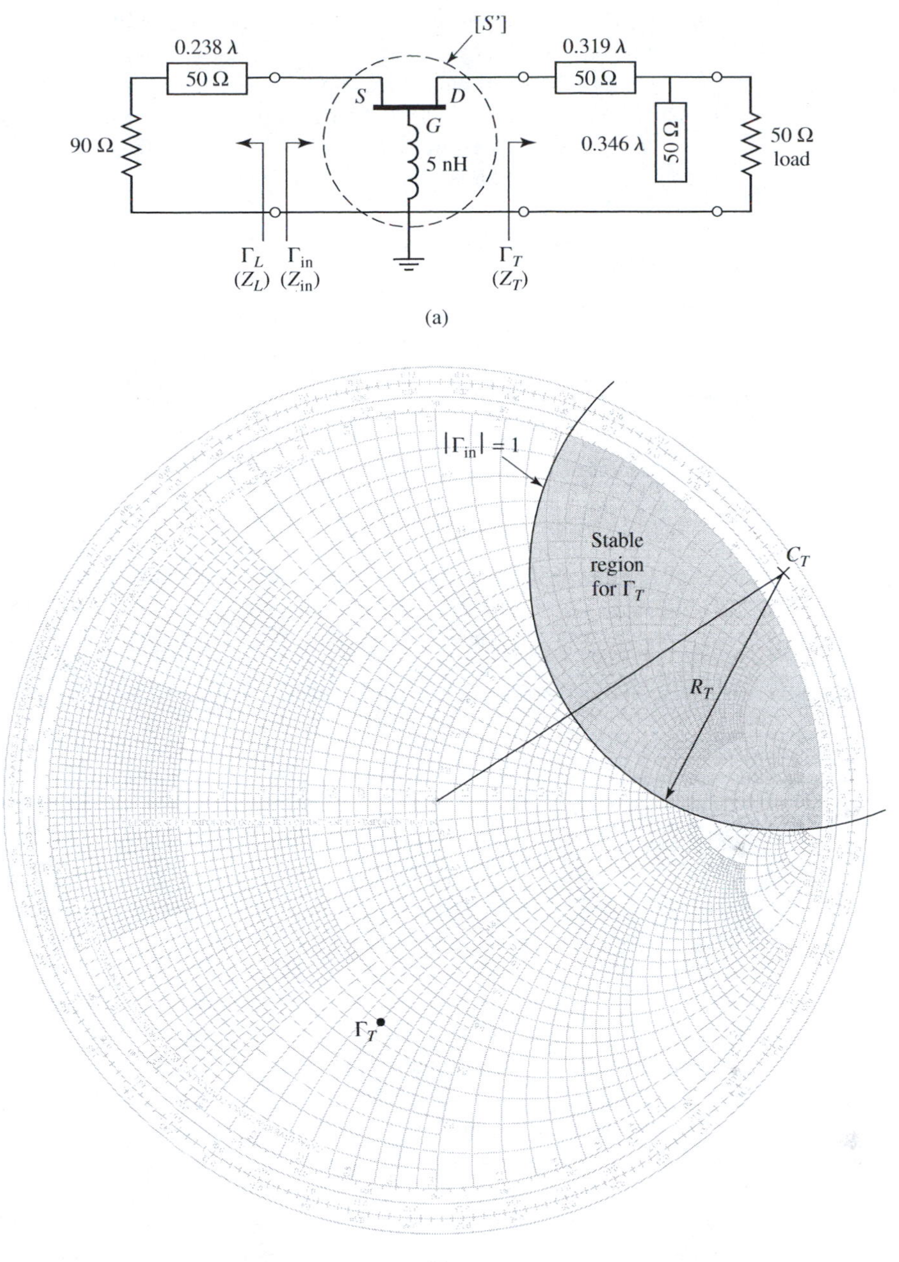

FIGURE 8.10 Circuit design for the transistor oscillator of Example 8.3. (a) Oscillator circuit. (b) Smith chart solution for finding Γ_T.

Calculating the output stability circle (Γ_T plane) parameters from (6.28) gives

$$C_T = \frac{(S'_{22} - \Delta' S'^{*}_{11})^{*}}{|S'_{22}|^2 - |\Delta'|^2} = 1.08\angle 33^\circ,$$

$$R_T = \left| \frac{S'_{12} S'_{21}}{|S'_{22}|^2 - |\Delta'|^2} \right| = 0.665.$$

Since $S'_{11} = 2.18 > 1$, the stable region is inside the stability circle, as shown in the Smith chart of Figure 8.10b.

There is a large amount of freedom in our choice for Γ_T, but one objective is to make $|\Gamma_{in}|$ large. Thus we try several values of Γ_T located on the opposite side of the chart from the stability circle, and select $\Gamma_T = 0.59\angle{-104°}$. Then we can design a single-stub matching network to convert a 50 Ω load to $Z_T = 20 - j35\ \Omega$, as shown in Figure 8.10a.

For the given value of Γ_T, we calculate Γ_{in} as

$$\Gamma_{in} = S'_{11} + \frac{S'_{12}S'_{21}\Gamma_T}{1 - S'_{22}\Gamma_T} = 3.96\angle{-2.4°},$$

or $Z_{in} = -84 - j1.9\ \Omega$. Then, from (8.31), we find Z_L as

$$Z_L = \frac{-R_{in}}{3} - jX_{in} = 28 + j1.9\ \Omega.$$

Using $R_{in}/3$ should ensure enough instability for reliable startup of oscillation. The easiest way to implement the impedance Z_L is to use a 90 Ω load with a short length of transmission line, as shown in Figure 8.10a. It is likely that the steady-state oscillation frequency will differ from 4 GHz because of the nonlinearity of the transistor parameters. ○

Dielectric Resonator Oscillators

As we saw from the result of (8.30), oscillator stability is enhanced with the use of a high Q tuning network. The Q of a resonant network using lumped elements or microstrip lines and stubs is typically limited to a few hundred [4], and while waveguide cavity resonators can have Qs of 10^4 or more, they are not well suited for integration in miniature microwave integrated circuitry. Another disadvantage of metal cavities is the significant frequency drift caused by dimensional expansion due to a variation in temperature. The dielectric cavity resonator [4] overcomes most of these disadvantages, as it can have an unloaded Q as high as several thousand, is compact and easily integrated with planar circuitry, and can be made from ceramic materials that have excellent temperature stability. For these reasons, transistor *dielectric resonator oscillators* (DROs) are becoming increasingly common over the entire microwave and millimeter wave frequency range.

A dielectric resonator is usually coupled to an oscillator circuit by positioning it in close proximity to a microstrip line, as shown in Figure 8.11a. The resonator operates in the $TE_{01\delta}$ mode, and couples to the fringing magnetic field of the microstrip line. The strength of coupling is determined by the spacing, d, between the resonator and microstrip line. Because coupling is via the magnetic field, the resonator appears as a series load on the microstrip line, as shown in the equivalent circuit of Figure 8.11b. The resonator is modeled as a parallel *RLC* circuit, and the coupling to the feed line is modeled by the turns ratio, N, of the transformer. Using the expression for the impedance of a parallel *RLC* resonator [4], we can express the equivalent series impedance, Z, seen by the microstrip line as

$$Z = \frac{N^2 R}{1 + j2Q\Delta\omega/\omega_0}, \tag{8.35}$$

where $Q = R/\omega_0 L$ is the unloaded resonator Q, $\omega_0 = 1/\sqrt{LC}$ is the resonant frequency, and $\Delta\omega = \omega - \omega_0$. The coupling factor between the resonator and the feed line is the ratio

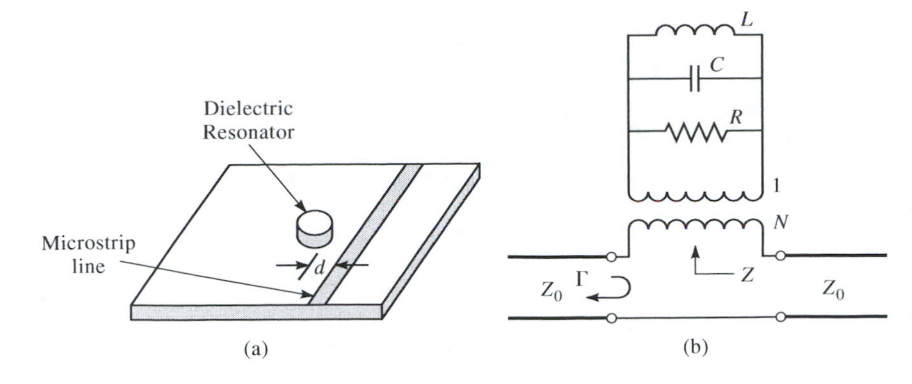

FIGURE 8.11 (a) Geometry of a dielectric resonator coupled to a microstripline; (b) Equivalent circuit.

of the unloaded to external Q [4], and can be found as

$$g = \frac{Q}{Q_e} = \frac{R/\omega_0 L}{R_L/N^2\omega_0 L} = \frac{N^2 R}{2Z_0}, \tag{8.36}$$

where $R_L = 2Z_0$ is the load resistance for a feed line with source and termination resistances Z_0. In some cases the feed line is terminated with an open-circuit $\lambda/4$ from the resonator to maximize the magnetic field at that point; in this case $R_L = Z_0$ and the coupling factor is twice the value given in (8.36).

The reflection coefficient seen on the terminated microstrip line looking toward the resonator can be written as

$$\Gamma = \frac{(Z_0 + N^2R) - Z_0}{(Z_0 + N^2R) + Z_0} = \frac{N^2R}{2Z_0 + N^2R} = \frac{g}{1+g}. \tag{8.37}$$

This allows the coupling coefficient to be found from $g = \Gamma/(1-\Gamma)$ after the simple procedure of measuring Γ at resonance; the resonant frequency and Q can also be found by measurement. Alternatively, these quantities can be calculated using approximate analytical solutions [5]. Note that this procedure leaves a degree of freedom between N and R, since only the product N^2R is uniquely determined.

There are many possible microwave oscillator configurations using common source (emitter), common gate (base), or common drain (collector) connections of either FET or bipolar transistors, in addition to the optional use of series or shunt elements to increase the instability of the device [1]–[3]. A dielectric resonator can be incorporated into the circuit to provide frequency stability using either the parallel feedback arrangement of Figure 8.12a, or the series feedback technique shown in Figure 8.12b. The parallel configuration uses a resonator coupled to two microstrip lines, functioning as a high-Q bandpass filter that couples a portion of the transistor output back to its input. The amount of coupling is controlled by the spacing between the resonator and the lines, and the phase is controlled by the length of the lines. The series feedback configuration is simpler, using only a single microstrip feed line, but typically does not have a tuning range as wide as that obtained with parallel feedback. Design of an oscillator using parallel feedback is most conveniently done using a microwave CAD package, but a dielectric resonator oscillator using series feedback can be designed using the same procedure that was discussed in the previous section on two-port oscillators.

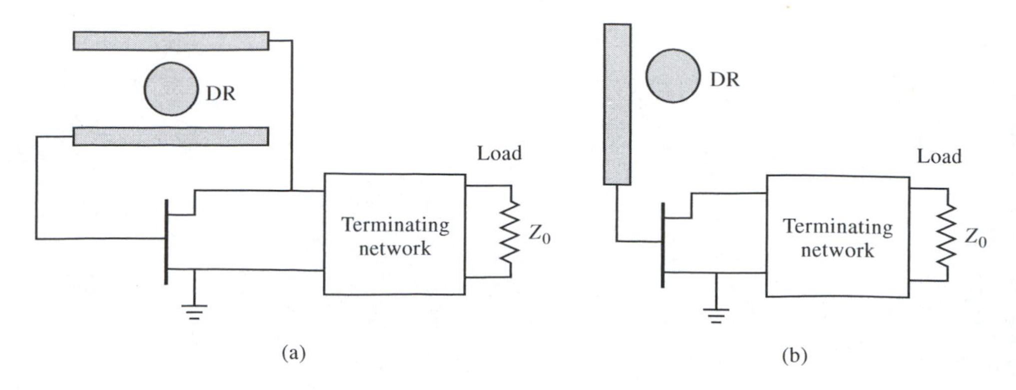

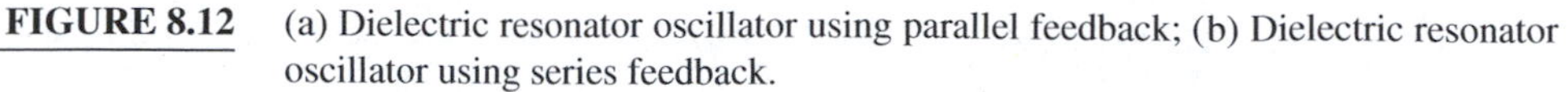
FIGURE 8.12 (a) Dielectric resonator oscillator using parallel feedback; (b) Dielectric resonator oscillator using series feedback.

EXAMPLE 8.4 DIELECTRIC RESONATOR OSCILLATOR DESIGN

A wireless local area network application requires a local oscillator operating at 2.4 GHz. Design a dielectric resonator oscillator using the series feedback circuit of Figure 8.12b with a bipolar transistor having the following S parameters ($Z_0 =$ 50 Ω): $S_{11} = 1.8\angle 130°$, $S_{12} = 0.4\angle 45°$, $S_{21} = 3.8\angle 36°$, $S_{22} = 0.7\angle -63°$. Determine the required coupling coefficient for the dielectric resonator, and a microstrip matching network for the termination network. The termination network should include the output load impedance. Plot the magnitude of Γ_{out} versus $\Delta f/f_0$, for small variations in frequency about the design value, assuming an unloaded resonator Q of 1000.

Solution

The DRO circuit is shown in Figure 8.13a. The dielectric resonator is placed $\lambda/4$ from the open end of the microstrip line; the line length ℓ_r can be adjusted to match the phase of the required value of Γ_L. In contrast to the oscillator of the previous example, the output load impedance for this circuit is part of the terminating network.

The stability circles for the load and termination sides of the transistor can be plotted if desired, but are not necessary to the design since we will begin by choosing Γ_L to provide a large value of $|\Gamma_{out}|$. From (8.34) we have

$$\Gamma_{out} = S_{22} + \frac{S_{12}S_{21}\Gamma_L}{1 - S_{11}\Gamma_L},$$

which indicates that we can maximize Γ_{out} by making $1 - S_{11}\Gamma_L$ close to zero. Thus we choose $\Gamma_L = 0.6\angle -130°$, which gives $\Gamma_{out} = 10.7\angle 132°$. This corresponds to an impedance

$$Z_{out} = Z_0\frac{1 + \Gamma_{out}}{1 - \Gamma_{out}} = 50\frac{1 + 10.7\angle 132°}{1 - 10.7\angle 132°} = -43.7 + j6.1\ \Omega.$$

Applying the analogous startup condition of (8.31) for the termination side gives the required termination impedance as

$$Z_T = \frac{-R_{out}}{3} - jX_{out} = 5.5 - j6.1\ \Omega.$$

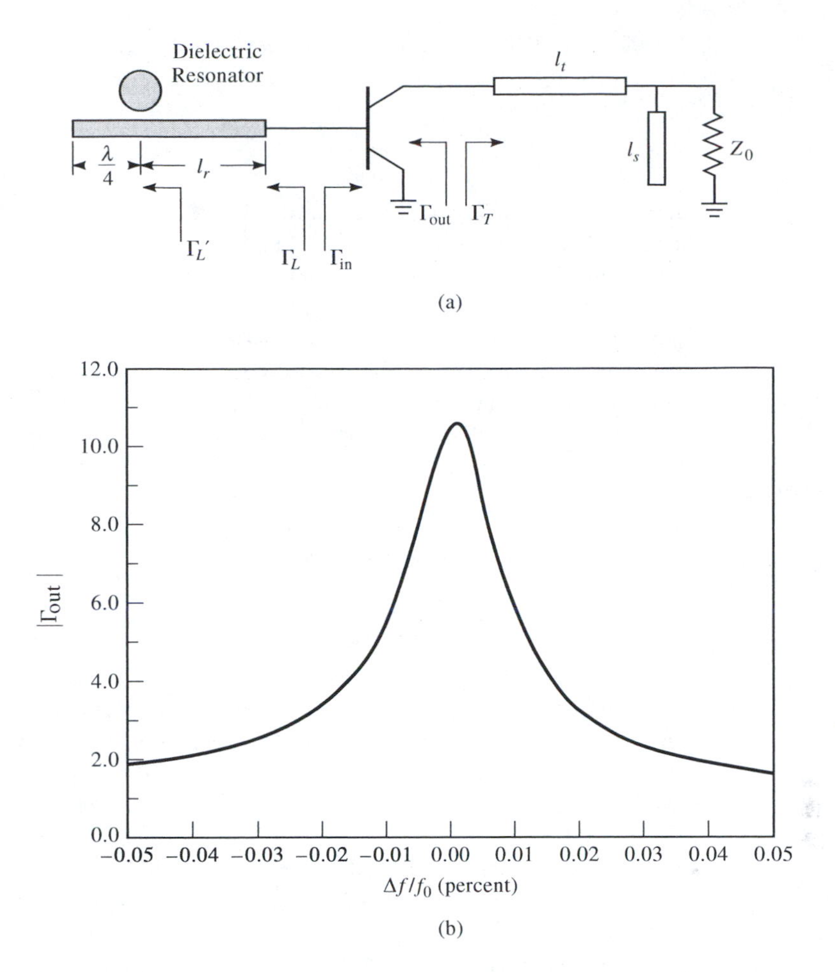

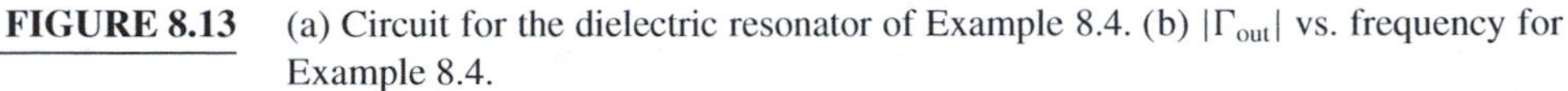

FIGURE 8.13 (a) Circuit for the dielectric resonator of Example 8.4. (b) $|\Gamma_{out}|$ vs. frequency for Example 8.4.

The termination matching network can now be designed using a Smith chart. The shortest transmission line length for matching Z_T to the load impedance Z_0 is $\ell_t = 0.481\lambda$, and the required open-circuit stub length is $\ell_s = 0.307\lambda$.

Next we match Γ_L to the resonator network. From (8.35) we know that the equivalent impedance of the resonator seen by the microstrip line is real at the resonant frequency, so the phase angle of the reflection coefficient at this point, Γ_L', must be either zero or 180°. For an undercoupled parallel *RLC* resonator, $R < Z_0$, so the proper phase will be 180°, which can be achieved by transformation through the line length ℓ_r. The magnitude of the reflection coefficient is unchanged, so we have the relation

$$\Gamma_L' = \Gamma_L e^{2j\beta\ell_r} = (0.6\angle -130^\circ)e^{2j\beta\ell_r} = 0.6\angle 180^\circ,$$

which gives $\ell_r = 0.431\lambda$. The equivalent impedance of the resonator at resonance is then

$$Z_L' = Z_0 \frac{1 + \Gamma_L'}{1 - \Gamma_L'} = 12.5\ \Omega.$$

The coupling coefficient can be found using (8.36), with a factor of two to account for the $\lambda/4$ stub termination, as

$$g = \frac{N^2 R}{Z_0} = \frac{12.5}{50} = 0.25.$$

The variation of $|\Gamma_{\text{out}}|$ with frequency will give an indication of the frequency stability of the oscillator. We can calculate Γ_{out} from (8.34), after first using (8.35) to compute Z'_L, Γ'_L and then transforming down the line of length ℓ_r to obtain Γ_L. The electrical line length can be approximated as constant for the small changes in frequency associated with this calculation. A short computer program or a microwave CAD package can be used to generate data for $-0.01 < \Delta f / f_0 < 0.01$, which is shown in the graph of Figure 8.13b. Observe that $|\Gamma_{\text{out}}|$ decreases rapidly with a change in frequency as small as a few hundredths of a percent, demonstrating the sharp selectivity that can be obtained with a dielectric resonator. ○

8.3 FREQUENCY SYNTHESIS METHODS

Frequency synthesizers provide a large number of precisely controlled frequencies derived from a stable oscillator. The stable reference source is usually a crystal-controlled oscillator, which may be housed in a temperature controlled environment for even greater stability. A frequency synthesizer eliminates the need for many independent crystal oscillators in a multichannel system, which would be expensive and occupy considerable space. Modern frequency synthesizers can be easily implemented using a variety of available integrated circuits, and are found in virtually all modern radios, cellular telephones, and wireless data equipment.

There are three basic methods that can be used for frequency synthesis. The oldest technique is called *direct synthesis*, and uses multiple stages of frequency mixing, division, and filtering to produce the desired product. Alternatively, a *phase-locked loop* can be used to derive multiples of a reference frequency. The newest method is *direct digital synthesis*, which uses a digital look-up table for the sine function and a digital-to-analog converter to construct a sine wave of arbitrary frequency. Each of these methods has its own advantages and disadvantages, which will be discussed next.

Direct Synthesis

Direct frequency synthesis is an analog method that uses mixers, frequency multipliers, dividers, bandpass filters, and switches to yield a large number of precisely controlled frequencies. The fundamental source is often a temperature-controlled crystal oscillator (TXCO), for a very stable output with excellent phase noise. The major drawback of this type of synthesis is that it is complicated and expensive, especially when wide frequency coverage with a fine resolution is required. For this reason, direct synthesis is usually reserved for RF and microwave test instrumentation.

The basic method of direct synthesis is shown in the example of Figure 8.14, where output frequencies ranging from 25 to 35 MHz in 1 MHz steps are derived from a 3 MHz reference source. The output of the 3 MHz source is first multiplied by 10, producing a 30 MHz signal. The reference is also divided by three, to produce a 1 MHz signal. Four additional multipliers produce 2 MHz, 3 MHz, 4 MHz, and 5 MHz signals. One of these five frequencies is selected by switch, and mixed with the 30 MHz signal, to produce 10 possible sum and difference products from 25 to 35 MHz. The mixer also passes part of the 30 MHz input signal, and produces higher order mixing products. A tunable bandpass filter is used

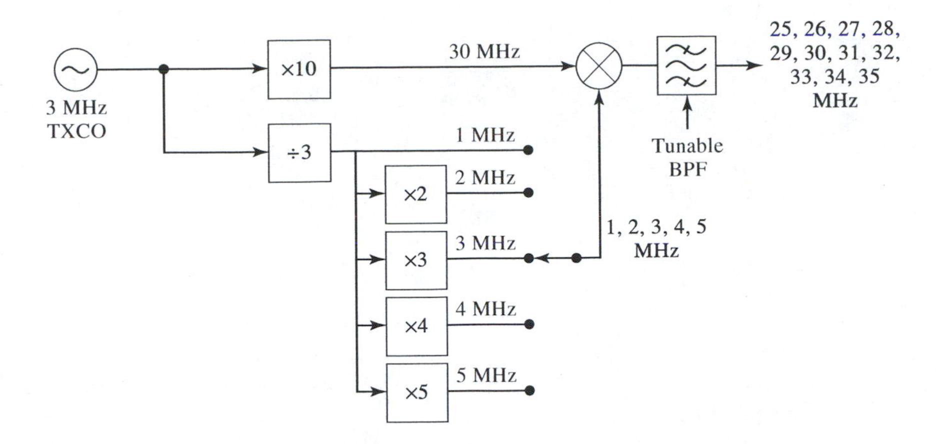

FIGURE 8.14 Example of direct frequency synthesis.

to select the desired output frequency. In order to have reasonable filter characteristics, the ratio of the two frequencies applied to the mixer should generally be less than 100.

A method of direct synthesis commonly used in commercial frequency synthesizers for test and measurement is the *double-mix-divide* circuit, shown in Figure 8.15. Here, an input frequency f_0 is mixed with the frequency f_1, and the upper sideband is selected by filtering to produce $f_0 + f_1$. This signal is then mixed with a signal at the frequency $f_2 + f^*$, where f^* is one of ten switch-selectable frequencies. Again the upper sideband is selected by filtering, and the output is divided by ten. If the frequencies f_1 and f_2 are selected so that

$$f_0 + f_1 + f_2 = 10 f_0, \tag{8.38}$$

then the output of the divider will be

$$f_{\text{out}} = \frac{f_0 + f_1 + f_2 + f^*}{10} = f_0 + \frac{f^*}{10}. \tag{8.39}$$

This method thus allows the input frequency to be incremented by $f^*/10$. This technique can be cascaded to achieve as many digits of frequency resolution as desired.

Digital Look-up Synthesis

In digital look-up synthesis, or *direct digital synthesis*, a digital-to-analog converter uses the output from a digital sine look-up table to generate a sinusoidal waveform. The block diagram for this type of synthesis is shown in Figure 8.16, and consists of a phase accumulator (an adder), a sine look-up table (read-only memory), a digital-to-analog converter (DAC), and a low-pass filter. Direct digital synthesis is extremely accurate and, because it is essentially a frequency dividing method, can provide frequency resolutions down to

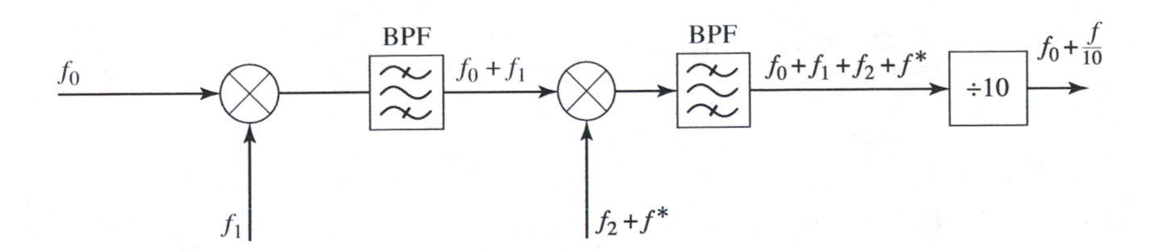

FIGURE 8.15 Direct frequency synthesis using the double-mix-divide method.

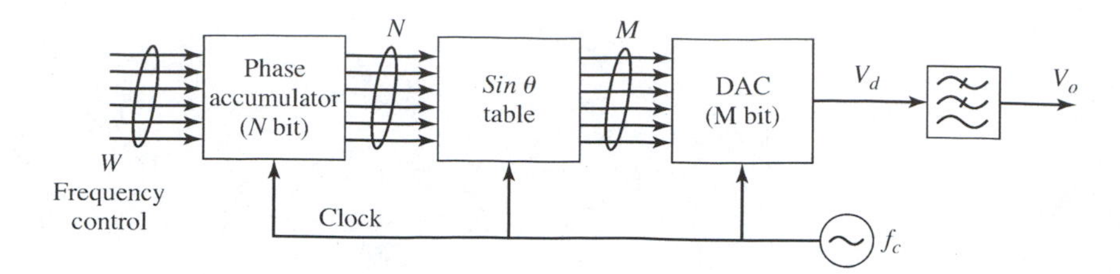

FIGURE 8.16 Block diagram of the direct digital synthesis technique.

small fractions of a Hertz. The main limitation of the method is that its upper frequency is typically limited to a few hundred MHz, primarily because of the speed limitations of available DACs.

At the lowest frequency of operation, the frequency control word W at the input to the phase accumulator is set to unity, so that the phase accumulator increments by one at each clock cycle. The sine look-up table stores 2^N uniformly spaced values of the sine function, and provides a digital output that is equal to the sine of the linearly increasing phase data from the accumulator. The DAC converts the digital values to a discretized sine waveform, which is low-pass filtered to provide an analog sine wave output. If the frequency control word is increased to a value $W > 1$, then the phase accumulator output will increment by W at each clock cycle, and every Wth value from the sine table will be sent to the DAC. This will result in an output that is W times higher in frequency. The cutoff frequency of the low-pass filter should be slightly greater than the highest frequency of operation, in order to suppress the clock frequency and its harmonics. The waveforms associated with this process are shown in Figure 8.17.

The output waveform can be expressed as

$$V_o = A \sin\left(\frac{2\pi f_c W t}{2^N}\right), \tag{8.40}$$

where f_c is the clock frequency. By the Nyquist sampling theorem a sine wave is uniquely determined if more than two samples are provided for each period of the waveform. In practice, direct digital synthesis usually employs a minimum of four samples per period, in which case the minimum clock frequency should be equal to four times the maximum output frequency of the synthesizer. Thus the maximum frequency of operation is given by

$$f_{\max} = \frac{f_c}{4}, \tag{8.41}$$

where f_c is the clock frequency. If N bits are used to address the sine look-up table containing

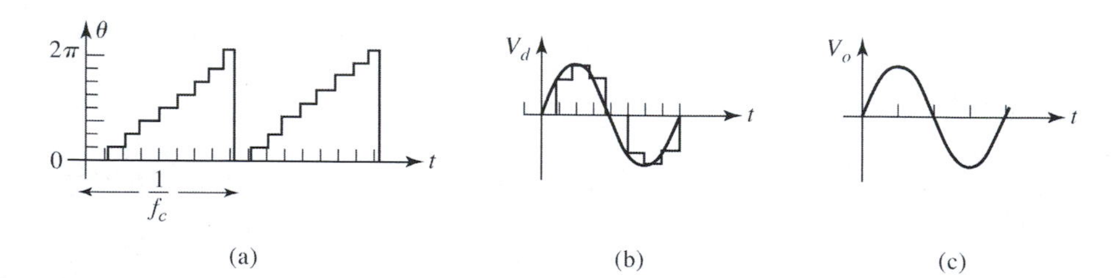

FIGURE 8.17 Waveforms in a direct digital frequency synthesizer. (a) Discretized linear phase values at output of phase accumulator. (b) Discretized sine waveform at output of digital to analog converter. (c) Sine wave output after low-pass filtering.

2^N sample values, then the minimum frequency obtainable from the synthesizer is

$$f_{\min} = \frac{f_c}{2^N}. \tag{8.42}$$

Because the output of the DAC is a discretized version of the analog waveform, the deterministic discretization error appears as noise, and in the worst case the associated noise power is given by

$$P_n = \frac{1}{2^{N-1}} = -6(M-1)\,\text{dB} \tag{8.43}$$

So, for example, if it is desired to have a total spurious noise less than 80 dB below the carrier, we should have $M \geq 15$. Some of this noise power will be suppressed by the output filter.

Phase-Locked Loops

A phase-locked loop (PLL) uses a feedback control circuit to allow a voltage-controlled oscillator to precisely track the phase of a stable reference oscillator, with the important feature that the output oscillator can be made to run at a multiple of the reference oscillator frequency. Phase-locked loops are used as FM demodulators, in carrier recovery circuits, and as frequency synthesizers for modulation and demodulation. Phase-locked loops have very good frequency accuracy and phase noise characteristics, but suffer from the fact that settling times (between changes in frequency) can be long.

The basic circuit of a phase-locked loop is shown in Figure 8.18. It consists of a reference oscillator, a phase detector that produces an output voltage proportional to the difference in phase of the inputs, a loop amplifier and filter, a voltage-controlled oscillator (VCO) operating at the desired output frequency, and a frequency divider. In operation, the output of the VCO is divided by N to match the frequency of the reference oscillator. The phase detector produces a voltage proportional to the difference in phase of these two signals, and is used to make small corrections in the frequency of the VCO in order to align the phase of the VCO with that of the reference source. The output of the phase-locked loop thus has a phase noise characteristic similar to that of the reference source, but operates at a higher frequency. If a programmable frequency divider is used, it is possible to synthesize a large number of closely spaced frequencies with a relatively simple circuit. This makes the phase-locked loop very useful for commercial wireless applications, especially those involving multiple channels. Phase-locked loops can be implemented in either digital or analog form, but we will only discuss analog PLLs because they are the only type capable of operating at RF and microwave frequencies.

There are several characteristics of phase-locked loops that are important in practice. The *capture range* is the range of input frequency for which the loop can acquire locking.

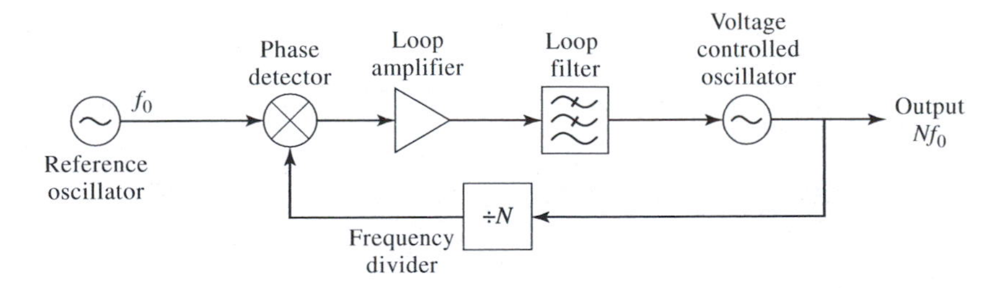

FIGURE 8.18 Block diagram of a phase-locked loop.

The *lock range* is the input frequency range over which the loop will remain locked; this is typically larger than the capture range. The *settling time* is the time required for the loop to lock on to a new frequency.

Practical Synthesizer Circuits

Practical RF and microwave frequency synthesizers for wireless systems often use variations and combinations of the above three basic synthesis methods. Frequently a phase-locked loop with a programmable divider, or a direct digital synthesizer, is used to obtain many closely spaced frequencies, followed by direct analog methods for up-conversion to the desired carrier or local oscillator frequency. While there are many possible circuits that can be studied, we consider one example here to illustrate the design of a practical synthesizer.

The AMPS cellular system requires a local oscillator in the 800 MHz band to receive one of several hundred voice channels having 30 kHz spacing. Using a standard phase-locked loop would require a reference source operating at 30 kHz and a VCO operating near 870 MHz, with a programmable divider providing a division ratio of more than 24,000. This would be impractical because of the large number of addresses required, as well as the high frequency at which the divider would have to operate. Instead, a phase-locked loop supplemented with a mixer and frequency multiplier is used [6], as shown in Figure 8.19.

In this synthesizer the VCO operates at the frequency f_0, which ranges from 217.5 to 222.5 MHz. The VCO output is frequency multiplied by four to achieve the desired synthesizer output in the range of 870 MHz. Part of the VCO output is mixed with a fixed reference crystal oscillator at $f_1 = 228.02250$ MHz. The filtered difference frequency of 6 to 11 MHz is low enough to be digitally divided with an inexpensive programmable counter. The division ratio is selected with a 10-bit address to lie between $737 \leq N \leq 1402$, according to the desired channel. The output of the divider is compared to a stable 7.5 kHz oscillator, f_2, and the phase error is used to control the VCO. When the loop is in lock, the output frequency is $f_{out} = 4(f_1 - Nf_2)$. Thus the output can be stepped in increments of $4f_2 = 30$ kHz. The stability of the output is set by the stability of the reference sources f_1 and f_2.

If it is desired to produce an output frequency of $f_{out} = 870.180$ MHz, for example, then we solve the equation

$$870.180 \text{ MHz} = f_{out} = 4(f_1 - Nf_2) = 4[228.02250 - N(0.0075)]$$

for N. This yields $N = 1397$, which is the required setting of the programmable divider.

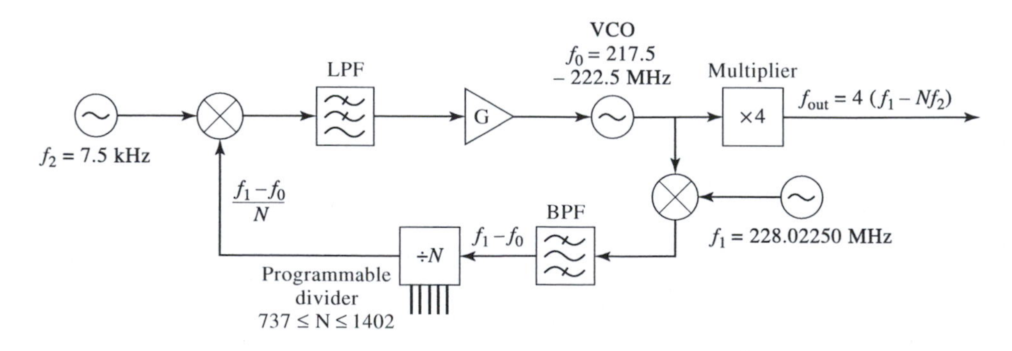

FIGURE 8.19 Block diagram of a frequency synthesizer used for AMPS service.

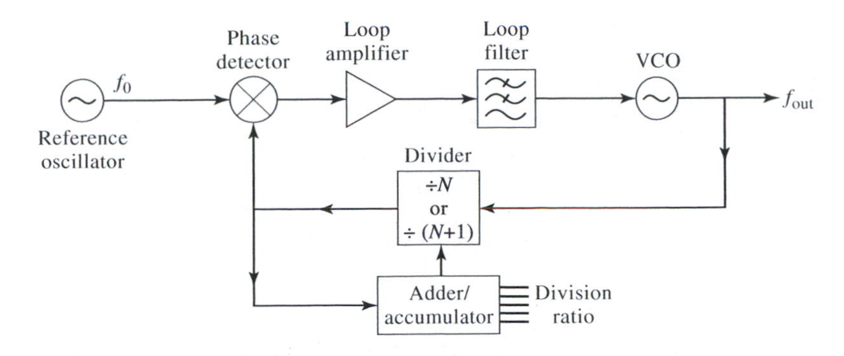

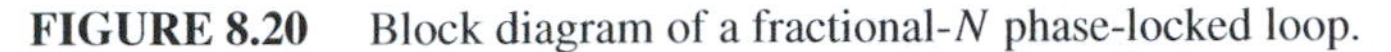

FIGURE 8.20 Block diagram of a fractional-N phase-locked loop.

Fractional-N Phase-Locked Loops

The conventional phase-locked loop shown in Figure 8.18 uses a frequency divider with an integral division ratio. This implies that the VCO output frequency must be an integral multiple of the reference oscillator frequency, and that the output frequency step size will be equal to the frequency of the reference oscillator. If the step size is very small, the division ratio will be large, and the loop settling time may be very long. Alternatively, if the division ratio can be effectively made to be a nonintegral value, then the VCO output frequency can operate at an integer-plus-fractional multiple of the reference oscillator frequency, rather than just integer multiples. This is called a *fractional-N phase-locked loop*.

There are several ways in which fractional-N synthesis can be achieved [3], but the basic conceptual block diagram of a fractional-N loop is shown in Figure 8.20. Digital dividers inherently provide integer division ratios, but a nonintegral division ratio between N and $N+1$ (where N is an integer) can be effectively obtained by dividing by N for a certain number of cycles, and dividing by $N+1$ for another number of cycles. Thus the frequency divider is constructed to switch between division ratios of N and $N+1$, under control of an adder that counts the number of cycles. In particular, if the VCO output frequency is divided by $N+1$ every R cycles, and divided by N for intervening cycles, then the average output frequency is given by

$$f_{\text{out}} = \left(N + \frac{1}{R}\right) f_0, \tag{8.44}$$

where f_0 is the frequency of the reference oscillator, N is an integer, and $1/R$ is the fractional part of the division ratio. For example, if $f_0 = 100$ MHz, and it is desired to generate an output frequency $f_{\text{out}} = 835$ MHz, then the required division ratio is 8.35. Thus $N=8$ and $1/R=0.35$, or $R=2.86$. Then the VCO output should be divided by $N+1=9$ every $R=2.86$ cycles, and divided by $N=8$ otherwise. Equivalently, every set of 100 cycles should be divided $100/2.86=35$ times by 9, and divided by 8 otherwise. This procedure can be implemented by adding 0.35 to the adder/accumulator for each cycle, and switching the divider from divide by 8 to divide by 9 whenever the accumulator overflows to unity or greater. The accumulator should retain only the fractional part of the summation, discarding the integer portion.

8.4 PHASE-LOCKED LOOP ANALYSIS

Because of the importance of phase-locked loops in modern wireless systems, we will focus in this section on an analysis of its basic operation. We begin with a discussion of the

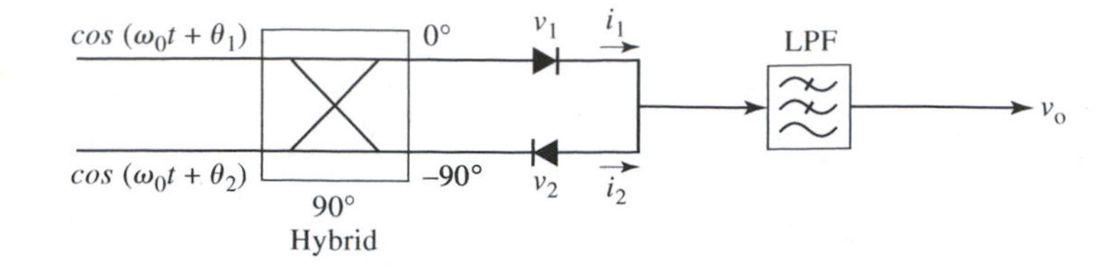

FIGURE 8.21 Circuit diagram for an analog phase detector.

phase detector, which is a key component in analog PLLs. A rigorous analysis of a phase-locked loop requires nonlinear analysis, which is best performed by computer, but we can gain much understanding by considering an approximate linearized model of the loop. The performance of the phase-locked loop will be seen to be dependent on the response of the loop filter. Our analysis will rely on the use of Laplace transforms; Appendix C lists several transform relationships that will be useful.

Phase Detectors

A *phase detector* provides an output voltage that is dependent on the phase difference between two input signals. As shown in Figure 8.21, the design of an analog phase detector is very similar to the balanced mixer described in Section 7.4. Two input signals of nominally the same frequency (ω_0), but different phases (θ_1 and θ_2), are applied to the input ports of a 90° hybrid coupler. The output voltages developed across the mixer diodes can be written as

$$\begin{aligned} v_1(t) &= \cos(\omega_0 t + \theta_1) + \cos(\omega_0 t + \theta_2 - 90^\circ) \\ &= \cos(\omega_0 t + \theta_1) - \sin(\omega_0 t + \theta_2) \end{aligned} \tag{8.45a}$$

$$\begin{aligned} v_2(t) &= \cos(\omega_0 t + \theta_2) + \cos(\omega_0 t + \theta_1 - 90^\circ) \\ &= \cos(\omega_0 t + \theta_2) - \sin(\omega_0 t + \theta_1) \end{aligned} \tag{8.45b}$$

If we assume a square-law response for the mixer diodes, and retain only the quadratic terms, the diode currents can be written as

$$\begin{aligned} i_1(t) &= K v_1^2(t) \\ &= K[\cos^2(\omega_0 t + \theta_1) - 2\cos(\omega_0 t + \theta_1)\sin(\omega_0 t + \theta_2) + \sin^2(\omega_0 t + \theta_2)] \end{aligned} \tag{8.46a}$$

$$\begin{aligned} i_2(t) &= -K v_2^2(t) \\ &= -K[\cos^2(\omega_0 t + \theta_2) - 2\cos(\omega_0 t + \theta_2)\sin(\omega_0 t + \theta_1) + \sin^2(\omega_0 t + \theta_1)] \end{aligned} \tag{8.46b}$$

The negative sign on i_2 accounts for the reversed diode polarity. After combining the diode currents and low-pass filtering, the output voltage can be expressed as

$$\begin{aligned} v_o(t) &= i_1(t) + i_2(t)|_{LPF} = K_d \sin(\theta_1 - \theta_2) \\ &\approx K_d(\theta_1 - \theta_2) \end{aligned} \tag{8.47}$$

This result shows that the output voltage of the phase detector is proportional to the sine of the difference in phase of the two input signals. If this difference is small, then the sine function can be approximated by its argument, so that the phase detector output is proportional to the phase difference. This is referred to as the *linearized* phase detector model. The constant K_d is the *phase detector gain factor*, and accounts for the diode square-law constants and current-to-voltage conversion. It has dimensions of volts/radian.

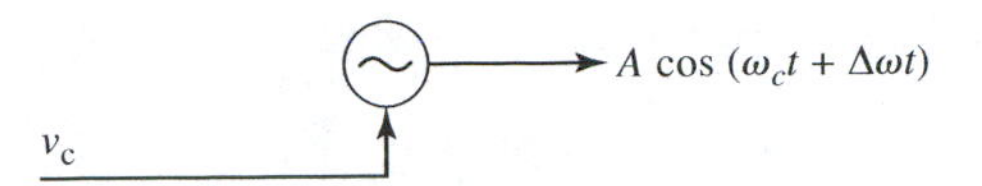

FIGURE 8.22 A voltage-controlled oscillator.

Transfer Function for the Voltage-Controlled Oscillator

In order to analyze a phase-locked loop system, we must model the voltage-controlled oscillator in terms of the transfer function between its control voltage and the phase of the output waveform. We can assume that the VCO has an output frequency ω_0 that is offset from its free-running frequency, ω_c, by an increment $\Delta\omega$:

$$\omega_0 = \omega_c + \Delta\omega = \omega_c + K_0 v_c, \tag{8.48}$$

where the offset frequency is controlled by the control voltage v_c applied to the VCO. The constant K_0 is the *VCO gain factor*, and has dimensions of Hz/V. Figure 8.22 illustrates this VCO model.

We define the phase of the offset frequency of the VCO as

$$\theta_o(t) = \Delta\omega t = K_0 v_c t. \tag{8.49}$$

Writing frequency as the time derivative of phase then gives

$$\frac{d\theta_o(t)}{dt} = \Delta\omega = K_0 v_c, \tag{8.50}$$

which can be inverted to express the output phase in terms of the control voltage as

$$\theta_o(t) = K_0 \int_{t'=0}^{t} v_c(t')\, dt' \tag{8.51}$$

In the Laplace transform domain (see Appendix C) this can be written as

$$\Theta_o(s) = \frac{K_0}{s} V_c(s), \tag{8.52}$$

which is the desired transfer function for the VCO.

Analysis of Linearized Phase-Locked Loop

With the above preliminaries we can now analyze the linearized analog phase-locked loop shown in Figure 8.23. We assume a reference input voltage given by

$$v_i(t) = \cos(\omega_0 t + \theta_i), \tag{8.53}$$

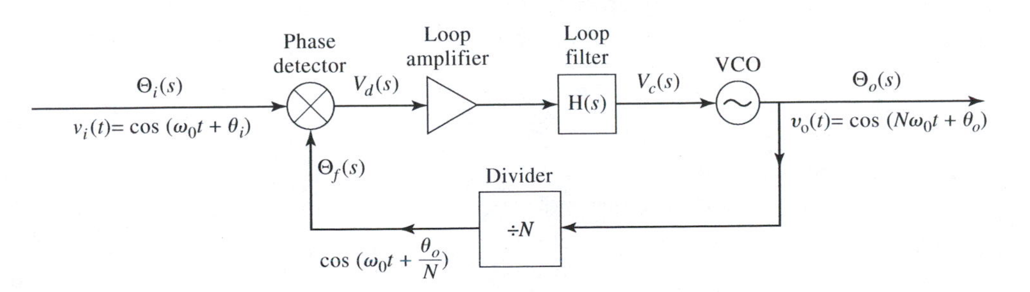

FIGURE 8.23 Analysis of a linearized phase-locked loop.

where θ_i is the phase of the input waveform. The output voltage of the VCO can be written as

$$v_o(t) = \cos(N\omega_0 t + \theta_o), \tag{8.54}$$

where θ_o is the phase of the output waveform. Note that the output frequency is N times the input (reference) frequency, due to the use of the frequency divider in the loop feedback path.

The phase detector output voltage given in (8.47) can be expressed in the Laplace transform domain as

$$V_d(s) = K_d[\Theta_i(s) - \Theta_f(s)], \tag{8.55}$$

where $\Theta_i(s)$ and $\Theta_o(s)$ are the Laplace transforms of $\theta_i(t)$ and $\theta_o(t)$, respectively. Since the divider divides the frequency by N, and phase is the derivative of frequency, the phase will also be divided by N. So the relation between the feedback phase, θ_f, and the output phase, θ_o, is

$$\Theta_f(s) = \frac{1}{N}\Theta_o(s). \tag{8.56}$$

The control voltage, $V_c(s)$, applied to the VCO is

$$V_c(s) = H(s)V_d(s), \tag{8.57}$$

where $H(s)$ is the transfer function of the loop filter. The effect of the loop amplifier can be assumed to be included in $H(s)$. The output phase is given in terms of the VCO control voltage by (8.52).

Combining (8.52), (8.55), (8.56), and (8.57) allows us to solve for the transfer function of the input and output phases as

$$\frac{\Theta_o(s)}{\Theta_i(s)} = \frac{K_0 K_d H(s)}{s + \dfrac{K_0 K_d}{N} H(s)} = \frac{NKH(s)}{s + KH(s)}, \tag{8.58}$$

where we have defined a new constant $K = K_0 K_d / N$. The VCO control voltage is then found as

$$V_c(s) = \frac{s}{K_0}\Theta_o(s) = \frac{sNKH(s)}{K_0[s + KH(s)]}\Theta_i(s) = \frac{sK_d H(s)}{s + KH(s)}\Theta_i(s). \tag{8.59}$$

We can define the *loop phase error* as

$$\varepsilon(s) = \Theta_i(s) - \Theta_f(s) = \left[1 - \frac{\Theta_o(s)}{N\Theta_i(s)}\right]\Theta_i(s) = \left[\frac{s}{s + KH(s)}\right]\Theta_i(s). \tag{8.60}$$

The above results can be used to find the output phase, the loop phase error, and the output frequency, once the input phase function is known. Depending on the application, the input phase function is often a step or a ramp function. The former occurs when the loop is used for phase synchronization, while the ramp corresponds to a step change in frequency. Here we limit our discussion to a step change $\Delta\omega$ in frequency, so that the input voltage is

$$v_i(t) = \cos(\omega_0 t + \theta_i) = \cos(\omega_0 t + \Delta\omega t). \tag{8.61}$$

The input phase function is then $\theta_i(t) = \Delta\omega t U(t)$, where $U(t)$ is the unit step function. From Appendix C the Laplace transform of $\theta_i(t)$ is

$$\Theta_i(s) = \frac{\Delta\omega}{s^2}. \tag{8.62}$$

We now consider two special cases of loop filters.

First-Order Loop

First consider the simplest case of a PLL with no loop filter. Then $H(s)=1$, and the VCO control voltage given in (8.59), for a step change in frequency, reduces to

$$V_c(s) = \frac{K_d \Delta\omega}{s(s+K)} = \frac{\Delta\omega K_d}{K}\left(\frac{1}{s} - \frac{1}{s+K}\right). \tag{8.63}$$

The last step in (8.63) is the result of applying a partial fraction expansion. Since the form of (8.63) involves poles of only first order, this is called a *first-order loop*.

Taking the inverse transform of (8.63) by using the table in Appendix C gives the following time-domain expression for the control voltage:

$$v_c(t) = \frac{\Delta\omega K_d}{K}(1 - e^{-Kt})U(t) \tag{8.64}$$

This result shows how the VCO control voltage varies in response to a step change in the input frequency. At $t=0$ $v_c(t)=0$, so by (8.48) the output frequency is $\omega_0=\omega_c$, the free-running VCO frequency. In the limit as $t \to \infty$ the control voltage exponentially converges to

$$v_c(\infty) = \frac{\Delta\omega K_d}{K} = \frac{\Delta\omega N}{K_0}. \tag{8.65}$$

Then the output voltage of the VCO is, after locking, given by (8.54) and (8.49):

$$v_o(t) = \cos(N\omega_0 t + \theta_o) = \cos N(\omega_0 + \Delta\omega)t, \tag{8.66}$$

which shows that the output frequency has tracked the input step in frequency, and is multiplied by N.

From (8.60) the loop phase error is

$$\varepsilon(s) = \frac{\Delta\omega}{s(s+K)} = \frac{\Delta\omega}{K}\left(\frac{1}{s} - \frac{1}{s+K}\right), \tag{8.67}$$

with the following time-domain expression:

$$\varepsilon(t) = \frac{\Delta\omega}{K}(1 - e^{-Kt})U(t). \tag{8.68}$$

Since $\varepsilon(\infty) = \Delta\omega/K \neq 0$, the loop is said to have a *static error*, even after locking has taken place.

The time required for the output of the PLL to respond to the step change in input frequency is called the *acquisition time*, and can be estimated as the time constant of the exponential in (8.64) or (8.68):

$$T_a = \frac{1}{K} = \frac{N}{K_0 K_d}. \tag{8.69}$$

These results show that both the static error and the acquisition time decrease as the loop gain, $K_0 K_d$, increases. Unfortunately, the 3 dB loop bandwidth, which is approximately equal to $1/T_a$, increases with the loop gain—this can lead to instability and increased phase noise. For this reason first-order loops are seldom used in practice.

EXAMPLE 8.5 RESPONSE OF A FIRST-ORDER PHASE-LOCKED LOOP

Typical values for a phase detector gain and the gain factor of a 200 MHz VCO are $K_d=2$ V/rad and $K_0 = 2$ MHz/V. If the reference frequency is 20 MHz, find the acquisition time and loop bandwidth for a first-order phase-locked loop.

Solution

For an output frequency of 200 MHz and an input frequency of 20 MHz, the value of the feedback divider must be $N = 10$. Then from (8.69) the acquisition time is

$$T_a = \frac{N}{K_0 K_d} = \frac{10}{(2 \times 10^6)(2)} = 2.5\ \mu S.$$

The loop bandwidth is approximately

$$B = \frac{1}{T_a} = \frac{1}{2.5 \times 10^{-6}} = 400 \text{ kHz.}$$ ○

Second-Order Loop

Next consider the use of a low-pass loop filter, with a transfer function given by

$$H(s) = \frac{1}{1 + s/\omega_f} = \frac{\omega_f}{s + \omega_f}, \tag{8.70}$$

where ω_f is the low-pass cutoff frequency. From (8.59) the Laplace transform of the control voltage, again with a step change in frequency, is

$$V_c(s) = \frac{K_d \omega_f \Delta\omega}{s(s + \omega_f)\left(s + \dfrac{K\omega_f}{s + \omega_f}\right)} = \frac{K_d \omega_f \Delta\omega}{s(s^2 + \omega_f s + K\omega_f)}. \tag{8.71}$$

Now define the variables

$$\omega_n^2 = K\omega_f,$$

and

$$2\zeta = \frac{\omega_n}{K} = \sqrt{\frac{\omega_f}{K}}.$$

Then (8.71) can be rewritten and expanded as

$$V_c(s) = \frac{K_d \omega_f \Delta\omega}{s\left(s^2 + 2\zeta\omega_n s + \omega_n^2\right)} = \frac{K_d \omega_f \Delta\omega}{\omega_n^2}\left[\frac{1}{s} - \frac{s + 2\zeta\omega_n}{s^2 + 2\zeta\omega_n s + \omega_n^2}\right]. \tag{8.72}$$

Using results from Appendix C gives the time domain form of the VCO control voltage as

$$v_c(t) = \frac{\Delta\omega N}{K_0}\left\{1 - \left[\cos\sqrt{1-\zeta^2}\,\omega_n t + \frac{\zeta}{\sqrt{1-\zeta^2}}\sin\sqrt{1-\zeta^2}\,\omega_n t\right]e^{-\zeta\omega_n t}\right\}U(t). \tag{8.73}$$

Since the transfer function now has a second-order pole, the loop is identified as a *second-order PLL*. The variable ω_n is seen to be the natural frequency of the second-order response, with ζ the damping constant. At $t = 0$ we see that $v_c = 0$, while for $t \to \infty$ the VCO control voltage converges to

$$v_c(\infty) = \frac{\Delta\omega N}{K_0}, \tag{8.74}$$

so again the output voltage of the VCO is

$$v_o(t) = \cos(N\omega_0 t + \theta_o) = \cos N(\omega_0 + \Delta\omega)t, \tag{8.75}$$

as desired. From (8.60) the Laplace transform of the loop phase error is

$$\varepsilon(s) = \frac{\Delta\omega(s+\omega_f)}{s(s^2+\omega_f s+K\omega_f)} = \frac{\Delta\omega(s+2\zeta\omega_n)}{s\left(s^2+2\zeta\omega_n s+\omega_n^2\right)}. \tag{8.76}$$

This can be converted to a time-domain expression, or the final value theorem for Laplace transforms can be used to show that the loop phase error for large t is

$$\varepsilon(t\to\infty) = \lim_{s\to 0} s\varepsilon(s) = \frac{\omega_f\Delta\omega}{K\omega_f} = \frac{\Delta\omega N}{K_0 K_d}, \tag{8.77}$$

indicating that the second-order loop also has a static phase error.

From these results it can be shown [2] that the 3 dB bandwidth of the loop is

$$B = \omega_n\left[1-2\zeta^2+\sqrt{2-4\zeta^2+4\zeta^4}\right]^{1/2}, \tag{8.78}$$

and the acquisition time is approximately

$$T_a = \frac{2.2}{B}. \tag{8.79}$$

The response of the second-order loop can be illustrated by considering the magnitude of the transfer function versus frequency. Using (8.70) in (8.58) with $s = j\omega$ gives the magnitude of the phase transfer function as

$$\left|\frac{\Theta_o(j\omega)}{\Theta_i(j\omega)}\right| = \frac{N}{\sqrt{(1-\omega/\omega_n)^2+(2\zeta\omega/\omega_n)^2}}. \tag{8.80}$$

This function is plotted in Figure 8.24, for $N = 1$ and various values of the damping constant ζ. Critical damping occurs for $\zeta = 0.707$.

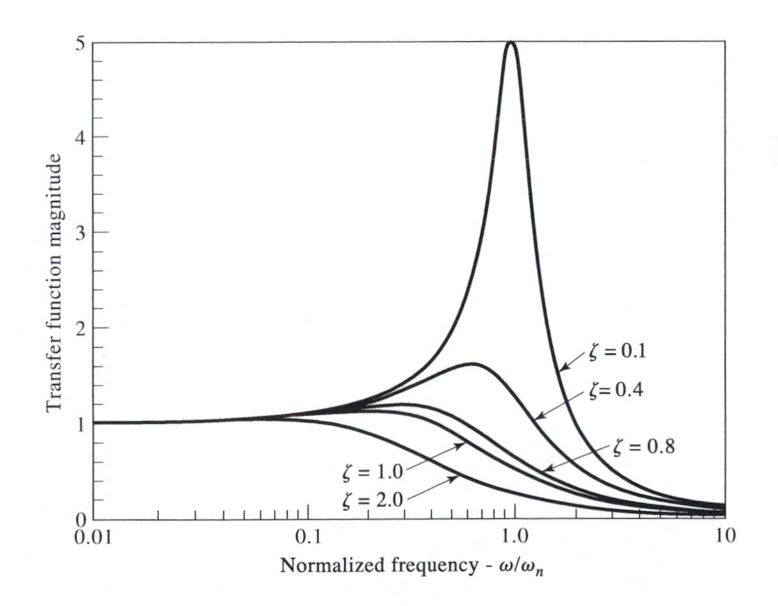

FIGURE 8.24 Magnitude of the phase transfer function of (8.80) for a second-order phase-locked loop ($N = 1$).

EXAMPLE 8.6 RESPONSE OF A SECOND-ORDER PHASE-LOCKED LOOP

A second-order PLL uses a low-pass filter with a cutoff frequency of $\omega_f/2\pi =$ 800 kHz. As in Example 8.5, the 200 MHz VCO has a gain factor of $K_d = 2$ V/rad and the phase detector has a gain factor $K_0 = 2$ MHz/V. If the reference frequency is 20 MHz, find the loop bandwidth and acquisition time.

Solution

As in Example 8.5, the feedback frequency divider ratio is 10. So we can calculate the natural frequency and damping constant as

$$K = \frac{K_0 K_d}{N} = \frac{(2 \times 10^6)(2)}{10} = 4 \times 10^5 \text{ Hz}$$

$$\omega_n = \sqrt{K\omega_f} = \sqrt{(2\pi)(4 \times 10^5)(2\pi)(800 \times 10^3)} = 3.55 \times 10^6 \text{ rad/S}$$

$$\zeta = \frac{\omega_n}{2K} = \frac{3.55 \times 10^6}{2(2\pi)(4 \times 10^5)} = 0.707$$

Note that we must be careful with conversions between Hertz and radians/second.

From (8.78) the loop bandwidth is

$$B = \omega_n\left[1 - 2\zeta^2 + \sqrt{2 - 4\zeta^2 + 4\zeta^4}\right]^{1/2}$$

$$= \frac{3.55 \times 10^6}{2\pi}\left[1 - 2(0.707)^2 + \sqrt{2 - 4(0.707)^2 + 4(0.707)^4}\right]^{1/2}$$

$$= 3.55 \times 10^6 \text{ rad/sec} = 565 \text{ kHz}$$

and from (8.79) the acquisition time is

$$T_a = \frac{2.2}{B} = \frac{2.2}{3.55 \times 10^6} = 0.62\ \mu S.$$

This is much faster than the acquisition time of 2.5 μS that was determined for the first-order loop of Example 8.5. ○

8.5 OSCILLATOR PHASE NOISE

The noise produced by a frequency synthesizer or oscillator is critically important in practice because it may severely degrade the performance of a wireless system. Besides adding to the noise level of the receiver, a noisy local oscillator will lead to down-conversion of undesired nearby signals, thus limiting the selectivity of the receiver and how closely adjacent channels may be spaced. *Phase noise* refers to the short-term random fluctuation in the frequency (or phase) of an oscillator signal. Phase noise also introduces uncertainty during the detection of digitally modulated signals, especially in the case of PSK or QAM modulation. Sources of noise in a phase-locked loop include leakage of the reference signal through the loop filter and VCO, harmonics of the VCO, and spurious frequencies generated by the phase detector. In addition, broadband noise generated by the active circuitry of the VCO sets a minimum noise floor at the output of the synthesizer. Probably most important, however, is the fact that noise within the passband of the oscillator feedback circuit will be amplified substantially, and therefore become the dominant source of phase noise.

An ideal oscillator would have a frequency spectrum consisting of a single delta function at its operating frequency, but a realistic oscillator will have a spectrum more like that shown in Figure 8.25. Spurious signals due to oscillator harmonics or mixer products appear as

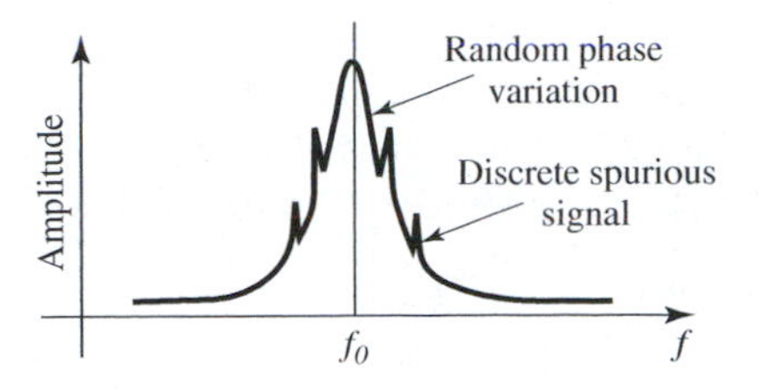

FIGURE 8.25 Output spectrum of a typical oscillator or frequency synthesizer.

discrete spikes in the spectrum. Phase noise, due to random fluctuations caused by thermal and other noise sources, appears as a broad continuous distribution localized about the output signal. Phase noise is defined as the ratio of power in one phase modulation sideband to the total signal power per unit bandwidth (one Hertz) at a given offset, f_m, from the signal frequency, and is denoted as $\mathcal{L}(f_m)$. It is usually expressed in decibels relative to the carrier power per Hertz of bandwidth (dBc/Hz). A typical oscillator phase noise specification for an FM cellular radio, for example, may be –110 dBc/Hz at 25 kHz from the carrier. In the following sections we show how phase noise may be represented, and present a widely used model for characterizing the phase noise of an oscillator.

Representation of Phase Noise

In general, the output voltage of an oscillator or synthesizer can be written as

$$v_o(t) = V_0[1 + A(t)]\cos[\omega_0 t + \theta(t)], \tag{8.81}$$

where $A(t)$ represents the amplitude fluctuations of the output, and $\theta(t)$ represents the phase variation of the output waveform. Of these, amplitude variations can usually be well controlled, and generally have less impact on system performance. Phase variations may be discrete (due to spurious mixer products or harmonics), or random in nature (due to thermal or other random noise sources). Note from (8.81) that an instantaneous phase variation is indistinguishable from a variation in frequency.

Small changes in the oscillator frequency can be represented as a frequency modulation of the carrier by letting

$$\theta(t) = \frac{\Delta f}{f_m}\sin\omega_m t = \theta_p \sin\omega_m t, \tag{8.82}$$

where $f_m = \omega_m/2\pi$ is the modulating frequency (see Chapter 9 for a more detailed discussion of frequency modulation). The peak phase deviation is $\theta_p = \Delta f/f_m$ (also called the modulation index). Substituting (8.82) into (8.81) and expanding gives

$$v_o(t) = V_0[\cos\omega_0 t\cos(\theta_p\sin\omega_m t) - \sin\omega_0 t\sin(\theta_p\sin\omega_m t)], \tag{8.83}$$

where we set $A(t) = 0$ to ignore amplitude fluctuations. Assuming the phase deviations are small, so that $\theta_p \ll 1$, the small argument expressions that $\sin x \cong x$ and $\cos x \cong 1$ can be used to simplify (8.83) to

$$\begin{aligned} v_o(t) &= V_0[\cos\omega_0 t - \theta_p\sin\omega_m t\sin\omega_0 t] \\ &= V_0\left\{\cos\omega_0 t - \frac{\theta_p}{2}[\cos(\omega_0+\omega_m)t - \cos(\omega_0-\omega_m)t]\right\} \end{aligned} \tag{8.84}$$

This expression shows that small phase or frequency deviations in the output of an oscillator result in modulation sidebands at $\omega_0 \pm \omega_m$, located on either side of the carrier signal at

ω_0. When these deviations are due to random changes in temperature or device noise, the output spectrum of the oscillator will take the form shown in Figure 8.25.

According to the definition of phase noise as the ratio of noise power in a single sideband to the carrier power, the waveform of (8.84) has a corresponding phase noise of

$$\mathcal{L}(f) = \frac{P_n}{P_c} = \frac{\frac{1}{2}\left(\frac{V_0\theta_p}{2}\right)^2}{\frac{1}{2}V_0^2} = \frac{\theta_p^2}{4} = \frac{\theta_{\mathrm{rms}}^2}{2}, \tag{8.85}$$

where $\theta_{\mathrm{rms}} = \theta_p/\sqrt{2}$ is the rms value of the phase deviation. The two-sided power spectral density associated with phase noise includes power in both sidebands:

$$S_\theta(f_m) = 2\mathcal{L}(f_m) = \frac{\theta_p^2}{2} = \theta_{\mathrm{rms}}^2. \tag{8.86}$$

White noise generated by passive or active devices can be interpreted in terms of phase noise by using the same definition. From Chapter 3 we know that the noise power at the output of a noisy two-port network is kT_0BFG, where $T_0 = 290$ K, B is the measurement bandwidth, F is the noise figure of the network, and G is the gain of the network. For a 1 Hertz bandwidth, the ratio of output noise power density to output signal power gives the power spectral density as

$$S_\theta(f_m) = \frac{kT_0F}{P_c}, \tag{8.87}$$

where P_c is the input signal power (note that the gain of the network cancels in this expression).

Leeson's Model for Oscillator Phase Noise

In this section we present Leeson's model for characterizing the power spectral density of oscillator phase noise [2], [7]. As in Section 8.1, we will model the oscillator as an amplifier with a feedback path, as shown in Figure 8.26. If the voltage gain of the amplifier is included in the feedback transfer function $H(\omega)$, then the voltage transfer function for the oscillator circuit is

$$V_o(\omega) = \frac{V_i(\omega)}{1 - H(\omega)}. \tag{8.88}$$

If we consider oscillators that use a high-Q resonant circuit in the feedback loop (e.g., Colpitts, Hartley, Clapp, and similar oscillators), then $H(\omega)$ can be represented as the

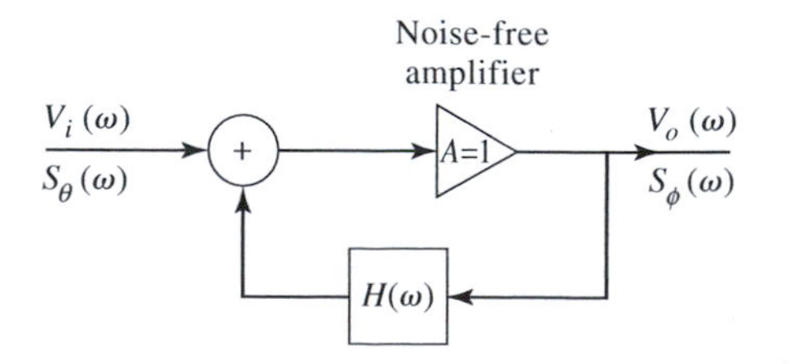

FIGURE 8.26 Feedback amplifier model for characterizing oscillator phase noise.

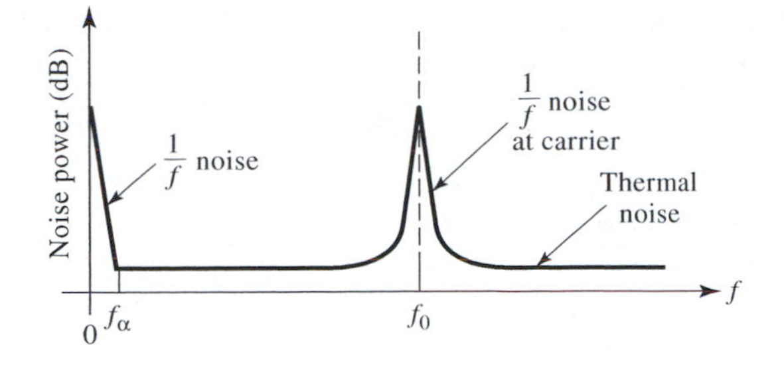

FIGURE 8.27 Noise power versus frequency for an amplifier with an applied input signal.

voltage transfer function of a parallel *RLC* resonator [4]:

$$H(\omega) = \frac{1}{1 + jQ\left(\dfrac{\omega}{\omega_0} - \dfrac{\omega_0}{\omega}\right)} = \frac{1}{1 + 2jQ\Delta\omega/\omega_0}, \tag{8.89}$$

where ω_0 is the resonant frequency of the oscillator, and $\Delta\omega = \omega - \omega_0$ is the offset relative to the resonant frequency. We know from Chapter 3 that the input and output power spectral densities are related by the square of the magnitude of the voltage transfer function, so we can use (8.88)–(8.89) to write

$$\begin{aligned} S_\phi(\omega) &= \left|\frac{1}{1 - H(\omega)}\right|^2 S_\theta(\omega) = \frac{1 + 4Q^2\Delta\omega^2/\omega_0^2}{4Q^2\Delta\omega^2/\omega_0^2} S_\theta(\omega) \\ &= \left(1 + \frac{\omega_0^2}{4Q^2\Delta\omega^2}\right) S_\theta(\omega) = \left(1 + \frac{\omega_h^2}{\Delta\omega^2}\right) S_\theta(\omega) \end{aligned} \tag{8.90}$$

where $S_\theta(\omega)$ is the input PSD, and $S_\phi(\omega)$ is the output PSD. In (8.90) we have also defined $\omega_h = \omega_0/2Q$ as the half-power (3 dB) bandwidth of the resonator.

The noise spectrum of a typical transistor amplifier with an applied sinusoidal signal at f_0 is shown in Figure 8.27. Besides *kTB* thermal noise, transistors generate additional noise that varies as $1/f$ at frequencies below the frequency f_α. This $1/f$, or *flicker*, noise is likely caused by random fluctuations of the carrier density in the active device. Due to the nonlinearity of the transistor, the $1/f$ noise will modulate the applied signal at f_0, and appear as $1/f$ noise sidebands around f_0. Since the $1/f$ noise component dominates the phase noise power at frequencies close to the carrier, it is important to include it in our model. Thus we consider an input power spectral density as shown in Figure 8.28, where $K/\Delta f$ represents the $1/f$ noise component around the carrier, and kT_0F/P_0 represents

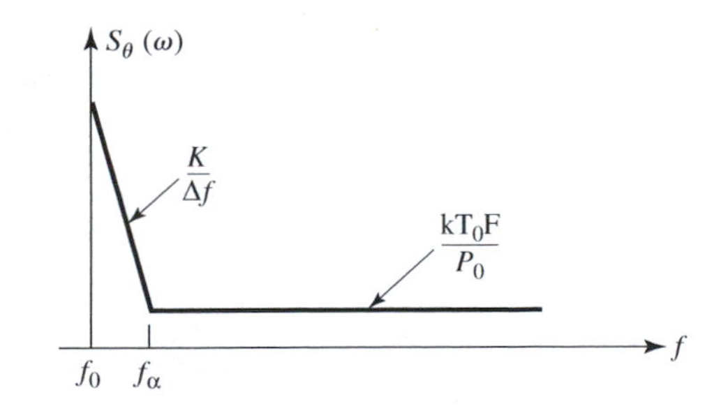

FIGURE 8.28 Idealized power spectral density of amplifier noise, including $1/f$ and thermal components.

thermal noise. Thus the power spectral density applied to the input of the oscillator can be written as

$$S_\theta(\omega) = \frac{kT_0F}{P_0}\left(1 + \frac{K\omega_\alpha}{\Delta\omega}\right), \tag{8.91}$$

where K is a constant accounting for the strength of the $1/f$ noise, and $\omega_\alpha = 2\pi f_\alpha$ is the *corner frequency* of the $1/f$ noise. The corner frequency depends primarily on the type of transistor used in the oscillator. Silicon junction FETs, for example, typically have corner frequencies ranging from 50 Hz to 100 Hz, while GaAs FETs have corner frequencies ranging from 2 to 10 MHz. Bipolar transistors have corner frequencies that range from 5 kHz to 50 kHz.

Using (8.91) in (8.90) gives the power spectral density of the output phase noise as

$$\begin{aligned} S_\phi(\omega) &= \frac{kT_0F}{P_0}\left(\frac{K\omega_0^2\omega_\alpha}{4Q^2\Delta\omega^3} + \frac{\omega_0^2}{4Q^2\Delta\omega^2} + \frac{K\omega_\alpha}{\Delta\omega} + 1\right) \\ &= \frac{kT_0F}{P_0}\left(\frac{K\omega_\alpha\omega_h^2}{\Delta\omega^3} + \frac{\omega_h^2}{\Delta\omega^2} + \frac{K\omega_\alpha}{\Delta\omega} + 1\right) \end{aligned} \tag{8.92}$$

This result is sketched in Figure 8.29. There are two cases, depending on which of the middle two terms of (8.92) is more significant. In either case, for frequencies close to the carrier at f_0, the noise power decreases as $1/f^3$, or –18 dB/octave. If the resonator has a relatively low Q, so that its 3 dB bandwidth $f_h > f_\alpha$, then for frequencies between f_α and f_h the noise power drops as $1/f^2$, or -12 dB/octave. If the resonator has a relatively high Q, so that $f_h < f_\alpha$, then for frequencies between f_h and f_α the noise power drops as $1/f$, or -6 dB/octave.

At higher frequencies the noise is predominantly thermal, constant with frequency, and proportional to the noise figure of the amplifier. A noiseless amplifier with $F = 1$ (0 dB) would produce a minimum noise floor of $kT_0 = -174$ dBm/Hz. In accordance with Figure 8.25, the noise power is greatest at frequencies closest to the carrier frequency, but (8.92) shows that the $1/f^3$ component is proportional to $1/Q^2$, so that better phase noise characteristics close to the carrier are achieved with a high Q resonator. Finally, recall from (8.86) that the single-sideband phase noise will be one-half of the PSD of (8.92).

The above results give a reasonably good model for oscillator phase noise, and quantitatively explain the roll-off of noise power with frequency offset from the carrier. The analysis can be extended to phase-locked loops, where similar results are obtained. Details can be found in reference [3].

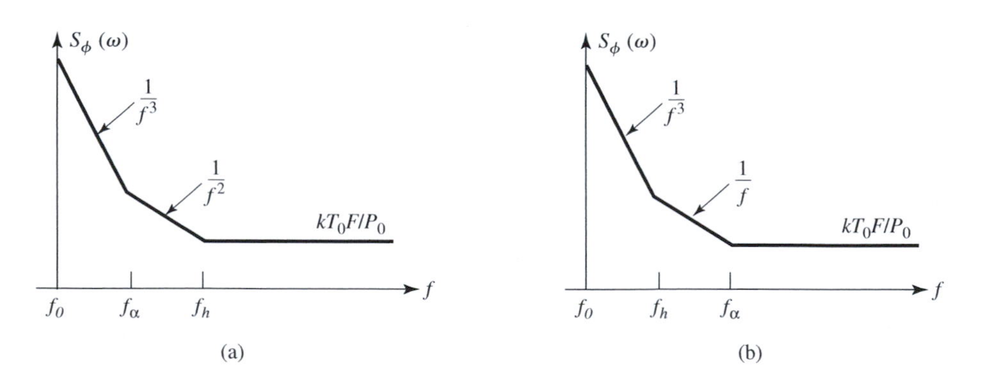

FIGURE 8.29 Power spectral density of phase noise at the output of an oscillator. (a) response for $f_h > f_\alpha$ (low Q). (b) response for $f_h < f_\alpha$ (high Q).

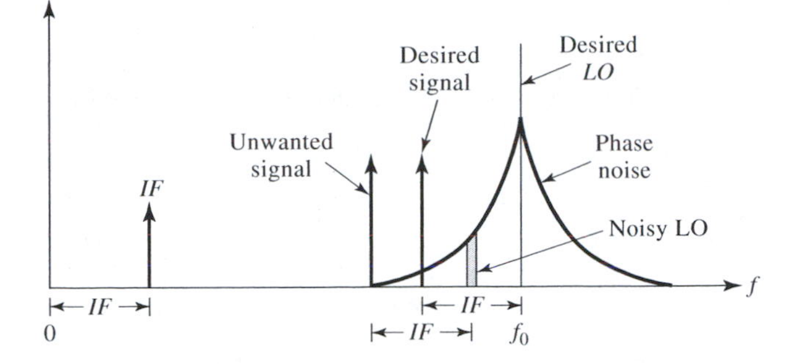

FIGURE 8.30 Illustrating how LO phase noise can lead to the reception of undesired signals adjacent to the desired signal.

Effect of Phase Noise on Receiver Performance

In the above analysis we studied the phase noise power produced by an oscillator, but an important question that remains to be answered is the effect of phase noise on receiver performance, and how much phase noise can be tolerated in a given system design. As mentioned earlier, phase noise can degrade both the bit error rate [1], and the selectivity of a receiver [8]. Of these, the impact of phase noise on selectivity is usually the most severe. (Because this material involves the topics of modulation and receiver design, the reader may wish to review this section after studying Chapters 9 and 10.)

As is discussed in Chapter 9, the probability of error for digital modulation methods depends on the variance of the noise power, σ^2. Thus the effect of phase noise on the bit error rate can be evaluated by converting the power spectral density of the phase noise to a total noise power. This can be done by integrating the power spectral density of the noise over the down-converted passband of the receiver. In general this is a complicated process that must be done experimentally, but a rough estimate can be obtained by approximating this power and using (8.85) to find the rms phase error. Then the expressions developed in Chapter 9 can be used to determine the expected bit error rate. As an approximate example, for a phase noise as high as -80 dBc/Hz over a 30 kHz bandwidth, the rms phase error is less than $1.5°$—this level of phase error will have a negligible effect for most modulation schemes.

Phase noise degrades receiver selectivity by causing down-conversion of signals located nearby the desired signal frequency. The process is shown in Figure 8.30. A local oscillator at frequency f_0 is used to down convert a desired signal to an IF frequency. Due to phase noise, however, an adjacent undesired signal can be down converted to the same IF frequency due to the phase noise spectrum of the LO. The phase noise that leads to this conversion is located at an offset from the carrier equal to the IF frequency from the undesired signal. This process is called *reciprocal mixing*. From this diagram, it is easy to see that the maximum phase noise required to achieve an adjacent channel rejection (or selectivity) of S dB ($S \geq 0$) is given by

$$\mathcal{L}(f_m) = C(\text{dBm}) - S(\text{dB}) - I(\text{dBm}) - 10\log(B), \quad (\text{dBc/Hz}) \tag{8.93}$$

where C is the desired signal level (in dBm), I is the undesired (interference) signal level (in dBm), and B is the bandwidth of the IF filter (in Hz).

EXAMPLE 8.7 GSM RECEIVER PHASE NOISE REQUIREMENTS

The GSM cellular standard requires a minimum of 9 dB rejection of interfering signal levels of -23 dBm at 3 MHz from the carrier, -33 dBm at 1.6 MHz from the carrier, and −43 dBm at 0.6 MHz from the carrier, for a carrier level of −99 dBm. Determine the required local oscillator phase noise at these carrier frequency offsets. The channel bandwidth is 200 kHz.

Solution
From (8.93) we have

$$\begin{aligned}\mathcal{L}(f_m) &= C(\text{dBm}) - S(\text{dB}) - I(\text{dBm}) - 10\log(B)\\ &= -99\,\text{dBm} - 9\,\text{dB} - I(\text{dBm}) - 10\,\log(2\times 10^5).\end{aligned}$$

The following table lists the required LO phase noise as computed from the above expression:

Frequency Offset f_m (MHz)	Interfering Signal Level (dBm)	$\mathcal{L}(f_m)$ dBc/Hz
3.0	−23	−138
1.6	−33	−128
0.6	−43	−118

This level of phase noise requires a phase-locked synthesizer. Bit errors in GSM systems are usually dominated by the reciprocal mixing effect, while errors due to thermal antenna and receiver noise are generally negligible [9]. ○

REFERENCES

[1] L. E. Larson, **RF and Microwave Circuit Design for Wireless Communications**, Artech House, Dedham, MA, 1996.

[2] J. R. Smith, **Modern Communication Circuits**, 2nd edition, McGraw-Hill, New York, 1998.

[3] U. L. Rohde, **Microwave and Wireless Synthesizers: Theory and Design**, Wiley Interscience, New York, 1997.

[4] D. M. Pozar, **Microwave Engineering**, 2nd edition, Wiley, New York, 1998.

[5] Y. Komatsu and Y. Murakami, "Coupling Coefficient Between Microstrip Line and Dielectric Resonator," *IEEE Trans. Microwave Theory and Techniques*, vol. MTT-31, pp. 34–40, January 1983.

[6] A. Mehrotra, **Cellular Radio Performance Engineering**, Artech House, Dedham, MA, 1994.

[7] D. B. Leeson, "A Simple Model of Feedback Oscillator Noise Spectrum," *Proc. IEEE*, vol. 54, pp. 329–330, 1966.

[8] M. K. Nezami, "Evaluate the Impact of Phase Noise on Receiver Performance," *Microwaves & RF Magazine*, pp. 1–11, June 1998.

[9] E. Ngompe, "Compute the LO Phase Noise Requirement in a GSM Receiver," *Applied Microwave & Wireless*, pp. 54–58, July 1999.

PROBLEMS

8.1 Derive the admittance matrix representation of the transistor oscillator circuit given in (8.3).

8.2 Derive the results in (8.20)–(8.22) for a Colpitts oscillator using a common emitter transistor with an inductor having a series resistance R.

8.3 Design a common emitter Colpitts oscillator operating at 30 MHz, using a transistor with $\beta = 40$ and $R_i = 800\ \Omega$. Select reasonable values for the inductor and the two capacitors. Determine the minimum value of the inductor Q in order to sustain oscillations.

8.4 A particular quartz crystal operating at 10 MHz has equivalent circuit parameters of $R = 30\ \Omega$, $C =$ 27 fF, and $C_0 = 5.5$ pF (1 fF $= 10^{-15}$ F). What is the value of the inductance in the equivalent circuit? What is the Q of this crystal? What is the percentage difference between the series and parallel resonant frequencies?

8.5 For either the one-port negative resistance oscillator of Figure 8.8, or the two-port transistor oscillator of Figure 8.9, show that $\Gamma_L \Gamma_{in} = 1$ for steady-state oscillation.

8.6 Repeat the oscillator design of Example 8.4 by replacing the dielectric resonator and microstrip feed line with a single-stub tuner to match Γ_L to a 50 Ω load. Find the Q of the tuner and 50 Ω load, then compute and plot $|\Gamma_{out}|$ versus $\Delta f / f_0$. Compare with the result in Figure 8.13b for the dielectric resonator case.

8.7 Repeat the dielectric oscillator design of Example 8.4 using a GaAs FET having the following S parameters: $S_{11} = 1.2\angle 150°$, $S_{12} = 0.2\angle 120°$, $S_{21} = 3.7\angle -72°$, $S_{22} = 1.3\angle -67°$.

8.8 Design a frequency synthesizer using direct synthesis to produce frequencies ranging from 10.0 MHz to 20.0 MHz, with a resolution of 0.1 MHz.

8.9 Design a direct digital frequency synthesizer operating from 1 kHz to 10 MHz. Assume that four samples are used to represent one cycle of the sinusoid at the highest operating frequency, and that it is desired to have a spectral purity of at least 40 dB. Determine the clock frequency, the size of the phase accumulator, the size of the look-up table, and the size of the DAC.

8.10 Verify the derivation of (8.47) from (8.46).

8.11 Verify the partial fraction expansion used in (8.72) for the second-order phase-locked loop control voltage.

8.12 Derive the result in (8.78) for the 3 dB loop bandwidth of a second-order phase-locked loop.

8.13 Consider a phase-locked loop using a lead-lag filter, having the transfer function

$$H(s) = \frac{1 + s\tau_1}{s\tau_2}.$$

Derive time-domain expressions for the VCO control voltage and the loop phase error. Evaluate the static phase error for a step change in input frequency.

8.14 An oscillator uses an amplifier with a noise figure of 6 dB and a resonator having a Q of 500, and produces a 100 MHz output at a power level of 10 dBm. If the measured f_α is 50 kHz, plot the spectral density of the output noise power, and determine the phase noise (in dBc/Hz) at the following frequencies: (a) at 1 MHz from the carrier. (b) at 10 kHz from the carrier. (assume $K = 1$).

8.15 Repeat Problem 8.14 for $f_\alpha = 200$ kHz.

8.16 Derive (8.93) giving the required phase noise for a specified receiver selectivity.

8.17 Find the necessary local oscillator phase noise specification if an 860 MHz cellular receiver with a 30 kHz channel spacing is required to have an adjacent channel rejection of 80 dB, assuming the interfering channel is at the same level as the desired channel. The final IF voice bandwidth is 12 kHz.

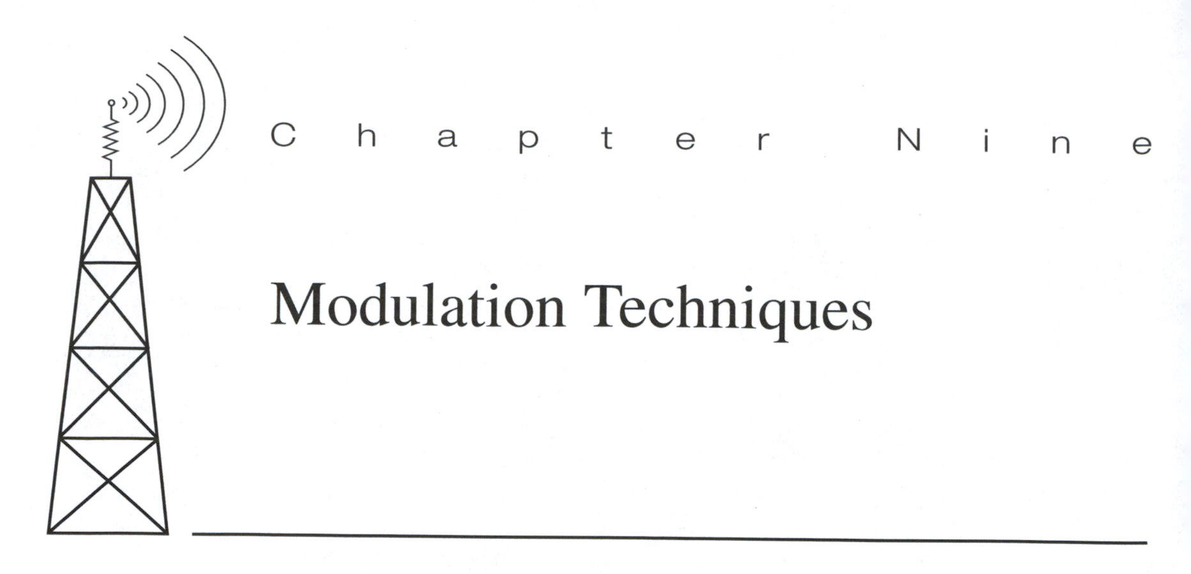

Modulation Techniques

While it is conceptually possible to transmit baseband signals directly, it is usually much more effective to transmit data by modulating a higher frequency carrier wave. This allows precise control of the radiated frequency spectrum, more efficient use of the allocated RF bandwidth, and flexibility in accommodating different baseband signal formats. In general, a sinusoidal carrier wave may be modulated by varying any of its three degrees of freedom: amplitude, frequency, or phase. Most modern wireless systems use digital modulation, where the modulation variables change in discrete steps, in contrast to older methods of analog modulation where carrier amplitude, frequency, or phase may vary continuously. In contrast to analog modulation, digital modulation makes more efficient use of the radio spectrum, and usually requires less prime power. In addition, digital modulation performs better over a fading communications channel, and is more compatible with the use of error-correcting codes. We will begin with a short review of analog modulation, then focus most of our attention on digital modulation techniques. Our objective is to study circuits for modulation and detection, the effect of noise on the detection process, and to derive expressions for the bit error rates of digital modulation methods in the presence of noise. These results will be useful for our study of receiving system performance in Chapter 10.

9.1 ANALOG MODULATION

We begin with a review of basic analog modulation techniques, including single-sideband and double-sideband amplitude modulation, and frequency modulation. We will describe circuits for modulation and demodulation, and derive expressions for the signal-to-noise ratio of these methods. Although some wireless systems employ other analog modulation methods, such as angle (phase) or pulse position modulation, we will not cover such topics here because of space limitations and the much greater importance of digital modulation methods. The subject of analog modulation, however, remains useful as

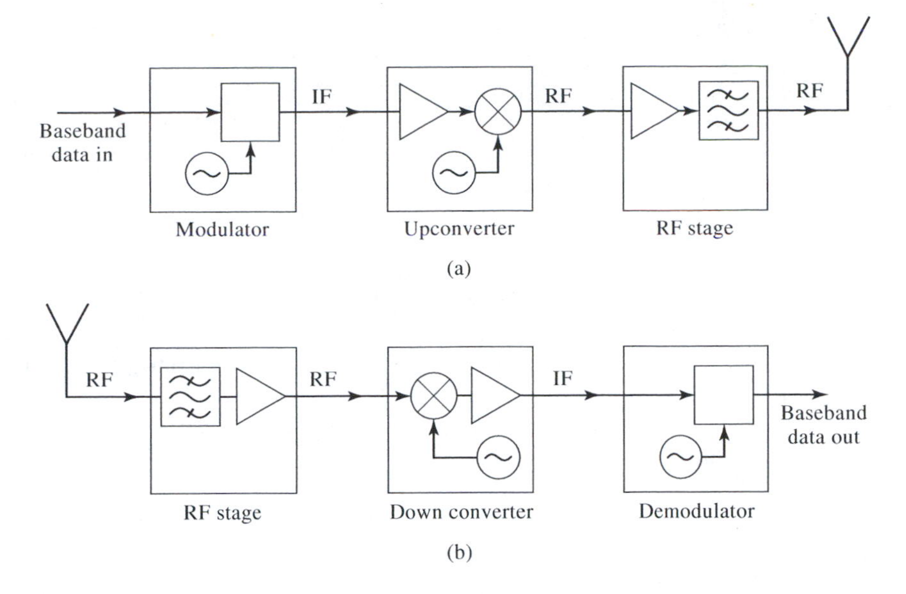

FIGURE 9.1 Illustrating the function of modulation and demodulation in transmitter and receiver systems: (a) modulator. (b) demodulator.

an introduction to the analysis of modulation methods, and as a comparison with digital modulation methods. References [1]–[3] can be consulted for further detail on these topics.

Receiver circuits were introduced in Chapter 1, and are discussed in more detail in Chapter 10, but it is useful here to review the basic heterodyne transmitter and receiver in order to place our discussion of modulation in its larger context. Figure 9.1 shows block diagrams for typical transmitter and receiver systems. The fundamental purpose of the transmitter is to modulate an RF carrier signal with the baseband data that is to be transmitted, while the receiver must recover the baseband data from the received modulated carrier wave. In both cases we see that the modulator/demodulator provides the interface between the baseband information and the IF signals. Because the IF signal in a receiver contains noise from the antenna and receiver circuitry, as well as the desired signal, the characteristics of the demodulator play a critical role in the overall performance of the wireless system.

In each case considered next we assume a bandlimited modulating waveform $m(t)$. This baseband signal may represent voice, music, computer data, or other information to be transmitted over the wireless channel, but for simplicity we assume $m(t)$ to be a single sinusoid of the form $m(t) = \cos 2\pi f_m t$. The maximum frequency of the modulating waveform is f_M, so that $0 \leq f_m \leq f_M$.

Single-Sideband Modulation

Figure 9.2 shows a block diagram for SSB modulation and demodulation. This diagram only shows the relevant IF and baseband stages, and omits the RF and other components that are required in a practical transmitter and receiver. The local oscillator of the SSB modulator is modulated by the bandlimited baseband signal $m(t)$, forming a DSB signal at the output of the mixer. The upper sideband is selected from the mixer output with a bandpass filter, although the lower sideband could be selected, if preferred. To avoid the requirement for the sharp-cutoff sideband filter, the single-sideband mixer discussed in Section 7.4 could instead be used to generate either a USB or LSB signal directly. In either

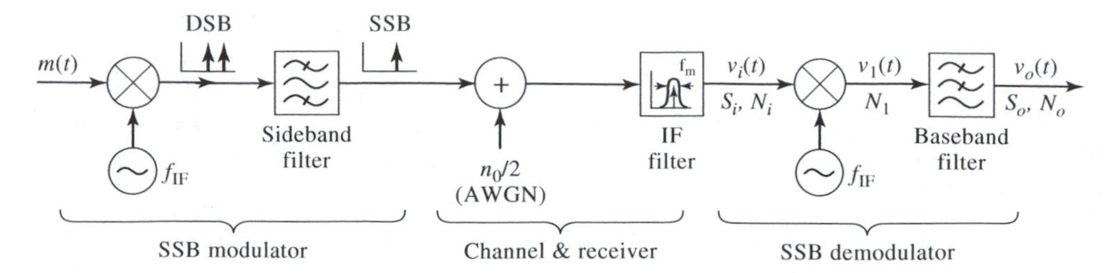

FIGURE 9.2 Block diagram of a SSB modulator and demodulator.

case the spectrum occupied by the SSB signal extends from f_{IF} to $f_{\mathrm{IF}} + f_M$, where f_M is the maximum frequency of $m(t)$.

Gaussian white noise, with a two-sided power spectral density $n_0/2$, is added to the *IF* signal to model noise contributed by the transmission channel, as well as noise generated by the input stages of the receiver. The IF bandpass filter has a passband extending from f_{IF} to $f_{\mathrm{IF}} + f_M$, which passes the received SSB signal as well as noise power within this band. The demodulator mixes the input SSB IF signal with a local oscillator of frequency f_{IF}. Note that the demodulator LO is identical in frequency and phase with the modulator LO; this is called a *synchronous*, or *coherent*, demodulator. The difference frequency from the mixer is low-pass filtered to recover the baseband signal, $m(t)$.

The SSB input voltage to the demodulator can therefore be expressed as

$$v_i(t) = A\cos(\omega_{\mathrm{IF}} + \omega_m)t + n(t), \tag{9.1}$$

where $\omega_{\mathrm{IF}} = 2\pi f_{\mathrm{IF}}$, $\omega_m = 2\pi f_m$, and $n(t)$ is gaussian white noise that has been passed through the IF bandpass filter. Since $n(t)$ is limited to a narrow frequency band near ω_{IF}, it can be expressed using the narrowband representation of noise discussed in Section 3.3. Thus we let $n(t) = x(t)\cos\omega_{\mathrm{IF}}t - y(t)\sin\omega_{\mathrm{IF}}t$, where $x(t)$ and $y(t)$ are the bandlimited quadrature components of the noise, and rewrite (9.1) as

$$v_i(t) = A\cos(\omega_{\mathrm{IF}} + \omega_m)t + x(t)\cos\omega_{\mathrm{IF}}t - y(t)\sin\omega_{\mathrm{IF}}t, \tag{9.2}$$

For a local oscillator signal $\cos 2\pi f_{\mathrm{IF}}t$, the output voltage of the mixer is proportional to

$$\begin{aligned} v_1(t) &= v_i(t)\cos\omega_{\mathrm{IF}}t \\ &= \frac{A}{2}\cos\omega_m t + \frac{A}{2}\cos(2\omega_{\mathrm{IF}} + \omega_m)t + \frac{1}{2}x(t)(1 + \cos 2\omega_{\mathrm{IF}}t) - \frac{1}{2}y(t)\sin 2\omega_{\mathrm{IF}}t \end{aligned} \tag{9.3}$$

Since our final result will involve the ratio of powers, we have ignored the mixer conversion constant. After low-pass filtering the final output voltage from the demodulator is

$$v_o(t) = \frac{A}{2}\cos\omega_m t + \frac{1}{2}x(t), \tag{9.4}$$

which contains the original modulating waveform and a bandpass noise signal. Note that, because it is orthogonal to the synchronous LO, the quadrature component of the noise voltage does not appear in the output of (9.4).

We now find the signal and noise powers in order to evaluate the output signal-to-noise ratio. The average input signal power of the SSB signal of (9.1) is found by time-averaging:

$$S_i = \frac{A^2}{2}. \tag{9.5}$$

Similarly, the average power of the output signal in (9.4) is

$$S_o = \frac{1}{2}\left(\frac{A}{2}\right)^2 = \frac{A^2}{8} = \frac{S_i}{4}. \tag{9.6}$$

The power of the narrowband input noise of (9.1) and (9.2) is given by

$$N_i = E\{n^2(t)\} = E\{x^2(t)\} = E\{y^2(t)\}, \tag{9.7}$$

according to (3.47). Then the noise power of the output voltage of (9.4) is

$$N_o = E\left\{\left[\frac{1}{2}x(t)\right]^2\right\} = \frac{1}{4}E\{x^2(t)\} = \frac{N_i}{4}. \tag{9.8}$$

So the output SNR is, from (9.6) and (9.8),

$$\frac{S_o}{N_o} = \frac{S_i}{4}\frac{4}{N_i} = \frac{S_i}{N_i}, \tag{9.9}$$

This result shows that the SSB demodulator does not degrade the input SNR. The frequency translation performed by the SSB demodulator affects the input signal and noise equally, leaving their ratio unchanged. Of course, practical mixers and filters have losses and generate noise, and this will lead to a degradation of the output SNR. But the process of SSB demodulation itself is inherently lossless in terms of the SNR.

While the above mathematical derivation provides the desired result, we may get a better understanding of the effect of mixing and filtering on the noise power by considering the power spectral density from a graphical point of view. We assume the power spectral density of the input white noise before the IF filter is $n_0/2$. As shown in Figure 9.3a,

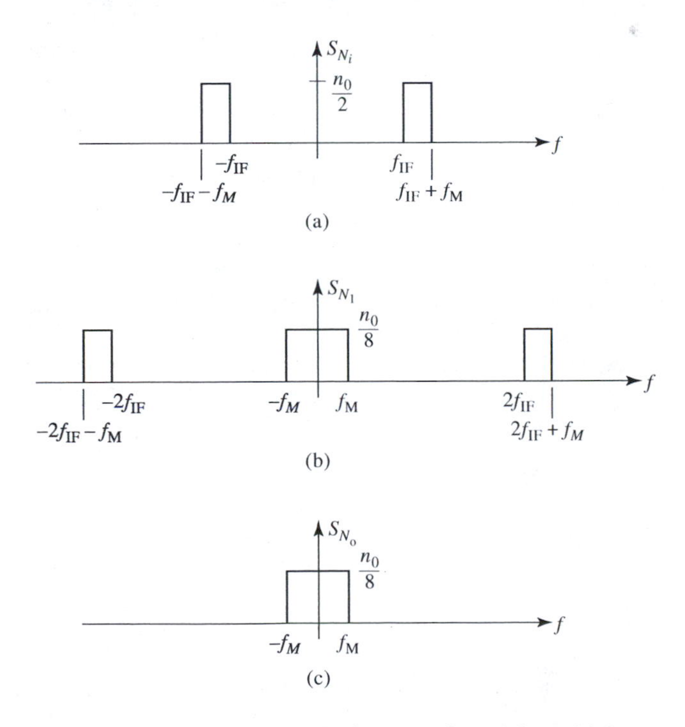

FIGURE 9.3 Power density spectrums of noise in a SSB demodulator. (a) Power spectral density of the noise at the demodulator input. (b) Power spectral density of the noise at the output of the mixer. (c) Power spectral density of the noise after low-pass filtering.

the IF bandpass filter passes noise power only over its passband, from f_{IF} to $f_{\text{IF}} + f_M$. Including the negative part of the frequency spectrum thus gives the input noise power to the demodulator as

$$N_i = 2f_M\left(\frac{n_0}{2}\right) = n_0 f_M. \tag{9.10}$$

Mixing the bandpass noise with the LO has the effect of shifting the noise spectrum up and down by f_{IF}, as shown in Figure 9.3b. As derived in Chapter 3, mixing noise has the effect of reducing its power by half. Thus, from (3.42), the power spectral density of the noise after the mixer must be $n_0/8$, so that the total noise power after the mixer is

$$N_1 = 4f_M\left(\frac{n_0}{8}\right) = \frac{n_0 f_M}{2}. \tag{9.11}$$

Finally, as shown in Figure 9.3c, the low-pass baseband filter selects only the noise spectrum from $-f_M$ to f_M, so the output noise power is

$$N_o = 2f_M\left(\frac{n_0}{8}\right) = \frac{n_0 f_M}{4} = \frac{N_i}{4}. \tag{9.12}$$

The last step follows from (9.10), and is in agreement with our earlier result of (9.8). Then we can rewrite the output SNR of (9.9) in terms of the power spectral density of the input white noise:

$$\frac{S_o}{N_o} = \frac{S_i}{n_0 f_M}. \tag{9.13}$$

Double-Sideband Suppressed-Carrier Modulation

Next we consider a double-sideband amplitude modulation system, with the block diagram shown in Figure 9.4. The mixer of the modulator forms the product of the modulating signal and the LO, producing a DSB waveform; unlike the SSB case, both sidebands are retained. If the mixer were ideal, the carrier frequency (f_{IF}) would not be present in the output, so this modulation is referred to as *double-sideband suppressed carrier* (DSB-SC). As we have seen in Chapter 7, realistic mixers produce virtually all combinations of the input frequencies, so filtering must be used to actually achieve a true DSB-SC waveform. The spectrum of the DSB-SC signal extends from $f_{\text{IF}} - f_M$ to $f_{\text{IF}} + f_M$, where f_M is again the maximum frequency of the modulating input. This is twice the bandwidth of the corresponding SSB waveform.

As in the previous case, we assume that gaussian white noise is added to the IF signal before demodulation. This noise is again bandlimited by the IF bandpass filter, which now has a bandwidth of $2f_M$, centered at f_{IF}, to pass both sidebands of the DSB signal. The

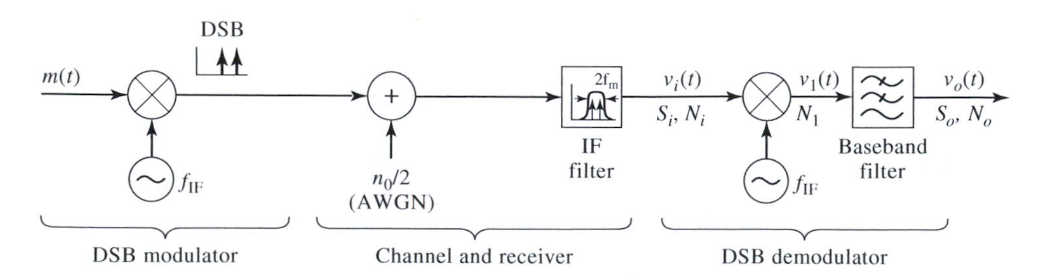

FIGURE 9.4 Block diagram of a DSB-SC modulator and demodulator.

input voltage to the demodulator can then be expressed as

$$v_i(t) = \frac{A}{\sqrt{2}} \cos(\omega_{\text{IF}} - \omega_m)t + \frac{A}{\sqrt{2}} \cos(\omega_{\text{IF}} + \omega_m)t + n(t). \tag{9.14}$$

We have normalized the input voltage so that the total DSB-SC input power is identical to the input power of the SSB case (corresponding to equal transmitter powers):

$$S_i = \frac{1}{2}\left(\frac{A}{\sqrt{2}}\right)^2 + \frac{1}{2}\left(\frac{A}{\sqrt{2}}\right)^2 = \frac{A^2}{2}. \tag{9.15}$$

Again using the narrowband representation for $n(t)$, we can rewrite (9.14) as

$$v_i(t) = \frac{A}{\sqrt{2}} \cos(\omega_{\text{IF}} - \omega_m)t + \frac{A}{\sqrt{2}} \cos(\omega_{\text{IF}} + \omega_m)t + x(t)\cos\omega_{\text{IF}}t - y(t)\sin\omega_{\text{IF}}t. \tag{9.16}$$

For an LO of $2\pi f_{\text{IF}}t$ the output of the mixer is

$$\begin{aligned} v_1(t) &= v_i(t)\cos\omega_{\text{IF}}t \\ &= \frac{A}{\sqrt{2}}\cos\omega_m t + \frac{A}{\sqrt{2}}\cos(2\omega_{\text{IF}} - \omega_m)t + \frac{1}{2}x(t)(1 + \cos 2\omega_{\text{IF}}t) - \frac{1}{2}y(t)\sin 2\omega_{\text{IF}}t \end{aligned} \tag{9.17}$$

After the low-pass baseband filter, the final output is

$$v_o(t) = \frac{A}{\sqrt{2}}\cos\omega_m t + \frac{1}{2}x(t). \tag{9.18}$$

Again, the quadrature component of the noise drops out because it is orthogonal to the synchronous LO. The output signal power is

$$S_o = \frac{1}{2}\left(\frac{A}{\sqrt{2}}\right)^2 = \frac{A^2}{4} = \frac{S_i}{2}. \tag{9.19}$$

Observe that we now have $S_o = S_i/2$, whereas in the SSB case we had $S_o = S_i/4$. This is because the two sidebands of the DSB signal add in phase after mixing with the LO, doubling the receive voltage, and increasing the output power by a factor of four. However, since each sideband of the DSB waveform has half the power of the SSB waveform, the overall increase in power is two, relative to the SSB case.

The noise power of the output voltage of (9.18) is

$$N_o = E\left\{\left[\frac{1}{2}x(t)\right]^2\right\} = \frac{N_i}{4}, \tag{9.20}$$

where $N_i = E\{n^2(t)\} = E\{x^2(t)\} = E\{y^2(t)\}$ as before. Then the output SNR is

$$\frac{S_o}{N_o} = \frac{S_i}{2}\frac{4}{N_i} = 2\frac{S_i}{N_i}. \tag{9.21}$$

This result suggests that the DSB-SC demodulator improves the input signal-to-noise ratio by a factor of two. This improvement is due to the effective doubling of signal power noted above. However, a more realistic result is found by expressing the input noise power in terms of the power spectral density of the white noise at the input to the IF stage. Since the IF bandwidth is $2f_M$, we have

$$N_i = 4f_M\left(\frac{n_0}{2}\right) = 2n_0 f_M, \tag{9.22}$$

where the factor of four accounts for positive and negative frequency portions of the filter response. Then the output SNR of (9.21) is rewritten as

$$\frac{S_o}{N_0} = \frac{S_i}{n_0 f_M}, \tag{9.23}$$

which is seen to be identical to the result for the SSB case given in (9.13). This implies that the coherent SSB and DSB-SC demodulators have the same SNR performance, when expressed in terms of a uniform input white noise level; this is a more meaningful comparison than that of (9.21).

EXAMPLE 9.1 SNR FOR A DSB-SC DEMODULATOR

Derive the output SNR for a DSB-SC demodulator with an arbitrary modulating signal, $m(t)$, where the input signal and noise is represented by

$$v_i(t) = m(t)\cos\omega_{\text{IF}}t + n(t).$$

Consider $m(t)$ to be a random process with $E\{m(t)\} = 0$ and $E\{m^2(t)\} = m^2/2$.

Solution

Assuming an LO voltage of $\cos\omega_{\text{IF}}t$, the output of the mixer is

$$\begin{aligned} v_1(t) &= v_i(t)\cos\omega_{\text{IF}}t \\ &= m(t)\cos^2\omega_{\text{IF}}t + x(t)\cos^2\omega_{\text{IF}}t - y(t)\cos\omega_{\text{IF}}t\sin\omega_{\text{IF}}t \\ &= \frac{1}{2}m(t)(1+\cos 2\omega_{\text{IF}}t) + \frac{1}{2}x(t)(1+\cos 2\omega_{\text{IF}}t) - \frac{1}{2}y(t)\sin 2\omega_{\text{IF}}t \end{aligned}$$

After low-pass filtering the output voltage is

$$v_o(t) = \frac{1}{2}m(t) + \frac{1}{2}x(t).$$

The average power of the input signal is found by time-averaging the expected value of the square of the input signal voltage:

$$S_i = \frac{1}{2}E\{m^2(t)\} = \frac{m^2}{4}.$$

The output signal power is

$$S_o = E\left\{\left[\frac{1}{2}m(t)\right]^2\right\} = \frac{m^2}{8} = \frac{S_i}{2}.$$

The noise power of the demodulator output is

$$N_o = E\left\{\left[\frac{1}{2}x(t)\right]^2\right\} = \frac{1}{4}E\{x^2(t)\} = \frac{N_i}{4}.$$

Then the output signal-to-noise ratio is

$$\frac{S_o}{N_o} = \frac{S_i}{2}\frac{4}{N_i} = 2\frac{S_i}{N_i},$$

which is the same result as was obtained for a sinusoidal modulating signal. ○

Double-Sideband Large-Carrier Modulation

If the double-sideband signal of the previous case is transmitted without suppression of the carrier wave, it is referred to as *double-sideband large-carrier* (DSB-LC) modulation. This is an advantage in that the carrier signal, even if much lower in amplitude than the sidebands, can be used as a reference signal to phase-lock the local oscillator to synchronization with the incoming signal. The block diagram of the DSB-LC modulator/demodulator is essentially the same as the DSB-SC system shown in Figure 9.4, but now the input signal is expressed as

$$\begin{aligned} v_i(t) &= A[1 + m(t)] \cos \omega_{\rm IF} t + n(t) \\ &= A \cos \omega_{\rm IF} t + \frac{mA}{2}[\cos(\omega_{\rm IF} - \omega_m)t + \cos(\omega_{\rm IF} + \omega_m)t] + n(t) \end{aligned} \tag{9.24}$$

where the modulating signal is $m(t) = m \cos \omega_m t$. The amplitude, m, of the modulating signal relative to the carrier is called the *modulation index*. The signal power associated with this input voltage is

$$S_i = \frac{A^2}{2} + \frac{m^2 A^2}{4}, \tag{9.25}$$

where the first term is the carrier power and the second is the power in the sidebands. The carrier power thus increases the total input power, but does not directly contribute to the output power after demodulation since it does not contain any modulation information.

Again we use the narrowband representation of noise to write $n(t) = x(t) \cos \omega_{\rm IF} t - y(t) \sin \omega_{\rm IF} t$. Mixing the input voltage of (9.24) with the LO and low-pass filtering gives the demodulator output voltage as

$$v_o(t) = \frac{mA}{2} \cos \omega_m t + \frac{1}{2} x(t). \tag{9.26}$$

Thus the output signal power is

$$S_o = \frac{m^2 A^2}{8} = \frac{m^2 A^2}{8} \frac{S_i}{\left(\dfrac{A^2}{2} + \dfrac{m^2 A^2}{4}\right)} = \frac{m^2 S_i}{2(2 + m^2)}, \tag{9.27}$$

where we used (9.25) to represent S_o in terms of S_i. Because the IF bandwidth for the DSB-LC case is the same as that for the DSB case, the input and output noise powers are the same. Using (9.20) and (9.22) with (9.27) gives the output SNR of the DSB-LC modulator as

$$\frac{S_o}{N_o} = \frac{S_i}{N_i} \frac{2m^2}{2 + m^2} = \frac{m^2 S_i}{n_0 f_M (2 + m^2)}. \tag{9.28}$$

Note that if $m \gg 1$, so that the carrier power is negligible relative to the sideband power, (9.28) reduces to (9.23). When the carrier power is not negligible, however, the output SNR is less than that for either SSB or DSB-SC. For example, when $m = 1$ (a modulation index of 100%), we have

$$\frac{m^2}{2 + m^2} = \frac{1}{3},$$

which implies a reduction in SNR of 4.8 dB.

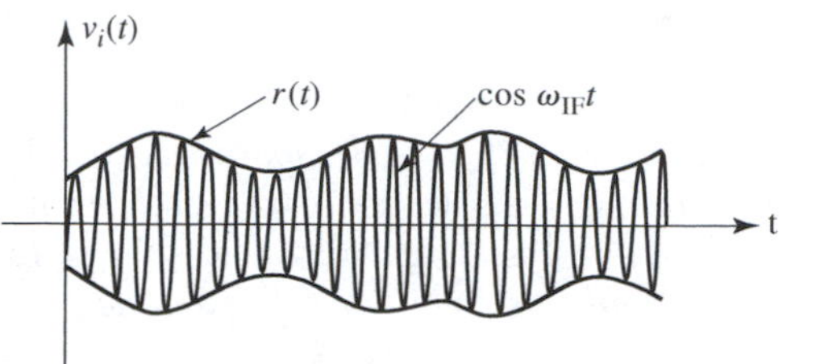

FIGURE 9.5 Modulation envelope of a double-sideband amplitude modulated carrier.

Envelope Detection of Double-Sideband Modulation

As we have seen, SSB and DSB-SC demodulators require a synchronous local oscillator for proper operation. An advantage of DSB with a carrier component (DSB-LC) is that detection can be done without a local oscillator and mixer, by using an *envelope detector*. Since it does not require knowledge of the phase of the incoming signal, an envelope detector is an example of a *noncoherent demodulator*. Because DSB-LC with an envelope detector results in a much simpler receiver circuit, it is the preferred method for broadcast AM radio, where it is desired to make the receiver as inexpensive as possible.

The DSB-LC signal waveform of (9.24) can be written as

$$v_i(t) = A[1 + m(t)] \cos \omega_{IF}t = r(t) \cos \omega_{IF}t, \tag{9.29}$$

where $r(t)$ is the amplitude variation, or envelope, of the carrier. Figure 9.5 shows a plot of the waveform of (9.29) for a typical modulating waveform $m(t)$. Note that $m(t)$ differs from $r(t)$ by a constant term.

The basic envelope detector circuit is shown in Figure 9.6a. When the DSB-LC signal of (9.29) is applied to the input of the circuit, the diode allows charging of the capacitor during the positive portions of the carrier cycle. During the negative portions the diode does not conduct, so the capacitor begins to discharge through the resistor. The resulting output voltage is sketched in Figure 9.6b, and is seen to approximate the actual envelope function. In reality, the carrier frequency is many times greater than the highest frequency component of the modulation, so the output of a properly designed detector can track the envelope much more closely than is indicated in the sketch. The RC time constant should be large enough so that the capacitor voltage does not decay too quickly before the next carrier peak arrives, but small enough so the output can track the envelope when it is decreasing. Also, the output of the envelope detector shown in Figure 9.6a has a DC level that must be removed with a series capacitor. This may limit the low frequency response of the detector.

Also note that $r(t)$ as defined in (9.29) may become negative if $|m| > 1$, a condition referred to as *overmodulation*. The synchronous demodulator discussed above will still correctly recover the modulating waveform in this case, but the envelope detector will not.

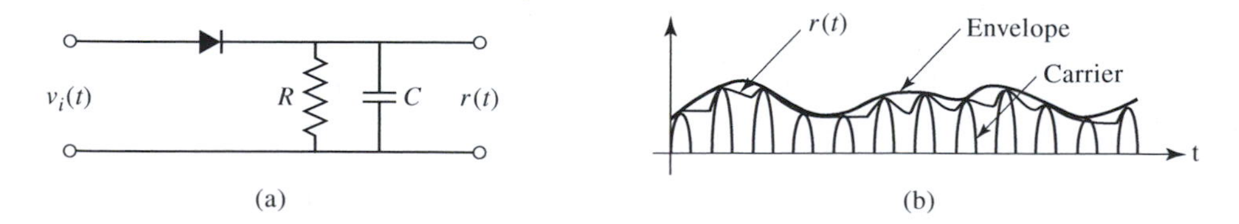

FIGURE 9.6 Envelope detection of an AM signal. (a) Basic envelope detector circuit. (b) Output waveform of the envelope detector.

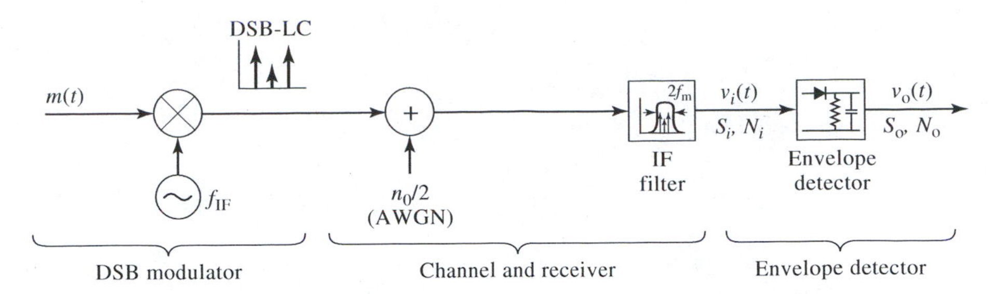

FIGURE 9.7 Block diagram of a DSB-LC modulator and demodulator using envelope detection.

This is because the output voltage of the envelope detector of Figure 9.6a is always positive, and cannot track a negative value of $r(t)$. Thus broadcast radio systems must ensure that the modulation index is always less than 100% in order to avoid signal distortion. This is also the reason why envelope detection cannot be used with a DSB-SC signal.

The block diagram of a DSB-LC system using envelope detection is shown in Figure 9.7. As in the case of the synchronous DSB demodulator, the IF bandwidth is $2f_M$, where f_M is the maximum frequency range of the modulating waveform, $m(t) = m \cos \omega_m t$. We can derive the output SNR of the envelope detector by writing the input voltage of (9.24) as

$$\begin{aligned} v_i(t) &= A[1 + m(t)] \cos \omega_{\text{IF}} t + n(t) \\ &= [A + Am \cos \omega_m t + x(t)] \cos \omega_{\text{IF}} t - y(t) \sin \omega_{\text{IF}} t \\ &= r(t) \cos[\omega_{\text{IF}} t + \theta(t)] \end{aligned} \tag{9.30}$$

where $r(t)$ is the envelope of the carrier, as defined by

$$r(t) = \sqrt{[A + Am \cos \omega_m t + x(t)]^2 + y^2(t)}. \tag{9.31}$$

The phase function $\theta(t)$ is defined by

$$\theta(t) = \tan^{-1} \frac{y(t)}{A + Am \cos \omega_m t + x(t)}. \tag{9.32}$$

The signal input power is

$$S_i = \frac{A^2}{2} + \frac{m^2 A^2}{4}, \tag{9.33}$$

and the output signal power is given by the time-average of the square of the envelope voltage with the noise terms set to zero:

$$S_o = A^2 + \frac{m^2 A^2}{2}. \tag{9.34}$$

The first term in (9.34) represents power of a DC component, which would generally be filtered with a DC block.

Evaluation of the noise power contained in the envelope voltage is complicated by the presence of the square root in (9.31). We can obtain an approximation for the special case when the input SNR is large by assuming $x(t) \ll A$ and $y(t) \ll A$. Then (9.31) reduces to

$$\begin{aligned} r(t) &= [A + Am \cos \omega_m t + x(t)] \sqrt{1 + \frac{y^2(t)}{A + Am \cos \omega_m t + y(t)}} \\ &\cong A + Am \cos \omega_m t + x(t) \end{aligned} \tag{9.35}$$

This gives the output noise power as

$$N_o = E\{x^2(t)\} = N_i, \tag{9.36}$$

so the output SNR becomes

$$\frac{S_o}{N_o} = \frac{S_i}{N_i}\frac{2m^2}{2+m^2}. \quad \text{(for large } S_i/N_i\text{)} \tag{9.37}$$

This is identical to the result of (9.28) for the synchronous demodulation of a DSB-LC signal, showing that no degradation in SNR occurs with the much simpler method of envelope detection for high input SNR.

When the input SNR is not large, however, the output SNR for envelope detection will be reduced. In fact, if we expand (9.31) as

$$r(t) = \sqrt{A^2 + 2A^2m(t) + 2Ax(t) + 2Am(t)x(t) + x^2(t) + A^2m^2(t) + y^2(t)}, \tag{9.38}$$

we see that the envelope voltage does not contain a term proportional to the modulating signal $m(t)$, but involves products of the signal and noise, the square of the signal, and the square root function. This means that the output of the envelope detector is seriously distorted and unusable for small input SNR. Figure 9.8 shows a graph of the output SNR versus input SNR for both envelope and synchronous demodulation. For an input SNR greater than about 10 dB, the two cases are comparable. As the input SNR drops below about 10 dB, the output SNR for envelope detection begins to decrease at a faster rate than the unity-slope response of the synchronous case. This is called the *threshold effect*.

Frequency Modulation

In an FM waveform the instantaneous frequency of the carrier wave is varied in proportion to the modulating signal. We will see that the SNR for FM radio can be much better than either SSB, DSB-SC, or DSB-LC, but at the expense of increased channel bandwidth. This improved performance has led to the application of frequency modulation in broadcast radio, television sound, two-way voice radio, and the AMPS cellular telephone system.

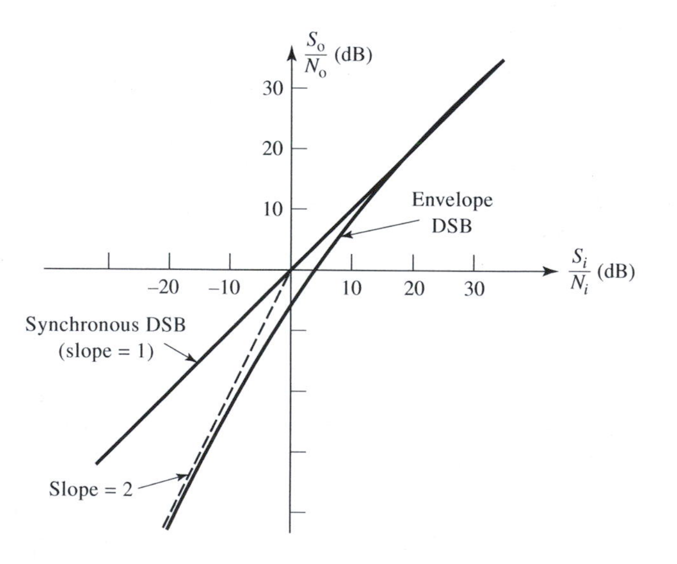

FIGURE 9.8 Output versus input SNR for synchronous and envelope detection of a DSB-LC AM signal.

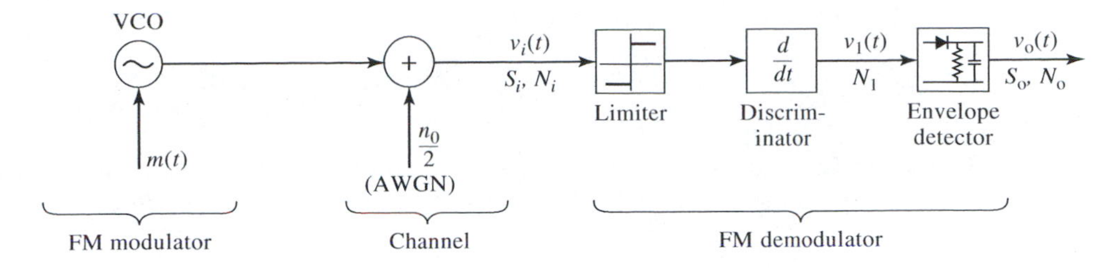

FIGURE 9.9 Block diagram of an FM modulator and demodulator.

Figure 9.9 shows a block diagram for an FM radio system. A direct method of generating an FM signal is to control the frequency of oscillation of a VCO with the modulating signal, $m(t)$. Demodulation can be accomplished by first passing the received (and downconverted) IF signal through an amplitude limiter, followed by a discriminator and an envelope detector. The operation of the demodulator will become clear when we analyze the SNR performance. A phase-locked loop can also be used to directly demodulate an FM signal, as discussed in Chapter 8.

An FM waveform has the general form

$$v(t) = A\cos\theta(t), \tag{9.39}$$

where the phase function $\theta(t)$ depends on the IF frequency and an arbitrary modulating waveform $m(t)$ as

$$\theta(t) = \omega_{\text{IF}}t + k\int_{\tau=0}^{t} m(\tau)\,d\tau. \quad \text{rad} \tag{9.40}$$

The instantaneous frequency (the rate of change of phase) of (9.39) is given by

$$\omega = \frac{d\theta(t)}{dt} = \omega_{\text{IF}} + km(t) \quad \text{rad/sec}. \tag{9.41}$$

For the special case of a sinusoidal modulating signal $m(t) = m\cos\omega_m t$, (9.40) reduces to

$$\theta(t) = \omega_{\text{IF}}t + \frac{km}{\omega_m}\sin\omega_m t. \tag{9.42}$$

If we define $\Delta\omega = \text{km}$ as the maximum frequency deviation, and define $\beta = \Delta\omega/\omega_m$ as the modulation index, then (9.39) can be written simply as

$$v(t) = A\cos(\omega_{\text{IF}}t + \beta\sin\omega_m t). \tag{9.43}$$

The FM signal of (9.43) has an infinite number of sidebands, in contrast to the finite-bandwidth AM signals discussed above. We can analyze the frequency spectrum of (9.43) by expanding the waveform into a Fourier series that is periodic in multiples of ω_m. The easiest way to do this is to write (9.43) as the real part of a complex exponential,

$$v(t) = A\,\text{Re}\{e^{j\theta(t)}\} = A\,\text{Re}\{e^{j\omega_{\text{IF}}t}e^{j\beta\sin\omega_m t}\}, \tag{9.44}$$

and then expand the second exponential as a complex Fourier series:

$$e^{j\beta\sin\omega_m t} = \sum_{n=-\infty}^{\infty} C_n e^{jn\omega_m t}.$$

The Fourier coefficients can be found as

$$C_n = \frac{1}{T}\int_{t=-T/2}^{T/2} e^{j\beta \sin \omega_m t} e^{-jn\omega_m t}\, dt, \tag{9.45}$$

where the period is $T = 2\pi/\omega_m$. Changing variables with $x = \omega_m t$, and using an integral identity for the Bessel function from Appendix B, reduces this result to

$$C_n = \frac{1}{2\pi}\int_{x=-\pi}^{\pi} e^{j(\beta \sin x - nx)}\, dx = J_n(\beta), \tag{9.46}$$

where $J_n(\beta)$ is the Bessel function of the first kind of order n. Then the Fourier series for (9.43) can be written as

$$v(t) = A\,\mathrm{Re}\left\{ e^{j\omega_{\mathrm{IF}} t} \sum_{n=-\infty}^{\infty} J_n(\beta) e^{jn\omega_m t} \right\} = A \sum_{n=-\infty}^{\infty} J_n(\beta)\cos(\omega_{\mathrm{IF}} + n\omega_m)t. \tag{9.47}$$

Thus there are sidebands spaced at multiples of ω_m on either side of the carrier at ω_{IF}, with amplitudes given by $AJ_n(\beta)$. Typical spectra for three values of β are shown in Figure 9.10 (note that we may have $0 \le \beta < \infty$). Observe that the sideband magnitude decreases for large n, so the practical bandwidth is not infinite. In fact, good transmission fidelity generally only requires those sidebands whose amplitude is 1% or larger relative to the unmodulated carrier (the $n = 0$ term); that is, for all n such that $|J_n(\beta)| \ge 0.01$. For large β, it can be shown that $n \cong \beta$ will satisfy this condition. In this case the required IF bandwidth is

$$B = 2n\omega_m \cong 2\beta\omega_m = 2\Delta\omega, \text{ (large } \beta)$$

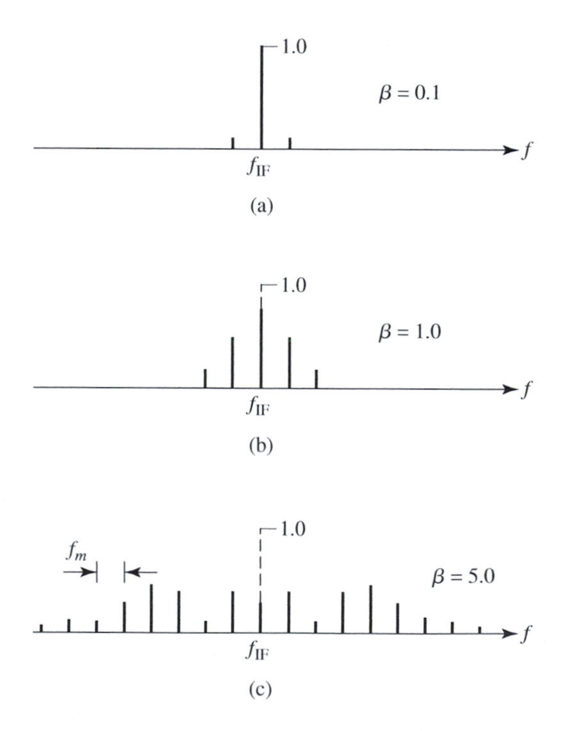

FIGURE 9.10 Magnitude of the spectrum of a frequency modulated signal, for various values of the modulation index β. (a) $\beta = 0.1$, (b) $\beta = 1.0$, and (c) $\beta = 5.0$. The center frequency is $f_{\mathrm{IF}} = \omega_{\mathrm{IF}}/2\pi$, and the spacing of all sidebands is $f_m = \omega_m/2\pi$.

since $\beta = \Delta\omega/\omega_m$. In the case of small β, $n \cong 1$ for the last required sideband, so the total IF bandwidth in this case is

$$B = 2n\omega_m \cong 2\omega_m. \quad \text{(small } \beta\text{)}$$

A convenient way to combine these two results is with the approximation known as *Carson's rule*:

$$B \cong 2(\Delta\omega + \omega_m) = 2\omega_m(1 + \beta). \tag{9.48}$$

We can now derive the SNR for FM modulation, assuming a sinusoidal modulating signal. The voltage at the input to the FM demodulator shown in Figure 9.9 can be written, for a sinusoidal modulating signal, as

$$v_i(t) = A\cos\theta(t) = A\cos(\omega_{\text{IF}}t + \beta\sin\omega_m t). \tag{9.49}$$

Then the input power of the carrier wave can be found from (9.49) with $\beta = 0$ as

$$S_i = A^2/2. \tag{9.50}$$

The purpose of the limiter is to remove any possible amplitude variation from the input to the discriminator, since this stage functions as a differentiator. If we assume the limiter constant is unity, then the output voltage of the discriminator can be written as

$$v_1(t) = \frac{d}{dt}v_i(t) = -(\omega_{\text{IF}} + \beta\omega_m\cos\omega_m t)\sin(\omega_{\text{IF}}t + \beta\sin\omega_m t). \tag{9.51}$$

After envelope detection (and DC blocking of the ω_{IF} term) the output voltage is

$$v_o(t) = \beta\omega_m\cos\omega_m t, \tag{9.52}$$

which is seen to be proportional to the original modulating signal, $m(t)$. The output signal power is

$$S_o = \beta^2\omega_m^2/2. \tag{9.53}$$

Next we consider the presence of noise, and set $\beta = 0$ to compute the output noise power. As we did for the case of envelope detection of AM, we express the noise voltage using the narrowband representation of noise and convert to polar form:

$$\begin{aligned} s_i(t) + n_i(t) &= A\cos\omega_{\text{IF}}t + x(t)\cos\omega_{\text{IF}}t - y(t)\sin\omega_{\text{IF}}t \\ &= r(t)\cos[\omega_{\text{IF}}t - \phi(t)] \end{aligned} \tag{9.54}$$

where the envelope function is

$$r(t) = \sqrt{[A + x(t)]^2 + y^2(t)},$$

and the phase function is

$$\phi(t) = \tan^{-1}\frac{y(t)}{A + x(t)}.$$

If we assume the noise voltage is small relative to the carrier amplitude, so that $x(t) \ll A$ and $y(t) \ll A$, then we can use a small argument approximation to write the phase term as

$$\phi(t) \cong \frac{y(t)}{A}. \tag{9.55}$$

The limiter will remove the effect of the time variation of $r(t)$, so the noise output of the discriminator will be

$$\begin{aligned} n_1(t) &= \frac{d}{dt}\cos[\omega_{\mathrm{IF}}t - \phi(t)] \\ &= -\left(\omega_{\mathrm{IF}} - \frac{d\phi(t)}{dt}\right)\sin[\omega_{\mathrm{IF}}t - \phi(t)] \end{aligned} \tag{9.56}$$

After envelope detection and DC blocking, the output noise voltage is

$$n_o(t) = \frac{d\phi(t)}{dt} = \frac{1}{A}\frac{dy(t)}{dt}. \tag{9.57}$$

Since $H(\omega) = j\omega$ for a differentiator (see Appendix B), the power spectral density of the output noise can be found as

$$S_{n_o}(\omega) = |H(\omega)|^2 S_y(\omega) = \frac{\omega^2}{A^2}n_0 \quad \text{for } |\omega| < \omega_m. \tag{9.58}$$

Then the output noise power over the IF passband from $-\omega_m$ to ω_m is

$$N_o = \frac{1}{2\pi}\int_{-\omega_m}^{\omega_m} S_{n_0}(\omega)\,d\omega = \frac{n_0\omega_m^3}{3\pi A^2}. \tag{9.59}$$

The input noise power over the same passband is

$$N_i = \frac{n_0}{2}(2\omega_m)\frac{1}{2\pi} = \frac{n_0\omega_m}{2\pi}. \tag{9.60}$$

Finally, the output SNR of the FM demodulator is

$$\frac{S_o}{N_o} = \frac{3\pi A^2\beta^2}{2n_0\omega_m} = \frac{3\pi\beta^2}{n_0\omega_m}S_i = \frac{3}{2}\beta^2\frac{S_i}{N_i}, \tag{9.61}$$

where (9.50) and (9.60) have been used to express this result in terms of S_i and N_i. Note that (9.61) indicates that the SNR of a demodulated FM signal can be improved by a factor of $3\beta^2/2$. We can compare this with the corresponding result for AM demodulation by using (9.28) with a 100% modulation index ($m = 1$),

$$\left.\frac{S_o}{N_o}\right|_{\mathrm{AM}} = \frac{2\pi S_i}{3n_0\omega_m},$$

and taking the following ratio:

$$\frac{\left.\frac{S_o}{N_o}\right|_{\mathrm{FM}}}{\left.\frac{S_o}{N_o}\right|_{\mathrm{AM}}} = \frac{9}{2}\beta^2. \tag{9.62}$$

This comparison is made on the basis of equal transmit power, since we have expressed each SNR in terms of S_i. The bandwidth of the FM signal, from $f_{\mathrm{IF}} \pm \beta f_m$, is

$$B_{\mathrm{FM}} = 2\beta f_m$$

while the bandwidth of the AM signal, from $f_{\mathrm{IF}} \pm f_m$, is

$$B_{\mathrm{AM}} = 2f_m.$$

This allows (9.62) to be written in terms of bandwidth as

$$\frac{\left.\dfrac{S_o}{N_o}\right|_{\mathrm{FM}}}{\left.\dfrac{S_o}{N_o}\right|_{\mathrm{AM}}} = \frac{9}{2}\left(\frac{B_{\mathrm{FM}}}{B_{\mathrm{AM}}}\right)^2. \tag{9.63}$$

The results of (9.62) and (9.63) clearly illustrate the main advantage of FM relative to AM, which is that FM allows an improvement in SNR at the expense of increased bandwidth, while AM does not. For example, with a modulation index of $\beta = 4$ the SNR for FM is 72 times better than that for AM, while the bandwidth is four times larger.

9.2 BINARY DIGITAL MODULATION

Digital modulation methods are quickly surpassing analog techniques for modern wireless systems because of their improved performance in the presence of noise and fading, lower transmit power requirements, and better suitability for transmission of digital data with error correction and encryption. In this section we consider binary modulation methods where a sinusoidal carrier wave is switched between one of two possible states according to the binary symbols "0" and "1." A carrier of the form $A\cos(\omega t + \phi)$ has three degrees of freedom: the amplitude (A), the frequency (ω), and the phase (ϕ). These correspond to three fundamental binary modulation methods:

- Amplitude shift keying (ASK)
- Frequency shift keying (FSK)
- Phase shift keying (PSK)

Figure 9.11 shows examples of waveforms for ASK, FSK, and PSK. In each case a carrier wave is either switched on or off (ASK), switched between two distinct frequencies

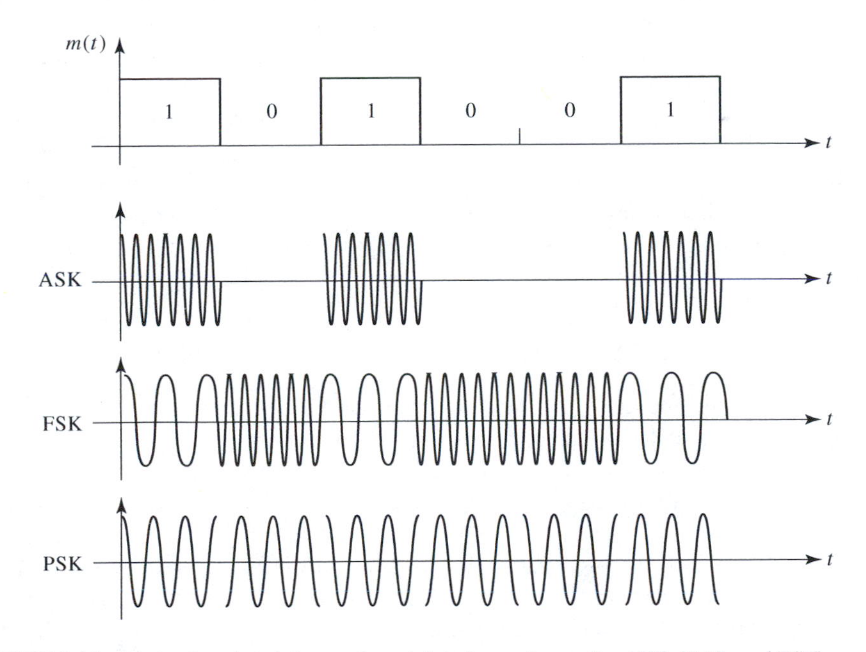

FIGURE 9.11 Binary baseband data and modulated waveforms for ASK, FSK, and PSK.

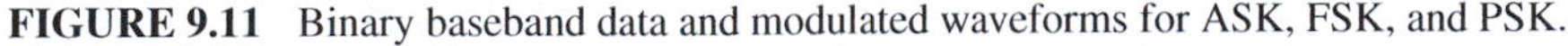

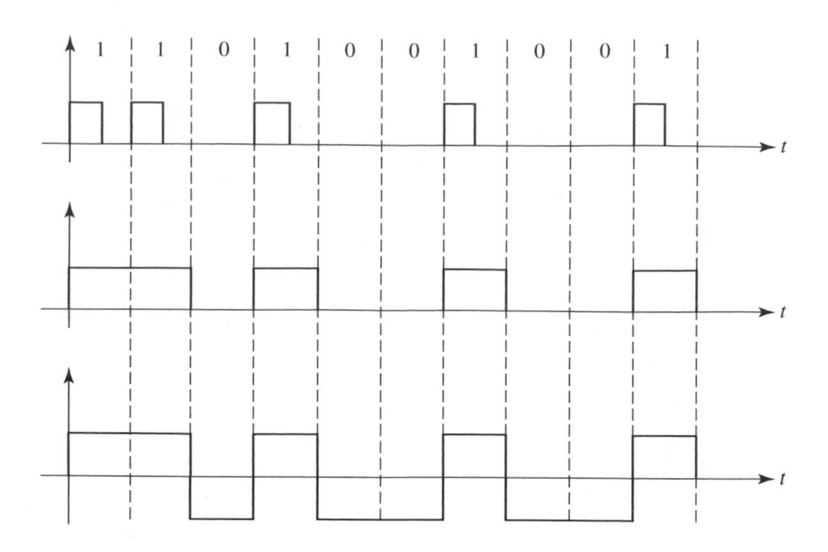

FIGURE 9.12 Binary signaling methods. (a) On-off, or return-to-zero (RZ), coding. (b) Non-return-to-zero (NRZ) coding. (c) Polar NRZ coding.

(FSK), or switched between two distinct phase states (PSK), according to the binary data. We will discuss methods of modulation and demodulation for each of these, and in the following section derive expressions for the error rates of these methods in the presence of noise. We will also briefly discuss the requirement for carrier recovery and synchronization.

Binary Signals

The baseband data source may be voice, fax, music, telemetry data, or computer data. In the case of an analog source, such as voice, music, or telemetry, an analog-to-digital converter can be used to obtain a digital representation. Several sources might be multiplexed together for transmission over a single channel, and a parallel-to-serial converter can be used to obtain sequential, or *line coded*, binary data. Thus we can assume baseband data consisting of a serial bit sequence as shown in Figure 9.12a. This representation involves a return of the signal voltage to zero before the end of the bit period, and is called *return-to-zero* (RZ) signaling. A method that uses less bandwidth is the *non-return-to-zero* technique of Figure 9.12b, where the signal voltage remains high for the duration of a binary "1." Another variation is to use the *polar non-return-to-zero* (polar NRZ) of Figure 9.12c, where a binary "0" is represented by a negative voltage. Polar NRZ thus has an average DC value of zero, which is an advantage in many practical implementations.

Amplitude Shift Keying

In amplitude shift keying the carrier wave is turned on and off according to the binary baseband data sequence. ASK is also known as *on-off keying*. As shown in Figure 9.13a, an ASK modulator can simply consist of a local oscillator and a mixer driven by NRZ binary data, and is virtually identical to the DSB-SC modulator discussed in the previous section. Alternatively, an ASK modulator can be made with an LO switched on and off with a switch controlled by NRZ data. The ASK waveform has a double-sideband suppressed-carrier spectrum. As in the case of DSB-SC modulation, ASK can be demodulated coherently using a synchronous local oscillator and mixer, but it is also possible to demodulate ASK with an envelope detector.

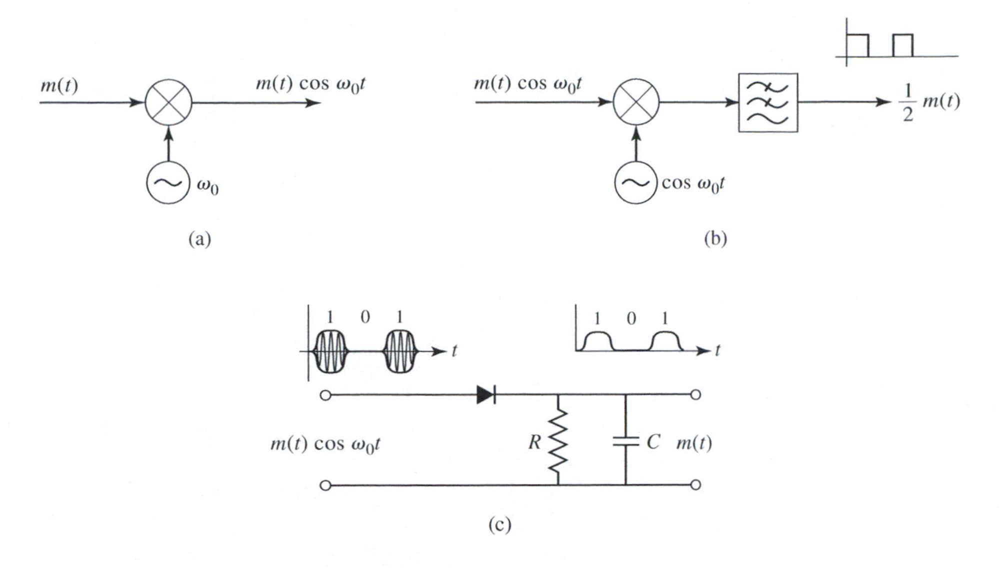

FIGURE 9.13 Modulation and demodulation of ASK signals. (a) ASK modulator. (b) Synchronous demodulation of ASK. (c) Envelope detection of ASK.

A coherent ASK demodulator is shown in Figure 9.13b. If we write the ASK modulated signal as

$$v(t) = m(t)\cos\omega_0 t, \tag{9.64}$$

where $m(t) = 0$ or 1, then the output of the mixer is

$$\begin{aligned} v_1(t) &= v(t)\cos\omega_0 t \\ &= \frac{1}{2}m(t)(1 + \cos 2\omega_0 t) \end{aligned}$$

where we again ignore the mixer conversion constant. After low-pass filtering this reduces to

$$v_o(t) = \frac{1}{2}m(t), \tag{9.65}$$

which is seen to be an exact reproduction of the original baseband signal. Of course, this requires the LO to have precisely the same phase and frequency as the incoming signal. When the local oscillator does not have perfect phase coherence or frequency synchronization, distortion may be introduced. This is explored further in Problem 9.7.

Because the envelope of an ASK signal varies according to the binary baseband data, it is possible to demodulate ASK noncoherently with an envelope detector, and thereby avoid the requirement for a coherent LO. The basic scheme is shown in Figure 9.13c. Although envelope detection cannot be used to demodulate a DSB-SC signal, it can be used with an ASK DSB-SC waveform because the modulating signal $m(t)$ is never negative. As discussed in Section 9.1, the envelope detector functions by extracting the envelope waveform from an AM signal. Equation (9.64) is of the form of an envelope $m(t)$ multiplying a high-frequency carrier, so the same argument used in Section 9.1 leads to the conclusion that the output of the envelope detector will be the envelope $m(t)$.

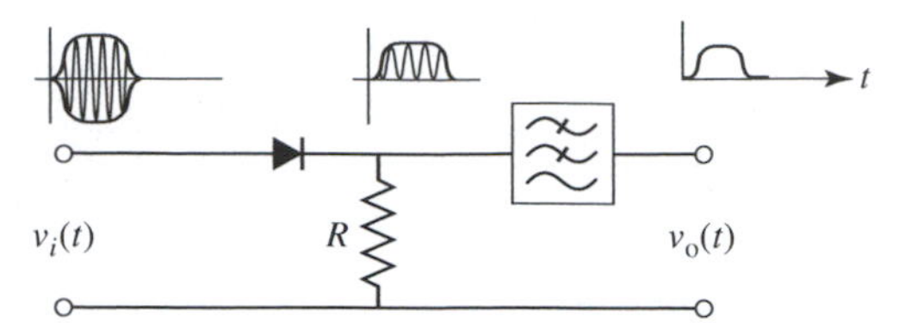

FIGURE 9.14 Circuit for a rectifier detector.

EXAMPLE 9.2 SQUARE-LAW DETECTOR

Another noncoherent demodulator is the rectifier, or square-law, detector shown in Figure 9.14. It is similar to the envelope detector, with the important difference that a capacitor is not used. If the diode is biased to operate in its square-law region, we can assume that the voltage across the resistor is proportional to the square of the input voltage. If the ASK waveform of (9.64) is applied to the input of the square-law detector, derive the output voltage from the low-pass filter.

Solution

Assuming a square-law dependence, the resistor voltage can be written as

$$\begin{aligned} v_R(t) &= Cv^2(t) = Cm^2(t)\cos^2\omega_0 t \\ &= \frac{C}{2}m^2(t)(1+\cos 2\omega_0 t) \end{aligned}$$

After low-pass filtering, we then have the output

$$v_o(t) = \frac{C}{2}m^2(t) = \frac{C}{2}m(t),$$

where the last step follows because $m^2(t) = m(t)$ when $m = 0$ or 1. ○

Frequency Shift Keying

Frequency shift keying involves switching a sinusoidal carrier wave between two frequencies, ω_1 and ω_2, as illustrated in Figure 9.11. While ω_1 and ω_2 can be selected arbitrarily, in practice the frequencies are usually symmetrically chosen so that

$$\begin{aligned} \omega_1 &= \omega_0 - \Delta\omega \\ \omega_2 &= \omega_0 + \Delta\omega \end{aligned}$$

where $\Delta\omega$ is known as the *frequency deviation*. The FSK modulated waveform can be written as

$$v(t) = \cos\omega t, \qquad \omega = \begin{cases} \omega_1 & \text{for } m(t) = 1 \\ \omega_2 & \text{for } m(t) = 0 \end{cases} \tag{9.66}$$

In general the spectrum of an FSK signal is complicated because of the essentially random switching between two frequency states. If $2(\Delta f) = 2(\Delta\omega/2\pi)$ is the difference in carrier frequencies, and T is the bit period of the binary data, then it can be shown that the effective bandwidth of an FSK signal is $B = 2(\Delta f + 2/T)$.

An FSK signal can be generated using the modulator circuit of Figure 9.15a, which employs a tunable oscillator to switch between ω_1 and ω_2. Coherent detection of FSK is

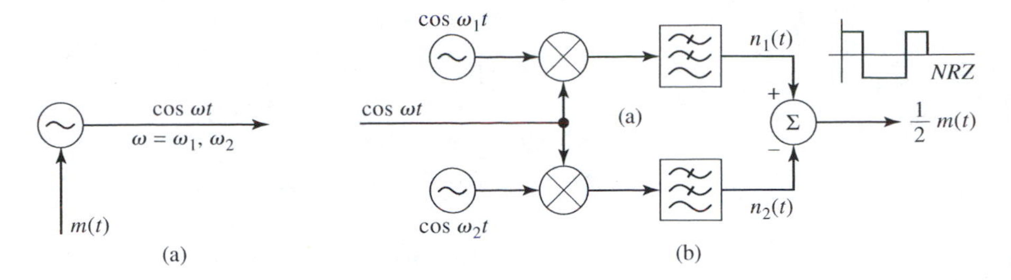

FIGURE 9.15 Modulation and demodulation of FSK signals. (a) FSK modulator. (b) Synchronous demodulation of FSK.

accomplished with the synchronous demodulator of Figure 9.15b, which requires two coherent local oscillators operating at ω_1 and ω_2. To analyze the operation of the demodulator, assume we have the incoming FSK waveform of (9.66) with $\omega = \omega_1$. Then the outputs of the top and bottom mixers are

$$v_1(t) = v(t)\cos\omega_1 t = \cos^2\omega_1 t = \frac{1}{2}(1 + \cos 2\omega_1 t) \tag{9.67a}$$

$$\begin{aligned} v_2(t) &= v(t)\cos\omega_2 t = \cos\omega_1 t\cos\omega_2 t \\ &= \frac{1}{2}[\cos(\omega_1 - \omega_2)t + \cos(\omega_1 + \omega_2)t] \end{aligned} \tag{9.67b}$$

After low-pass filtering only the DC term from (9.67a) remains, resulting in a positive pulse at the output of the summer, indicating that a "1" has been received. Similarly, if the incoming FSK signal has $\omega = \omega_2$, the outputs of the mixers will be reversed from that of (9.67), and the output of the summer will be a negative pulse, indicating that a "0" has been received.

FSK can also be demodulated using an envelope detector, thereby avoiding the requirement for two coherent local oscillators. The FSK envelope demodulator circuit is shown in Figure 9.16a. Operation is based on the fact that bandpass filtering can be used to decompose the FSK signal into two ASK components, as shown in Figure 9.16b. These ASK signals can then be individually demodulated using envelope or square- law detectors. The required bandpass filtering is usually done at the IF stage, as opposed to RF filtering, to ease the filter cutoff requirements. The outputs of the envelope detectors are combined with a summer to form a polar NRZ output.

Phase Shift Keying

In PSK modulation the phase of the carrier wave is switched between two states, usually 0° and 180°. In this case the PSK waveform can be written as

$$v(t) = m(t)\cos\omega_0 t, \tag{9.68}$$

where $m(t) = 1$ or -1. The PSK waveform can be generated by mixing a polar NRZ version of the binary data with a local oscillator, as shown in Figure 9.17a. The spectrum of the PSK waveform is relatively wide in bandwidth due to the sharp transitions caused by phase reversal. These are usually smoothed by filtering, but the resulting bandwidth is usually still wide enough that PSK is impractical for multichannel wireless systems.

Synchronous demodulation of PSK can be accomplished using the demodulator of Figure 9.17b. After mixing the PSK signal of (9.68) with the local oscillator, the output of

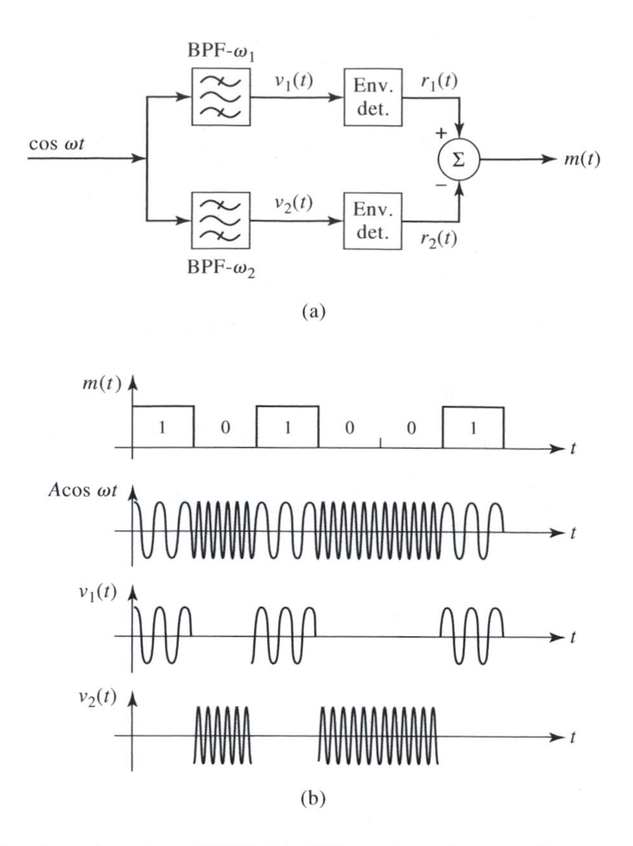

FIGURE 9.16 Envelope detection of FSK. (a) FSK envelope detector. (b) Decomposition of FSK signal into two ASK signals.

the mixer is

$$v_1(t) = v(t)\cos\omega_0 t = \frac{1}{2}m(t)[1 + \cos 2\omega_0 t],$$

so, after low-pass filtering, the output voltage is

$$v_o(t) = \frac{1}{2}m(t), \tag{9.69}$$

which is proportional to the original polar-NRZ data. Since the PSK waveform has a constant envelope, it cannot be demodulated with an envelope detector. Note that it is possible to add $\cos\omega_0 t$ to the PSK waveform of (9.68) to obtain an ASK waveform, which could then be demodulated with an envelope detector. But this procedure offers no advantage over coherent detection because it still requires a synchronous local oscillator.

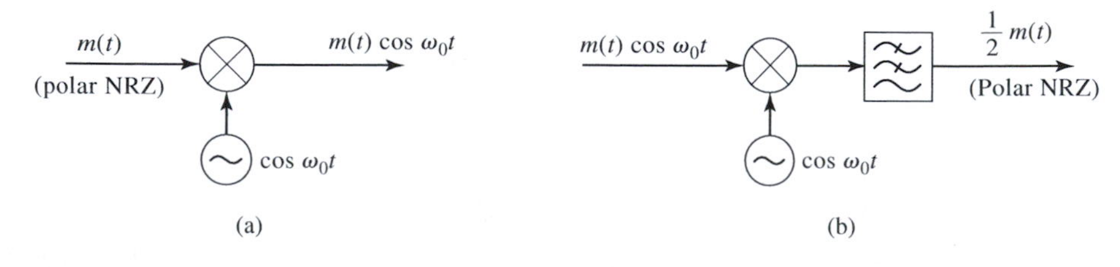

FIGURE 9.17 Modulation and demodulation of PSK signals. (a) PSK modulator, (b) Synchronous demodulation of PSK.

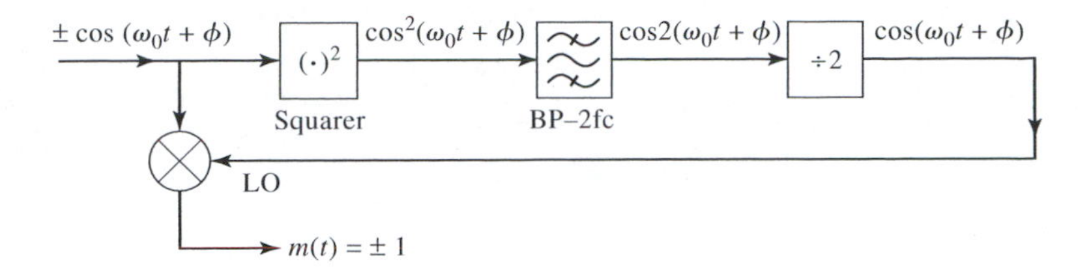

FIGURE 9.18 A carrier-recovery circuit suitable for synchronous PSK demodulation.

Carrier Synchronization

We have seen that ASK, FSK, and PSK can each be demodulated using coherent detectors, where the local oscillator is in synchronism with the incoming carrier. Synchronism implies that both frequency and phase are identical, a result that is generally difficult to achieve in practice. The effect of a phase error $\Delta\phi$ is that the output signal is reduced in amplitude by $\cos \Delta\phi$, while an error $\Delta\omega$ in frequency introduces a factor of $\cos \Delta\omega t$ (see Problem 9.7). As an example, if we set a criteria of requiring less than 45° phase error at a carrier frequency of 1 GHz, synchronization of the LO to the carrier must be better than $T/8 = 0.125$ nS. A free-running local oscillator will virtually never exhibit such synchronism because of frequency drift, Doppler effects, and the arbitrary (and sometimes variable) distance between the transmitter and receiver. An advantage of ASK and FSK is that they can be demodulated without a synchronous LO by using envelope detection, but we will see that a penalty is paid in these cases, since the bit error rates are not as good as those obtained with coherent detection.

In general, there are two ways in which a local oscillator can be synchronized with an incoming carrier wave: transmit a pilot carrier, or use a carrier-recovery circuit. Transmitting a low-level carrier is probably the easiest way, as this signal can be used to phase-lock the local oscillator. The transmitted carrier passes through the same propagation delays as the modulated signal, and so automatically arrives at the receiver with the same phase and frequency.

Carrier recovery circuits use phase or frequency information from the received signal to synchronize the local oscillator. Many of the variations of these circuits use a phase-locked loop, which we study in detail in Chapter 8, but a clever alternative carrier-recovery circuit for PSK demodulation is shown in Figure 9.18. A PSK signal of the form of (9.68), with $m(t) = \pm 1$, is applied to the input of the circuit. The phase ϕ of the waveform accounts for the arbitrary phase delay between transmitter and receiver. A squarer circuit (such as a diode operating in its square-law region) squares the input and effectively removes the phase modulation. A bandpass filter is used to select the $2\omega_0$ component of the squared signal, and a divide-by-two circuit yields an output of $\cos(\omega_0 t + \phi)$, which is the desired local oscillator signal synchronized to the incoming PSK waveform. Mixing with the PSK signal then produces a polar-NRZ version of $m(t)$.

Modern digital demodulators usually operate at the IF stage, and employ digital signal processing (DSP) circuits to perform all functions of signal conditioning, carrier recovery and synchronization, demodulation, and signal formatting.

9.3 ERROR PROBABILITIES FOR BINARY MODULATION

In an ideal situation a receiver will detect the same binary digit that was transmitted, but the presence of noise in a communication channel introduces the possibility that errors will be made during the detection process. Here we derive expressions for the probability of error for the binary modulations schemes discussed in the previous section. We will do

TABLE 9.1 PCM Signals for ASK, FSK, and PSK

Modulation Type	$s_1(t)$ "1"	$s_2(t)$ "0"	Detection Threshold
ASK	V	0	V/2
FSK	V	−V	0
PSK	V	−V	0

this for synchronous detection of ASK, FSK, and PSK, and envelope detection of ASK and FSK. Although envelope detection is simpler to implement than synchronous detection, a price is paid in the error rates for envelope detection because of the absence of phase information.

PCM Signals and Detectors

ASK, FSK, and PSK signals can all be considered as special cases of *pulse code modulation* (PCM). In general, PCM codes a binary "1" as a signal voltage $s_1(t)$, and a binary "0" as a signal voltage $s_2(t)$, each of bit duration T. In the transmitter these PCM signals are used to modulate a carrier by amplitude, frequency, or phase modulation. In the receiver, as we have shown, synchronous demodulation or envelope detection (for ASK or FSK) can be used to recover $s(t) = s_1(t)$ or $s_2(t)$. The PCM signals for the cases of ASK, FSK, and PSK modulation can be summarized as shown in the Table 9.1. The demodulated signal is then used to make the decision as to whether a binary "1" or "0" has been received. In the absence of noise this can simply be done by setting a *detection threshold*, where "1" or "0" is decided using a comparator circuit to determine whether $s(t)$ is greater or less than the threshold value. For FSK and PSK, which use symmetric voltages of $\pm V$, the optimum threshold is obviously zero, while for ASK the optimum threshold is $V/2$. The fact that the optimum threshold depends on the actual value of the received signal is a serious disadvantage of ASK.

When noise is present the problem of detection is more difficult. It can be shown in this case that optimum detection of PCM signals can be made with a *correlation receiver*, consisting of an integrator (or low-pass filter) followed by a sampler and a comparator. (It can be shown that this circuit is equivalent to the optimum matched filter for PCM signals.) The basic circuit is shown in Figure 9.19a. As illustrated in Figure 9.19b, the integrator integrates the incoming signal and noise over the bit duration, T. The output of the integrator is sampled at the end of the bit period, and a comparator is used to determine whether the output voltage is above or below the threshold value. The result is a close approximation of the original binary message $m(t)$. Although the effect of the (zero mean) noise voltage is minimized by averaging over the bit period, the random nature of the noise results in occasional errors, and the likelihood of an error increases with the power of the input noise. The likelihood that a single bit is received incorrectly is called the *probability of error*, or the *bit error rate* (BER).

If we assume that the input signal and gaussian noise voltage is of the form $s(t) + n(t)$, where $s(t) = s_1(t)$ or $s_2(t)$, then the sampled output of the integrator in Figure 9.19a can be written as

$$v_o(T) = \int_{t=0}^{T} [s(t) + n(t)]\, dt = s_o(T) + n_o(T), \tag{9.70}$$

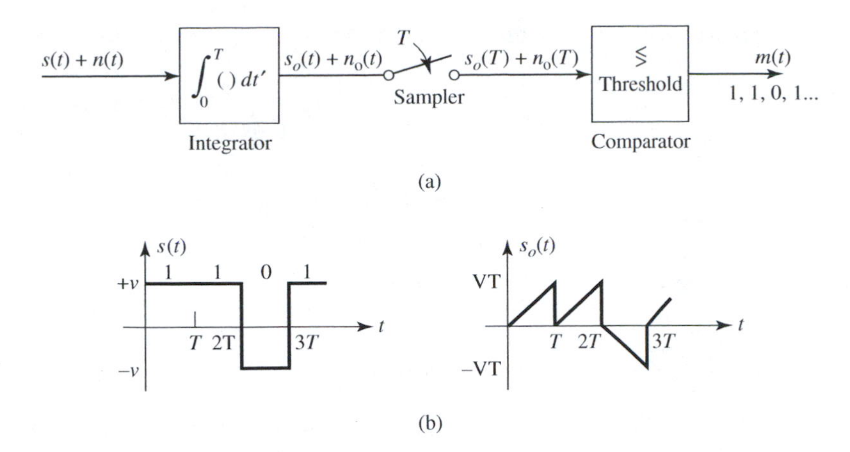

FIGURE 9.19 Optimum correlation receiver for PCM detection. (a) Correlation detector circuit. (b) Input and output signal voltage waveforms.

where the integrated signal voltage is

$$s_o(T) = \int_{t=0}^{T} s(t)\, dt = \begin{cases} VT & \text{if } s(t) = V \\ -VT & \text{if } s(t) = -V \\ 0 & \text{if } s(t) = 0 \end{cases} \tag{9.71}$$

and the integrated noise voltage is

$$n_o(T) = \int_{t=0}^{T} n(t)\, dt. \tag{9.72}$$

We further define the *bit energy* as the energy of the (non-zero) signal voltage over one bit period:

$$E_b = \int_{t=0}^{T} s^2(t)\, dt = V^2 T. \tag{9.73}$$

The noise power output from an integrator with a bandlimited white noise input was derived in (3.38) as

$$N_o = \frac{n_0 T}{2} = \sigma^2, \tag{9.74}$$

where $n_0/2$ is the two-sided power spectral density of the white noise, and σ^2 is the variance of the gaussian probability distribution function (don't confuse $n_o(T)$ with n_0). We can now derive the probability of error for the various binary modulation methods separately.

Synchronous ASK

Assume that a binary "0" has been transmitted in an ASK system, in the presence of bandlimited gaussian white noise. Thus we have $s(t) = s_2(t) = 0$ and $s_o(T) = 0$, with a detection threshold of $VT/2$ (the threshold must also be integrated over the bit period). Correct detection occurs when the output signal and noise voltage from the sampler is less than the threshold. Conversely, an error will occur if the output signal and noise voltage is greater than the threshold. Thus we can write the probability of error when a "0" is sent as

$$P_e^{(0)} = P\left\{s_o(T) + n_o(T) > \frac{VT}{2}\right\} = P\left\{n_o(T) > \frac{VT}{2}\right\}$$
$$= \int_{n_0=\frac{VT}{2}}^{\infty} \frac{e^{-n_0^2/2\sigma^2}}{\sqrt{2\pi\sigma^2}}\, dn_0 \tag{9.75}$$

since $s_o(T) = 0$. This integral can be expressed in terms of the complementary error function by using the change of variable $x = n_0/\sqrt{2}\sigma$:

$$P_e^{(0)} = \frac{1}{\sqrt{\pi}} \int_{x=\frac{VT}{2\sqrt{2}\sigma}}^{\infty} e^{-x^2}\, dx = \frac{1}{2} erfc\left(\frac{VT}{2\sqrt{2}\sigma}\right).$$

The argument of the complementary error function can be expressed in terms of E_b, the bit energy, and the power spectral density of the white noise by using (9.73) and (9.74):

$$\frac{VT}{2\sqrt{2}\sigma} = \sqrt{\frac{E_b}{T}}\frac{T}{2\sqrt{2}}\sqrt{\frac{2}{n_0 T}} = \sqrt{\frac{E_b}{4n_0}}.$$

Then the final expression for the probability of error when a "0" is sent via ASK is

$$P_e^{(0)} = \frac{1}{2} erfc\left(\sqrt{\frac{E_b}{4n_0}}\right). \tag{9.76}$$

It is left as a problem to show that the probability of error when a "1" is transmitted, $P_e^{(1)}$, is equal to $P_e^{(0)}$ when the threshold is set at $VT/2$. In this case the total probability of error of a particular bit in a message having equal occurrences of "0" and "1" bits is

$$P_e = \frac{1}{2}P_e^{(0)} + \frac{1}{2}P_e^{(1)} = P_e^{(0)} = P_e^{(1)} = \frac{1}{2} erfc\left(\sqrt{\frac{E_b}{4n_0}}\right). \tag{9.77}$$

Synchronous PSK

Now assume that a binary "0" is transmitted in a PSK system, in the presence of bandlimited gaussian white noise. Then we have $s(t) = s_2(t) = -V$ and $s_o(T) = -VT$, with a detection threshold of zero. An error will occur if the sampled output voltage from the integrator is greater than the threshold, so we can write

$$P_e^{(0)} = P\{s_o(T) + n_o(T) > 0\} = P\{n_o(T) > VT\}$$
$$= \int_{n_0=VT}^{\infty} \frac{e^{-n_0^2/2\sigma^2}}{\sqrt{2\pi\sigma^2}}\, dn_0 = \frac{1}{2} erfc\left(\frac{VT}{\sqrt{2}\sigma}\right) = \frac{1}{2} erfc\left(\sqrt{\frac{E_b}{n_0}}\right) \tag{9.78}$$

where the change of variable $x = n_0/\sqrt{2}\sigma$ was again used to evaluate the integral, and (9.73)–(9.74) were used to express the argument of the complementary error function in terms of the bit energy and the power spectral density of the noise. Due to the symmetry of the PSK signal and the demodulator, $P_e^{(1)} = P_e^{(0)}$.

Observe that the bit energy-to-noise ratio in the argument of the complementary error function for the PSK case differs by a factor of four (6 dB) compared to the result for ASK. This implies that, for the same probability of error, PSK requires only one-fourth the power of an ASK system. Since an ASK signal is off half the time, however, the average transmit power of an ASK system is half that of a PSK system, for the same peak power (same signal voltage, V). Thus, in terms of average transmit power, the PSK result is better by a factor of two (3 dB), compared with ASK.

Synchronous FSK

Next we consider a synchronous FSK receiver, and assume that a binary "0" is transmitted in the presence of bandlimited gaussian white noise. Then we have $s(t) = s_2(t) = -V$ and $s_o(T) = -VT$, with a detection threshold of zero. These signal levels are similar to the PSK case, but in the synchronous FSK demodulator output noise consists of the difference between noise that has passed through both the ω_1 and ω_2 channels. As shown in Figure 9.15b, if $n_1(t)$ and $n_2(t)$ are the noise voltages due to the ω_1 and ω_2 channels, respectively, then the total output noise voltage from the demodulator is

$$n(t) = n_1(t) - n_2(t). \tag{9.79}$$

The noise voltages $n_1(t)$ and $n_2(t)$ are both gaussian with zero mean and variance σ^2, but are uncorrelated because they are taken over different filter passbands from a white noise spectrum. The variance of $n(t)$ can be computed as

$$\begin{aligned} E\{n^2(t)\} &= E\left\{n_1^2(t) - 2n_1(t)n_2(t) + n_2^2(t)\right\} \\ &= E\left\{n_1^2(t)\right\} + E\left\{n_2^2(t)\right\} = 2\sigma^2 \end{aligned} \tag{9.80}$$

since $E\{n_1(t)n_2(t)\} = E\{n_1(t)\}E\{n_2(t)\} = 0$ because $n_1(t)$ and $n_2(t)$ are uncorrelated and have zero mean. The result of (9.80) shows that the total noise power of the FSK demodulator is doubled relative to the synchronous ASK or PSK demodulator.

The probability of error for synchronous FSK is calculated in the same way as for PSK, but with σ^2 replaced with $2\sigma^2$:

$$\begin{aligned} P_e^{(0)} &= P\{s_o(T) + n(T) > 0\} = P\{n_1(T) - n_2(T) > VT\} \\ &= \int_{n_0=VT}^{\infty} \frac{e^{-n_0^2/4\sigma^2}}{\sqrt{4\pi\sigma^2}}\, dn_0 = \frac{1}{2} erfc\left(\frac{VT}{2\sigma}\right) = \frac{1}{2} erfc\left(\sqrt{\frac{E_b}{2n_0}}\right) \end{aligned} \tag{9.81}$$

The change of variable $x = n_0/2\sigma$ was used to evaluate the integral, and (9.73)–(9.74) were used to express the argument of the complementary error function in terms of the bit energy and the power spectral density of the noise. As before, $P_e^{(1)} = P_e^{(0)}$.

Observe that synchronous FSK requires 3 dB more signal power than an equivalent PSK system for the same probability of error, and 3 dB less power than an ASK system on a peak power basis. FSK and ASK, however, have equal error rates when compared in terms of average transmit power, since an ASK system transmits power only half the time.

Envelope Detection of ASK

Although envelope detection of ASK is simpler than synchronous detection, the derivation of the probability of error is more difficult because of the effect of the nonlinearity of the envelope detector on the noise. Let the incoming ASK signal and noise voltage be written as

$$v(t) = m(t) \cos \omega_0 t + n(t), \tag{9.82}$$

where $m(t) = V$ or 0, for a transmitted "1" or "0," respectively. The bandpass gaussian white noise $n(t)$ can be expressed using the narrowband representation of (3.43) as $n(t) = x(t) \cos \omega_0 t - y(t) \sin \omega_0 t$. Then (9.82) can be expressed as an envelope function $r(t)$ times the carrier:

$$v(t) = [m(t) + x(t)]\cos\omega_0 t - y(t)\sin\omega_0 t$$
$$= r(t)\cos[\omega_0 t + \theta(t)] \tag{9.83}$$

The envelope and phase functions are defined as

$$r^2(t) = [m(t) + x(t)]^2 + y^2(t), \tag{9.84}$$

$$\tan\theta(t) = \frac{y(t)}{m(t) + x(t)}. \tag{9.85}$$

As shown in Figure 9.13c, the output of the envelope detector will be the envelope function $r(t)$. The data $m(t)$ can be recovered from the output of the envelope detector by using the PCM threshold detector of Figure 9.19a. The probability of error thus depends on the statistics of $r(t)$, which are not gaussian and, in fact, depend on whether $m(t) = 0$ or V.

First consider the case where $m(t) = 0$. Since $x(t)$ and $y(t)$ are independent gaussian random variables, we can form the joint probability distribution function as

$$f_{xy}(x, y) = f_x(x)f_y(y) = \frac{e^{-(x^2+y^2)/2\sigma^2}}{2\pi\sigma^2}, \tag{9.86}$$

for $-\infty < x,y < \infty$. Since in this case $r^2(t) = x^2(t) + y^2(t)$, we use the change of variables

$$x(t) = r(t)\cos\theta(t)$$
$$y(t) = r(t)\sin\theta(t)$$

to convert from rectangular to polar coordinates. The differential element is $dxdy = rdrd\theta$. Applying this transformation to (9.86) gives the joint pdf of $r(t)$ and $\theta(t)$ as

$$f_{r\theta}(r, \theta) = f_{xy}(x, y) = \frac{e^{-r^2/2\sigma^2}}{2\pi\sigma^2}, \tag{9.87}$$

where $0 \le r < \infty$. Then the pdf of $r(t)$ is computed as

$$f_r(r) = \int_{\theta=0}^{2\pi} f_{r\theta}(r, \theta) r\, d\theta = \frac{re^{-r^2/2\sigma^2}}{\sigma^2}, \tag{9.88}$$

for $0 \le r < \infty$. This is the Rayleigh probability distribution function, which occurs in many problems of probability that involve circular symmetry; a common example is that of throwing darts at a circular target. The probability that a particular sample lies in the annular ring from r to $r + dr$ is given by $f_r(r)dr$. We also used the Rayleigh pdf in Section 4.5 to describe the statistics of fading in a multipath propagation environment. If the envelope detection threshold is set at r_0, the probability of error when sending $m(t) = 0$ is

$$P_e^{(0)} = P\{r(t) > r_0\} = \int_{r=r_0}^{\infty} f_r(r)\,dr = -e^{-r^2/2\sigma^2}\Big|_{r=r_0}^{\infty} = e^{-r_0^2/2\sigma^2}. \tag{9.89}$$

Now consider transmission of $m(t) = V$, and let $x' = x + V$. Then x' is a gaussian random variable with mean V, and the pdf for x' is

$$f_{x'}(x') = \frac{e^{-(x'-V)^2/2\sigma^2}}{\sqrt{2\pi\sigma^2}}. \tag{9.90}$$

The envelope and phase functions in this case are

$$r^2(t) = [V + x(t)]^2 + y^2(t) = x'^2(t) + y^2(t), \tag{9.91}$$

$$\tan\theta(t) = \frac{y(t)}{V + x(t)} = \frac{y(t)}{x'(t)}. \tag{9.92}$$

If we use the change of variables

$$x'(t) = r(t)\cos\theta(t)$$

$$y(t) = r(t)\sin\theta(t),$$

then the joint probability distribution function can be found as

$$\begin{aligned} f_{r\theta}(r,\theta) = f_{x'y}(x',y) = f_{x'}(x')f_y(y) &= \frac{e^{-[(x'-V)^2+y^2]/2\sigma^2}}{2\pi\sigma^2} \\ &= \frac{e^{-A^2/2\sigma^2}e^{-(r^2-2r\cos\theta)/2\sigma^2}}{2\pi\sigma^2} \end{aligned} \tag{9.93}$$

since $(x' - V)^2 + y^2 = (r\cos\theta - V)^2 + r^2\sin^2\theta = V^2 + r^2 - 2Vr\cos\theta$. Then the pdf of $r(t)$ can be calculated as

$$f_r(r) = \int_{\theta=0}^{2\pi} f_{r\theta}(r,\theta) r\,d\theta = \frac{re^{-V^2/2\sigma^2}}{2\pi\sigma^2}\int_{\theta=0}^{2\pi} e^{-Vr\cos\theta/2\sigma^2}\,d\theta. \tag{9.94}$$

The integral in (9.94) can be written in terms of the modified Bessel function of zero order, $I_0(x)$ (see Appendix B):

$$f_r(r) = \frac{r}{\sigma^2}e^{-(V^2+r^2)/2\sigma^2} I_0\left(\frac{Vr}{\sigma^2}\right), \tag{9.95}$$

for $0 \le r < \infty$. This is known as the *rician* probability distribution function, named after S. O. Rice, of Bell Labs, who in the 1940s derived this result in connection with early communications systems. The peak of (9.95) occurs when $r = V$. Note that the rician distribution reduces to the Rayleigh pdf when $V = 0$, as expected; it is left as a problem to show that the rician pdf approaches a gaussian distribution when the argument is large.

The probability of error when sending $m(t) = V$ is then found as

$$P_e^{(1)} = P\{r(t < r_0)\} = \int_{r=0}^{r_0} f_r(r)\,dr = \int_{r=0}^{r_0} \frac{r}{\sigma^2}e^{-(V^2+r^2)/2\sigma^2} I_0\left(\frac{Vr}{\sigma^2}\right)dr, \tag{9.96}$$

where r_0 is the detection threshold. The integral in (9.96) cannot be evaluated in closed form, but must be calculated numerically.

The expressions for $P_e^{(0)}$ and $P_e^{(1)}$ given in (9.89) and (9.96) differ because the statistics of the noise are different for $m(t) = 0$ and $m(t) = V$. If P_0 is the probability that a given transmitted message bit is a "0," and P_1 is the probability of a "1," then the overall probability of error for envelope detected ASK is

$$P_e = P_0 P_e^{(0)} + P_1 P_e^{(1)}. \tag{9.97}$$

The bit error probabilities $P_e^{(0)}$ and $P_e^{(1)}$ depend on the detection threshold r_0, as well as the ratio V/σ; these expressions can be written in terms of the (peak) bit energy-to-noise ratio using the relation that $V^2/\sigma^2 = 2E_b/n_0$.

Figure 9.20 shows a graphical interpretation of the detection process. The Rayleigh and rician probability distribution functions are plotted versus $r(t)$, and the error probabilities when sending a "0" or a "1" are shown as the appropriate areas under these curves. For

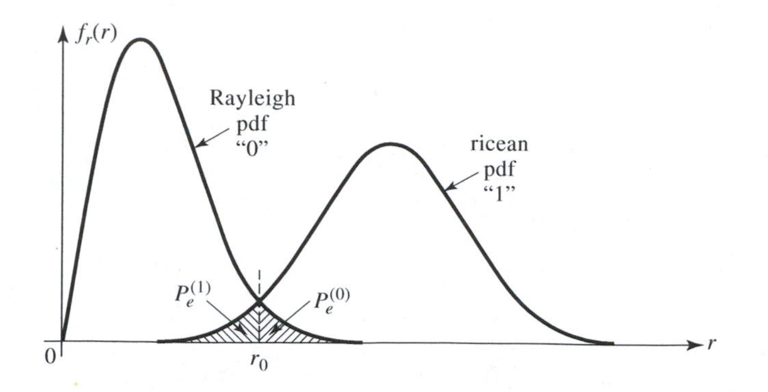

FIGURE 9.20 Envelope detection of ASK. The optimum detection threshold occurs at the intersection of the Rayleigh and rician pdf.

the usual case where $P_0 = P_1 = 0.5$, the probability of error will be minimized when the detection threshold is chosen at the intersection point of the probability distribution functions for $P_e^{(0)}$ and $P_e^{(1)}$ (see Problem 9.20). In general, the optimum threshold is a function of V and σ, and must be found numerically, but for the case where the input SNR is very large it can be shown that an approximate value is $r_0 = V/2$. The fact that the threshold depends on the received signal level is a disadvantage of ASK, particularly when fading is present.

Envelope Detection of FSK

Derivation of the probability of error for noncoherent FSK is similar to the procedure used for noncoherent ASK, since the two channels of the FSK demodulator of Figure 9.16a essentially decompose the FSK signal into two ASK signals. If frequency ω_1 is transmitted for a "1," then the output of the envelope detector for ω_1 will consist of the detected signal plus noise:

$$r_1(t) = V + n_1(t), \tag{9.98a}$$

while the output of the envelope detector for ω_2 will consist of only noise:

$$r_2(t) = n_2(t). \tag{9.98b}$$

The converse results apply when frequency ω_2 is transmitted for a "0." This symmetry implies that the error probabilities are the same for sending a "1" or a "0" and, since the output of the demodulator is formed as $r_1(t) - r_2(t)$, the detection threshold can be set at zero.

The probability of error when a "1" is sent can be written as

$$P_e^{(1)} = P\{r_1(t) - r_2(t) < 0\} = P\{r_2(t) > r_1(t)\}. \tag{9.99}$$

In this expression both $r_1(t)$ and $r_2(t)$ are random variables. Since $r_1(t)$ consists of signal plus noise, it has a rician pdf; $r_2(t)$ has a Rayleigh pdf since it consists of noise only. If we temporarily assume a fixed value of $r_1(t)$, the probability of error can be computed as

$$\int_{r_2=r_1}^{\infty} f_{r_2}(r_2)\, dr_2.$$

We must now integrate this result over all possible values of r_1, weighted by the pdf of $r_1(t)$, to find the overall probability of error:

$$P_e^{(1)} = \int_{r_1=0}^{\infty} f_{r_1}(r_1) \int_{r_2=r_1}^{\infty} f_{r_2}(r_2)\, dr_2\, dr_1$$

$$= \int_{r_1=0}^{\infty} \frac{r_1}{\sigma^2} e^{-(V^2+r_1^2)/2\sigma^2} I_0\left(\frac{Vr_1}{\sigma^2}\right) \int_{r_2=r_1}^{\infty} \frac{r_2}{\sigma^2} e^{-r_2^2/2\sigma^2}\, dr_2\, dr_1 \tag{9.100}$$

where we have substituted the Rayleigh and rician density functions from (9.88) and (9.95). While this expression might appear formidable, in fact it can be reduced to a very simple result. First, the inner integral can be evaluated directly as

$$\int_{r_2=r_1}^{\infty} \frac{r_2}{\sigma^2} e^{-r_2^2/2\sigma^2}\, dr_2 = e^{-r_1^2/2\sigma^2},$$

which reduces (9.100) to

$$P_e^{(1)} = \int_{r_1=0}^{\infty} \frac{r_1}{\sigma^2} e^{-V^2/2\sigma^2} e^{-r_1^2/\sigma^2} I_0\left(\frac{Vr_1}{\sigma^2}\right) dr_1. \tag{9.101}$$

Now use the change of variable $x = \sqrt{2}r_1$, which gives

$$P_e^{(1)} = \frac{1}{2} e^{-V^2/4\sigma^2} \int_{x=0}^{\infty} \frac{x}{\sigma^2} e^{-(V^2/2+x^2)/2\sigma^2} I_0\left(\frac{Vx}{\sqrt{2}\sigma^2}\right) dx. \tag{9.102}$$

The integrand of (9.102) is identical to the rician pdf of (9.95), if V^2 is replaced with $V^2/2$. Thus the integral in (9.102) must be unity. This gives the final result that

$$P_e^{(1)} = P_e^{(0)} = \frac{1}{2} e^{-V^2/4\sigma^2} = \frac{1}{2} e^{-E_b/2n_0}. \tag{9.103}$$

Bit Rate and Bandwidth Efficiency

Each of the above expressions for probability of error is expressed in terms of E_b/n_0, the ratio of bit energy to noise power spectral density. The dimensions of E_b are W-sec, while the dimensions of n_0 are W/Hz, so the ratio is dimensionless. In practice it is more convenient to write the bit energy in terms of signal power and the data rate. Let R_b be the bit rate of the binary message signal, with dimensions of bits per second (bps). Then the signal power is $S = R_b E_b$, in Watts. The bit energy-to-noise density ratio can then be written as

$$\frac{E_b}{n_0} = \frac{S}{n_0 R_b}. \tag{9.104}$$

Thus the probability of error is determined solely by the carrier power, the bit rate, and the PSD of the input noise. Since all our expressions for probability of error are monotonically decreasing with an increase in E_b/n_0, (9.104) shows that the error rate will increase with an increase in bit rate, for a fixed noise level. Note that (9.104) is independent of the receiver bandwidth.

We can also express the bit energy-to-noise density ratio in terms of the SNR of the receiver. If the receiver has an IF bandwidth Δf, then the noise power is $N = (2\Delta f)(n_0/2) = n_0 \Delta f$, for noise with a two-sided PSD of $n_0/2$. Then we have

$$\frac{E_b}{n_0} = \frac{S}{N}\frac{\Delta f}{R_b}. \tag{9.105}$$

Depending on the type of modulation, the required receiver bandwidth may range from one to several times the bit rate.

Each of the above binary modulation methods transmit one bit during each bit period, and are therefore said to have a *bandwidth efficiency* of 1 bps/Hz. Thus, for example, an analog telephone circuit having a bandwidth of 2400 Hz (600 Hz–3 kHz) is limited to a maximum data rate of (2400 Hz)(1 bps/Hz) = 2400 bps when used with binary modulation. This rate can be approached with FSK, but we will see later in this chapter that substantially greater data rates can be achieved with multilevel modulation methods that transmit more than one bit per period. Such methods have bandwidth efficiencies greater than 1 bps/Hz. Thus the terms *data rate* and *bandwidth* are not synonymous, in spite of the often-heard (but incorrect) use of the word "bandwidth" when referring to high data rates.

Comparison of ASK, FSK, and PSK Systems

Now that we have derived expressions for the probability of error for ASK, FSK, and PSK systems, we can compare the relative advantages of each. Figure 9.21 shows a comparison of bit error rates for ASK, FSK, and PSK. Note that the lowest error rate occurs for the coherent detection of PSK, but the price paid for this performance is the need for a synchronized LO and a relatively wide signal bandwidth. Next in order of performance is coherent FSK, which requires 3–4 dB more power than PSK for the same error rate. The error rate for noncoherent (envelope detected) FSK is only slightly worse (about 1 dB) than for coherent FSK, and is better than coherent ASK. This conclusion only applies, however, when using the peak power for ASK; when the average power of the ASK signal is used the error rate for coherent ASK and coherent FSK are identical. The worst performance is obtained with noncoherent ASK.

The error rate that is required in practice depends on the application, as well as trade-offs between bandwidth, transmitter power, range, and receiver complexity and cost. It may range from 10^{-2} (image data from early space probes) to 10^{-8} for high-data rate computer networks. A typical error rate that is often used as a goal for modern wireless

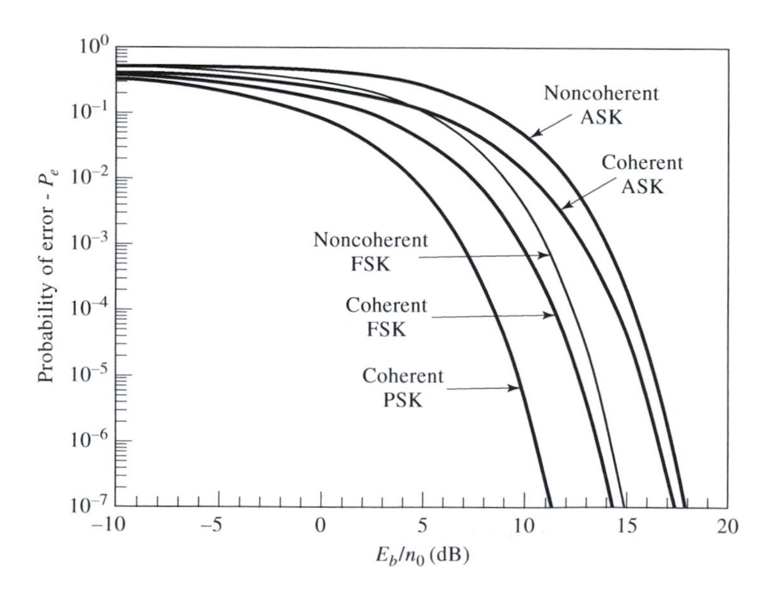

FIGURE 9.21 Comparison of bit error rates for coherent (ASK, FSK, and PSK) and noncoherent (ASK and FSK) demodulation. The results for ASK are based on peak bit energy.

TABLE 9.2 E_b/n_0 for $P_e = 10^{-5}$ for Various Modulation Methods

Modulation Method	E_b/n_0 for $P_e = 10^{-5}$
coherent PSK	9.6 dB
coherent FSK	12.6 dB
noncoherent FSK	13.4 dB
coherent ASK	15.6 dB
noncoherent ASK	16.4 dB

systems is 10^{-5}; the Table 9.2 lists the required E_b/n_0 for various binary modulation methods. ASK transmitters are very simple, and efficient since no power is radiated when no data is being sent. ASK receivers are also simple, if envelope detection is used. Because the bit error rate is poor in comparison to other modulation methods, ASK is limited to low data rates. In addition, the fact that the detection threshold depends on the received signal level makes ASK performance very poor in a fading environment. For these reasons, ASK applications are usually limited to short-range, low-cost telemetry and RFID.

FSK systems can use a zero threshold, regardless of signal strength. An FSK transmitter is only slightly more complicated than for ASK, and an FSK receiver using envelope detection can be made simply and inexpensively. In addition, the error rate for noncoherent FSK is comparable to coherent FSK, and much better than noncoherent ASK. Because of these features, noncoherent FSK has found widespread historical application in a wide variety of both baseband and modulated data transmission systems, such as data modems, teletype, and fax.

PSK gives better error rate performance than any other binary modulation method, but requires synchronous detection. PSK has a constant envelope, and uses a zero detection threshold. This eases requirements on the transmitter power amplifiers, and makes the performance of PSK in a fading environment better than ASK or FSK. PSK also requires a relatively wide signal spectrum, typically ranging from $2R_b$ to $4R_b$. For these reasons, applications of PSK are generally limited to high performance systems such as space and satellite communications, such as interplanetary space missions and the *Global Positioning Satellite* (GPS) system.

EXAMPLE 9.3 COMPARISON OF FSK AND ASK MODULATION

A wireless local area network operating at 2.44 GHz transmits data at the rate of 1.15 Mbps, with a desired error rate of 10^{-5}. The transmitter and receiver each use a monopole antenna with a gain of 4.5 dB. The transmit power is 0.5 W, and the receiver and receive antenna have a combined system noise temperature of 600 K. Compare the maximum possible operating range for coherent FSK and noncoherent ASK modulation

Solution

From Table 9.2, $E_b/n_0 = 12.6$ dB $= 18.2$ for coherent FSK with $P_e = 10^{-5}$. The wavelength is

$$\lambda = \frac{300}{2440} = 0.123 \text{ m}.$$

Using (9.104) to find the required receive carrier power gives

$$P_r = \frac{E_b}{n_0}(n_0 R_b) = \frac{E_b}{n_0} k T_{\text{sys}} R_b = (18.2)(1.38 \times 10^{-23})(600)(1.15 \times 10^6)$$
$$= 1.73 \times 10^{-13}\text{ W}$$

Then the Friis formula gives the range as

$$R = \sqrt{\frac{P_t G_t G_r \lambda^2}{(4\pi)^2 P_r}} = \sqrt{\frac{(0.5)(2.82)^2(0.123)^2}{(4\pi)^2(1.73 \times 10^{-13})}} = 47\text{ km. (coherent FSK)}$$

For the case of noncoherent ASK, $E_b/n_0 = 16.4$ dB $= 43.6$ for $P_e = 10^{-5}$. The required receive carrier power is now

$$P_r = \frac{E_b}{n_0}(n_0 R_b) = \frac{E_b}{n_0} k T_{\text{sys}} R_b = (43.6)(1.38 \times 10^{-23})(600)(1.15 \times 10^6)$$
$$= 4.15 \times 10^{-13}\text{ W}$$

Then the maximum range is

$$R = \sqrt{\frac{P_t G_t G_r \lambda^2}{(4\pi)^2 P_r}} = \sqrt{\frac{(0.5)(2.82)^2(0.123)^2}{(4\pi)^2(4.15 \times 10^{-13})}} = 30\text{ km. (noncoherent ASK)}$$

We see that the use of noncoherent ASK modulation leads to a reduction of about 25% in maximum range, compared with coherent FSK. Because of less than ideal propagation conditions, such as blockage, attenuation, and fading effects, these ranges would not likely be achieved in practice. ○

9.4 EFFECT OF RAYLEIGH FADING ON BIT ERROR RATES

As we discussed in Section 4.5, the presence of multiple signal paths between transmitter and receiver lead to Rayleigh fading. This causes sharp reductions in received signal power over short time intervals, or with movement over small distances, due to phase cancellation. Fading has the effect of significantly increasing the error rate for digital modulation or, equivalently, decreasing the usable range of a wireless system (for a fixed EIRP, error rate, and system noise level). Although the average receive power is not greatly reduced, fading has a major impact on error rates because the presence of brief but deep drops in received power results in very large error rates for short periods, which has the effect of greatly increasing the average error rate.

Figure 9.22 shows an idealized model for a propagation channel with fading and additive noise. The input signal is $s_i(t)$, and the multiplicative factor $r(t)$, with $0 \le r(t) < \infty$, represents the change in amplitude due to Rayleigh fading. White gaussian noise is added after the fading factor, and may include noise received by the receiver antenna as well as noise generated in the receiver itself. As derived in Section 4.5, $r(t)$ is a random variable

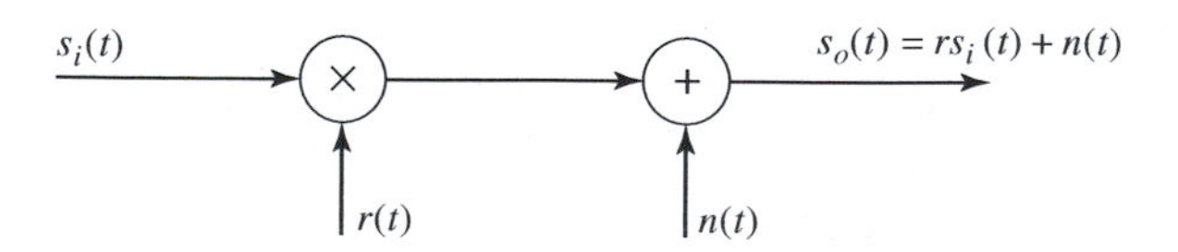

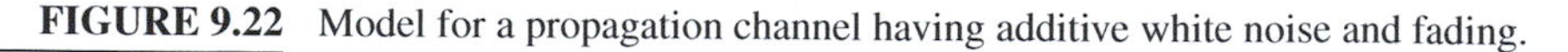
FIGURE 9.22 Model for a propagation channel having additive white noise and fading.

with a Rayleigh PDF given by

$$f_r(r) = \frac{r}{\alpha^2} e^{-r^2/2\alpha^2}, \tag{9.106}$$

for $0 \leq r(t) < \infty$. The rms value of the distribution of $r(t)$ is $\sqrt{2}\alpha$ (we use α here to avoid confusion with σ, the rms noise voltage). We can modify the expressions for probability of error derived in the previous section to account for Rayleigh fading by considering the conditional probability of error for a fixed signal amplitude, then integrating over the Rayleigh probability distribution function for this amplitude. This can be done for coherent ASK, FSK, and PSK, and for noncoherent FSK. The results show that realistic Rayleigh fading can significantly increase the average bit error rate, and thus have a major impact on the performance of wireless systems.

Effect of Rayleigh Fading on Coherent PSK

First consider the case of coherent phase shift keying with Rayleigh fading. If we consider $r(t)$ to be a fixed multiplier, then the probability of error derived in (9.78) can be modified to give

$$P_e^{(0)}(E_b \mid r) = P\{n_o(T) > rVT\} = \frac{1}{2} erfc\left(\sqrt{\frac{r^2 E_b}{n_0}}\right). \tag{9.107}$$

The overall probability of error is now found by weighting (9.107) with the Rayleigh PDF of (9.106), and integrating over the range of r:

$$P_e^{(0)} = \int_{r=0}^{\infty} P_e^{(0)}(E_b \mid r) f_r(r)\, dr = \frac{1}{2} \int_{r=0}^{\infty} erfc\left(\sqrt{\frac{r^2 E_b}{n_0}}\right) \frac{r}{\alpha^2} e^{-r^2/2\alpha^2}\, dr. \tag{9.108}$$

To evaluate this integral we first use the change of variable $u = r^2 E_b / n_0$, and the definition of the complementary error function

$$erfc(x) = \frac{2}{\sqrt{\pi}} \int_x^{\infty} e^{-x^2}\, dx,$$

to rewrite (9.108) as

$$\begin{aligned} P_e^{(0)} &= \frac{n_0}{4E_b\alpha^2} \int_{u=0}^{\infty} erfc(\sqrt{u}) e^{-un_0/2\alpha^2 E_b}\, du \\ &= \frac{n_0}{2E_b\alpha^2\sqrt{\pi}} \int_{u=0}^{\infty} \int_{x=\sqrt{u}}^{\infty} e^{-x^2} e^{-un_0/2\alpha^2 E_b}\, dx\, du \end{aligned} \tag{9.109}$$

Next we define

$$\Gamma = \frac{2\alpha^2 E_b}{n_0}, \tag{9.110}$$

which is the average received bit energy-to-noise power spectral density ratio of the faded received signal. Then (9.109) can be simplified to

$$P_e^{(0)} = \frac{1}{\sqrt{\pi}\Gamma} \int_{u=0}^{\infty} \int_{x=\sqrt{u}}^{\infty} e^{-x^2} e^{-u/\Gamma}\, dx\, du. \tag{9.111}$$

The inner integral can be evaluated by using integration by parts. Let $U = \int_{x=\sqrt{u}}^{\infty} e^{-x^2}\, dx$, and $dV = e^{-u/\Gamma}\,du$, then $\int U dV = UV - \int V dU$ reduces (9.111) to

$$P_e^{(0)} = \frac{1}{\sqrt{\pi}\Gamma}\left\{-\Gamma e^{-u/\Gamma}\int_{x=\sqrt{u}}^{\infty} e^{-x^2}\,dx\Bigg|_{u=0}^{\infty} - \Gamma\int_{u=0}^{\infty}\frac{e^{-u}}{2\sqrt{u}}e^{-u/\Gamma}\,du\right\}$$

$$= \frac{1}{\sqrt{\pi}\Gamma}\left\{\frac{\sqrt{\pi}\Gamma}{2} - \Gamma\int_{u=0}^{\infty}\frac{e^{-(1+1/\Gamma)u}}{2\sqrt{u}}\,du\right\}$$

$$= \frac{1}{2}\left[1 - \sqrt{\frac{\Gamma}{1+\Gamma}}\right] \tag{9.112}$$

The last integral in (9.112) is a standard integral, and is given in Appendix B. Because of the similarity of (9.76), (9.78), and (9.81), virtually identical derivations can be made for the cases of coherent ASK and FSK; these are left as problems.

Equation (9.112) is the final result for the probability of error for PSK with Rayleigh fading. Note that this result is expressed in terms of Γ, defined in (9.110) as the average energy-to-noise ratio. While the probability of error for large E_b/n_0 for the nonfaded case decreases exponentially (see Problem 9.17), (9.113) decreases much more slowly. Also observe that (9.112) does not reduce to the non-faded result of (9.78) for limiting values of $\Gamma = 0$ or 1; this is because Rayleigh fading does not include a line-of-sight term, which is the assumption for the nonfaded case. When the fading environment includes a strong line-of-sight component the fading statistics become rician distributed, and in this case the expression for probability of error will reduce to the nonfaded result when $V = 0$ [4]–[5].

Effect of Rayleigh Fading on Noncoherent FSK

Next consider the effect of Rayleigh fading on the error rate of noncoherent ASK. In this case the derivation is easier because the non-faded expression for probability of error in (9.113) is simpler.

Following the same procedure as in the previous case, the conditional probability of error for noncoherent FSK for a fixed amplitude factor r is

$$P_e(E_b\,|\,r) = \frac{1}{2}e^{-r^2E_b/n_0}. \tag{9.113}$$

Weighting this result with the Rayleigh PDF of (9.106) and integrating over the range of r gives

$$P_e = \int_{r=0}^{\infty} P_e(E_b\,|\,r)f_r(r)\,dr = \frac{1}{2}\int_{r=0}^{\infty}\frac{r}{\alpha^2}e^{-r^2E_b/2n_0}e^{-r^2/2\alpha^2}\,dr$$

$$= \frac{1}{2\left(1+\dfrac{\alpha^2E_b}{n_0}\right)} = \frac{1}{2+\Gamma} \tag{9.114}$$

The required integral in (9.114) is listed in Appendix B. As before, we define Γ in (9.110) as the average bit energy-to-noise ratio.

Comparison of Faded and Nonfaded Error Rates

Figure 9.23 shows the effect of Rayleigh fading on the bit error rates of coherent PSK and noncoherent FSK modulation. The results are plotted versus $\Gamma = 2\alpha^2E_b/n_0$, the rms average of the received bit energy-to-noise power density ratio. The error rates for the faded

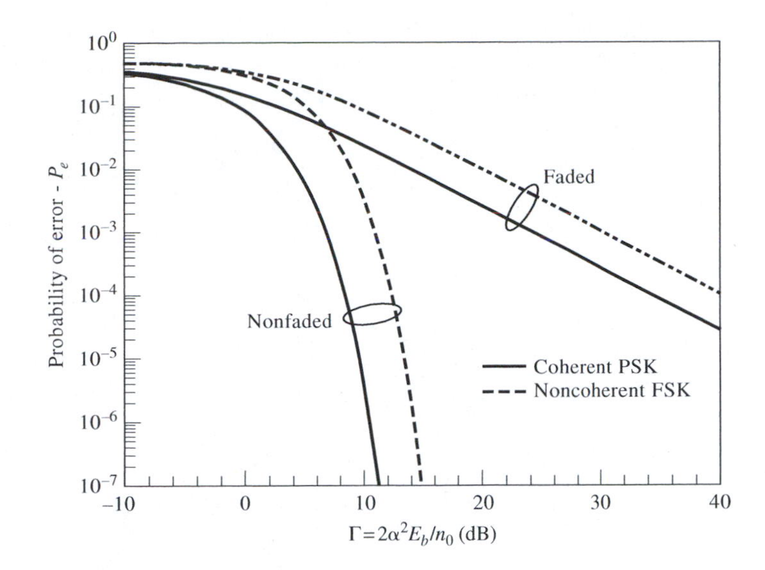

FIGURE 9.23 The effect of Rayleigh fading on the bit error rates of coherent PSK and noncoherent FSK.

cases are given by (9.112) and (9.114). For the nonfaded cases we let $2\alpha^2 = 1$, so that $\Gamma = E_b/n_0$, and use the expressions given in (9.78) and (9.103).

Note the dramatic increase in the probability of error when fading is present. For example, for $P_e = 10^{-5}$, fading has the effect of increasing the required bit energy-to-noise ratio by approximately 30 dB for both PSK and FSK. Since transmit power is usually limited by regulation, fading has the ultimate effect of reducing the usable range of a wireless system, for a given data rate, error rate, and noise level. Fortunately, because most of the errors on a fading channel occur in short bursts, error-correcting codes can be used very effectively to improve the error rate.

EXAMPLE 9.4 EFFECT OF RAYLEIGH FADING

A cellular phone base station using BPSK transmits at 882 MHz with an EIRP of 50 W, and an antenna height of 50 m. The mobile receive antenna is at a height of 2 m, with a gain of -1.0 dBi and a noise temperature of 300 K. Find the maximum operating range if the required error rate is 10^{-5}, for a nonfaded and a flat Rayleigh faded channel with $2\alpha^2 = -10$ dB. Assume a data rate of 24.3 kbps, and a receiver noise figure of 8 dB.

Solution

The equivalent noise temperature of the receive antenna and the receiver is

$$T_{sys} = T_A + (F - 1)T_0 = 300 + (6.3 - 1)(290) = 1837 \text{ K}$$

since the receiver noise figure is $F = 8 \text{ dB} = 6.3$.

Without fading, $E_b/n_0 = 9.6 \text{ dB} = 9.12$ for PSK with $P_e = 10^{-5}$. So from (9.104) the required receiver input power is

$$P_r = \frac{E_b}{n_0}(n_0 R_b) = \frac{E_b}{n_0}(kT_{sys}R_b) = (9.12)(1.38 \times 10^{-23})(1837)(24.3 \times 10^3)$$
$$= 5.6 \times 10^{-15} \text{ W}$$

Since the propagation channel involves multipath scattering, we will use the link formula for ground reflections to include a realistic path loss factor. So from (3.58) we calculate the maximum range as

$$R = \left[\frac{P_t G_t G_r h_1^2 h_2^2}{P_r}\right]^{1/4} = \left[\frac{(50)(0.8)(50)^2(2)^2}{5.6 \times 10^{-15}}\right]^{1/4} = 9.2 \times 10^4 \text{ m}.$$

For the faded case, (9.112) gives $\Gamma = 44$ dB for $P_e = 10^{-5}$. Since $2\alpha^2 = -10$ dB, we have $E_b/n_0 = 44 + 10$ dB $= 54$ dB $= 2.51 \times 10^5$. Then the receiver input power is

$$\begin{aligned} P_r &= (2.51 \times 10^5)(1.38 \times 10^{-23})(1837.)(24.3 \times 10^3) \\ &= 1.55 \times 10^{-10} \text{ W} \end{aligned}$$

and the maximum range is now

$$R = \left[\frac{(50)(0.8)(50)^2(2)^2}{1.55 \times 10^{-10}}\right]^{1/4} = 7.1 \times 10^3 \text{ m},$$

which is reduced by a factor of more than 10 from the nonfaded result. ○

9.5 M-ary DIGITAL MODULATION

As we have seen in the preceding sections, binary modulation methods transmit one bit per signaling interval, with a bandwidth efficiency of 1 bps/Hz. Here we consider the more general case of *M-ary* modulation methods, where more than one bit may be transmitted per signaling interval. This allows greater bit rates for the same bandwidth, at the expense of a more complex system.

If we transmit $M = 2^n$ symbols for each signaling interval, a bandwidth efficiency of n bps/Hz can be achieved. This can be done by using multiple discrete amplitude levels, or with multiple phase states, or with a combination of these. If the symbol rate is R_s, then the effective bit rate is $R_b = nR_s$. In the binary case we have $n = 1$, with $M = 2$ symbols (0 and 1). Figure 9.24 shows an example for $n = 5$. In this case there are $M = 2^5 = 32$ distinct symbols that can be transmitted, each consisting of $n = 5$ bits. If the symbol transmission rate is R_s, then the bit rate is $R_b = 5R_s$. The symbol period is $T = 1/R_s = 5/R_b$, and the bit period is $1/R_b$. Note that if an error is made upon receiving a symbol, it is possible that all the bits in that symbol may be in error.

We will first consider the $M = 4$ extension of PSK, then discuss M-ary phase shift keying and the general case of quadrature amplitude modulation.

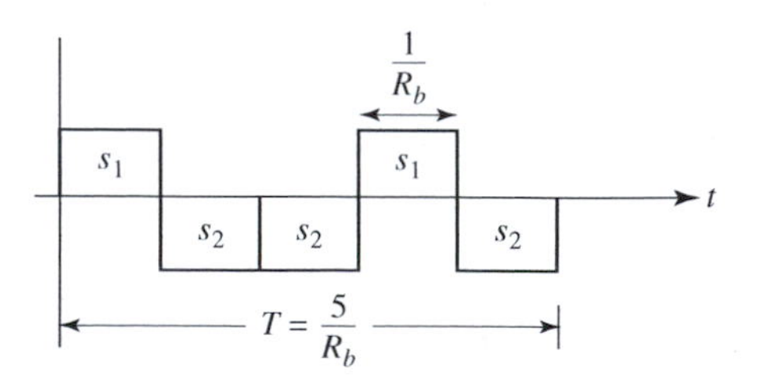

FIGURE 9.24 Example of an M-ary signal with $n = 5$. There are $M = 2^5 = 32$ symbols, each consisting of 5 bits. The symbol period is $T = 5/R_b$, and the bit period is $T/5 = 1/R_b$.

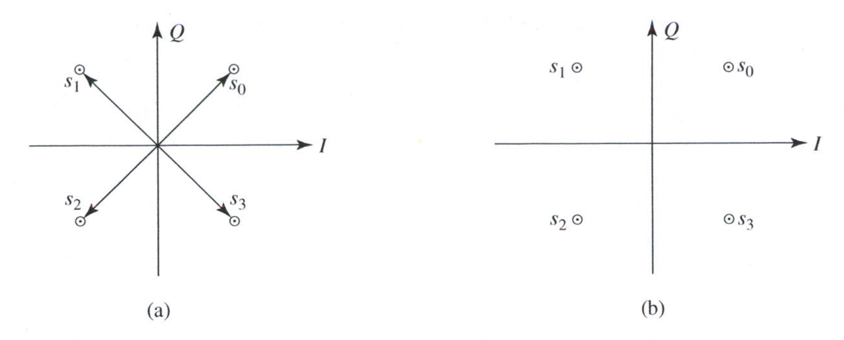

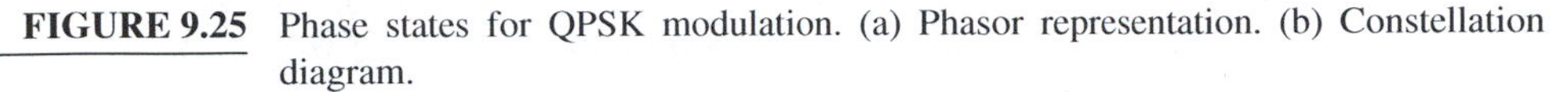

FIGURE 9.25 Phase states for QPSK modulation. (a) Phasor representation. (b) Constellation diagram.

Quadrature Phase Shift Keying

It is easy to see that phase shift keying can be generalized to divide the allowable phase states of the carrier into more than the two values (0° and 180°) used in BPSK. If we use four states, where $n = 2$ and $M = 4$, then we can transmit two bits, or four symbols, for each signaling interval. This is called *quadrature phase shift keying* (QPSK).

The four phase states of a QPSK modulated carrier can be written as

$$s_0(t) = A\cos(\omega_0 t + 45^\circ) \tag{9.115a}$$

$$s_1(t) = A\cos(\omega_0 t + 135^\circ) \tag{9.115b}$$

$$s_2(t) = A\cos(\omega_0 t - 135^\circ) \tag{9.115c}$$

$$s_3(t) = A\cos(\omega_0 t - 45^\circ) \tag{9.115d}$$

These can be written more succinctly as

$$s_i(t) = A\cos(\omega_0 t + \phi_i),\ \phi_i = (2i+1)\frac{\pi}{4}, \quad \text{for } i = 0, 1, 2, 3. \tag{9.116}$$

These four phase states can be represented in phasor form, as in Figure 9.25a. Assuming cosine-based phasors, the horizontal axis represents the in-phase (I) component, while the vertical axis represents the quadrature (Q), component that is shifted 90° in phase. Each of the four signals of (9.115) is represented by a separate phasor, with constant magnitude and a phase of ±45° or ±135°. A related graphical representation is the *constellation diagram*, shown in Figure 9.25b. In this case only the endpoint of the phasor is shown. Constellation diagrams can be useful for visualizing M-ary modulation methods with either amplitude or phase changes. The in-phase and quadrature components of the QPSK signal set can be more clearly seen by using a trigonometric identity to expand (9.116):

$$s_i(t) = A_I \cos\omega_0 t + A_Q \sin\omega_0 t, \tag{9.117}$$

where A_I and A_Q are given in Table 9.3. In this table we have ignored a common factor of $\sqrt{2}/2$ in the values of A_I and A_Q. The above results suggest that the QPSK signal can be generated by using binary NRZ data to modulate the in-phase (cosine) and quadrature (sine) components of the carrier wave. Each QPSK phase state can then be used to represent two bits of data, as shown in the above table. The incoming serial binary data stream must be multiplexed into groups of two bits, which separately modulate the I and Q components of the carrier. The QPSK modulator thus takes the form shown in Figure 9.26. A 90° hybrid divider can be used to provide the I and Q components of the LO from a single oscillator. The low-pass filters serve to limit the bandwidth of the radiated signal; this filtering is

TABLE 9.3 I and Q Components of a QPSK Signal

i	ϕ_i	A_I	A_Q	Binary Data
0	45°	1	1	1,1
1	135°	−1	1	0,1
2	−135°	−1	−1	0,0
3	−45°	1	−1	1,0

not easily performed after up-conversion due to filter limitations. Because the average transition between phase states is 90°, the bandwidth of the QPSK spectrum is narrower than the spectrum of a BPSK signal. The output of the QPSK modulator is a double sideband suppressed carrier signal.

Figure 9.27 shows a typical input binary data stream and the resultant I and Q data for the QPSK modulator. Each pair of input bits is coded into distinct I and Q symbols according to Table 9.3. These symbols are then used to form the QPSK signal as shown in the figure. If the bit rate of the input data is R_b bits per second, the output symbol rate is $R_s = R_b/2$ symbols per second. Thus QPSK requires only half the channel bandwidth of BPSK to transmit the same data rate, and therefore has a bandwidth efficiency of 2 bps/Hz. Note that the QPSK output is a constant envelope signal.

Demodulation of QPSK requires coherent detection. A block diagram of a QPSK demodulator is shown in Figure 9.28. The demodulator uses two mixers with quadrature LO components to recover the I and Q signals. PCM detectors, each consisting of an integrator and a sampler, are used to detect the I and Q NRZ data, which is then decoded with a parallel-to-serial converter to provide serial binary output. Since QPSK is a constant envelope modulation, the detectors can use a zero threshold.

To analyze the operation of the QPSK demodulator, assume that the QPSK signal $s_i(t)$ of (9.116) is applied to the input. Then the I and Q outputs of the demodulator are

$$V_I = \int_{t=0}^{T} A^2 \cos(\omega_0 t + \phi_i) \cos \omega_0 t \, dt$$

$$= \frac{A^2}{2} \int_{t=0}^{T} [\cos(2\omega_0 t + \phi_i) + \cos \phi_i] \, dt = \frac{A^2 T}{2} \cos \phi_i \tag{9.118a}$$

$$V_Q = -\int_{t=0}^{T} A^2 \cos(\omega_0 t + \phi_i) \sin \omega_0 t \, dt$$

$$= \frac{-A^2}{2} \int_{t=0}^{T} [\sin(2\omega_0 t + \phi_i) - \sin \phi_i] \, dt = \frac{A^2 T}{2} \sin \phi_i \tag{9.118b}$$

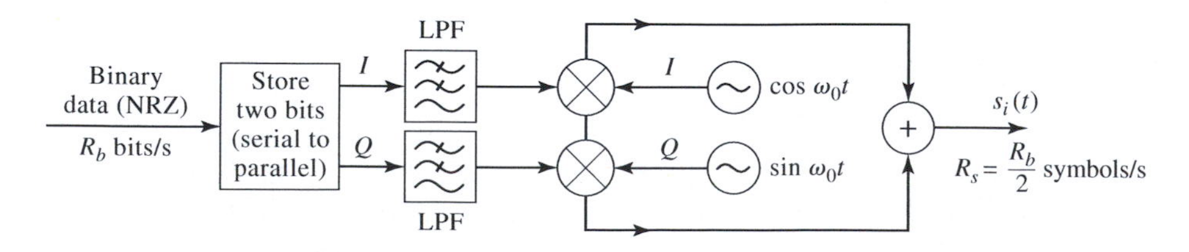

FIGURE 9.26 Block diagram of a QPSK modulator.

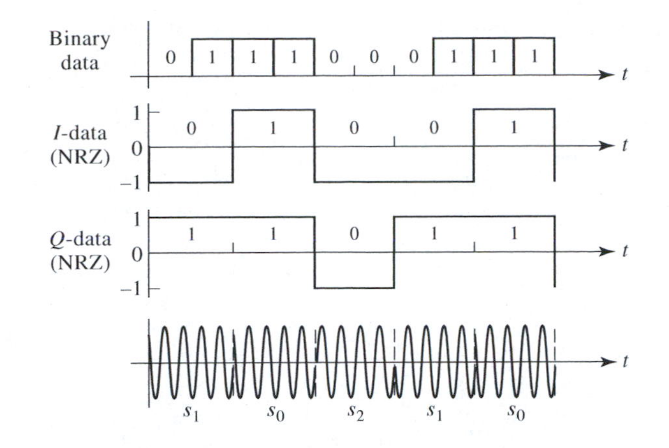

FIGURE 9.27 Data and timing diagrams for the QPSK modulator.

where T is the symbol period, and ϕ_i is the phase of the ith QPSK signal. Table 9.4 lists the specific outputs for the four possible input symbols, where the constant C is defined as $A^2T/2\sqrt{2}$. Comparison with Table 9.3 shows that the correct binary data is recovered by the QPSK demodulator.

The assignment of bits pairs to symbols given in Table 9.3 is not unique, as we could have chosen to assign the bit pair 1,1 to s_1, s_2, or s_3, instead of s_0, for example. But the above choice has the very useful property that when an error occurs in the detection of a symbol, it is most likely that only one of the bits will be in error, rather than both bits. This is because an error is most likely to result in a shift from the correct phase state to the immediately adjacent phase, rather than the diametrically opposite phase (see Figure 9.25), and the assignment of bits has been made so that there is a difference of only one bit between consecutive QPSK phase states. This is called *Gray coding*, and is an example of how judicious coding of the transmitted data can lead to improved bit error rates.

For example, if the binary input is 0,1 the s_1 symbol is sent. The presence of noise may cause the demodulator to incorrectly indicate reception of either s_0 (for the bit pair 1,1), or s_2 (for the bit pair 0,0). In either case there is only one bit that is in error. It is possible that an error is severe enough so that the detector indicates s_3, for which both bits would be in error, but this is much less likely than either of the first two events.

Probability of Error for QPSK

We can derive the probability of error for QPSK using the same method as was used for BPSK. By symmetry the error rates for all s_i are equal, so let us assume that s_1 is received

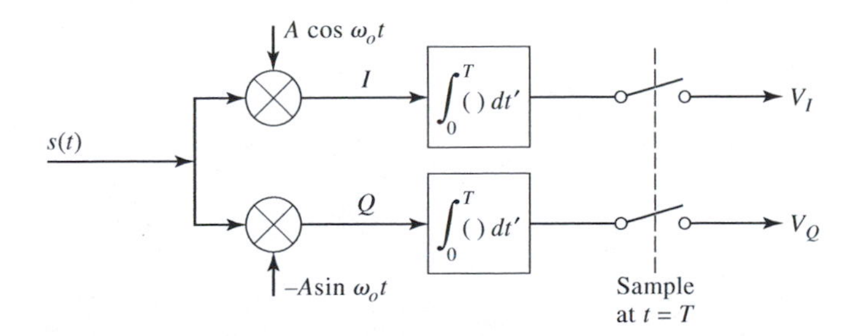

FIGURE 9.28 Coherent QPSK demodulator.

TABLE 9.4 I and Q Outputs of a QPSK Demodulator

$s_i(t)$	ϕ_i	V_I	V_Q	Binary Output
s_0	45°	C	C	1,1
s_1	135°	−C	C	0,1
s_2	−135°	−C	−C	0,0
s_3	−45°	C	−C	1,0

by the demodulator along with additive gaussian noise having a power spectral density of $n_0/2$:

$$s(t) + n(t) = s_1(t) + n(t) = A\cos(\omega_0 t + 135°) + n(t). \tag{9.119}$$

Then from (9.118) the I and Q outputs from the detector are

$$V_I(T) = \frac{-A^2T}{2\sqrt{2}} + n_o(T) = \frac{-E_s}{\sqrt{2}} + n_o(T) \tag{9.120a}$$

$$V_Q(T) = \frac{A^2T}{2\sqrt{2}} + n_o(T) = \frac{E_s}{\sqrt{2}} + n_o(T) \tag{9.120b}$$

where $E_s = \frac{A^2T}{2}$ is the symbol energy. The noise component of the outputs is

$$n_o(T) = \int_{t=0}^{\infty} An(t)\cos\omega_0 t\, dt, \tag{9.121}$$

which is a gaussian random variable. We can find the variance of $n_o(T)$ as follows:

$$\begin{aligned}\sigma^2 = E\{n_o^2(T)\} &= E\left\{\int_{t=0}^{T}\int_{s=0}^{T} A^2 n(t)n(s)\cos\omega_0 t\cos\omega_0 s\, dt\, ds\right\}\\
&= A^2\int_{t=0}^{T}\int_{s=0}^{T} E\{n(t)n(s)\}\cos\omega_0 t\cos\omega_0 s\, dt\, ds\\
&= \frac{A^2 n_0}{2}\int_{t=0}^{T}\int_{s=0}^{T}\delta(t-s)\cos\omega_0 t\cos\omega_0 s\, dt\, ds\\
&= \frac{A^2 n_0}{2}\int_{s=0}^{T}\cos^2\omega_0 s\, ds = \frac{A^2 T n_0}{4} = \frac{E_s n_0}{2}\end{aligned} \tag{9.122}$$

The detection threshold for both detectors is zero, so the probability of a symbol error at the I correlator is

$$\begin{aligned}P_e^{(I)} &= P\{V_I > 0\} = p\left\{n_o(T) > \frac{E_s}{\sqrt{2}}\right\}\\
&= \int_{E_s/\sqrt{2}}^{\infty}\frac{e^{-n_0^2/2\sigma^2}}{\sqrt{2\pi\sigma^2}}\, dn_0 = \frac{1}{\sqrt{\pi}}\int_{E_s/2\sigma}^{\infty} e^{-x^2}\, dx\\
&= \frac{1}{2}erfc\left(\frac{E_s}{2\sigma}\right) = \frac{1}{2}erfc\left(\sqrt{\frac{E_s}{2n_0}}\right)\end{aligned} \tag{9.123a}$$

since from (9.122) $\frac{E_s}{2\sigma} = \sqrt{\frac{E_s}{2n_0}}$. By symmetry we will also obtain the same expression for the probability of a symbol error at the Q correlator:

$$P_e^{(Q)} = \frac{1}{2} erfc\left(\sqrt{\frac{E_s}{2n_0}}\right). \tag{9.123b}$$

Since the overall probability that a symbol is received correctly is the product of the probabilities that each correlator operates correctly, the overall probability of error for a symbol is

$$\begin{aligned} P_e^{(s)} &= 1 - \left(1 - P_e^{(I)}\right)\left(1 - P_e^{(Q)}\right) = 1 - \left[-2P_e^{(I)} + \left(P_e^{(I)}\right)^2\right] \\ &\cong 2P_e^{(I)} = erfc\left(\sqrt{\frac{E_s}{2n_0}}\right) \end{aligned} \tag{9.124}$$

where we made the approximation that $P_e \ll 1$.

If, as is the usual case in practice, we use QPSK with Gray coding, then we can assume that a symbol error is most likely to cause only a single bit error. Then since each symbol contains two bits, the bit error rate for QPSK will be one-half the symbol error rate:

$$P_e^{(e)} = \frac{1}{2} erfc\left(\sqrt{\frac{E_s}{2n_0}}\right). \tag{9.125}$$

Note that because the symbol period T is twice the bit period, $E_s = 2E_b$, where E_b is the bit energy. This shows that the expression of (9.125) is equivalent to the probability of error for BPSK as given in (9.78). Thus with QPSK it is possible to achieve twice the data rate as for BPSK, with the same bandwidth and error rate. This fact has led to the extensive use of QPSK modulation (and its several variations) in a wide variety of applications, including CDMA-PCS telephone systems, the Iridium LEO satellite telephone system, and the direct broadcast television system (DBS).

EXAMPLE 9.5 COMPARISON OF BPSK AND QPSK MODULATION

A QPSK system transmits data at 20 Mbps over a radio link with an average transmit power of 2 W. The total link loss is 110 dB, and the two-sided power spectral density of the noise is $n_0/2 = 4 \times 10^{-20}$ W/Hz. If Gray coding is used, find the bit error rate, and compare with a similar system using BPSK.

Solution

The receive power is $P_r = 2 \times 10^{-110/10} = 2 \times 10^{-11}$ W. For QPSK the symbol energy-to-noise ratio is

$$\frac{E_s}{n_0} = \frac{P_r}{n_0 R_s} = \frac{2P_r}{n_0 R_b} = \frac{2(2 \times 10^{-11})}{(8 \times 10^{-20})(20 \times 10^6)} = 25.$$

The probability of a bit error is given by (9.125)

$$P_e = \frac{1}{2} erfc\left(\sqrt{\frac{E_s}{2n_0}}\right) = \frac{1}{2} erfc\left(\sqrt{\frac{25}{2}}\right) = \frac{1}{2} erfc(3.535) = 2.87 \times 10^{-7}.$$

For BPSK the bit energy is

$$\frac{E_b}{n_0} = \frac{P_r}{n_0 R_b} = \frac{2 \times 10^{-11}}{(8 \times 10^{-20})(20 \times 10^6)} = 12.5,$$

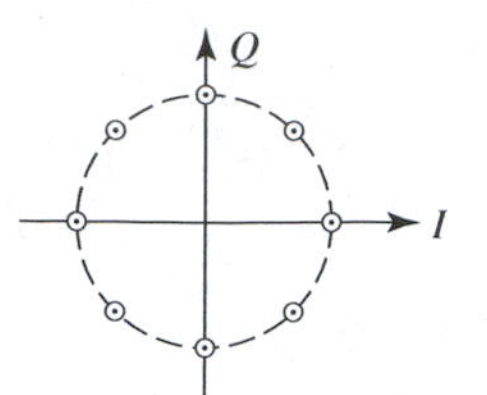

FIGURE 9.29 Constellation diagram for 8-PSK.

and the probability of a bit error is given by (9.78):

$$P_e = \frac{1}{2}erfc\left(\sqrt{\frac{E_b}{n_0}}\right) = \frac{1}{2}erfc(\sqrt{12.5}) = \frac{1}{2}erfc(3.535) = 2.87 \times 10^{-7}.$$

As expected, the bit error rates for QPSK and BPSK are identical. ○

M-ary Phase Shift Keying

We can generalize phase shift keying to M different phase states. By extension of (9.116), an M-PSK signal can be defined as

$$s_i(t) = A\cos(\omega_0 t + \phi_i), \tag{9.126}$$

with $\phi_i = \frac{2\pi i}{M}$, for $i = 0, 1, 2, \ldots M - 1$. This definition divides the unit circle into M equally spaced phase states. Thus $M = 2$ corresponds to BPSK, while $M = 4$ corresponds to QPSK. Figure 9.29 shows a constellation diagram for 8-PSK. If $M = 2^n$, then each of the M M-PSK symbols corresponds to n bits of binary data. If R_s is the symbol rate of M-PSK, then the effective bit rate will be $R_b = nR_s$. Thus the bandwidth efficiency of M-PSK is n bps/Hz.

It is straightforward to show that the probability of a symbol error for M-PSK is approximately given by

$$P_e^{(s)} = erfc\left(\sqrt{\frac{E_s}{n_0}\sin^2\frac{\pi}{M}}\right). \tag{9.127}$$

This result is valid for $M > 2$ and $P_e < 10^{-3}$. If Gray coding is used for M-PSK, the bit error rate is given to a close approximation by

$$P_e = \frac{1}{n}P_e^{(s)} = \frac{1}{n}erfc\left(\sqrt{\frac{E_s}{n_0}\sin^2\frac{\pi}{M}}\right). \tag{9.128}$$

Quadrature Amplitude Modulation

We can further generalize M-ary modulation by allowing the amplitudes of the I and Q components of the carrier as given in (9.117) to vary arbitrarily. Thus we let

$$s_i(t) = a_i\cos\omega_0 t + b_i\sin\omega_0 t. \tag{9.129}$$

This is called *quadrature amplitude modulation* (QAM), because the quadrature components of the carrier are independently controlled. This representation encompasses ASK and M-PSK modulations. For example, for $M = 4$, QPSK modulation will result if we select the

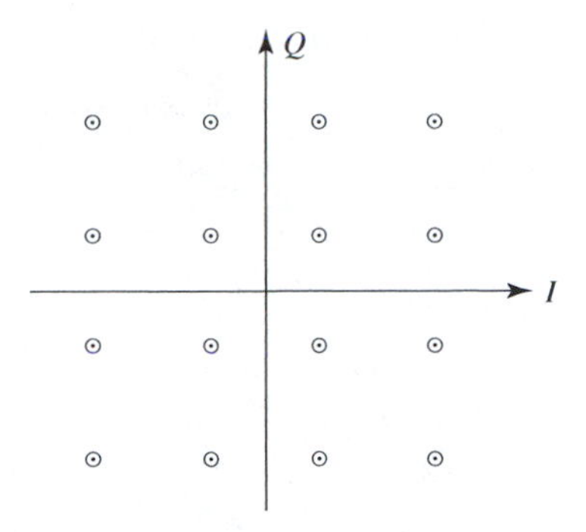

FIGURE 9.30 Constellation diagram for 16-QAM.

coefficients as

$$(a_0, b_0) = \left(\frac{1}{\sqrt{2}}, \frac{1}{\sqrt{2}}\right)$$

$$(a_1, b_1) = \left(\frac{-1}{\sqrt{2}}, \frac{1}{\sqrt{2}}\right)$$

$$(a_2, b_2) = \left(\frac{-1}{\sqrt{2}}, \frac{1}{\sqrt{2}}\right)$$

$$(a_3, b_3) = \left(\frac{1}{\sqrt{2}}, \frac{-1}{\sqrt{2}}\right)$$

We can therefore say that 4-QAM is identical to QPSK, if the signal amplitudes are constant. Although it is not necessary, the a_i and b_i coefficients are usually chosen to form symmetrical signal sets. The constellation diagram for 16-QAM is shown in Figure 9.30. It can be shown that an approximate expression for the symbol error rate of 16-QAM is given by

$$P_e = \frac{3}{2} erfc\left(\sqrt{\frac{2E_b}{5n_0}}\right). \tag{9.130}$$

Because of the high bandwidth efficiency that can be obtained with QAM, it is increasingly being used in modern wireless systems, including point-to-point microwave radios, LMDS systems, and the DVB-C digital video cable broadcasting system. Table 9.5 summarizes the ideal performance of several types of coherent digital modulation methods.

Channel Capacity

We have seen in the above sections that bit error rates decrease exponentially fast with an increase in E_b/n_0. This trend continues with higher-order M-ary modulation methods, so it is possible to achieve as low an error rate as is desired once E_b/n_0 is above a critical value. Since $E_b = S/R_b$, this also implies that, for a fixed signal power S, there is a critical value of the data rate R_b for which the error rate can be made as small as desired. This particular value is called the *channel capacity*, and is given by a formula derived by Claude Shannon:

$$C = B \log_2\left(1 + \frac{S}{n_0 B}\right), \tag{9.131}$$

TABLE 9.5 Summary of Performance of Various Digital Modulation Methods

Modulation Type	E_b/n_0 (dB) for $P_e = 10^{-5}$	Bandwidth Efficiency
binary ASK	15.6	1
binary FSK	12.6	1
binary PSK	9.6	1
QPSK (4-QAM)	9.6	2
8-PSK	13.0	3
16-PSK	18.7	4
16-QAM	13.4	4
64-QAM	17.8	6

where C is the maximum data rate capacity of the channel (bps), B is the bandwidth (Hz), S is the signal power (W), and $n_0/2$ is the two-sided power spectral density of the gaussian noise (W/Hz). The Shannon channel capacity formula gives the upper bound on the maximum data rate that can be achieved for a given channel in the presence of additive gaussian noise. Practical modulation methods usually perform at only a fraction of this value, but the use of error correcting codes can provide performance close to the Shannon limit.

REFERENCES

[1] F. G. Stremler, **Introduction to Communication Systems**, Addison-Wesley, Reading, MA, 1977.

[2] B. P. Lathi, **Modern Digital and Analog Communications Systems**, 3rd edition, Oxford University Press, New York, 1998.

[3] M. Schwartz, **Information Transmission, Modulation, and Noise**, 3rd edition, McGraw-Hill, New York, 1980.

[4] A. Mehrotra, **Cellular Radio Performance Engineering**, Artech House, Dedham, MA, 1994.

[5] W. C. Jakes, **Microwave Mobile Communications**, IEEE Press, Piscataway, NJ, 1974.

PROBLEMS

9.1 An SSB demodulator has an IF bandwidth of 30 kHz, and an input noise power spectral density (two-sided) of 10^{-8} W/Hz. If the required output SNR is 25 dB, find the minimum input signal power and the output noise power. Repeat for a DSB-SC demodulator.

9.2 The DSB-SC demodulator requires a local oscillator that is synchronized in phase and frequency with the carrier of the received signal. Consider the effect of an LO phase error $\Delta\phi$ by letting the local oscillator voltage be $\cos(\omega_{\text{IF}}t + \Delta\phi)$, and deriving an expression for the demodulated signal voltage. Will this error distort the received signal?

9.3 Repeat Problem 9.2 for a local oscillator having an error $\Delta\omega$ in frequency. Let the local oscillator voltage be $\cos(\omega_{\text{IF}} + \Delta\omega)t$, and derive an expression for the demodulated signal voltage. Will this error distort the received signal?

9.4 A broadcast radio station transmits a DSB-LC signal with a total power of 30 kW, using a modulation index of 70%. Find the power in the carrier and sidebands of the radiated signal.

9.5 A square wave is applied to an envelope detector circuit, as shown below. Assuming an ideal diode characteristic, sketch the output voltage waveform.

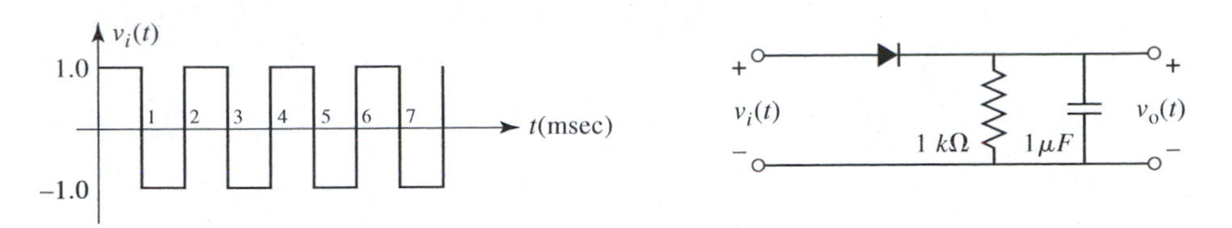

9.6 Repeat Example 9.1 for a synchronous DSB-LC demodulator.

9.7 Derive the output voltage for synchronous demodulation of ASK, FSK, and PSK when the local oscillator has a phase error $\Delta\phi$, and a frequency error Δf. That is, let the local oscillator be represented as $\cos[(\omega + \Delta\omega)t + \Delta\phi]$.

9.8 Can the square-law detector described in Example 9.2 be used to demodulate a DSB-LC amplitude modulated signal? Derive a result to justify your answer.

9.9 An FM radio transmitter uses a carrier frequency of 90 MHz, with a sinusoidal modulating signal at a frequency of $f_m = 20$ kHz. The voltage amplitude of the carrier is 30 V (peak), the modulation index is $\beta = 5$, and the transmitter drives a 50 Ω load. (a) What is the maximum frequency deviation from the carrier frequency? (b) What is the total power delivered by the transmitter? (c) What fraction of the total power is generated at the carrier frequency? (d) What is the approximate bandwidth according to Carson's rule? (e) What fraction of the total power is generated within this bandwidth?

9.10 Consider the special case of an FSK signal generated from a square-wave modulating waveform, with $f_1 = m/T$ and $f_2 = n/T$, where T is the bit period. Find the Fourier spectrum of the FSK signal, and sketch the form of the spectrum.

9.11 The coherent FSK demodulator circuit shown below avoids the need for two mixers and filters, as used in the demodulator of Figure 9.15b. Analyze this circuit to verify that it operates as shown.

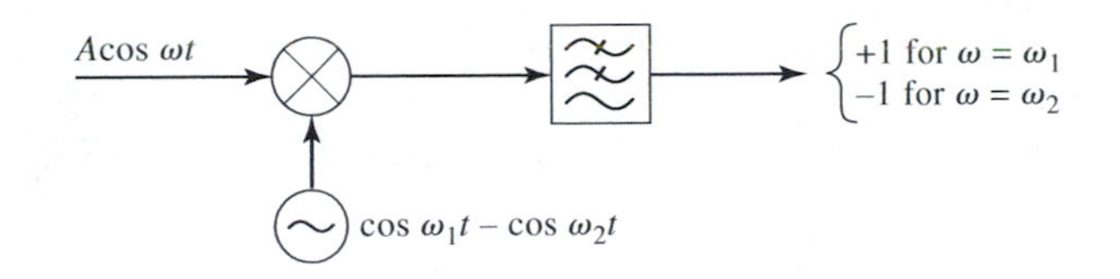

9.12 Derive an expression for $P_e^{(1)}$, the probability of error when a "1" is transmitted, for synchronous ASK, PSK, and FSK demodulators.

9.13 Calculate the E_b/n_0 ratio, in dB, that is required for $P_e = 10^{-2}$ and for $P_e = 10^{-8}$, for binary ASK, FSK, and PSK modulation using coherent detection.

9.14 Binary data is received by a coherent FSK receiver. The input signal power is 1 μW, and the two-sided PSD of the input noise is 10^{-14} W/Hz. Find the probability of error if the bit rate is 5 Mbps. How does this result change if noncoherent FSK is used?

9.15 The Mariner 10 satellite used to explore the planet Mercury in 1974 used PSK with $P_e = 0.05$ ($E_b/n_0 = 1.4$ dB) to transmit image data back to Earth (a distance of about 1.6×10^8 km). The satellite transmitter operated at 2.295 GHz, with an antenna gain of 27.6 dB and a carrier power of 16.8 W. The ground station had an antenna gain of 61.3 dB, and an overall system noise temperature of 13.5 K. Compute the maximum possible data rate.

9.16 The probability of error for coherent detection of ASK, FSK, and PSK was derived by assuming that noise was added after down converting to the IF stage. For the down-converter and PSK demodulator shown below, show that the same expression for P_e is obtained when noise is added before down conversion (as actually occurs in practice). Let the input signal be of the form $s(t) = m(t) \cos \omega_0 t$, where $m(t) = \pm V$, and let the two-sided PSD of the noise be $n_0/2$. Note that the energy of the input signal is now given as $E_b = V^2 T/2$, where T is the bit period.

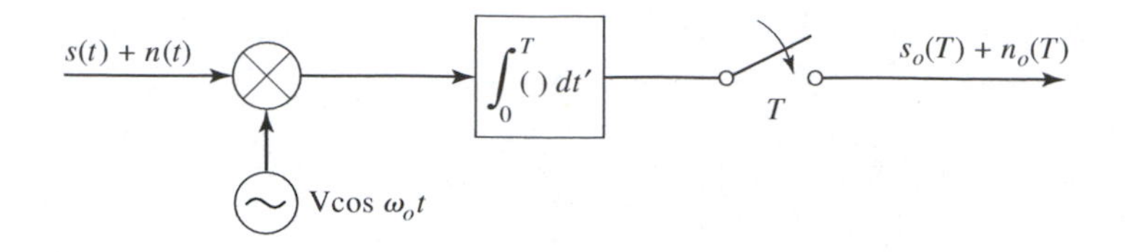

9.17 Use the large argument form of $erfc(x)$ given in Appendix D to derive expressions for the probability of error for coherent ASK, FSK, and PSK modulation when E_b/n_0 is very large. Compare your results with exact values for $E_b/n_0 = 5$ dB and 15 dB. Give an estimate for the range of E_b/n_0 for which the large argument expressions for P_e give accurate results.

9.18 For the Rayleigh distribution function of (9.88) show that $\int_{r=0}^{\infty} f_r(r)\,dr = 1$, and that the maximum value of $f_r(r)$ occurs at $r = \sigma$.

9.19 Use the large argument form for $I_0(x)$ given in Appendix B to show that the rician pdf approaches a gaussian distribution when $rV \gg \sigma^2$.

9.20 The general expression for the probability of error for binary modulation is given by

$$P_e = P_0 \int_{x_0}^{\infty} f_0(x)\,dx + P_1 \int_{-\infty}^{x_0} f_1(x)\,dx,$$

where P_0 and P_1 are the probabilities of occurrence for a "0" or "1," respectively, and f_0 and f_1 are the probability density functions for the occurrence of an error when receiving a "0" or "1." Minimize this expression to find the optimum value of the detector threshold, x_0. Show that for the usual case where $P_0 = P_1 = 0.5$, the result reduces to $f_0(x_0) = f_1(x_0)$. Hint: use the fact that

$$\frac{\partial}{\partial x} \int_{a}^{x} f(y)\,dy = f(x).$$

9.21 Consider a PSK receiver using a local oscillator with $v_{LO}(t) = \cos(\omega_0 t + \phi)$, where ϕ represents departure from ideal phase coherence with the incoming signal. (a) derive an expression for the probability of error as a function of the phase error ϕ. (b) consider ϕ as a random variable distributed uniformly from $-\pi$ to π, and derive an expression for the average probability of error. Hint: use the even and odd properties of $erfc(x)$ given in Appendix D.

9.22 Follow the procedure in Section 9.4 to derive expressions for the error rates of coherent ASK and FSK with Rayleigh fading.

9.23 Evaluate the required value of Γ for $P_e = 10^{-2}$, 10^{-5}, and 10^{-8} for faded and nonfaded BPSK.

9.24 An IS-54 cellular phone system uses QPSK (Gray coded) and operates in a propagation environment with a ground reflection. The base station antenna transmits at 882 MHz with an EIRP of 30 W, with an antenna that is 40 m high. The receive antenna has a gain of –1.0 dB, and is 1.5 m high. The receive antenna noise temperature is 200 K, and the receiver noise figure is 7 dB. The channel bandwidth is 30 kHz, and the data rate is 46.6 kbps. If the required bit error rate is 10^{-5}, find the maximum operating range of the system.

9.25 Reconsider the interplanetary communication problems of Chapter 4 with the use of digital modulation methods. Let a microwave signal at 2 GHz be transmitted to the nearest star (Alpha Centuri), with a transmitter power of 1 kW. Assume large dish antennas with $G = 60$ dB for transmit and receive, a 4 K background noise temperature, and a receiver bandwidth of 1 kHz. If coherent FSK is used with $P_e = 10^{-5}$ is used, what is the maximum data rate?

9.26 Calculate the required E_b/n_0 ratio for M-PSK with $P_e = 10^{-5}$, for $M = 2, 4, 8, 16, 32$, and 64.

9.27 A standard wired telephone circuit has a usable frequency range extending from 600 Hz to 3000 Hz, with a signal-to-noise ratio of about 30 dB. According to the Shannon channel capacity expression, what is the maximum data rate that can be achieved with this system?

9.28 Evaluate the Shannon channel capacity for an IS-54 cellular phone system. The channel bandwidth is 30 kHz. Assume a receive signal power of –60 dBm and $n_0/2 = 1 \times 10^{-18}$ W/Hz.

Receiver Design

The receiver is often the most critical component of a wireless system, having the overall purpose of reliably recovering the desired signal from a wide spectrum of transmitting sources, interference, and noise. Here we discuss some of the fundamental principles of radio receiver design, beginning with the evolution of receivers to provide progressively improved performance. The receiver must be sensitive enough to detect signal levels that may be as low as −110 dBm, while not being overloaded by much stronger signals, so we will review concepts such as minimum detectable signal and dynamic range, and discuss the need for automatic gain control. Another consideration is the requirement for filtering at various stages in the receiver, in order to eliminate spurious response from mixers, as well as provide rejection of image frequencies. The Globalstar LEO satellite telephone handsets shown in Figure 10.1 exemplify the sophistication of state-of-the-art receiver design.

10.1 RECEIVER ARCHITECTURES

Here we present an overview of some of the most important types of receiver architectures. Receiver design has evolved from the simple circuits used in the early days of radio in order to provide improved performance, ultimately allowing more efficient use of the radio spectrum for more users, communication over larger distances, and the use of lower transmit powers [1]–[2].

Receiver Requirements

The well-designed radio receiver must provide the following requirements:

- *High gain* (~100 dB) to restore the low power of the received signal to a level near its original baseband value.

FIGURE 10.1 Subscriber handsets for the Globalstar low-earth-orbit satellite telephone system. Courtesy of F. Dietrich, Globalstar, Inc., San Diego, CA.

- *Selectivity*, in order to receive the desired signal while rejecting adjacent channels, image frequencies, and interference.
- *Down-conversion* from the received RF frequency to an IF frequency for processing.
- *Detection* of the received analog or digital information.
- *Isolation* from the transmitter to avoid saturation of the receiver.

Because the typical signal power level from the receive antenna may be as low as –100 to –120 dBm, the receiver may be required to provide gain as high as 100 to 120 dB. This much gain should be spread over the RF, IF, and baseband stages to avoid instabilities and possible oscillation; it is generally good practice to avoid more than about 50–60 dB of gain at any one frequency band. The fact that amplifier cost generally increases with frequency is a further reason to spread gain over different frequency stages.

In principle, selectivity can be obtained by using a narrow bandpass filter at the RF stage of the receiver, but the bandwidth and cutoff requirements for such a filter are usually impractical to realize at RF frequencies. It is more effective to achieve selectivity by down-converting a relatively wide RF bandwidth around the desired signal, and using a sharp-cutoff bandpass filter at the IF stage to select only the desired frequency band. In addition, many wireless systems use a number of narrow but closely spaced channels which must be selected using a tuned local oscillator, while the IF passband is fixed. The alternative of using an extremely narrow band electronically tunable RF filter is not practical.

As discussed in Chapter 1, wireless systems may be simplex, half-duplex, or full-duplex, but most systems today require full-duplex operation where transmission and reception can occur simultaneously. Full-duplex communications systems usually use separate frequency bands for transmit and receive, thus avoiding the difficult (but not impossible) problem of isolating incoming and outgoing radiation at the same frequency. In addition, it is often preferred to use a single antenna for both transmit and receive. In this case it is necessary to use a *duplexing filter* to provide isolation between the transmitter and receiver, while still providing a signal path with the antenna.

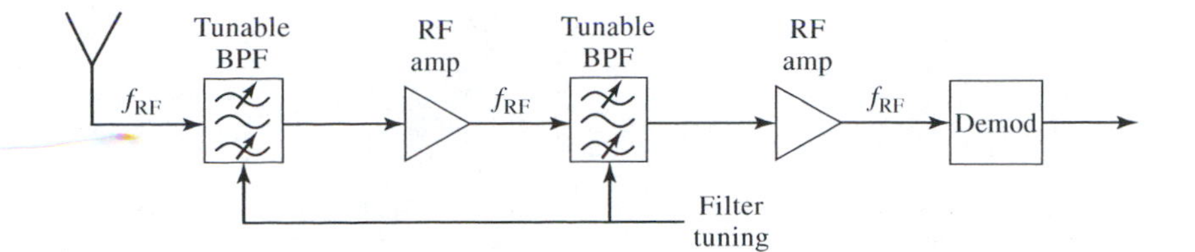

FIGURE 10.2 Block diagram of a tuned radio frequency receiver.

Tuned Radio Frequency Receiver

One of the earliest types of receiving circuits to be developed was the *tuned radio frequency* (TRF) receiver. As shown in Figure 10.2, a TRF receiver employs several stages of RF amplification along with tunable bandpass filters to provide high gain and selectivity. Alternatively, filtering and amplification may be combined by using amplifiers with a tunable bandpass response. At relatively low broadcast radio frequencies, such filters and amplifiers have been historically tuned using mechanically variable capacitors or inductors. But tuning is very difficult because of the need to tune several stages in parallel, and selectivity is poor because the passband of such filters is fairly broad. In addition, all the gain of the TRF receiver is achieved at the RF frequency, limiting the amount of gain that can be obtained before oscillation occurs, and increasing the cost and complexity of the receiver. Because of these drawbacks TRF receivers are seldom used today, and are an especially bad choice for higher RF or microwave frequencies.

Direct Conversion Receiver

The *direct conversion* receiver, shown in Figure 10.3, uses a mixer and local oscillator to perform frequency down-conversion with a zero IF frequency. The local oscillator is set to the same frequency as the desired RF signal, which is then converted directly to baseband. For this reason, the direct conversion receiver is sometimes called a *homodyne* receiver. For AM reception the received baseband signal would not require any further detection. The direct conversion receiver offers several advantages over the TRF receiver, as selectivity can be controlled with a simple low-pass baseband filter, and gain may be spread through the RF and baseband stages (although it is difficult to obtain stable high gain at very low frequencies). Direct conversion receivers are simpler and less costly than superheterodyne receivers, since there is no IF amplifier, IF bandpass filter, or IF local oscillator required for final down conversion. Another important advantage of direct conversion is that there is no image frequency, since the mixer difference frequency is effectively zero, and the sum frequency is twice the LO and easily filtered. But a serious disadvantage is that the LO must have a very high degree of precision and stability, especially for high RF frequencies, to avoid drifting of the received signal frequency. This type of receiver

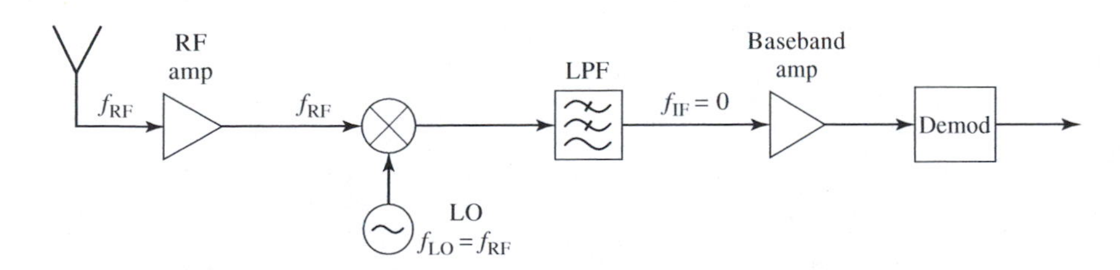

FIGURE 10.3 Block diagram of a homodyne receiver.

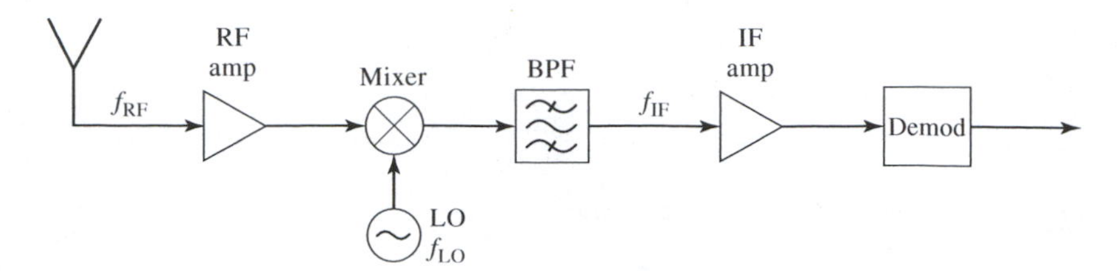

FIGURE 10.4 Block diagram of a single-conversion superheterodyne receiver.

is often used with Doppler radars, where the exact LO can be obtained from the transmitter, but a number of newer wireless systems are being designed with direct conversion receivers.

Superheterodyne Receiver

By far the most popular type of receiver used today is the *superheterodyne* circuit, shown in Figure 10.4. The block diagram is similar to the direct conversion receiver, but the IF frequency is now nonzero, and generally selected to be between the RF frequency and baseband. A midrange IF allows the use of sharper cutoff filters for improved selectivity, and higher IF gain through the use of an IF amplifier. Tuning is conveniently accomplished by varying the frequency of the local oscillator so that the IF frequency remains constant. The superheterodyne receiver represents the culmination of over 50 years of receiver development, and is used in the majority of broadcast radios and televisions, radar systems, cellular telephone systems, and data communications systems.

At microwave and millimeter wave frequencies it is often necessary to use two stages of down conversion to avoid problems due to LO stability. The *dual-conversion* superheterodyne receiver of Figure 10.5 employs two local oscillators and mixers to achieve down-conversion to baseband with two IF frequencies.

Duplexing

If a single antenna is to be used for both transmit and receive in a duplex system, a *duplexer* must be used to allow both the transmitter and receiver to be connected to the antenna, while preventing the transmit signal from directly entering the receiver. Isolation between the transmitter and receiver is usually required to be greater than 100 dB. The function of a duplexer is illustrated in Figure 10.6.

If the system is half-duplex, where the transmitter and receiver are not operating simultaneously, duplexing can be achieved with a T/R switch. A single-pole double-throw switch can connect the antenna to either the transmitter or the receiver. Solid-state switches

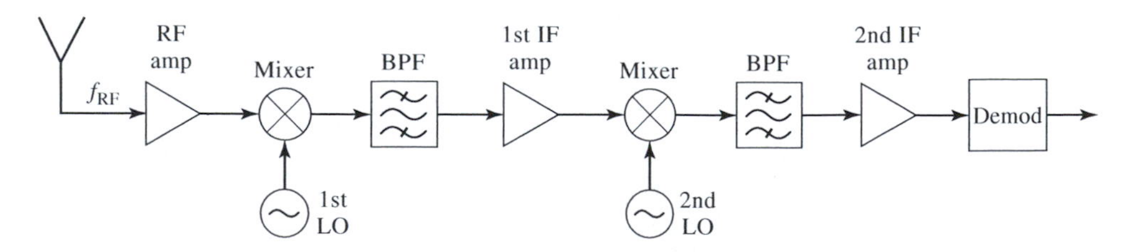

FIGURE 10.5 Block diagram of a double-conversion superheterodyne receiver.

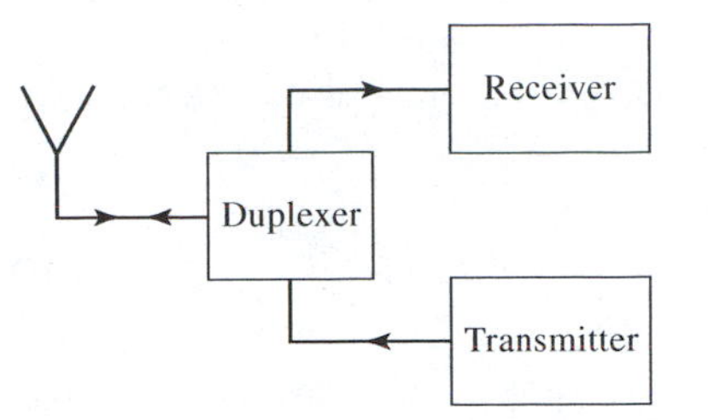

FIGURE 10.6 Use of a duplexer to connect a common antenna to a transmitter and a receiver, while providing isolation between the two.

can switch in microseconds, and provide isolation of 40 dB or more; limiters or filters may also be required to increase the isolation.

Full-duplex systems usually use separate transmit and receive frequency bands with bandpass filters to provide duplexing. Figure 10.7 shows a typical set of bandpass filter responses for duplexing. One filter is used to connect the antenna to the transmitter, while the other connects the antenna to the receiver. Duplexing filters are usually made as a single unit, with a crossover point at the 3 dB level. They can also provide some preselective filtering on receive, and attenuate spurious out-of-band signals from the transmitter. Duplexing filters often have insertion losses on the order of 1–3 dB, however, which degrades the noise figure of the receiver.

A related component is a *diplexer*, a term generally used to refer to a device that combines two or more frequency components into a single channel. Examples include frequency selective surfaces used in dual band antennas, hybrid junctions used to combine the RF and LO ports in a mixer, and filters used for frequency combining. Since a duplexing filter used with different transmit and receive frequency bands fits this definition, it is sometimes referred to as a diplexing filter.

EXAMPLE 10.1 FREQUENCY DUPLEXING

The AMPS and IS-54 cellular telephone system mobile units transmit over the frequency range of 824–849 MHz, and receive over the frequency range of 869–894 MHz. These two bands are each divided into 832 channels, each 30 kHz wide, to provide full-duplex communication. The first IF frequency of the receiver is 88 MHz. Compare the fractional bandwidths of the IF bandpass filter and a hypothetical tunable bandpass filter used at the RF stage. If we would like to have a minimum rejection of 50 dB between transmit and receive bands, what is the required order of the duplexing filter?

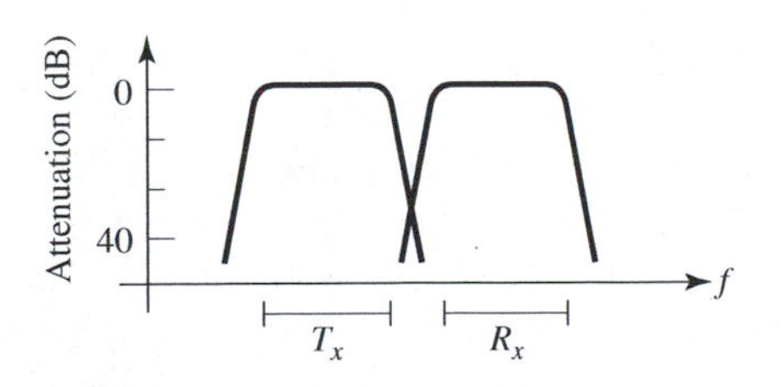

FIGURE 10.7 Transmit and receive passband responses of a duplexing filter.

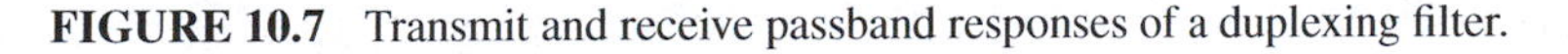

Solution
At the first IF frequency of 88 MHz, the fractional bandwidth of the 50 kHz IF filter is

$$\frac{\Delta f}{f} = \frac{0.05}{88} = 0.06\%,$$

whereas the same passband at the (midband) RF receive frequency of 882 MHz would be

$$\frac{\Delta f}{f} = \frac{0.05}{882} = 0.006\%.$$

Fractional bandwidths of 0.06% can be achieved with crystal or surface acoustic wave (SAW) filters, whereas bandwidths of 0.006% are too narrow to be achievable in practice.

To find the required order of the duplexer filter, we must first transform the transmit bandpass filter response to a low-pass prototype response. The worst-case isolation will occur for the receive frequency that is closest to the transmit band: $f = 869$ MHz. We thus require a minimum of 50 dB attenuation at 869 MHz from the bandpass filter used for the transmit band. The transmit band bandpass filter has a center frequency of

$$f_0 = \frac{824 + 849}{2} = 836.5 \text{ MHz},$$

and a fractional bandwidth of

$$\Delta = \frac{\Delta f}{f_0} = \frac{849 - 824}{836.5} = 0.03.$$

Using results from Chapter 5, the receive frequency of 869 MHz maps to a low-pass prototype (normalized to a low-pass filter with a cutoff frequency of 1 Hz) of

$$f' = \frac{1}{\Delta}\left(\frac{f}{f_0} - \frac{f_0}{f}\right) = \frac{1}{0.03}\left(\frac{869}{836.5} - \frac{836.5}{869}\right) = 2.54$$

The filter design graphs in Chapter 5 show that a filter of order $N = 4$ is required for 50 dB attenuation. ○

10.2 DYNAMIC RANGE

The definitions of linear dynamic range and spurious free dynamic range introduced in Chapter 3 are useful in the context of characterizing an individual component. Here we introduce a third definition of dynamic range that is more relevant for receiving systems. This definition involves the minimum detectable signal power, which is dependent on the type of modulation used in the receiving system, as well as the noise characteristics of the antenna and receiver. We will also see that the very large gain that most receivers are required to provide to detect weak signals can lead to saturation for higher level incoming signals; this problem is alleviated with the use of automatic gain control. High signal levels may also lead to distortion if the compression or third-order intercept points are exceeded.

TABLE 10.1 Typical Minimum SNR for Various Applications

System	SNR (dB)
Analog voice	5–10
Analog telephone	25–30
Analog television	45–55
AMPS cellular	18
AM-PCM	30–40
QPSK ($P_e = 10^{-5}$)	10

Minimum Detectable Signal

We have seen in Chapter 9 that reliable communication requires a receive signal power at or above a certain minimum level, which we call the *minimum detectable signal* (MDS). For a given system noise power, the MDS determines the minimum signal-to-noise ratio (SNR) at the demodulator of the receiver. The usable SNR depends on the application, with some typical values listed in Table 10.1.

The minimum SNR at the output of a receiver is sometimes expressed in terms of the *signal plus noise and distortion-to-noise and distortion ratio* (SINAD). The SINAD is related to SNR as

$$\text{SINAD} = \frac{S+N}{N} = 1 + \frac{S}{N}. \tag{10.1}$$

Knowing the minimum SNR or SINAD and the noise characteristics of the receiving system allows us to calculate the minimum detectable signal power. Figure 10.8 shows the block diagram of a general receiving system. The antenna delivers an input signal level S_i to the receiver, along with noise that is characterized by the antenna noise temperature T_A. The receiver block has an equivalent noise temperature T_e, and overall power gain G. Alternatively, we can use the receiver noise figure F, which is related to the receiver noise temperature as $T_e = (F-1)T_0$. The signal-to-noise ratio at the output of the receiver is S_o/N_o, which is applied to the input of the demodulator.

With these definitions we can write the output signal power as

$$S_o = GS_i, \tag{10.2}$$

and the total output noise power as

$$N_o = kBG(T_A + T_e), \tag{10.3}$$

where B is the bandwidth of the receiver (usually set by the IF bandpass filter). Then the

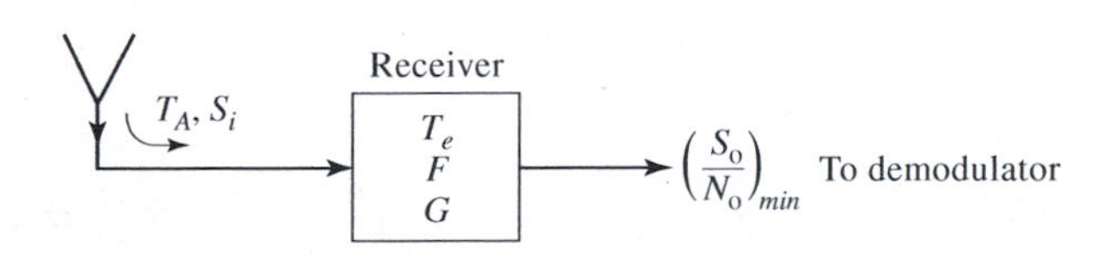

FIGURE 10.8 Block diagram of receiving system for the determination of minimum detectable signal.

minimum detectable input signal level is

$$S_{i_{\min}} = \frac{S_{o_{\min}}}{G} = \left(\frac{N_o}{G}\right)\left(\frac{S_o}{N_o}\right)_{\min} = kB(T_A + T_e)\left(\frac{S_o}{N_o}\right)_{\min}$$
$$= kB[T_A + (F-1)T_o]\left(\frac{S_o}{N_o}\right)_{\min} \quad (10.4)$$

This is an important result, relating the minimum detectable signal power at the input of the receiver to the noise characteristics of the receiving system and the minimum SNR required for that application. This equation provides the interface between the radio link equation (e.g., the Friis equation or ground reflection link equation) and the SNR or error rate equations of Chapter 9, thereby allowing characterization of the complete wireless system. For digital modulation, recall from (9.105) that the bit energy to noise power spectral density, E_b/n_0, is related to the SNR and bit rate R_b as follows:

$$\frac{E_b}{n_0} = \frac{S_o}{N_o}\frac{B}{R_b}. \quad (10.5)$$

For the special case where $T_A = T_0$, (10.4) reduces to the following result:

$$S_{i_{\min}} = kBT_0F\left(\frac{S_o}{N_o}\right)_{\min}. \quad (10.6)$$

This equation can be conveniently expressed in dB:

$$S_{i_{\min}}(dB) = 10\log(kT_0) + 10\log B + F(dB) + \left(\frac{S_o}{N_o}\right)_{\min}(dB)$$
$$= -174\,\text{dBm} + 10\log B + F(dB) + \left(\frac{S_o}{N_o}\right)_{\min}(dB) \quad (10.7)$$

Although (10.6)–(10.7) are sometimes used in place of (10.4), it is important to realize that (10.6)–(10.7) are only valid when the antenna temperature is 290 K; this situation is seldom true in practice. In either case, note that the minimum detectable signal level does not depend on the gain of the receiver, since both signal and noise are increased equally.

EXAMPLE 10.2 MINIMUM DETECTABLE SIGNAL

An IS-54 PCS telephone system uses QPSK with a bit rate of 46.6 kbps. The receiver has a bandwidth of 30 kHz, and a noise figure of 8 dB. The receive antenna has a noise temperature of 900 K. If the bit error rate is 10^{-5}, find the minimum detectable signal level.

Solution
Assuming Gray coded QPSK, $E_b/n_0 = 10$ dB $= 10$. Then the minimum SNR is found from (10.5) as

$$\left(\frac{S_o}{N_o}\right)_{\min} = \frac{R_b}{B}\frac{E_b}{n_0} = \frac{46.6\times 10^3}{30\times 10^3}(10) = 15.5$$

Using (10.4) gives the MDS as

$$S_{i_{\min}} = kB[T_A + (F-1)T_o]\left(\frac{S_o}{N_o}\right)_{\min}$$
$$= (1.38\times 10^{-23})(30\times 10^3)[900 + (6.3-1)(290)](15.5)$$
$$= 1.57\times 10^{-14}\text{ W} = 1.57\times 10^{-11}\text{ mW} = -108\text{ dBm}.$$

This value indicates the need for a total receiver gain on the order of 100 dB.

As a comparison, the (incorrect) use of (10.7) gives

$$\begin{aligned} S_{i_{\min}} &= -174\ \text{dBm} + 10\log B + F(dB) + \left(\frac{S_o}{N_o}\right)_{\min} \quad (dB) \\ &= -174 + 10\log(30 \times 10^3) + 8 + 15.5 \\ &= -105.7\ \text{dBm}. \end{aligned}$$

Note that this result is in error by several dB. ○

Sensitivity

The minimum detectable signal power can be converted to a minimum detectable signal voltage, for a given receiver input impedance. This quantity is called the *receiver voltage sensitivity*, usually shortened to simply the receiver *sensitivity*. Receiver specifications often list the sensitivity using this quantity. If the impedance of the receiver's antenna input is Z_0 (often 50 Ω), then the receiver voltage sensitivity is related to the minimum detectable signal power as

$$V_{i_{\min}} = \sqrt{2Z_0 S_{i_{\min}}} \cdot \text{V (rms)}. \tag{10.8}$$

The voltage sensitivity is usually expressed in microvolts. Sensitivity is sometimes expressed in terms of power, in which case it is identical to the minimum detectable signal level.

Dynamic Range

In contrast to the very low power of a received signal at or near the MDS level, the receiver must also be able to accept signals when the receiver is close to the transmitter. This signal level can be estimated by using the Friis link formula or, in the case of ground reflection or multipath, the link formula for ground reflection. This maximum allowable signal level and the minimum detectable signal level then set the *receiver dynamic range*, DR_r, defined as

$$DR_r = \frac{\text{maximum allowable signal power}}{\text{minimum detectable signal power}}. \tag{10.9}$$

Note that the receiver dynamic range depends on the noise characteristics of the receiver as well as the type of modulation being used, and the required minimum SNR. The maximum allowable signal power could alternatively be defined by the third-order intercept point, P_3, at the input to the receiver, as this would be the maximum input power before intermodulation distortion becomes unacceptable.

EXAMPLE 10.3 DYNAMIC RANGE

Consider the AMPS cellular system, transmitting from the base station at 880 MHz with a transmit power of 20 dBm. If the transmit and receive antennas have gains of 1 dB, find the receive power versus distance under the assumption of free-space conditions. If we assume the minimum detectable signal level from Example 10.2 of –108 dBm, and a minimum distance of 10 m between the transmit and receive antennas, what is the required dynamic range of the receiver? If the third-order intercept point at the input of the receiver is –15 dBm, how close can the receiver be to the transmit antenna before third-order intermodulation distortion becomes severe?

Solution

The Friis equation can be used to calculate the received power level versus distance R between the transmitter and receiver:

$$P_r = \frac{G_r G_t P_t \lambda^2}{(4\pi R)^2}.$$

The antenna gains are $G_r = G_t = 1.26$, and the wavelength is $\lambda = 300/880 = 0.341$ m. The received power versus range is listed in the next table:

R (m)	P_r (dBm)
0.5	−3.3
1	−9.3
10	−29.3
100	−49.3
1000	−69.3
10,000	−89.3

Observe that the received signal level decreases by 6 dB with a doubling in range, and decreases by 20 dB for an increase of 10 in the range. (In the case of ground reflections, the received signal level would decrease by 12 dB for a doubling in range.)

If we limit the minimum distance between transmitter and receiver to 10 m, we see that we can expect a maximum received power level of about −29.3 dBm. For a minimum detectable signal level of −108 dBm, this gives a receiver dynamic range of

$$DR_r = (-29.3\,\text{dBm}) - (-108\ \text{dBm}) = 78.7\,\text{dB}.$$

This table shows that when $R = 1$ m, the received power is about −9 dBm. Doubling that distance to $R = 2$ m will decrease the received power to about −15 dBm, which is the level of the third-order intercept at the input to the receiver. ○

Automatic Gain Control

The above examples show the need for about 80–100 dB of receiver gain to raise the minimum detectable signal to a usable level of approximately 10 mW (about 1 V peak at 50 Ω). As discussed previously, this much gain should be distributed throughout the RF, IF, and baseband stages to avoid oscillation, with most of the gain occurring at the IF stage to take advantage of the fact that amplifiers and other components are generally cheaper at lower frequencies. Another reason to avoid very high levels of gain at RF frequencies is that high input signal levels may exceed the 1 dB compression point (P_1), or the third-order intercept point (P_3), of the front-end components if the gain of the early stages is too high. On the other hand, it is useful to have a moderate level of gain at the RF stage to set a good noise figure for the receiver system.

At the output of the receiver the detected baseband signal often drives a digital signal processing (DSP) circuit, or a digital to analog converter (DAC), where the input voltage range is typically 1 mV to 1 V. For example, in a digital PCS telephone receiver the input signal is demodulated to recover digitized data, and then converted to an analog voice signal with a DAC. A 10-bit DAC with a maximum output voltage of 1 V has a resolution of $1/2^{10} = 1/1024 \approx 1$ mV, and provides a dynamic range of $20 \log 1000 = 60$ dB. Thus the

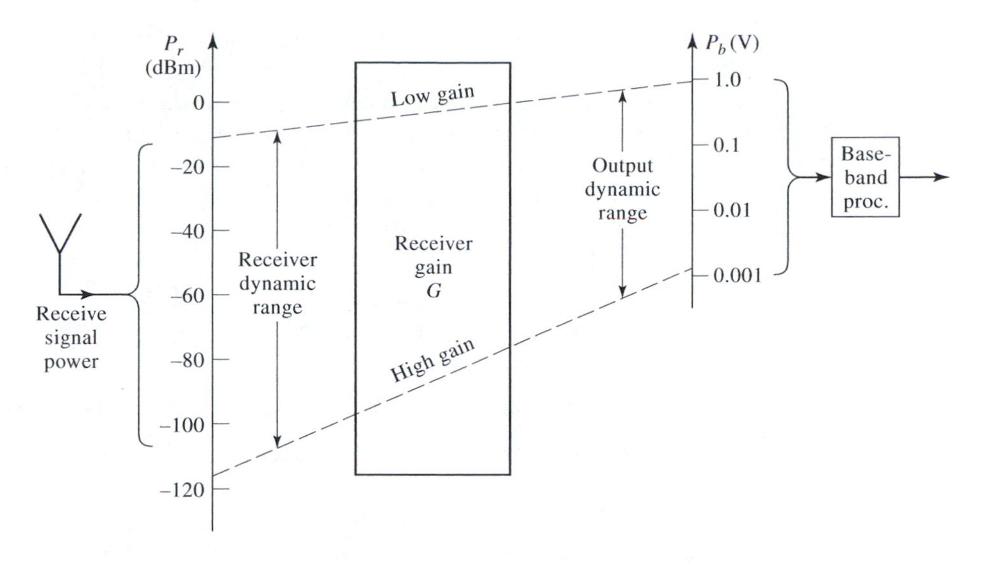

FIGURE 10.9 Diagram illustrating the change in power levels between the input and output of a typical receiver.

dynamic range at the output of the receiver is much smaller than the 80–100 dB dynamic range at the receiver input.

Figure 10.9 shows a graphical view of the input and output dynamic ranges of a typical receiver. The dynamic range of the input signal from the antenna is on the order of 100 dB, while the output dynamic range of the receiver is typically about 60 dB. The power gain through the receiver must therefore vary as a function of the input signal strength in order to fit the input signal range into the baseband processing range, for a wide range of input signal levels. A receiver gain on the order of 60 dB may be required for low level signals, while a gain of only 20–30 dB may be required for high-level signals. This variable-gain function is accomplished with an *automatic gain control* (AGC) circuit. AGC is most often implemented at the IF stage, but some receivers may use AGC at the RF stage as well. Virtually all modern communications, broadcast radio, and television receivers use AGC to control signal levels.

A typical IF AGC circuit is shown in Figure 10.10. It consists of a variable voltage-controlled attenuator (or variable gain amplifier) with a detector to convert a sample of the

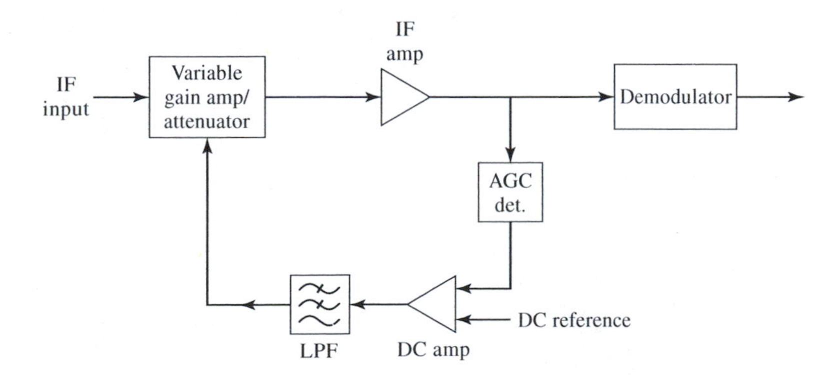

FIGURE 10.10 Block diagram of an automatic gain control circuit at the IF stage.

IF voltage to a DC value. The rectified signal is then compared with a reference level, and passed through a low-pass filter to provide a time constant long enough to avoid having the AGC following low-frequency components of the modulated signal. The output of the filter controls the variable attenuator, so that an increase in control voltage increases the attenuation. The dynamic range of the AGC is the ratio of the expected maximum and minimum input powers; this sets the required range of attenuation, which typically ranges from 20 to 30 dB. This loss will increase the thermal noise of the IF stage, but presumably this will not be a problem when the signal level is already large enough that attenuation is required. If properly designed, and spread over two or more stages of amplification, an AGC circuit can provide improved SNR as the input signal level increases.

Compression and Third-order Intermodulation

As discussed in Chapter 2, power levels that exceed the 1 dB compression point P_1 of an amplifier will cause harmonic distortion, and power levels in excess of the third-order intercept point P_3 will cause intermodulation distortion. Thus it is important to track power levels through the stages of the receiver to ensure that P_1 and P_3 are not exceeded. This can be conveniently done with a graph of the form shown in Figure 10.11. This diagram tracks

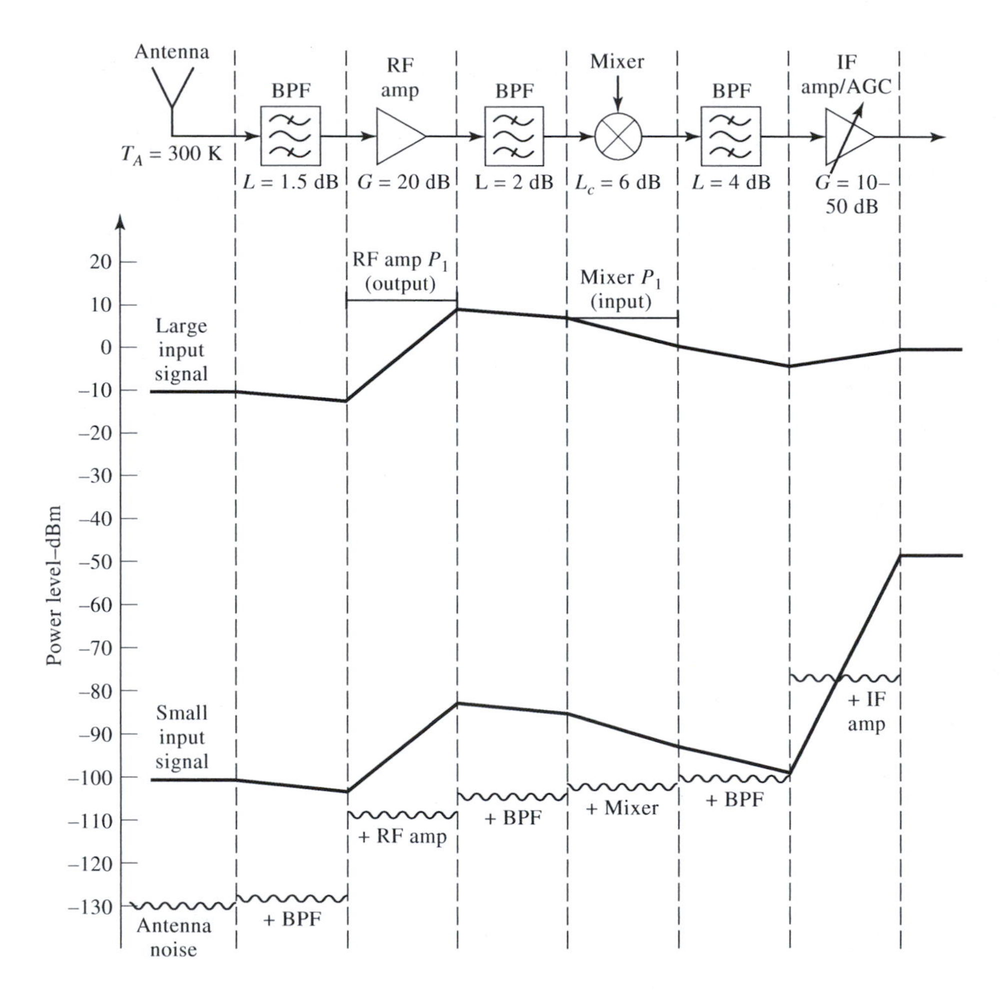

FIGURE 10.11 Diagram of power and noise levels at consecutive stages of a receiver.

the power level of small and large input signals, along with the noise level (noise power or noise figure may be plotted), through the front-end stages of the receiver. The compression and third-order intercept points of the amplifiers and mixers can be plotted on the graph, making it easy to see if these limits are exceeded, and to determine the effect of changing component specifications or position in the receiver circuit.

Placing an AGC attenuator in the RF stage may reduce the possibility of saturating the RF amplifier with a large input signal, but will degrade the noise performance of the receiver, even for a low attenuator setting. Since the power output of an amplifier is limited by the power supply voltage, an RF amplifier without AGC is limited in the gain it can provide without distortion. For example, if $S_i = -10$ dBm, an RF amplifier with a gain of 40 dB would theoretically yield an output power of 30 dBm = 1 W (about 7.1 V rms at 50 Ω), which would likely lead to saturation for a small-signal amplifier. Thus, a condition on the received signal power level, S_i, to avoid compression can be written as

$$S_{i_{\max}} + G_{\mathrm{RF}} < P_1, \tag{10.10}$$

where G_{RF} is the gain of the RF stage, and P_1 is the 1 dB compression point at the output of the RF amplifier. If the third-order intercept point is less than the compression power, P_3 can be substituted for P_1 in (10.10).

10.3 FREQUENCY CONVERSION AND FILTERING

In this section we will look at some details related to frequency down-conversion in a superheterodyne receiver, including the selection of the IF frequency, and the important role of filtering in the RF and IF stages.

Selection of IF Frequency

A key decision in the design of a superheterodyne receiver is the choice of IF frequency. As we have seen, the IF frequency is related to the RF and LO frequencies by

$$f_{\mathrm{IF}} = |f_{\mathrm{RF}} - f_{\mathrm{LO}}|. \tag{10.11}$$

While it is possible to use a local oscillator either above or below the RF signal frequency, most receivers use the lower sideband, so that the LO frequency is given by

$$f_{\mathrm{LO}} = f_{\mathrm{RF}} - f_{\mathrm{IF}}, \tag{10.12}$$

(where positive values are assumed). The mixer also responds to an RF image frequency separated by twice the IF frequency:

$$f_{\mathrm{IM}} = f_{\mathrm{RF}} - 2f_{\mathrm{IF}}. \tag{10.13}$$

This result assumes a lower sideband IF, with LO given by (10.12); if the upper sideband IF is used, the sign must be changed in (10.13). Because the image signal is often removed by filtering (as opposed to using an image reject mixer), using a large IF frequency eases the cutoff requirements of the image filter. In addition, to ensure that the image frequency is outside the RF bandwidth of the receiver, it is necessary to have

$$f_{\mathrm{IF}} > \frac{B_{\mathrm{RF}}}{2}, \tag{10.14}$$

where B_{RF} is the RF bandwidth of the receiver. This result follows directly from (10.13),

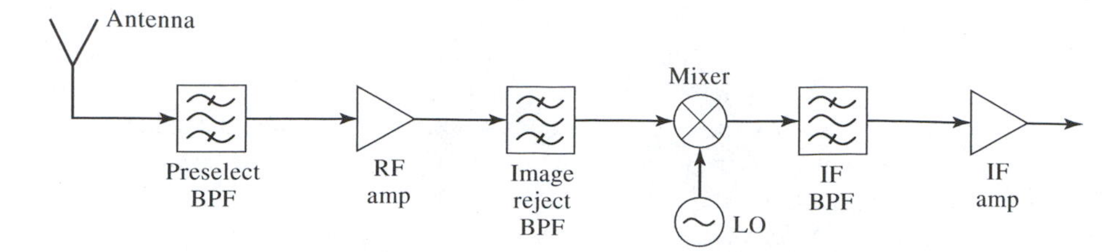

FIGURE 10.12 RF and IF filtering in a superheterodyne receiver

and the fact that the separation between the RF and image frequencies must be greater than the bandwidth of the system in order to filter the image without affecting the RF response.

Finally, it is usually helpful to use an IF frequency less than 100 MHz, because of component cost and availability considerations, as well as the fact that filters can be obtained in this frequency range with reasonable size and good cutoff characteristics. These are usually crystal, ceramic, or surface acoustic wave filters.

Filtering

Filtering is required in a superheterodyne receiver to provide interference rejection, image rejection, selectivity, and suppression of LO radiation. Figure 10.12 shows the usual location of these filters in the RF and IF stages of a typical receiver.

A *preselect* filter is usually placed ahead of the first RF amplifier (or mixer if an RF amplifier is not used). This is a bandpass filter set to the RF tuning range of the receiver. It rejects out-of-band interference, which is particularly important for preventing strong interference signals from saturating the RF amplifier or mixer. In order to keep the noise figure as low as possible, this filter should have a low insertion loss. This implies that its cutoff characteristics will not be very sharp, so this filter generally does not provide much rejection of the image frequency.

The *image reject* filter is usually placed after the RF amplifier, where the insertion loss associated with a filter having a sharp cutoff will have less effect on the noise figure of the receiver. This filter is often a ceramic dielectric resonator type. The image reject filter may also reduce the effect of possible harmonic distortion from the RF amplifier.

Because the local oscillator frequency is separated from the RF frequency only by the IF frequency, it often lies in the RF passband of the receiver, and may pass back through the RF stages to be radiated by the antenna. Such radiation may interfere with other users, and therefore must be attenuated to a very low level. This is usually accomplished by the combined attenuation of the preselect and image reject filters, the LO-RF isolation of the mixer, and the reverse attenuation of the RF amplifier. Because the LO is only one IF frequency away from the RF frequency, while the image is twice the IF frequency away, it is sometimes more difficult to meet the requirement for low LO radiation than it is for image rejection.

The IF bandpass filter sets the overall noise bandwidth of the receiver, as well as removing most unwanted mixer products. These include the nf_{LO} and mf_{RF} frequencies, as well as many intermodulation products of the form $nf_{\mathrm{LO}} \pm mf_{\mathrm{RF}}$. Some of these products may be within the passband of the IF filter, as discussed in more detail next.

Spurious-free Range

The nonlinear action of the mixer produces the sum and difference frequencies of the input signals, along with smaller levels of power at the intermodulation products at frequencies given by

$$f = |mf_{RF} - nf_{LO}|, \tag{10.15}$$

where m and n are positive integers. Most of these products are far outside the passband of the IF stage, but some may fall within the IF band. These are called *spurious responses* (or "spurs"), and are a problem because the receiver will respond to undesired signals at RF frequencies within its tuning range that produce spurs within the IF passband. This is especially a concern with multichannel wireless receivers. Because the amount of power contained in a particular intermodulation product decreases with the order of the product, it is usually sufficient to specify that the order of spurious responses within the IF passband be greater than a value in the range of 6–10.

In order to check the spurious response, the following procedure can be used. Because of the large number of combinations to check, it is best to write a simple computer program to evaluate the intermodulation product frequencies.

- Divide the RF tuning range of the receiver into K frequencies, spaced by the IF bandwidth.
- For each RF frequency, compute the required LO frequency from (10.12).
- Compute the intermodulation frequency using (10.15), for $1 \leq m \leq M$ and $1 \leq n \leq N$, where M and N are the upper limits of the maximum order to be searched.
- A spurious response lies within the IF passband if the value of (10.15) is less than the IF bandwidth

The number of combinations that must be checked is $K \times M \times N$.

While (10.15) gives all possible spurious frequency products, mixers often will inherently suppress some products due to symmetries and phase cancellations in the mixer circuit. Double-balanced mixers, for example, will reject all spurious responses where either m or n is even. Depending on their design, singly balanced mixers may reject some, but not all, products with m or n even.

EXAMPLE 10.4 SPURIOUS RESPONSES

Consider an AMPS receiver operating from 869 to 894 MHz, with a first IF of 88 MHz, and an IF bandwidth of 50 kHz. Find the possible spurious responses of order less than 10.

Solution

A short computer program was written to implement the above procedure, checking over 50,000 combinations in less than one second. The following products resulted in frequencies within the IF passband:

f_{RF} (MHz)	f_{LO} (MHz)	m	n	$\|mf_{RF} - nf_{LO}\|$ (MHz)
all	all	1	1	88
880.	792.	8	9	88

The first row, with $m = n = 1$, occurs for any RF frequency in the range from 869 to 894 MHz, and represents the desired down-conversion to the IF. The second row represents a spurious response with $m = 8$ and $n = 9$. This product may be suppressed if a doubly balanced mixer is used. ○

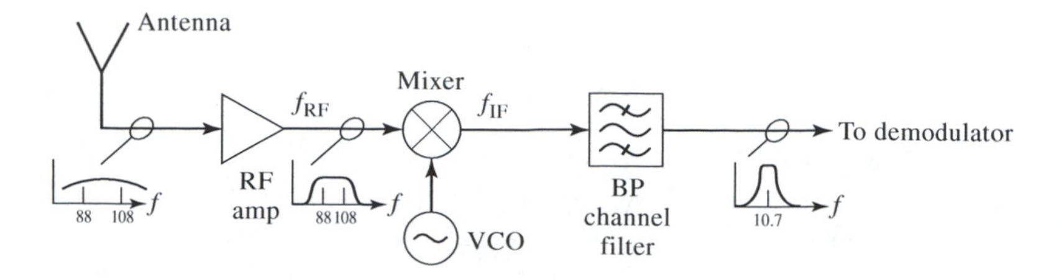

FIGURE 10.13 Block diagram of an FM broadcast receiver.

10.4 EXAMPLES OF PRACTICAL RECEIVERS

We have seen that the sensitivity, selectivity, and image rejection performance of a receiver depends on its overall design and layout. In this section we consider some practical examples of receivers to see how these problems have been solved in specific cases, and to study some of the trade-offs in performance and design.

FM Broadcast Receiver

We first consider the design of an FM broadcast receiver. Broadcast FM radio in the United States operates over the frequency band from 88 to 108 MHz, with channels spaced 200 kHz apart. Each FM broadcast station uses a maximum frequency deviation of ± 75 kHz, with receiver sensitivity on the order of 10 μV.

The block diagram of a traditional FM receiver is shown in Figure 10.13. The antenna receives signals over a relatively wide frequency range, and the RF amplifier usually provides a small amount of bandpass filtering to reject signals far outside the FM band. The desired channel cannot be tuned at the RF stage, however, because the required tunable filter bandwidth of 150 kHz/100 MHz = 0.15% cannot be inexpensively realized in practice. Thus the voltage-controlled oscillator is used to mix the RF signal down to an IF frequency of 10.7 MHz. Selectivity is then provided by the IF bandpass filter, with a fixed center frequency of 10.7 MHz. The fractional bandwidth of this filter is about 150 kHz/10.7 MHz = 1.4%, which can be easily realized.

If the lower sideband is used, the VCO must tune over the range of $f_{LO} = f_{RF} - f_{IF} = 77.3$ to 97.3 MHz. This is a tuning ratio of $97.3/77.3 = 1.3$, which can be easily implemented with a varactor-tuned Colpitts oscillator, or with a digital synthesizer. The image frequency is separated from the RF frequency by twice the IF frequency, which therefore lies in the band from 66.6 to 86.6 MHz. Thus the image frequency will always be outside the FM band, and attenuated mainly by the passband of the RF amplifier. The image frequency will always lie outside the receiver band if the IF frequency is selected to be at least $(108 - 88)/2 = 10$ MHz.

This FM receiver design has been used successfully for over thirty years, with little variation. Most of the components shown in Figure 10.13 can be miniaturized using integrated circuit technology. In particular, the RF and IF amplifiers, the VCO, and the mixer can be integrated on a single chip, as well as the lower frequency demodulator and audio amplifier circuits [4]. In addition, integrated circuits with increased use of transistors could be used, such as active filters, double-balanced Gilbert mixers, phase-locked loops, and balanced amplifiers. This trend improves radio performance and reduces cost. But miniaturization is ultimately limited by the IF bandpass filter, whose size cannot be reduced significantly without accepting seriously degraded performance, and cannot be replaced with integrated passive or active filters.

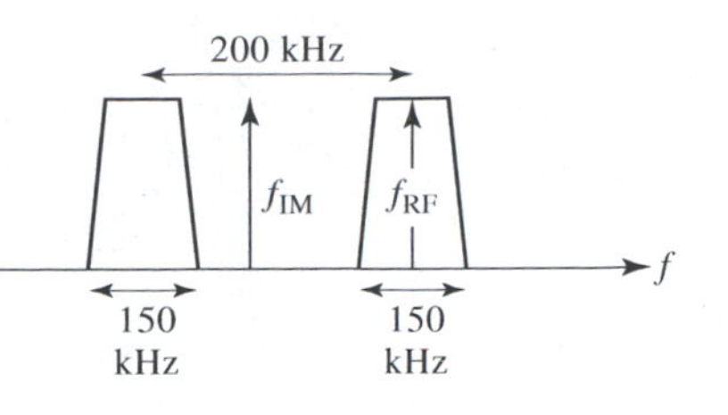

um of two FM channels and the image frequency for an FM receiver with a
IF frequency.

ther miniaturization of FM broadcast receivers led engineers at
propose the innovative idea of lowering the IF frequency from
is converts the desired FM channel down to baseband (0 to
vity can be obtained with a low-pass filter, which can be easily
grated circuitry. The image frequency in this case is located
desired RF channel, placing it between the selected channel
hown in Figure 10.14. The image signal therefore is not
preselect filter (if one is used). Generally there should
M channels, but the noise from the image bandwidth
dB. This is a trade-off in performance for a very high
eivers of this type can be made as small as a wrist-

f an IS-54 digital PCS telephone. The mobile
uency range of 824–849 MHz and receives
and receive bands are each divided into
ls through the use of frequency division
o the AMPS cellular system, so much
can be easily upgraded to the digital
without frequency reuse (within a
ate of 48.6 kbps, and time-division
share the same transmit and receive

IS-54 dual-conversion receiver. The
it and receive bands. This is followed
lter to provide further interference and
d to provide up to 20 dB of attenuation
ading the receiver. This is followed by
ain over two stages interspersed with
ltering without severely degrading the
ceptably low level.
controlled oscillator to the necessary
lies that the image frequency will be
(87) = 695 to 720 MHz. This is well
the duplexing and first RF bandpass
SAW, or ceramic resonator filter to
d IF is fixed at 87.5 MHz to produce

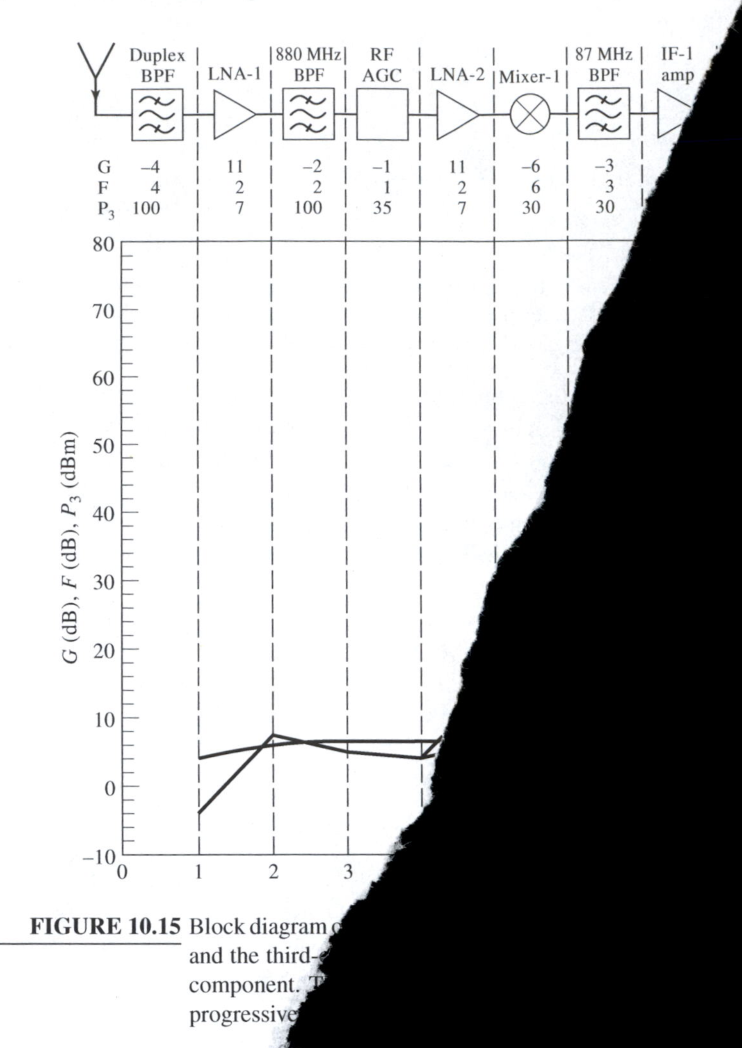

FIGURE 10.15 Block diagram o
and the third-
component. T
progressive

The gain, noise f
in Figure 10.15. Usin
gain, noise figure, and
stage, versus position th
about 7 dB within a few
the receiver. The gain of
filters and first mixer, w
receiver gain. The third-
receiver.

Millimeter Wave Poi

An example of a m
Figure 10.16. Such radi
base stations to mobile te
network (PSTN), and fo
data rates, and a wireles

FIGURE 10.14 Specti
70 kHz

The demand for fu
Philips in the 1980s to p
10.7 MHz to 70 kHz. Th
70 kHz), so channel selecti
implemented with active inte
$2f_{IF} = 140$ kHz away from the c
and the next lower channel, as s
attenuated by the RF amplifier o
not be radiated signals between F
will reduce the receiver SNR by 3
degree of miniaturization—FM rec
watch.

Digital Cellular Receiver

Next we consider the receiver section o
unit of this U.S. system transmits over the fre
over the range of 869–894 MHz. The transmit
30 kHz channels, providing 832 full-duplex chann
duplexing. These details of the system are identical
of the AMPS infrastructure and spectrum allocations
IS-54 PCS. While the AMPS system uses analog FM
cell), the IS-54 system uses QPSK with a channel bit r
multiple access (TDMA) to allow up to three users to
channels simultaneously.

Figure 10.15 shows a block diagram of a typical
antenna feeds a duplexing filter, to separate the transm
by an RF low-noise amplifier, and another bandpass fi
image rejection. A switchable AGC attenuator is use
at the RF stage to prevent strong signals from overlo
another RF low-noise amplifier. Spreading the RF g
bandpass filters provides the required RF gain and fi
noise figure, and prevents P_3 from falling to an unac

The desired channel is tuned by setting a voltage
LO frequency. The first IF is at 87 MHz, which imp
located in the frequency range from (869–894) – 2
outside the receive band, and can be attenuated by
filters. The first IF bandpass filter may be a crystal,
provide a sharp cutoff response. The LO for the secon
a second IF of 455 kHz.

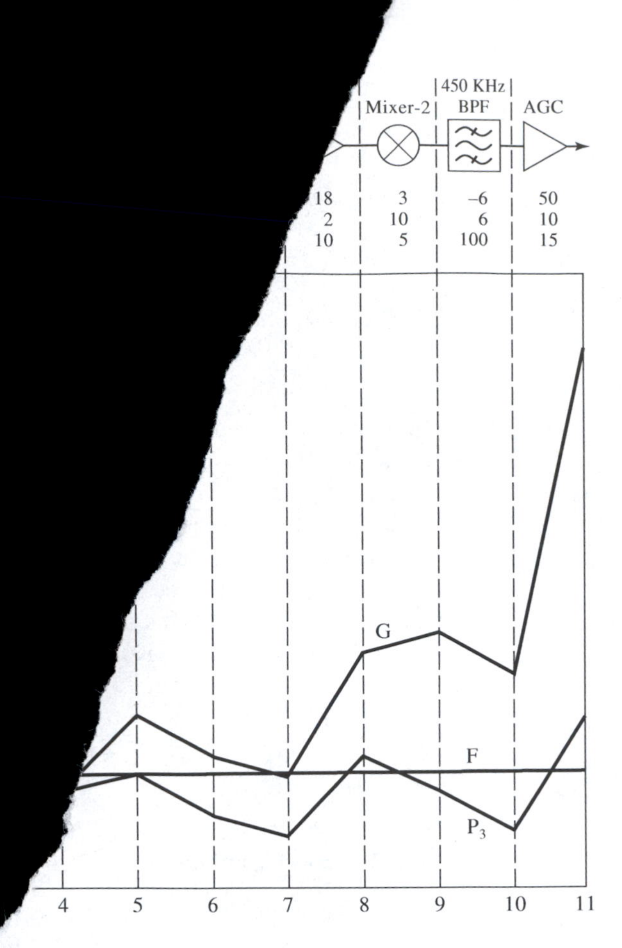

…f an IS-54 digital PCS receiver. The gain (in dB), noise figure (in dB), …order intercept point (in dBm, reference at output) are listed for each … he cascaded gain, noise figure, and third-order intercept are plotted …ly through the system.

…gure, and third-order intercept point are listed for each component … the cascade formulas for noise figure and third-order intercept, the …third-order intercept are plotted in Figure 10.15 at the output of each …rough the receiver. Note that the noise figure approaches a value of … stages of the input, and changes very little throughout the rest of … the RF amplifiers essentially compensates for the losses of the RF …ith the first and second IF amplifiers providing most of the overall …order intercept point remains near or above 0 dBm throughout the

…t-to-Point Radio Receiver

…illimeter wave point-to-point radio receiver front end is shown in …s are commonly used to provide backhaul service from cellular …lephone switching offices (MTSO) and the public switched phone …r private data network connections. Such applications require high …s link is often more economical than laying fiber or coaxial cables. A

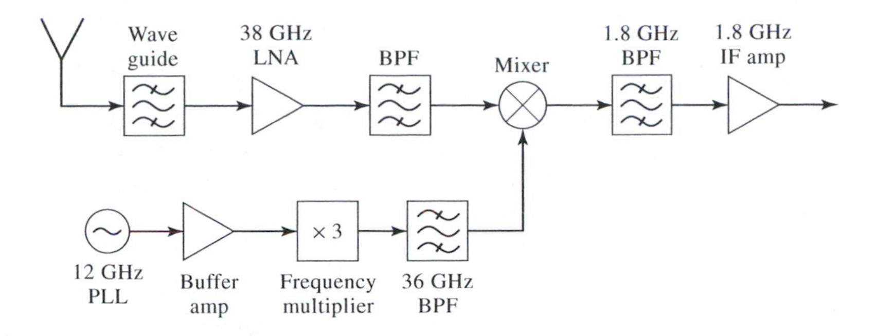

FIGURE 10.16 Block diagram of the front-end of a 38 GHz point-to-point radio receiver.

photograph of a set of commercial millimeter wave transmit and receive modules is shown in Figure 10.17.

The block diagram of Figure 10.16 shows only the first conversion to the first IF frequency of 1.8 GHz. A high gain antenna and waveguide feed provide some bandpass filtering over the waveguide band. The local oscillator frequency of 36 GHz is derived from a phase-locked loop at 12 GHz, followed by a frequency tripler. A 36 GHz bandpass filter removes the spurious products generated by the frequency multiplier. The first IF of 1.8 GHz is converted to a lower second IF in later stages. Specifications for the receiver

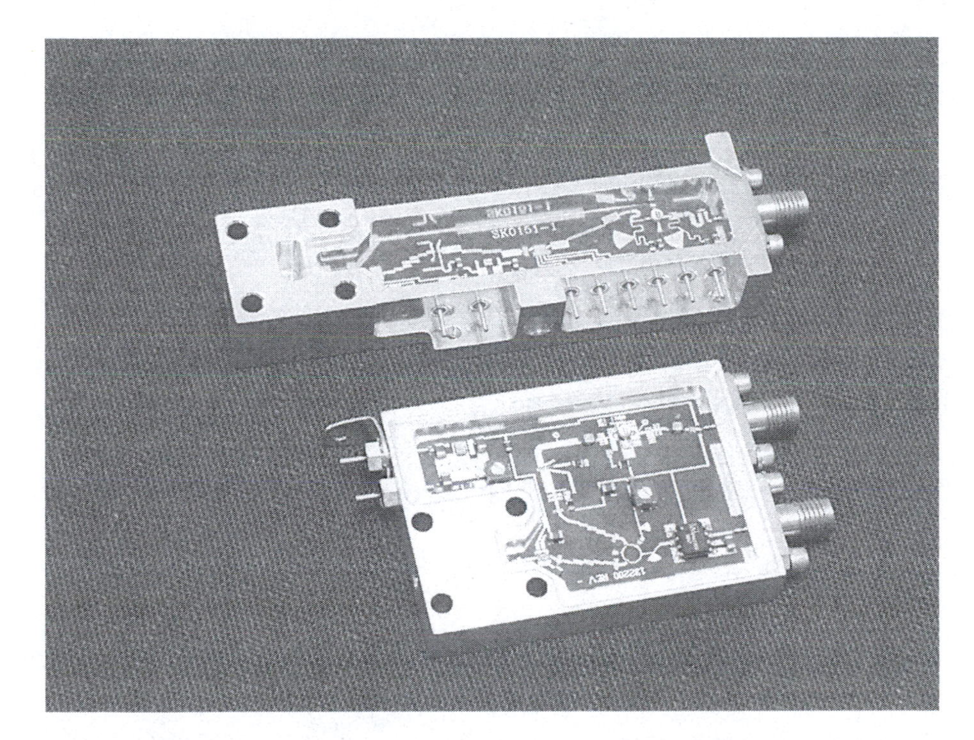

FIGURE 10.17 Photograph of transmit (top) and receive (bottom) modules for a 38 GHz point-to-point radio. Each unit contains all necessary millimeter wave low-noise/power amplifiers, mixers, frequency multipliers, filters, and IF amplifiers. Courtesy of Ar-com, Inc., Salem, NH.

components are:

38 GHz waveguide transition:	insertion loss = 1.0 dB
38 GHz low-noise amplifier:	gain = 20 dB noise figure = 3.5 dB third-order intercept = 15 dBm
38 GHz band pass filter:	insertion loss = 4 dB
first mixer:	conversion loss = 7 dB noise figure = 7 dB third-order intercept = 10 dBm
1.8 GHz IF amplifier:	gain = 13 dB noise figure = 2.5 dB third-order intercept = 25 dBm

Application of the cascade formula gives the following overall characteristics of the receiver front end:

receiver gain:	21 dB
receiver noise figure:	4.9 dB
receiver intercept point:	−5.5 dBm (at input)

The 36 GHz local oscillator has a phase noise level that is 85 dB below the carrier at an offset of 100 kHz.

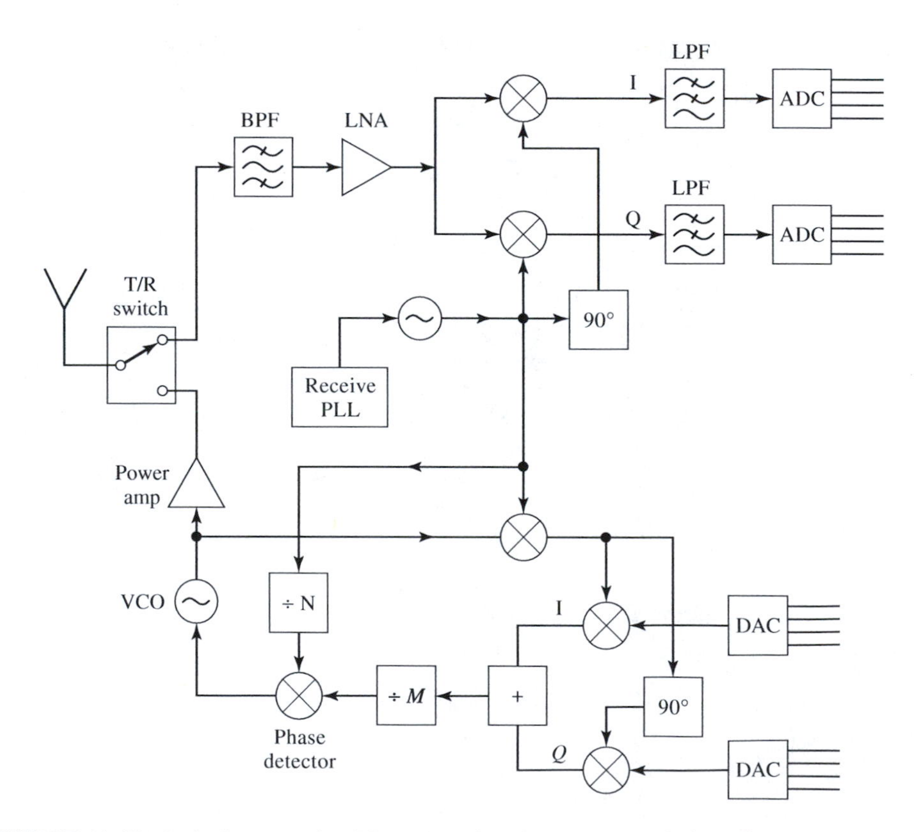

FIGURE 10.18 Block diagram of a GSM transceiver front end, consisting of a transmitter and a direct conversion receiver.

Direct-Conversion GSM Receiver

Because of the very competitive nature of the cellular/PCS telephone market, there is a strong demand to reduce the parts count, size, weight, and cost of transceiver handsets. Direct conversion receivers are therefore of significant interest, since the IF filters, amplifiers, and IF LOs are eliminated with this type of receiver design.

Figure 10.18 shows the block diagram of a proposed second-generation GSM transceiver, where direct conversion is used for the receiver. GSM is used throughout Europe with a transmit frequency band of 880–915 MHz, and a receive band of 925–960 MHz. Many of the components in this circuit can be integrated into two or three chips, with a total parts count of less than 90. Duplexing is performed with a transmit/receive switch, since multiplexing in GSM is done via TDMA.

The receiver uses a preselect bandpass filter, followed by a low-noise amplifier, and down conversion to baseband. The I and Q mixer outputs are low-pass filtered, then fed to separate analog-to-digital converters for baseband processing. A fractional-N phase-locked loop synthesizer, derived from a crystal controlled source, is used for the local oscillator for both transmitter and receiver.

REFERENCES

[1] P. J. Nahin, **The Science of Radio**, AIP Press, Greenwich, CT, 1996.

[2] S. J. Erst, **Receiving Systems Design**, Artech House, Dedham, MA, 1984.

[3] D. M. Pozar, **Microwave Engineering**, 2nd edition, Wiley, New York, 1998.

[4] L. E. Larson, **RF and Microwave Circuit Design for Wireless Communications**, Artech House, Dedham, MA, 1996.

PROBLEMS

10.1 The AMPS mobile telephone receiver has a noise figure of about 8 dB, with an IF bandwidth of 50 kHz. If the antenna temperature is 1000 K, and the required minimum SNR at the output of the receiver is 20 dB, find the minimum detectable signal level at the input to the receiver. If the maximum allowable signal level at the input to the receiver is -20 dBm, what is the dynamic range of the receiver?

10.2 A wireless data network operates at 900 MHz, with a transmit power of 100 mW, and a transmit antenna gain of 3 dB. The receive antenna has a gain of 1 dB, with a background temperature of 100 K, and a radiation efficiency of 70%. The system data rate is 1.6 Mbps, using Gray-coded QPSK with a bit error rate of 10^{-5}. The receiver noise figure is 12 dB. If free-space propagation is assumed, what is the maximum range of this system?

10.3 A receiver has a sensitivity of $3\,\mu$V (rms), for a 50 Ω input impedance, and an IF bandwidth of 30 kHz. If the minimum SINAD at the output of the receiver is 12 dB, what is the required receiver noise figure?

10.4 The input signal power to an AGC circuit ranges from -60 dBm to 0 dBm. The input noise power is fixed at -80 dBm. If the AGC is set to provide a constant output power of -60 dBm, calculate the effect of the noise generated by the attenuator on the output SNR. Plot the output SNR versus input power level. Assume the attenuator is at room temperature, and that the bandwidth is 1 MHz.

10.5 Consider the receiver front-end block diagram shown below, with the given parameters for each component. The 1 dB compression points and the third-order intercept points are referred to the output of the amplifiers, while for the mixer these quantities are referred to the input. Determine the noise figure of the receiver. If the required output SNR is 12 dB, find the minimum detectable signal power and the receiver voltage sensitivity, assuming an antenna noise temperature of 400 K, an IF bandwidth of 50 kHz, and a receiver impedance of 50 Ω. Plot a power level diagram similar to that in Figure 10.11, for input signal levels of −90 dBm and −30 dBm. Is P_1 or P_3 exceeded for any component?

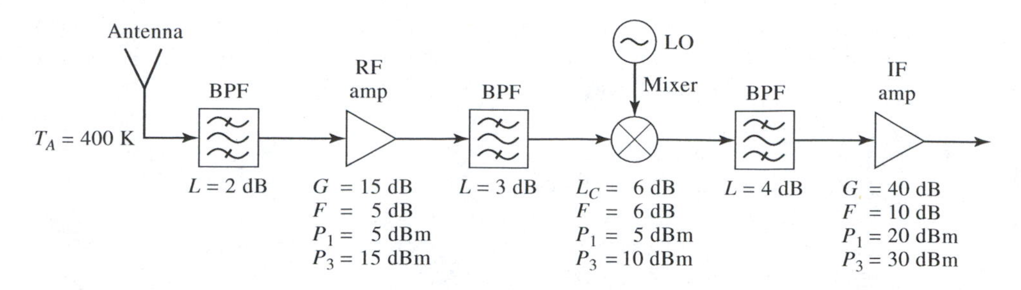

10.6 For the receiver circuit of Problem 10.5, plot the output SINAD versus input signal power S_i, for -120 dBm $\leq S_i \leq -20$ dBm.

10.7 A radio receiver is tuned to receive a signal at 880 MHz. It uses an IF frequency of 88 MHz. What is the frequency of the image frequency that could be received by this system? Describe three methods that could be used to minimize reception of an image signal.

10.8 A multi-user radio system uses three possible channel frequencies of 900 MHz, 910 MHz, and 920 MHz. The channel bandwidth is 1 MHz, and the receiver IF frequency is 10 MHz. (a) Assuming the receiver input is bandpass filtered from 899.5 to 920.5 MHz, will the receiver pick up any image frequencies? (verify this by calculating all possible image frequencies) (b) Will the mixer produce any spurs of order 6 or less that fall within the IF passband? (Calculate at least several examples to justify your answer.)

10.9 Consider the first five stages of the IS-54 receiver shown in Figure 10.15 (the duplexing BPF through LNA-2). If LNA-2 is moved to a position directly following LNA- 1, calculate and plot the gain, noise figure, and third-order intercept for this new configuration. Discuss the performance of this configuration relative to that of Figure 10.15.

10.10 Verify the gain, noise figure, and third-order intercept point for the 38 GHz point-to-point millimeter wave receiver of Figure 10.16, using the component characteristics listed in the text.

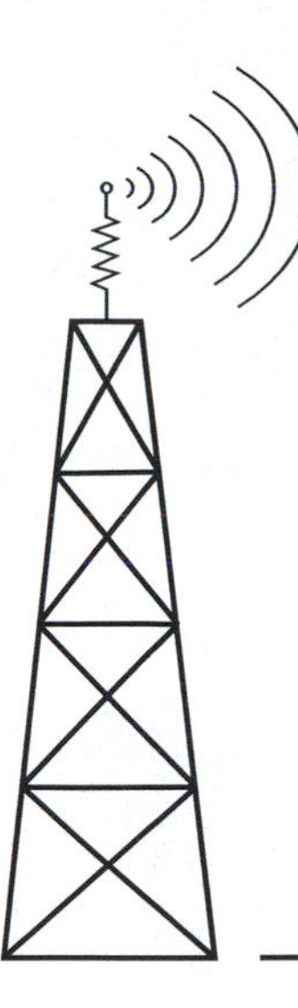

Appendices

Appendix A WIRELESS SYSTEM FREQUENCY BANDS

Radio and Microwave Frequency Bands:

Medium frequency (MF)	300 kHz to 3 MHz
High frequency (HF)	3 MHz to 30 MHz
Very high frequency (VHF)	30 MHz to 300 MHz
Ultra high frequency (UHF)	300 MHz to 3 GHz
L band	1–2 GHz
S band	2–4 GHz
C band	4–8 GHz
X band	8–12 GHz
Ku band	12–18 GHz
K band	18–26 GHz
Ka band	26–40 GHz

Commercial Radio and Television Frequencies:

AM broadcast radio	535–1605 kHz
FM broadcast radio	88–108 MHz
VHF television (channels 2–4)	54–72 MHz
VHF television (channels 5–6)	76–88 MHz
UHF television (channels 7–13)	174–216 MHz
UHF television (channels 14–83)	470–890 MHz

Other system frequencies:

AMPS (mobile unit)	824–849 MHz (T), 869–894 MHz (R)
European GSM (mobile unit)	880–915 MHz (T), 925–960 MHz (R)
PCS (mobile unit)	1710–1785 MHz (T), 1805–1880 MHz (R)
Paging	931–932 MHz
GPS	1575 MHz (L1), 1227 MHz (L2)
DBS	11.7–12.5 GHz
ISM bands	902–928 MHz
	2.400–2.484 GHz
	5.725–5.850 GHz

Appendix B USEFUL MATHEMATICAL RESULTS

Useful Integrals:

$$\int_{-\infty}^{\infty} \frac{\sin^2 x}{x^2} dx = \pi$$

$$\int_{-\infty}^{\infty} e^{-x^2} dx = \frac{\sqrt{\pi}}{2}$$

$$\int_{0}^{\infty} x e^{-ax^2} dx = \frac{1}{2a}$$

$$\int_0^\infty x^2 e^{-x^2} dx = \frac{\sqrt{\pi}}{4}$$

$$\int_0^\infty \frac{e^{-au}}{\sqrt{u}} du = \sqrt{\frac{\pi}{a}}$$

$$\int_0^\infty \frac{dx}{1+x^2} = \frac{\pi}{2}$$

$$\int_0^\pi \sin^3\theta d\theta = \frac{4}{3}$$

Taylor Series:

$$\sin x \cong x - \frac{1}{3!}x^3 + \cdots \text{ for small } x$$

$$\cos x \cong 1 - \frac{1}{2!}x^2 + \cdots \text{ for small } x$$

$$\sqrt{1+x} \cong 1 + \frac{1}{2}x + \cdots \text{ for small } x$$

Bessel Functions:

$$J_n(x) = \frac{1}{2\pi}\int_{-\pi}^{\pi} e^{j(x\sin\theta - n\theta)} d\theta$$

$$I_n(x) = \frac{1}{2\pi}\int_0^{2\pi} e^{x\cos\theta}\cos n\theta d\theta$$

$$I_n(x) \cong \frac{e^x}{\sqrt{2\pi x}} \text{ for large } x$$

Appendix C FOURIER AND LAPLACE TRANSFORMS

Fourier Transforms:

$f(t)$	$F(\omega)$
$f(t) = \frac{1}{2\pi}\int_{-\infty}^{\infty} F(\omega)e^{j\omega t} d\omega$	$F(\omega) = \int_{-\infty}^{\infty} f(t)e^{-j\omega t} dt$
$\delta(t)$	1
1	$2\pi\delta(\omega)$
$e^{j\omega_0 t}$	$2\pi\delta(\omega - \omega_0)$
$\cos\omega_0 t$	$\pi[\delta(\omega - \omega_0) + \delta(\omega + \omega_0)]$
$\sin\omega_0 t$	$j\pi[\delta(\omega + \omega_0) - \delta(\omega - \omega_0)]$
$f(t) = \begin{cases} 1 & \text{for } \lvert t\rvert < \text{T} \\ 0 & \text{for } \lvert t\rvert > \text{T} \end{cases}$	$2T\frac{\sin\omega T}{\omega T}$
$\frac{W}{\pi}\frac{\sin Wt}{Wt}$	$\begin{cases} 1 & \text{for } \lvert\omega\rvert < W \\ 0 & \text{for } \lvert\omega\rvert > W \end{cases}$
$\frac{df(t)}{dt}$	$j\omega F(\omega)$

Laplace Transforms:

$f(t)$	$F(s)$
$f(t) = \frac{1}{2\pi j}\int_{\alpha-j\infty}^{\alpha+j\infty} F(s)e^{st}ds$	$F(s) = \int_0^{\infty} f(t)e^{-st}dt$
$U(t) = \begin{cases} 0 & \text{for } t < 0 \\ 1 & \text{for } t \geq 0 \end{cases}$	$\frac{1}{s}$
$tU(t)$	$\frac{1}{s^2}$
$e^{-at}U(t)$	$\frac{1}{s+a}$
$\frac{\sin\sqrt{1-\zeta^2}\omega_n t}{\sqrt{1-\zeta^2}\omega_n} e^{-\zeta\omega_n t}$	$\frac{1}{s^2+2\zeta\omega_n s+\omega_n^2}$
$\left[\cos\sqrt{1-\zeta^2}\omega_n t - \frac{\zeta}{\sqrt{1-\zeta^2}}\sin\sqrt{1-\zeta^2}\omega_n t\right]e^{-\zeta\omega_n t}$	$\frac{s}{s^2+2\zeta\omega_n s+\omega_n^2}$
$\frac{d}{dt}f(t)$	$sF(s) - f(0^+)$
$\int_0^t f(t)dt$	$\frac{F(s)}{s}$

Appendix D THE COMPLEMENTARY ERROR FUNCTION

Definition of error function and complementary error function (for $x \geq 0$):

$$erf(x) = \frac{2}{\sqrt{\pi}}\int_0^x e^{-u^2}du \qquad erfc(x) = \frac{2}{\sqrt{\pi}}\int_x^{\infty} e^{-u^2}du = 1 - erf(x)$$

Symmetry properties:

$$erf(x) = -erf(-x) \qquad erfc(x) = 2 - erfc(-x)$$

Special values of complementary error function:

$$\begin{aligned} erfc(0) &= 1 \\ erfc(1) &= 1.573 \times 10^{-1} \\ erfc(2) &= 4.678 \times 10^{-3} \\ erfc(3) &= 2.209 \times 10^{-5} \\ erfc(\infty) &= 0 \end{aligned}$$

Closed-form approximations for complementary error function:

$$erfc(x) \cong \begin{cases} 1 - \dfrac{3.372x}{3+x^2} & \text{for } x \leq 0.96 \\ \dfrac{1.132xe^{-x^2}}{1+2x^2} & \text{for } x \geq 2.71 \end{cases}$$

Large-argument form of complementary error function:

$$erfc(x) \cong \frac{e^{-x^2}}{\sqrt{\pi}x}$$

BASIC code for computing complementary error function:

```
Rem compute complementary error function erfc(x) for x=>0
    pi = 3.14159265
    If x < 1.5 Then
       j = 3 + Int(9 * x)
       erfc = 1.
1      erfc = 1. + erfc * x * x * (0.5 - j) / (j * (j+0.5))
       j = j - 1
       If j <> 0 Then GoTo 1
       erfc = 1. - 2. * erfc * x / Sqr(pi)
    Else
       j = 3 + Int(32. / x)
       erfc = 0.
2      erfc = 1. / (j * erfc + Sqr(2. * x * x))
       j = j - 1
       If j <> 0 Then GoTo 2
       erfc = erfc * Sqr(2. / pi) * Exp(-x * x)
    End If
    Print ''erfc(x)='' ,erfc
    End
```

Appendix E CHEBYSHEV POLYNOMIALS

The nth-order Chebyshev polynomial is a polynomial of degree n, and is denoted by $T_n(x)$. The first four Chebyshev polynomials are

$$
\begin{aligned}
T_1(x) &= x, \\
T_2(x) &= 2x^2 - 1, \\
T_3(x) &= 4x^3 - 3x, \\
T_4(x) &= 8x^4 - 8x^2 + 1.
\end{aligned}
$$

Higher-order polynomials can be found by using the following recurrence formula:

$$T_n(x) = 2xT_{n-1}(x) - T_{n-2}(x)$$

The first four Chebyshev polynomials are plotted in Figure E.1; note that for $-1 \leq x \leq 1$,

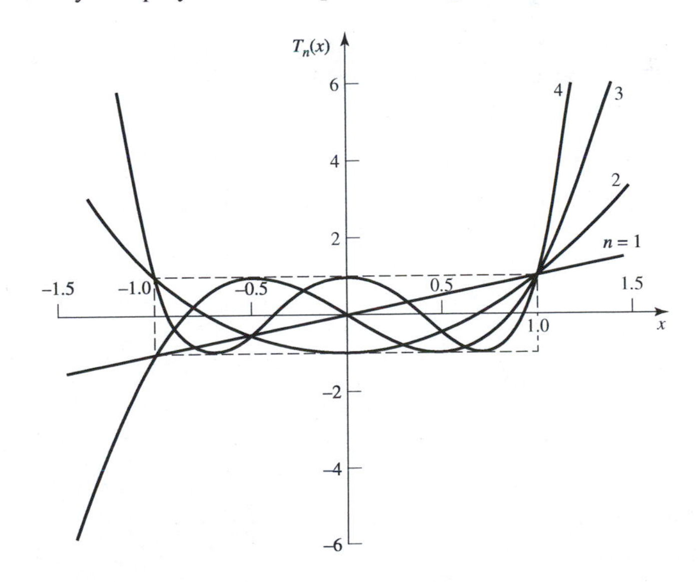

FIGURE E.1

$|T_n(x)| \leq 1$. This *equal ripple* property makes the Chebyshev polynomials very useful for problems in approximation and synthesis.

Appendix F DECIBELS AND NEPERS

Often the ratio of two power levels, P_1 and P_2, in an RF or microwave system is expressed in decibels (dB) as

$$10 \log \frac{P_1}{P_2} \text{ dB}.$$

Thus, a power ratio of 2 is equivalent to 3 dB, while a power ratio of 0.1 is equivalent to -10 dB. Using power ratios in dB makes it easy to calculate power loss or gain through a series of components, since multiplicative loss or gain factors can be accounted for by adding the loss or gain in dB for each stage. For example, a signal passing through a 20 dB amplifier followed by a 6 dB attenuator will have an overall gain of $20 - 6 = 14$ dB.

Decibels are only used to represent power ratios, but if $P_1 = V_1^2/R_1$ and $P_2 = V_2^2/R_2$, then the result in terms of voltage ratios is

$$10 \log \frac{V_1^2 R_2}{V_2^2 R_1} = 20 \log \frac{V_1}{V_2} \sqrt{\frac{R_2}{R_1}} \text{ dB},$$

where R_1 and R_2 are the load resistances, and V_1 and V_2 are the voltages appearing across these loads. If the load resistances are equal, then this result simplifies to

$$20 \log \frac{V_1}{V_2} \text{ dB}.$$

The ratio of voltages across equal load resistances can also be expressed in terms of nepers (Np) as

$$\ln \frac{V_1}{V_2} \text{ Np}.$$

The corresponding expression in terms of powers is

$$\frac{1}{2} \ln \frac{P_1}{P_2} \text{ Np},$$

since voltage is proportional to the square root of power. Transmission line attenuation is often expressed in nepers. Since 1 Np corresponds to a power ratio of e^2, the conversion between nepers and decibels is

$$1 \text{ Np} = 10 \log e^2 = 8.686 \text{ dB}.$$

Absolute powers can also be expressed in decibel notation if a reference power is assumed. If we let $P_2 = 1$ mW, for example, then the power P_1 can be expressed in dBm as

$$10 \log \frac{P_1}{1 \text{ mW}} \text{ dBm}.$$

Thus a power level of 1 mW is equivalent to 0 dBm, while a power level of 1 W is equivalent to 30 dBm. If the reference power is taken to be 1 W, then the power P_1 can be expressed in dBW as

$$10 \log \frac{P_1}{1 \text{ W}} \text{ dBW}.$$

Thus a power level of 1 W is equivalent to 0 dBW, while a power level of 4 W is equivalent to 6 dBW.

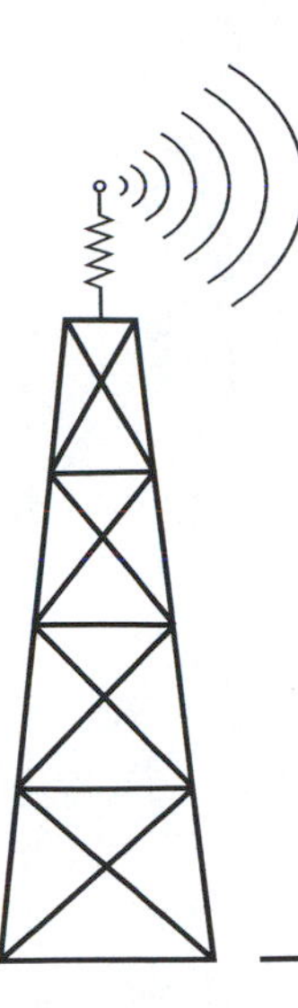

Index